Preface

Soft Computing has emerged as an important approach towards achieving intelligent computational paradigms where key elements are learning from experience in the presence of uncertainties, fuzzy belief functions, and evolution of the computing strategies of the learning agent itself. Fuzzy, neural and evolutionary computing are the three major themes of soft computing. The book presents original research papers dealing with the theory of soft computing and its applications in engineering design and manufacturing. The methodologies have been applied to a large variety of real life problems. Application of soft computing has provided the opportunity to integrate human like 'vagueness' and real life 'uncertainty' to an otherwise 'hard' computer programme. Now, a computer programme can learn, adapt, and evolve using soft computing. The book identifies the strengths and limitations of soft computing techniques, particularly with reference to their engineering applications. The applications range from design optimisation to scheduling and image analysis. Goal optimisation with incomplete information and under uncertainty is the key to solving real-life problems in design and manufacturing. Soft computing techniques presented in this book address these issues. Computational complexity and efficient implementation of these techniques are also major concerns for realising useful industrial applications of soft computing. The different parts in the book also address these issues.

The book contains 9 parts, 8 of which are based on papers from the '2nd On-line World Conference on Soft Computing in Engineering Design and Manufacture (WSC2)'. The conference was hosted on the Internet by five different universities from four different countries. Internet has already started to revolutionise many parts of our life, and now, it is going to modernise the way researchers, academics, and practitioners communicate between each other. This series of on-line events in Soft Computing started a couple of years ago at Nagoya University, Japan under the chairmanship of Prof. Takeshi Furuhashi. Every event was successful and raised more interest. The growing popularity of these events provided the motivation for WSC2. The conference has removed the financial burden from the authors and participants. Though there were some troubles with the Internet connections, WSC2 has successfully attracted participants from all over the World.

Publication of this book was possible due to the successful organisation of WSC2. Thus we would like to thank the organising committee, the authors, the reviewers, the sponsors, and the advertisers of the conference. This book is the final result of hard work of many people across the World. We hope the book will promote the research and development in soft computing for engineering design and manufacturing and justify the hard work.

Dr. Pravir K. Chawdhry
Dr. Rajkumar Roy
Prof. R. K. Pant

Cyberspace
18 August 1997

Table of Contents

Introduction

Soft Computing has emerged as an important field in computing. Fuzzy Technology, Neural Computing, and Evolutionary Computing are the three major players in soft computing. The methodologies have been successfully applied to a large variety of real life problems. Application of soft computing has provided the opportunity to integrate human like 'vagueness' and real life 'uncertainty' to an otherwise 'hard' computer program. Now a computer programme can learn, adapt, and evolve using soft computing. Engineering design and manufacturing are the two major application areas of soft computing. This book identifies the strengths and weaknesses of soft computing methodologies for the engineering applications. The book also identifies some of the issues involved in real life engineering problems. The book is divided into 9 parts. The papers presented in the next eight parts of this book are the papers presented in the 2nd On-line World Conference on Soft Computing in Engineering Design and Manufacturing (WSC2). The last part summarises discussions about the papers during the conference. This section briefly introduces different areas covered in the book, and includes introduction about different papers presented in the next eight parts.

1. Evolutionary Computing

Charles Darwin, in his book 'The Origin of Species' in 1859, championed the principle of 'Evolution through natural selection'. Evolutionary Computing (EC) gleans concepts from the 'natural evolution' to perform optimisation, search and synthesis of information. According to the nature of the algorithm, there are several categories of evolutionary computing, such as, evolutionary strategies, genetic algorithms, and evolutionary programming. Rechenberg [1] introduced 'evolutionary strategies' as an optimisation technique. He started with a 'population' of two individuals, one parent and one offspring. The offspring was produced by a random change (mutation) in the parent. Genetic Algorithms were invented by John Holland [2], where he used a number of genetic operators, like 'crossover', 'mutation', and 'inversion'. He also introduced a large population (more than two members) based genetic algorithm. Fogel, Owens, and Walsh [3] developed 'evolutionary programming', an optimisation technique where candidate solutions were represented as finite-state machines. New individuals were evolved by mutating the state-transition diagrams the parent individuals. Among these three techniques genetic algorithms have recently become very popular. Goldberg [4] has largely contributed to the development and popularity of genetic algorithms. More recently, Koza [5] has developed 'genetic programming' technique that works in a similar way to genetic algorithms but evolves better computer programmes. Genetic programming is also gaining wide popularity in optimisation, system identification, and symbolic regression.

This first part of the book contains six papers, and gives a quick overview about some of the trends in 'evolutionary computing' research. The first paper, by Falco et al., is about the use of evolutionary algorithms, in this case a 'breeder' genetic algorithm, in the optimisation of neural network structure. The performance of the 'breeder' genetic algorithm is compared with standard genetic algorithms. This paper also exhibits how fuzzy logic concepts can be utilised to develop an efficient genetic operator. 'Fitness Causes Bloat' by Langdon and Poli is one of the most discussed paper during the conference. One of the challenges faced by 'genetic programming' researchers is how to control the growth of the computer programmes (bloat). Often programmes produced are very large and with a lot of redundancy. The authors argue that such growth is inherent in using a fixed evaluation function with a discrete but variable length representation. The paper also mentions that the redundancy can also be very useful in some applications. The paper has increased the understanding about the mechanism behind genetic programming. The next paper by Bapi is very different from others. He tries to relate the concepts of evolutionary computing with the emergence of intelligence in human. This conceptual understanding is essential for the development of 'Intelligent Systems'. He has borrowed the concept of 'triune-brain' as introduced by Paul MacLean to develop his viewpoint. The author has placed symbolic processing (rational and logical thinking) on the top layer and quantitative methods of statistics at the instinctual layer (bottom layer). Evolutionary computing along with other soft computing techniques is placed in the intermediate layer. He argues that the interaction between these three layers is crucial for theories of intelligent computation. Roy and Parmee have presented an overview of evolutionary computing in multimodal function optimisation. In real life, problems often have more than one solution to a task. The paper discusses strengths and weaknesses of evolutionary computing techniques to optimise such problems. The major focus is to identify the issues involved in real life problems. A new technique (ARTS), developed by the authors, is briefly described. The paper mentions how ARTS can optimise in the case of multimodal real life problems. The next paper is about evolving 'cellular automata' to perform 'pattern association' and develop 'associative memory'. The paper exhibits how a large set of simple interacting elements with little information produce a complex co-ordinated information processing behaviour. The final paper in this part discusses a possible use of column tables in genetic programming. Column tables are used to define a chromosome. The authors argue that column tables inherently code directed acyclic graphs. The implementation of crossover is simple and needs neither traversing nor checking of integrity of resulting data structures and should therefore be more efficient. The paper demonstrates the use of column tables with examples of symbolic regression.

2. Neural Networks

Artificial neural networks, as their name suggests, have their origins in mathematical models proposed by neurobiologists in the 1940s and 1950s to simulate neuron behaviour in the brain. The associationist abilities evident in these early networks provided a plausible learning mechanism, and so excited the interest of the artificial intelligence and cognitive science research communities. However, the networks' reasoning at a numerical level also had the effect of precipitating a split between those who believed that intelligence exists at such a low level of granularity and those who maintain that it necessarily involves the manipulations of symbols at a higher level of cognitive abstraction. The predominance of the latter faction, along with the supposed limitations of the networks when learning certain classes of problems, leads to neural networks being all but ignored for well over a decade. However, interest began to be re-awakened in the early 1980s, and the seminal work of the Parallel Distributed Processing (PDP) group, under James McClelland and David Rumelhart, at the University of California, San Diego in the mid-1980s was instrumental in rehabilitating neural networks. This work had a significant impact in a number of disparate fields and subject areas, with, in particular, the back-propagation network algorithm of McClelland and Rumelhart, along with Geoffrey Hinton, providing a powerful learning metaphor which provides a suitable riposte to the criticism which lead to the initial abandonment of research into neural networks.

There are now many different 'varieties' of neural network algorithms (which, however, no longer make any claim for biological plausibility). The fundamental element (the 'neuron') of most of these is a computational unit that takes a number of numerical inputs, sums them, and, according to some internal function, produces a numerical response, the output of the neuron [6]. A number of such neurons are connected into a network in such a way that the outputs of some feed the inputs of others. Associated with each connection is a numerical weighting, which serves to amplify or else inhibit the signal passing along it. Some of the neurons are designated as input units, others as outputs. Provided with a set of training information, consisting of corresponding pairs of input and output patterns, which together form some coherent description of some concept, the 'learning' task becomes one of modifying the weightings (which have been initialised to random values) so that each input pattern, when propagated through the network produces the correct output pattern. When this is achieved (and depending on the quality of the training data), the network and weightings embody a distributed representation of the concept, sufficiently generalised to provide a correct output response to input data outside that in the training set. To some, such behaviour is akin to induction forming processes in humans; to others, it is merely statistical pattern recognition. Nevertheless, neural network theory is now one of the core constituents of soft computing, displaying many of the qualities typical of such techniques: neural networks are robust, being able to cope with imprecision in the data; they can handle and 'reason' with uncertainty in the data; networks can produce concise representations of complex problems in a tractable manner, these representations emerging from an initial state of stochastic chaos.

The work of the PDP group has lead to widespread acceptance and utilisation of neural networks; they are now an established weapon in the armoury of many an engineer for tackling analytical problems. Among the following papers, that of Stavroulakis et al describes their application of neural networks to the task of producing an analytical model of semi-rigid connections in steel structures based on empirical data. The work of Blumenstein and Verma, using networks to comprehend zip code information, represents a fairly typical problem, involving a degree of uncertainty and 'fuzziness', to which soft computing offers a solution where hard computing has proved inadequate. However, despite finding practical applications, aspects of neural theory are still open research issues. For example, a criticism often levied is that the design of the network topology so as to best suit the problem in hand is itself a difficult task; an inappropriate choice can lead to a poor representation being constructed, or in some cases, in a failure to construct a representation at all. This has prompted a branch of research which attempts to overcome these problems by applying another of the core paradigms of soft computing to the task, that of evolutionary computation. The concept of producing neural networks suited to a particular problem through evolutionary imperatives is a very appealing analogue to the adaptation of brains in living creatures, but is not without its difficulties. Two of the following papers address this idea, albeit with differing approaches. Zhang and Joung employ genetic programming as their evolutionary model and physiological data as their test domain. They attempt to combine sub-networks ('neural trees') to produce a suitable network structure. Tsui and Plumbley use a classifier system to manipulate a network at the level of its connections. They use standard neural theory benchmark data to measure the worth of their approach. The final paper in this section, that of Kovacs, is not concerned with neural networks per se, but it shares the theme of classifier systems with the work of Tsui and Plumbley.

3. Fuzzy Logic

Fuzzy Logic is one of the main areas in Soft Computing. The technique allows us to model a system using 'vague' or 'ambiguous' terms (as we often use in real life). In real life many decisions can not be just 'yes' or 'no', there is always an overlap where the decision is 'grey'. Capturing and representing that 'grey' areas in the knowledge and reason with them are the major challenges for classical computing techniques. Fuzzy Logic captures and represents the 'grey' knowledge using a set of 'fuzzy terms' or 'linguistic variables'. The definition of the fuzzy terms was first proposed by Lukasiewicz in the 1920s. Prof. Lotfi A. Zadeh developed a new logic tool [7] for representing and manipulating fuzzy terms, called fuzzy logic. He proposed graded memberships in sets which is to say that an element could be, say, 20% element of set A and 80% element of A¢ (i.e. complement of A). For example, if ambient temperature is 24 °C, for many people it is medium temperature whereas others may consider that as hot. Thus, the temperature can be represented as having memberships of 60% in the medium set and 40% in the hot set. The terms, medium and hot, are referred to as fuzzy terms. At that time the concept was very radical, and even many did not accept the idea at all. The concept of fuzzy logic showed first indications of success only after almost two decades of research. Engineering design and manufacturing use a vast amount of domain knowledge, expressing the knowledge is very difficult, and also often experts use vague terms to express the knowledge. Thus many engineering design and manufacturing problems have been addressed using fuzzy logic. Successful applications of fuzzy logic include control of chemical plants, control of camcorders, washing machines, knowledge capturing for engineering designs etc. For a good account on application of fuzzy logic in engineering problems, please refer to Mendel [8]. Expert Systems (Durkin [9]) have been the most obvious recipients of the benefits of fuzzy logic, since their application domain is often inherently fuzzy. Expert systems that utilise fuzzy logic concepts are termed 'fuzzy expert systems' (Roy et al. [10]). This part of the book presents research in some of the present issues in fuzzy logic, and describes some applications. The last paper presents application of Chaos Engineering in Industrial Applications.

The first paper by Kelly and Painter describes how Hypertrapezoidal Fuzzy Membership Functions (HFMFs) that is a multi-dimensional fuzzy membership function can be useful for decision support. This is a very interesting research, but requires further investigation to address some of the issues like 'can we extract very complex multi-dimensional knowledge from human experts for a HFMF', 'how do we determine the representative point(s) for a HFMF', etc. ? The paper presents successful application of HFMFs for Flight Mode Interpreter. The next paper on 'Fuzzy System Design' describes a methodology to obtain smaller but equally efficient fuzzy rule set using neural networks and genetic algorithms. The paper proposes an integrated approach to rule structure and parameter identification for fuzzy systems. The next paper by Jantzen and Dotoli presents some of their experience in promoting fuzzy logic education through Internet. People of similar interest should find it interesting. Patki et al. has presented their research in the area of developing 'fuzzy operating systems'. This research is very useful for corporate computing, and for improving usability of any operating system. The developed system, FUZOS, has fuzzy decision making ability in a context sensitive fashion. The final paper is not directly related to fuzzy logic it is about industrial application of chaos engineering. The paper is a very good overview in the area, and is recommended for the beginners in chaos.

4. Genetic Algorithms

Genetic Algorithms (GA) are based on the Darwinian model of survival of the fittest. According to this theory, an individual with the most favourable genetic characteristics is more likely to survive and produce offspring. Genetic Algorithms were originally developed by Holland [2] and have been analysed and developed further by many researchers such as De Jong [11] and Goldberg [4]. Genetic algorithms have been successfully employed for various applications related to artificial intelligence such as machine learning & game theory, and also in several widely differing areas such as image processing, manufacturing, economics, political science, linguistics, psychology, and immunology, to name a few. However, the most widely reported application of these methods is in the area of search and optimisation related to complex engineering problems, for which the conventional optimisation methods are found to be inadequate. GA is proved to be a robust search method for a wide range of optimisation problems.

The first step in GAs is to create a pool of individuals (called the initial population), which could be generated based on judgement and intuition, or on a purely random basis. Each member of the population represents a possible set of design variables corresponding to the optimum solution. New offspring are produced by carrying out three different processes, such as: reproduction, crossover and mutation. In reproduction, members of the population are selected randomly with a view to create new individuals, but with a bias towards fitter individuals. Some measure of fitness has to be developed for these purposes, and in most cases it is directly related to the objective function value corresponding to the set of design variables represented by the individual. This strategy, through which fitter individuals have a higher change of surviving through the next generation, ensures propagation of genetic information across generations.

This part of the book on Genetic Algorithms consists of six papers. The first paper by Moriwaki et al. discusses a genetic method for evolutionary agents in a co-evolutionary environment. This is a numerical simulation of an evolutionary environment in which a herbivore, a carnivore and plants co-exist and co-evolve. A new gene expression (n-BDD) is proposed for investigating the environment, and a food-chain relation has been observed. Using this method, the herbivore was seen to more rapidly acquiring its ability to survive from a carnivore compared to the finite state automation. One of the problems confronting the GA user is the appropriate choice of GA related parameters, which can greatly influence the efficiency and efficacy of the algorithm. The second paper by Williams and Crossley provides empirical guidelines on the choice of population size and mutation rates for a GA employing Uniform Crossover, based on a study involving several parameter combinations on four mathematical functions and one engineering design problem. In the third paper, Vasiljevik and Golobic have discussed the application of various evolutionary algorithms in the optimisation of a simple optical system viz. a two lens system called Doublet. A description of various evolutionary strategies and their relative merits and de-merits are provided, which include an adaptive steady state GA without duplicates, 1+1 evolutionary strategy and a μ, λ evolutionary strategy with and without

duplicates. The performances of these strategies have been compared with the classical dumped least squares method. Many research ideas from Genetic Algorithms find application in other areas of Evolutionary Computing, and vice versa. An example of this is provided in the next paper, in which Poli and Langdon discuss the application of a new crossover operator and (its variant) for Genetic Programming, which is quite similar to the One-Point Crossover in GA. Experimental evidence is provided, demonstrating that this form of crossover compares well with the standard crossover in Genetic Programming. Several researchers have also developed hybrid optimisation methods, in which various techniques are judiciously combined to derive maximum benefit. One such method is the Evolutionary Tabu Search (ETS), which is discussed in the paper by Chai, Jiang and Ma. They have formulated the geometric primitive extraction method used in content-based image retrieval and other problems related to computer vision as a cost function minimisation problem, and have used ETS for shape extraction in images. A comparison of ETS with three global optimisation methods such as, simulated annealing, genetic algorithms and tabu search showed that ETS can yield good near-optimal solutions and has better convergence speed.

Parallelism is an inherent feature in GAs, and many researchers have used this to develop GA implementations on parallel computers. One such method is illustrated in the paper by Goodman et al., which discusses the application of parallel Genetic Algorithms in the optimisation of composite structures. In most composite structures, the search space is highly discontinuous and multi-modal, with the presence of many local sub-optimal solutions or even singular extrema, which makes them very much suitable for application of GAs. Three applications have been discussed viz. design of energy absorbing laminated beams, airfoils with tailored bending-twisting coupling, and flywheel structures. Laminated beams were designed using a special FEM by optimising material stacking sequences to maximise the mechanical energy absorbed before fracture. The optimum stacking sequence of idealised aerofoil was determined to produce a desired twisting response while minimising weight, maximising in-plane stiffness, and maintaining acceptable stress levels. GA was used to select various flywheel shapes and material sequences along the radius to optimise the rotational energy per unit mass.

5. Decision Support, Constraints and Optimisation

Decision support is one of the main and complex activities in engineering. The advent of new technologies, especially computer based tools has helped engineers to design and manufacture a product more efficiently. The new technologies are mostly useful in automating routine tasks involved in the design and manufacturing processes. Decision making is still very much left to the engineers. The ever growing competition in the market place and increasing expectation of the users are adding many more dimensions to the decision making process. Thus decision making is becoming increasingly complex. Engineering decision making involves a lot of heuristic knowledge and also uncertainties. Soft computing provides the right technology required for the decision support. Fuzzy Expert Systems, and Neural Networks have been used to model engineering design and manufacturing knowledge (Parsaei and Jamshidi [12]). The models can then be used for decision support. Evolutionary computing has been used as a search technique to identify multiple good solutions to a problem, and thus help in decision support (Roy et al. [13]). Evolutionary computing is also heavily used in optimisation (Goldberg [4]). The technique can also optimise in the presence of constraints (Michalewicz [14]). This part of the book mostly presents application of evolutionary computing techniques for engineering decision support, and how they can effectively optimise constrained search space.

The first paper in this part, by De Falco et al., describes the development of a new mutation technique (Mijn) for a simple evolutionary algorithm. The technique is applied to the Inverse Aerofoil Design Optimisation problem and the performance is compared with that of a classical genetic algorithm. The next paper by Marian Mach presents engineering design as the class of problems with linear constraints. The paper introduces a concept of 'soft' constraints. Genetic algorithm (GA) has been applied for the constraint optimisation problem. The GA uses penalty function to represent the constraints. The penalty function can be modified easily to reflect different combinations of 'soft' constraints. Optimisation in case of multiple objectives is a major challenge. Bentley and Wakefield describe the problem of using a genetic algorithm to converge on a small, user-defined subset of acceptable solutions to multi-objective problems, in the Pareto-optimal (P-O) range. The paper tries to answer some of the questions like 'why is multi-objective optimisation difficult?', 'how can a set of acceptable solutions be defined?', etc. The next paper, by Valkovsky et al., describes the approach and prototype of the tool for handling uncertainty in probabilistic interpretation in the framework of Constraint Logic Programming (CLP). The research helps to integrate the real life uncertainties involved in scheduling, planning and manufacturing problems with the relevant decision making. The developed technique is applied in a network based project planning problem as a case study. The paper also discusses possible techniques to make Logical Inference with Probability more efficient by reducing the unnecessary trials. Baron et al. utilises domain knowledge and the knowledge about genetic algorithms to design suitable genetic operators for a shape optimisation problem. The paper examines n-dimensional pixel based shape representation for a beam design problem. The paper can be a good example to demonstrate the need for further development in an evolutionary algorithm to address specific problems. The next paper, by Myung-Ju and Chi-Geun, is about multiobjective optimisation. The authors have tried three different crossover techniques for Rural Postman Problem with Time Windows (RPPTW). The research identifies a Pareto-optimal set of solutions to the problem. Experiments show that a modified Order Exchange (OX) crossover technique, Modified Order Exchange (MOX), is more efficient for RPPTW. The last paper in this part, by Pant and Kalker-Kalkman, is about optimisation of 'commuter aircraft configurations'. The authors have applied simulated annealing (SA), Monte-Carlo method (MC), and Genetic Algorithms (GA) to the problem. The optimum configurations obtained by the three stochastic methods are then compared with those obtained earlier with a modified univariate method (MUV). The comparison shows SA, MC, and GA performed better than MUV, and the best results are achieved by the simulated annealing.

6. Engineering Design

In the past two decades, engineering design has begun to emerge as a discipline that tests the limits of the established computing techniques, originally developed for analytical and scientific applications. Design requires the intersection of scientific, mathematical and AI techniques to solve numerical and symbolic problems involving large search spaces. One distinction between design synthesis and the analytical approaches is that design rarely has right and wrong answers - it can admit good or bad solutions, optimal or suboptimal solutions, feasible or infeasible solutions. Creativity, described as a trait leading to novel solutions to a design problem, is often seen essential to any computational approach to design. This is particularly true for conceptual design where the problem is ill-defined, information is incomplete and past experience in the problem domain is often vital. The under-constrained set of specifications leads to a very large search space that requires fast pruning to reduce the search to only the feasible solutions. It is important to remember that due to ill-defined nature of the problem, optimality becomes a subjective rather than objective choice, thus defying the application of established optimisation methods in this case. These characteristics of design call for the application of soft computing techniques. Genetic algorithms have been successfully used to generate a feasible search space of an arbitrary size. Simulated annealing has been applied to reduce a search space beginning with a minimal set of constraints and still reach towards good solutions, if not optimal. Neural networks have been used to deliver the capability for learning from experience in a given domain. Fuzzy techniques have been applied to enable decisions based on ill-defined parameters. Naive physics has been applied for reasoning about the qualitative behaviour of a conceptual system. It is obvious that no single technique from the above categories will alone meet the challenge posed by the design problem. The various soft computing techniques need to be deployed jointly to carry out a design with the aid of a thinking machine.

As feasible design concepts are established and detailed further, more and more analysis is introduced in design. The search space becomes much smaller and the underlying the computational problem is better defined. This transition in turn lends the problem to relatively hard computational techniques. Still, heuristics continue to play a major role even in the detailed design stage as further choices have to be made such as in component selection, in building safety and reliability into the product for minimum cost, as well as in applying the criteria based on manufacturability and serviceability. The major computational problem at this stage is that of distributed information management and communication of this information to all interested design agents. Clearly, no algorithmic solution nor a universal information model exists to solve the generic detailed design problem, and again domain-based soft computing techniques have to be applied.

Seven papers have been selected in this section to highlight the major computational issues in design and the application of soft techniques to address some of these issues. Hopper and Turton review the application of GAs to a variety of packing problems in two and three dimensions. The authors have placed particular emphasis on the underlying genetic representation and the genetic operators. Bentley and Wakefield have applied the genetic theme to design evolution where a range of conceptual designs are able to evolve. The authors demonstrate this technique for the design of solid objects such as tables and car bodies. A key contribution here is a study of the issues that need investigated for further advance in automated design evolution. The theme of design evolution is carried further in a paper by Fogarty, Miller and Thomson whose object is the digital logic circuit design. The authors are able to carry out functional design of circuits and their implementation in a given family of integrated circuit devices thus dealing with function, component placement and routing in a single step. Again, the GA approach to circuit evolution has been favoured by the authors, and the techniques would be applicable to design in other domains such as mechanical and hydraulic systems with similar issues. The subsequent paper by Wu et al. addresses the communication issues between a team during the conceptual design stage. The authors treat the conceptual design process as an 'idea evolution' from an initial set of design parameters. Several design agents participate in an evolving dialogue based on population strings in a GA-based approach. Feasible solutions are thus jointly generated and evaluated by the team. The approach is illustrated on a microwave appliance. Potter et al. continue this theme to design emergence with the knowledge and learning in circuit configuration. Instead of 'hard-wiring' design rules into a knowledge base, they examine the approach of machine learning from known configurations. In this case, neural networks are to the task of automating the configuration design. The authors consider the use of neural networks to be particularly agreeable in a design context, due to some emergent properties associated with aspects of design creativity. The technique has been demonstrated on the configuration design of fluid power systems. Airframe structure evolution by means of GAs is the focus of the paper by Zarubin et al. The problem of aircraft structural optimisation has been defined in terms of a vector of design variables. The authors present the solution as an evolution of the concept, where the design process is an iterative one, where the sequence of models is used and values of design variables are verified. The method has been numerically tested for a short-range passenger aircraft. Experience in simulated evolution and adaptive search in engineering design is discussed by Greene. The author argues in favour of the application of these methods in the routine design of engineering systems and artefacts. Their flexibility in constraint-handling and expression of complex design goals requires designers who are experienced in the task domain. The author presents a compact and accessible package of robust, user-friendly black-box optimisers for use both by designers in industry and the students at the university.

7. Scheduling, Manufacturing and Robotics

With increasing customer demand and competition in the market, terms like 'intelligent manufacturing', 'agile manufacturing', 'manufacturing intelligence', 'manufacturing automation' are becoming very popular. Soft computing provides underlying technology to make manufacturing more responsive, more intelligent and overall more flexible. One of the major areas in manufacturing is control. There is a need to develop efficient (in terms of performance and time), and robust control systems. Neural networks (Nguyen and Widrow [15]) and fuzzy logic (Sugeno [16]) have been successfully applied to develop such control systems. Neural networks are good in learning about a system from data, but where knowledge is involved people often use Expert Systems, an artificial intelligence technique, to capture and reuse the knowledge. Recently there is a trend to integrate expert systems with neural networks (Bapi et al. [17]) to utilise the strengths of both the techniques in order to model a real life system. The other major areas in manufacturing where soft computing is very popular are scheduling, diagnostics, and decision support. Optimisation techniques like Genetic Algorithms are widely used to identify the best schedule for a manufacturing process. Neural networks and fuzzy logic are used to identify patterns in a set of data, and integrate domain knowledge to provide assistance in fault diagnosing or automatic monitoring (Hou and Lin [18]) of a manufacturing system. This part of the book presents research results on soft computing applications in different areas of manufacturing.

The first paper by Visweswaran and Anvekar investigates the application of a genetic algorithm (GA) to the Hybrid Channel Allocation (HCA) scheme in mobile cellular communication systems. This is a constrained optimisation problem. The paper reports a comparison between the optimisation based approach and the corresponding Fixed Channel Assignment scheme. The next paper, by Norenkov and Goodman, discusses how a multistage scheduling problem can be optimised using genetic algorithms. The paper describes a Heuristics Combination Method (HCM) that optimises the choice and sequence of application of a set of heuristic rules. A hybrid evolutionary-genetic method is reported with experimental results. The third paper is about optimising the geometrical parameters of drill point by combining GA and Gradient Method. The paper deals with a new evaluation method for complete geometry of conical reground drill point, aiming to develop a Computer Aided Inspection (CAI) system. The experimental investigation shows that a combined method of Steepest Gradient and Genetic Algorithm was the best in terms of computational time and accuracy of results. The paper by Michaud is very different from the previous three papers. This paper describes how Behaviour Exploitation can be used to influence motives using a simulated environment for mobile Robots, and to acquire knowledge about the World using a Pioneer 1 mobile robot. The research develops an adaptive control system for the robot. The paper discusses issues related to the design of a general control architecture for Intelligent Systems. The next paper by Tanaka et al. presents an application of genetic algorithm to optimise a time tabling problem. The authors have developed a more user friendly approach to solve the time tabling problem. The clients use 'application slips' to express their preferred time slots for each lecture. GA is used to optimise the order of the application slip to be used. This is an example of combinatorial optimisation problem with many hard and/or soft constraints. Pham and Chan use a self-organising neural network for pattern classification. The paper uses a new and better type of firing criterion that takes into account the individual components of the patterns to be clustered. The neural network is applied to recognise patterns in control charts. This research can contribute in developing an automatic 'statistical process control' system for a manufacturing plant.

8. Dynamic Systems, Identification and Control

System modelling and control were amongst the earliest applications for soft computing, particularly the fuzzy logic and neural networks although GAs has seen relatively fewer applications in this area. Fuzzy control techniques allow to describe a range of system parameter values through membership functions for fuzzy variables. Thus a dynamic system can be said to be in a fuzzy state with partial membership. They allow uncertainties in systems dynamics to be represented without the need to introduce statistical techniques such as probabilities.

In this section, five papers have been included, demonstrating the application of fuzzy, neural and genetic techniques to dynamic systems and their control. Hyotyniemi presents automatic structuring of unknown dynamic systems using a novel algorithm CCHA, which extracts linearly additive features from the system data. The ANN-based technique can be applied to the dynamic systems for both the recurrent and non-recurrent model structures. The models can be used for associative prediction tasks, so that different kinds of soft sensors can be realised. The author recognises that an explicit method for model structure identification does not exists and an exhaustive search still needs to be employed.

Coelho and Coelho utilise evolutionary computation in process identification and control. Evolution techniques of particular interest are the hybrid algorithms composed from genetic algorithms with simulated annealing. These are applied to the tuning of design parameters of a mono-variable PID controller for a non-linear mono-tank level and temperature processes composed of coupled twin-tanks. Nuernberger et al. deal with the optimisation of fuzzy controller which is usually tuned by 'trial-and-error' methods. The neuro-fuzzy approaches can simplify the design and optimisation process via learning techniques. The authors describe an updated version of the neuro-fuzzy model NEFCON, which is able to learn and to optimise the rule base of a Mamdani-like fuzzy controller on-line by a reinforcement learning algorithm that uses a fuzzy error measure. Authors present an implementation of the model and an application example under the MATLAB/SIMULINK development environment. Kim and Cho have developed an optimal COG (Centre Of Gravity) defuzzification method that improves the control performance of a fuzzy logic controller. The defuzzification method incorporates the membership values and the effective widths of membership functions in calculating a crisp value. An optimal effective width is determined automatically by the genetic algorithm through the training of some typical examples. Simulation results over the truck backer-upper control problem show that the proposed optimal COG defuzzifier reduces the average tracing distance by 23.8% compared with the conventional COG defuzzifier. Abonyi et al. focus on the temperature control of a batch polymerisation reactor where difficulty arises due to the physical and chemical properties of the contents vary between and within runs. The Takagi-Sugeno fuzzy logic controllers (FLC) are shown to be capable of providing good overall system performance for the non-linear operating range.

References

[1] Rechenberg, I., 1965, *Cybernetic Solution Path of an Experimental Problem*, Ministry of Aviation, Royal Aircraft Establishment (U.K.).

[2] Holland, J. H., 1975, *Adaptation in Natural and Artificial Systems*, University of Michigan Press, (second edition by: MIT Press, 1992).

[3] Fogel, L. J., Owens, A. J., and Walsh, M. J., 1966, *Artificial Intelligence through Simulated Evolution*, Wiley.

[4] Goldberg, D. E., 1989, *Genetic Algorithms in Search, Optimisation, and Machine Learning*, Addison-Wesley.

[5] Koza, J.R., 1992, *Genetic Programming: On the Programming of Computers by Means of Natural Selection*, MIT Press.

[6] Schalkoff, R. J., 1997, *Artificial Neural Networks*, The McGRAW-HILL Companies, Inc.

[7] Zadeh, L. A., 1965, Fuzzy Sets, *Information and Control*, v. **8**.

[8] Mendel, M. J., 1995, Fuzzy logic systems for engineering: a tutorial, *Proceedings of the IEEE*, v. 83, n. 3, pp. 345-376.

[9] Durkin, J., 1994, *Expert systems: design and development*, Prentice-Hall International Inc.

[10] Roy, R., Parmee, I. C., and Purchase, G., 1996, Qualitative evaluation of engineering designs using fuzzy logic, *CD-Rom Proceedings of the ASME DETC-Design Automation Conference, 18-22 August, Irvine, CA (USA)*, 96-DETC/DAC-1449.

[11] De Jong, K. A., 1975, *Analysis of the Behavior of a class of Genetic Adaptive Systems*, Doctoral Dissertation, university of Michigan, Dissertation Abstract International, 36(10), 5140B, USA.

[12] Parsaei, H. R., and Jamshidi, M. (Eds.), 1995, *Design and Implementation of Intelligent Manufacturing Systems*, Prentice Hall P T R, New Jersey (USA).

[13] Roy, R., Parmee, I. C., and Purchase, G., 1996, Integrating the genetic algorithms with the preliminary design of gas turbine blade cooling systems, *Proceedings of ACEDC'96 Conference, 26-28 March, Plymouth (UK), University of Plymouth*, pp. 228-235.

[14] Michalewicz, Z., 1995, A survey of constraint handling techniques in evolutionary computation methods, *Proceedings of the Evolutionary Programming IV, 1-3 March, CA (USA), MIT Press*, pp. 135-155.

[15] Nguyen, D. H., and Widrow, B., 1990, Neural networks for Self-Learning Control Systems, *IEEE Control Systems*, pp. 18-23.

[16] Sugeno, M. (Ed), 1985, *Industrial Applications of Fuzzy Control*, North-Holland, Amsterdam.

[17] Bapi, R.S., McCabe, S.L., and Roy, R., 1996, Artificial Intelligence and Neural Networks: Steps toward Principled Integration, *Book Review, Neural Networks*, vol. 9, no. 3, pp. 545-548.

[18] Hou, T. -H,, and Lin, L, 1995, Using Neural Networks for the automatic monitoring and recognition of Signals in Manufacturing Processes, In the book: *Design and Implementation of Intelligent Manufacturing Systems, Parsaei, H. and Jamshidi, M. (Eds.), Prentice Hall P T R*, pp. 141-160.

Part 1: Evolutionary Computing

Papers:

Artificial Neural Networks Optimization by means of Evolutionary Algorithms

I. De Falco[1], **A. Della Cioppa**[1], **P. Natale**[2] and **E. Tarantino**[1]

[1] *Research Institute on Parallel Information Systems*
National Research Council of Italy (CNR)
Via P. Castellino 111, Naples, ITALY.
[2] *Department of Mathematics and Applications, University of Naples "Federico II"*
Monte S. Angelo via Cintia, Naples, ITALY.

Keywords: Evolutionary Algorithms, Breeder Genetic Algorithms, Artificial Neural Networks.

Abstract

In this paper Evolutionary Algorithms are investigated in the field of Artificial Neural Networks. In particular, the Breeder Genetic Algorithms are compared against Genetic Algorithms in facing contemporaneously the optimization of (i) the design of a neural network architecture and (ii) the choice of the best learning method for nonlinear system identification. The performance of the Breeder Genetic Algorithms is further improved by a fuzzy recombination operator. The experimental results for the two mentioned evolutionary optimization methods are presented and discussed.

1. Introduction

EVOLUTIONARY Algorithms have been applied successfully to a wide variety of optimization problems. Recently a novel technique, the Breeder Genetic Algorithms (BGAs) [1, 2, 3] which can be seen as a combination of Evolution Strategies (ESs) [4] and Genetic Algorithms (GAs) [5, 6], has been introduced. BGAs use *truncation selection* which is very similar to the (μ, λ)-strategy in ESs and the search process is mainly driven by recombination making BGAs similar to GAs.

In this paper the ability of BGAs in the field of Artificial Neural Networks (ANNs) [7] has been investigated. Differently from GAs, which have been widely applied to design these networks [8], BGAs have not been examined in this task. The most popular neural network, the Multi–Layer Perceptron (MLP) [9], for which the best known and successful training method is the Back Propagation (BP) [10], has been considered.

A response must be given on how to construct the network and reduce the learning time. The BP method is a gradient descent method making use of derivative information to descend along the steepest slope and reach a minimum. This technique has two drawbacks: it may get stuck in local minima and requires the specification of a number of parameters so that the training phase can take a very long time. In fact, learning neural network weights can be considered a hard optimization problem for which the learning time scales exponentially becoming prohibitive as the problem size grows [10]. Besides in any new problem a lot of time can be wasted to find an appropriate network architecture. It seems natural to devote attention to heuristic methods capable of facing satisfactorily both problems.

Several heuristic optimization techniques have been proposed. In most cases the heuristic techniques have been used for the training process [11, 12, 13]. Whitley has faced the above tasks separately [14]. A different approach in which GAs have been utilized both for the architecture optimization and to choose the best variant of the BP method among four proposed in literature is followed in [15]. In [16] the GAs and the Simulated Annealing have been employed with the aim to optimize the neural network architecture and train the network with a particular model of BP.

Our approach is to use BGAs to yield the network architecture optimization and contemporaneously choose the best technique to update the weights in the BP optimizing the related parameters. In this way, an automatic procedure to train the neural network and to handle the network topology optimization at the same time is provided. Nonlinear system identification has been chosen as test problem to verify the efficiency of the approach proposed. This test is representative of a very intriguing class of problems for control systems in engineering and allows to effect significant experiments in combining evolutionary methods and neural networks.

The paper is organized as follows. In section 2 BGAs are briefly described. Section 3 is dedicated to the MLP and the BP methods, while in section 4 an explanation of the aforementioned identification problem is reported. In section 5 some implementation details are outlined. In section 6 the experimental results are presented and discussed. Section 7 contains final remarks and prospects of future work.

2. Breeder Genetic Algorithms

BGAs are a class of probabilistic search strategies particularly suitable to deal with continuous parameter optimization. They, differently from GAs which model natural evolution, are based on a rational scheme 'driven' by the breeding selection mechanism. This consists in the selection, at each generation, of the λ best elements within the current population of μ elements (λ is called *truncation rate* and its typical values are within the range 10% to 50% of μ).

The selected elements are let free to mate (self–mating is prohibited) so that they generate a new population of $\mu - 1$ individuals. The former best element is then inserted in this new population (*elitism*) and the cycle of life continues. In such a way the best elements are mated together hoping that this can lead to a fitter population. These concepts are taken from other sciences and mimic animal breeding.

A wide set of appropriate genetic recombination and mutation operators has been defined to take into account all these topics. Typical recombination operators are the Discrete Recombination (DR), the Extended Intermediate Recombination (EIR) and the Extended Line Recombination (ELR) [2]. As concerns mutation operators, the continuous and discrete mutation schemes (CM) are considered. However, a comprehensive explanation of these operators can be found in [2].

Moreover, in our approach the fuzzy recombination operator described below has also been considered [17]. Let us consider the following genotypes $\mathbf{x} = \{x_1, \ldots, x_n\}$ and $\mathbf{y} = \{y_1, \ldots, y_n\}$, where the generic x_i and y_i are real variables. The probability to obtain the i–th value of the offspring $\mathbf{z}$ is given by a bimodal distribution $p(z_i) = \{\psi(x_i), \psi(y_i)\}$ where $\psi(r)$ is a triangular probability distribution with modal values x_i and y_i with

$$x_i - d \mid y_i - x_i \mid \le r \le x_i + d \mid y_i - x_i \mid \quad \text{and} \quad y_i - d \mid y_i - x_i \mid \le r \le y_i + d \mid y_i - x_i \mid$$

for $x_i \le y_i$ and d is usually chosen in the range $[0.5, 1.0]$. For our aims the following triangular probability distribution is introduced:

$$\psi_s^b(r) = 1 - \frac{2 \mid s - r \mid}{b}$$

where s is the centre and b is the basis of the distribution.

A BGA can be formally described by:

$$BGA = (\mathcal{P}^0, \mu, \lambda, \mathcal{R}, \mathcal{M}, \mathcal{F}, \mathcal{T})$$

where $\mathcal{P}^0$ is the initial random population, μ the population size, λ the truncation threshold, $\mathcal{R}$ the recombination operator, $\mathcal{M}$ the mutation operator, $\mathcal{F}$ the fitness function and $\mathcal{T}$ the termination criterion. A general scheme of a BGA is outlined in the following:

```
Procedure Breeder Genetic Algorithm
begin
    randomly initialize a population of μ individuals;
    while (termination criterion not fulfilled) do
        evaluate goodness of each individual;
        save the best individual in the new population;
        select the best λ individuals;
        for i = 1 to μ - 1 do
            randomly select two elements among the λ;
            recombine them so as to obtain one offspring;
            perform mutation on the offspring;
        od
        update variables for termination;
    od
end
```

3. Artificial Neural Networks

ANNs represent an important area of research which opens a variety of new possibilities in different fields including control systems in engineering.

The MLP are ANNs consisting of a number of elementary units arranged in a hierarchical layered structure, with each internal unit receiving inputs from all the units in the previous layer and sending outputs to all the units in the following layer. In the reception phase of the unit the sum of the inputs of the unit is evaluated. Except than in the input layer, each incoming signal is the synaptic weighted output of another unit in the network. Let us consider that the MLP is composed of $\mathcal{L}$ layers with each layer l containing $\mathcal{N}^l$ neurons. Moreover, let us suppose that the neurons in the first layer contain $\mathcal{N}^0 = m$ inputs each, every such neuron receiving the same number of inputs and the output layer contains $\mathcal{N}^{\mathcal{L}} = n$ neurons. The total activation of the i-th neuron in the hidden layer l, denoted by h_i^l, is calculated as follows:

$$h_i^l(w_i^l, y^{l-1}) = \sum_{j=1}^{\mathcal{N}^{l-1}} w_{ij}^l y_j^{l-1} + \Theta_i \qquad i = 1, \ldots, \mathcal{N}^l \quad l = 1, \ldots, \mathcal{L} \tag{1}$$

where w_{ij}^l represents the connection weight between the j-th neuron in layer $l-1$ and the i-th neuron in layer l, y_j^{l-1} is the input arriving from the layer $l-1$ to the i-th neuron, and Θ_i is the threshold of the neuron. In the transmission phase, the output of the neuron is an activation value $y_i^l = f(h_i^l)$ computed as a function of its total activation. In the experiments performed three forms have been used for this function; in particular the 'sigmoid':

$$y_i^l = f(h_i^l) = \frac{1}{1 + e^{-kh_i^l}} \quad k > 0 \tag{2}$$

the related form symmetric about the h-axis:

$$y_i^l = f(h_i^l) = \tanh\left(h_i^l\right) \tag{3}$$

and a further semi-linear function here introduced:

$$y_i^l = \begin{cases} 1 & \text{if } h_i^l > 1 \\ h_i^l & \text{if } -1 \leq h_i^l \leq 1 \\ -1 & \text{if } h_i^l < -1 \end{cases} \tag{4}$$

The learning phase consists in finding an appropriate set of weights from the presentation of the inputs and the corresponding outputs. The most popular method for this phase is the BP. The system is trained by using a set containing p example patterns. Therefore, when the single pattern is presented to the network, the output of the first layer is evaluated. Then the neurons in the hidden layers will compute their outputs propagating them until the last layer is reached. The output of the network will be given by the neurons at this level, i.e. $y_i^{\mathcal{L}}$ for $i = 1, \ldots, \mathcal{N}^{\mathcal{L}}$. In the supervised learning, the goal is to minimize, with respect to the weight vector, the sum E of the squares of the errors observed at the output units after the presentation of the chosen pattern:

$$E = \sum_{s=1}^{p} E^s = \frac{1}{2} \sum_{s=1}^{p} \sum_{i=1}^{n} (y_i^{s,\mathcal{L}} - d_i^s)^2 \tag{5}$$

where d_i^s represents the desired output at the i-th output unit in response to the s-th training case. The weight changes are chosen so as to reduce the output error by an approximation to gradient descent until an acceptable value is attained. The last step is to "change" the weights by a small amount in the direction which causes the error, i.e. the direction opposed to the partial derivative of the (5) with respect to the weights in the hidden layers. Hence the following update procedure is introduced:

$$w_{ij}^l(t+1) = w_{ij}^l(t) - \eta \frac{\partial E(t)}{\partial w_{ij}^l(t)} \tag{6}$$

where η is the training balance which is conventionally chosen in the interval $[0,1]$. There are a number of other heuristic modifications to the basic approach which may speed up the training time or enable completion of the learning process. These approaches propose different ways to proceed in the change of the weights. The rationale behind the modifications suggested can be found in the reported references. In [9] a momentum term $\alpha \in [0,1]$ is suggested to be included in (6) so the formula becomes:

$$\Delta w_{ij}^l(t+1) = -\eta \frac{\partial E(t)}{\partial w_{ij}^l(t)} + \alpha \Delta w_{ij}^l(t) \tag{7}$$

where $\Delta w_{ij}^l(t)$ is the previous weight change.

In [18] an alternative strategy, known as exponential smoothing, is proposed which modifies the (6) in the following way:

$$\Delta w_{ij}^l(t+1) = (1-\eta) \frac{\partial E(t)}{\partial w_{ij}^l(t)} + \eta \Delta w_{ij}^l(t) \tag{8}$$

Besides, other modifications to the basic backpropagation method have been introduced. A statistical technique presented in [18] leads to the following equation:

$$\Delta w_{ij}^l(t+1) = \mu((1-\eta) \frac{\partial E(t)}{\partial w_{ij}^l(t)} + \eta \Delta w_{ij}^l(t)) + (1-\mu)\omega \tag{9}$$

where ω is determined by a Cauchy distribution and its usage is established by the Simulated Annealing technique. Thus, an initial temperature t_0 must be fixed together with the value μ which simply extends the dimension of the training parameter space.

Since the values of the parameters involved depend on the problem under examination, it is not possible to take an *a priori* decision by a qualitative analysis about which of the proposed variants is the best.

4. Neural Networks for System Identification

Neural networks, thanks to their ability to learn, approximate and classify examples, represent an effective response to the modern demand for complex control systems with a high precision degree. It is desirable that a trained network should produce correct response not only for the training patterns but also for hitherto unseen data. Therefore, in the simplest case, to assess the generalization ability of a trained network, the training data set is calculated in the same way but for different data points and the unseen patterns are assumed to be located in the region of the data evaluated. In system identification for modeling the input–output behavior of a dynamic system, the network is trained using the input–output data and the weights adjusted by the BP algorithm. The network is provided with information related to the system history so that it can represent adequately the dynamic behavior of the system in fixed ranges of a particular application. The training can be performed by observing the input–output behavior of the system with the neural network which receives the same input than the system and a fixed number of delayed inputs and outputs. The system output is the desired output of the network. The system and the network output are compared to allow the weight update so as to reduce the error until the required precision is reached. Fig. 1 illustrates the principle of modeling a nonlinear SISO (Single Input Single Output) system by using a neural network assuming that the output depends only on the input and the output at the previous time step. Thus, one assumes that the unknown system is discrete in time and continuous with respect to the input.

It is evident that the identification is performed off–line because the neural network operation is relatively slow.

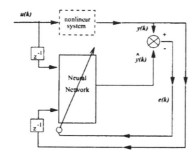

Fig. 1. *Modeling a nonlinear system with the MLP (z^{-1} denotes the time delay of a unit).*

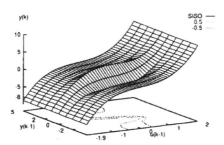

Fig. 2. *Response surface of the nonlinear system (10) to be modeled by the neural network.*

This problem has been chosen as a benchmark to compare GAs and BGAs both for its applicative interest and because the large size of the search space allows a statistically significant test of the relative performance of different algorithms.

To verify the performance of a neural network optimized by GAs and BGAs for a nonlinear system identification, the following test problem has been chosen:

$$y(k) = 2.5y(k-1)\sin(\pi e^{(-u^2(k-1)-y^2(k-1))}) + u(k-1)[1 + u^2(k-1)] \tag{10}$$

where u and y represent the input and the output respectively [19]. Eq. (10) evidences that the output $y(k)$ depends only on the previous input $u(k-1)$ and the previous output $y(k-1)$. This problem can be represented in a three-dimensional space as illustrated in Fig. 2.

5. Multi–Layer Perceptron Optimization by Evolutionary Algorithms

The basic idea is to provide an automatic procedure to find the most appropriate neural network structure. The optimization of an MLP which must be trained to solve a problem P is characterized by the need to determine :

1. **The architecture**

    (a) *The number of hidden layers* NHL,
    (b) *The number of nodes* NHl *for each layer l.*

2. **The activation function f_i^l to be used for each layer l.**
3. **Presence or absence of the bias b.**
4. **The most appropriate technique for weight change TRM and the training balance η.**
    On the basis of this technique the following parameters are to be fixed:

    (a) *the momentum term α,*

(b) *the initial temperature t_0,*

(c) *the statistical training term μ.*

We wish to point out that in our approach the training phase is effected with the BP algorithm. Analytical procedures to determine the exact configuration of these factors for a given application do not exist. Each time it is necessary to fix a particular configuration on the basis of empirical considerations, then to train the network with the patterns and to evaluate its quality. The algorithmical analysis demonstrates that this search presents an exponential complexity if performed with an exaustive technique. Furthermore, in addition to the time complexity, there is the possibility that the training fails because it runs into a configuration which is a local minimum. We have utilized heuristic methods to reduce the search time and the probability to get stuck in local minima. Namely, GAs and BGAs have been used to determine the appropriate set of parameters listed above. The aim is to provide the most general possible technique to determine the network structure. At the end of the evolutionary process not only the best network architecture for a particular application but also the trained network will be provided.

5.1. Encoding

The neural network is defined by a "genetic encoding" in which the genotype is the encoding of the different characteristics of the MLP and the phenotype is the MLP itself. Therefore, the genotype contains the parameters related to the network architecture, i.e. NHL and NHl, and other genes representing the activation function type f_i^l, the different BP methods and the related parameters.

For both the Evolutionary Algorithms considered, the chromosome structure $\mathbf{x} = (x_1, \ldots, x_n)$, constituted by 17 loci, is reported in Table 1. Each allele is defined in the subset A_i, $i \in \{1, \ldots, 7\}$

Locus	x_1	x_2	x_3	x_4	x_5	x_6	x_7	x_8	x_9	x_{10}	x_{11}	x_{12}	x_{13}	x_{14}	x_{15}	x_{16}	x_{17}
Gene	NHL	η	α	b	TRM	μ	t_0	f_i^0	NH1	f_i^1	NH2	f_i^2	NH3	f_i^3	NH4	f_i^4	f_i^5
Set	A_1	A_2	A_2	A_3	A_4	A_2	A_5	A_7	A_6	A_7	A_6	A_7	A_6	A_7	A_6	A_7	A_7

Table 1.

reported in the third row of the Table 1. Since it is sufficient in most applications that a network have few hidden layers, the value of NHL has been allowed to vary from 1 to 4 while the maximum value for NHl has been fixed equal to 20. Namely, the loci are defined within the following subsets:

$A_1 = \{1, \ldots, 4\}; A_2 = [0, 1] \subset \mathbb{R}; A_3 = \{0, 1\};$

$A_4 = \{1, 2, 3\}$ with $\{1 \equiv$ momentum, $2 \equiv$ exponential smoothing, $3 \equiv$ statistical technique$\}$;

$A_5 = [0, 100] \subset \mathbb{R}; A_6 = \{1, \ldots, 20\}; A_7 = \{1, 2, 3\}$ with $\{1 \equiv f_1, 2 \equiv f_2, 3 \equiv f_3\}$

where the locus A_4 individuates the way to change the weights, while the genes which individuate the activation function used can assume the value f_1 for the sigmoid, f_2 for tanh and f_3 for the semi–linear; it is worth noting that the activation function can be different for each layer.

5.2. The fitness function

To evaluate the goodness of an individual, the network is trained with a fixed number of patterns and then evaluated according to determined parameters. The parameters which seem to describe better the goodness of a network configuration are the mean square error E at the end of the training and the number of epochs ep needed for the learning. Clearly it is desirable to attain for both the parameters E and ep values as low as possible: in fact, a neural network must learn as fast as possible (small values for ep), and with a good approximation of the output desired (small values for E).

It is necessary to establish an upper limit $ep_{\max}$ to the number of epochs ep utilized by the network for the training, thus $0 < ep \le ep_{\max}$. Moreover, it is desirable that $0 \le e_{\min} \le E$ where $e_{\min}$ represents the minimum error required. Since the heuristic techniques are implemented for the minimization, the fitness function has been chosen as a function increasing when ep and E increase. Specifically the fitness function is:

$$\mathcal{F}(\mathbf{x}) = \frac{ep}{ep_{\max}} + E \qquad (11)$$

The choice of this fitness function is justified as it takes into account both E and the learning speed ep, weighting these contributions in a dynamic manner. Note that $e_{\min}$ is generally chosen equal to 10^{-a} with $a > 2$.

5.3. The optimization algorithm for the neural network structure

Let $\mathrm{MLP}(\mathbf{x}_i)$ be the algorithm for training the MLP related to the individual $\mathbf{x}_i$ representing a neural network configuration. The general schema for the BGA is the following[1]:

Given a pattern set P for the network training;
Procedure *Breeder Genetic Algorithm*
begin
    randomly initialize a population of μ neural network structures;
    while (termination criterion not fulfilled) **do**
        train $\mathbf{x}_i$ by means of $\mathrm{MLP}(\mathbf{x}_i)$ on P;
        evaluate the trained network $\mathbf{x}_i$;
        save the best trained network in the new population;
        select the best λ neural network configurations;
        for $i = 1$ **to** $\mu - 1$ **do**
            randomly select two structures among the λ;
            recombine them so as to obtain one offspring;
            perform mutation on the offspring
        od
        update variables for termination;
    od
end

The procedure is constituted by an evolutionary algorithm, a BGA, and by a procedure for training the MLP encapsulated in the main program. This procedure returns the values for the fitness evaluation. Its action is completely transparent to the main program which can only evaluate the goodness of the result.

6. Experimental results

Due to the fact that the choice of the best selection method and genetic operators would require very high execution times if preliminary tests were effected directly for the non linear system identification, we have decided to use as benchmark a subset of the classical test functions known as F6, F7 and F8 [20].

For the GA these experiments have been performed to establish the best selection method. In particular, the truncation selection has resulted to have better performance than that achieved by the proportional, the tournament and the exponential selections.

For the BGA the objective of the tests has been to determine the best combination among mutation and recombination operators. The experimental results have proved that the discrete mutation in combination with the fuzzy recombination operator has led to the best performance. It has to be pointed out that the fuzzy recombination allows to obtain a reduction of about 40% in terms of the convergence time with respect to the extended intermediate and the extended line recombination operators. Other preliminary tests for the MLP have been conducted by using as application the well-known "exclusive OR" (XOR) function. This task is commonly considered an initial test to evaluate the network ability to classify data.

The GA has a population of 30 individuals and it utilizes the truncation selection with a 1–elitist strategy. The percentage of the truncation is set equal to 30%. The algorithm has been executed 10 times fixing as termination criteria $ep_{\max} = 500$ and $e_{\min} = 10^{-8}$. A one-point crossover operator with probability $p_c = 0.6$ and a mutation operator with probability $p_m = 0.06$ have been employed. The

---

[1] For the GA there are two offspring so the cycle **for** is to be repeated $\frac{\mu-1}{2}$ times.

minimum value for the error has been obtained at the second generation demonstrating the effectiveness of the approach proposed. This result is due to the presence of good individuals in the initial population. This can be explained by the fact that the number of individuals fixed is probably large for the problem under consideration. Moreover, the result achieved implies the presence of neural network structures in the population able to learn in a number of epochs lower than ep_{max}. The best individual in the population has learned in 80 epochs. The architecture of this individual is 2 16 5-1, without bias and the momentum technique with $\eta = 4.72656 \cdot 10^{-1}$ and $\alpha = 4.375 \cdot 10^{-1}$ has resulted to be the best. In the ninety per cent of the cases the technique with momentum and the absence of the bias has resulted to be the best result. The statistical and the exponential techniques produce individuals with bad fitness values which disappear during the evolution. For the XOR we have only established that the network has learned the patterns as the network has been trained with the complete space of four possible solutions. The application of BGAs for this problem has provided results not much different because the test is very simple.

For this reason we consider a more complex problem to compare GAs and BGAs. The problem derives from a real-world application in engineering and it is the nonlinear system identification described by (10). The first problem is to determine the dimension of the pattern set to train the network. Being the (10) a continuous function with respect to u, the number of samples sufficient to approximate it with the desired precision must be found. In fact, a small set of patterns cannot be sufficient while a large set involves a high time to effect the training phase.

The dimension of the pattern set A has been fixed equal to 100 which has turned out to be a number of samples sufficient to train the network. The system has been excited by setting u to be a random signal uniformly distributed between -2.0 and 2.0. Considering that the output at the time $k = 0$ is equal to zero, Eq. (10) has been evaluated on the set of patterns. In this way, we have obtained the triples $(u(k), y(k), y(k + 1))$ in which the first two elements represent the input and the previous output of the system and the third the desired output.

Moreover, a pattern set B constituted by 500 samples different from the previous ones has been determined in order to verify how much the network has learned about the SISO system. To evaluate the goodness attained by the trained networks, the network has been excited by the first two elements of the triples of the set B and the output $\hat{y}(k + 1)$ computed by the network has been compared with the desired output evaluating the value of the error.

For the GA the number of individuals in the population has been fixed equal to 50 and as termination criterium we have established a value of 100 for the maximum number of generations. It utilizes the truncation selection with a 1-elitist strategy and the percentage of the truncation equal to 30%. The operators are the one-point crossover with $p_c = 0.8$ and the mutation with $p_m = 0.06$. The value of the error e_{min} has been set equal to 10^{-3}. We have executed ten trials of the algorithm. It has been necessary to train the neural networks for 1000 epochs to allow the GA to converge in a not too high number of generations, so as to reduce the network error at the end of the training phase. As concerns the techniques for the weight variation during the training phase, the best method found is that based on the momentum.

The best architecture achieved is 2-19-19-18-1, the technique for changing the weights is that of the momentum with $\eta = 1.95312 \cdot 10^{-2}$, and $\alpha = 4.98047 \cdot 10^{-2}$ and the bias present. The value of the error obtained is $7.89309 \cdot 10^{-3}$. The activation functions determined by the evolution from the first to the last level are respectively: sigmoid, semi-linear, tanh, tanh and semi-linear. In Fig. 3 it is possible to observe the response surface of the network MLP achieved by the evolutionary process, after the verification of the network effected on the pattern set B. Furthermore, the error surface computed as the difference between the network output $\hat{y}(k + 1)$ and the system output $y(k + 1)$ is shown.

The results obtained with the BGA are better than those achieved by the GA. The number of individuals in the population and the termination criterion for the BGA have been the same used for the GA. The truncation rate λ is 30% and $e_{min} = 10^{-4}$. The discrete mutation with $p_m = 0.06$ has been employed. It has to be pointed out that the performance of the BGA has allowed to choose a value of e_{min} lower than that of the GA. The evolution terminates after 50 generations, half of those needed by the GA. In particular, the results for the BGA have been improved in terms of convergence speed by using the fuzzy recombination instead of the extended intermediate and the extended line recombination operators. The best architecture achieved is 2-20-16-17 1, the technique for changing the weights is that of the momentum with $\eta = 1.57313 \cdot 10^{-2}$ and $\alpha = 7.03549 \cdot 10^{-1}$. The value of the error obtained is $5.21883 \cdot 10^{-4}$. The activation functions determined by the evolution from the first to the last level are

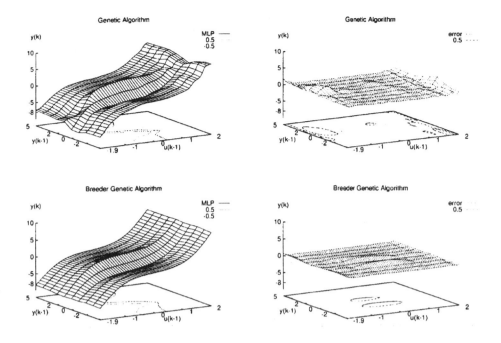

Fig. 3. *The response surface of the MLP on the pattern set B and the error surface obtained by the GA and the BGA.*

respectively: sigmoid, tanh, tanh, sigmoid and semi-linear. It is possible to note that the BGA converges faster than the GA so as to reduce the time to complete the evolutionary process. Suffice it to consider that the GA has needed on average six days to terminate its execution on a RISC 6000 station, while the BGA has taken on average five days when the discrete and extended line recombination have been employed and three days with the fuzzy operator. In figure 3 it is also possible to observe the approximation of function (10) performed by the neural network at the end of the evolutionary process of the BGA. As it can be noted, in such case, the error is much lower than that achieved by the GA.

Since the set B utilized to verify the effectiveness of the neural network optimization performed by the BGA is larger than the set A used for the training, it is evident that not only has the trained network learned the patterns, also it has shown ability to generalize being able to appropriately simulate the characteristics of the system to be identified.

7. Conclusions

In this paper the effectiveness of the Evolutionary Algorithms in the field of ANNs has been demonstrated. The BGAs are compared against the GAs in facing the simultaneous optimization of the design of the architecture and the choice of the best learning method. In particular, the fuzzy recombination operator, which improves the performance of the BGA, has been employed. The problem test used is a nonlinear system identification. The experimental results have proved the superiority of the BGA with respect to the GA in terms of both solution quality and speed of convergence.

The BGA method can also be easily distributed among several processors as it operates on populations of solutions that may be evaluated concurrently. Further work will be concerned with the implementation of a parallel version of the evolutionary approach proposed with the aim at improving the performance.

References

[1] Mühlenbein, H., Schlierkamp–Voosen, D., 1993, Analysis of Selection, Mutation and Recombination in Genetic Algorithms, *Neural Network World*, **3**, pp. 907–933.

[2] Mühlenbein, H., Schlierkamp-Voosen, D., 1993, Predictive Models for the Breeder Genetic Algorithm I. Continuous parameter optimization, *Evolutionary Computation*, **1**(1), pp. 25–49.

[3] Mühlenbein, H., Schlierkamp-Voosen, D., 1994, The Science of Breeding and its Application to the Breeder Genetic Algorithm, *Evolutionary Computation*, **1**, pp. 335–360.

[4] Bäck, T., Hoffmeister, F., Schwefel H.P., 1991, A survey of evolution strategies. *Proceedings of the Fourth International Conference on Genetic Algorithms, San Mateo CA, USA*, Morgan Kauffmann, pp. 2–9.

[5] Holland, J.H., 1975, *Adaptation in Natural and Artificial Systems*, University of Michigan Press, Ann Arbor.

[6] Goldberg, D. E., 1989, *Genetic Algorithms in Search, Optimization and Machine Learning*, Addison-Wesley, Reading, Massachussets.

[7] Hertz, J., Krogh, A., Palmer, R.G., 1991, *Introduction to the Theory of Neural Computation*, Addison-Wesley Publishing.

[8] Kuscu, I., Thornton, C., 1994, Designing Neural Networks using Genetic Algorithms: Review and Prospect, *Cognitive and Computing Sciences*, University of Sussex.

[9] Rumelhart, D. E., Hinton, G. E., Williams, R. J, 1986, Learning Internal Representations by Error Propagation, *Parallel Distributed Processing: Explorations in the Microstructure of Cognition*, **VIII**, MIT Press.

[10] Rumelhart, D. E., McLelland, J. L., 1986, *Parallel Distributed Processing*, **I-II**, MIT Press.

[11] Montana, D. J., Davis, L., 1989, Training Feedforward Neural Networks using Genetic Algorithms, *Proceedings of the Eleventh International Joint Conference on Artificial Intelligence*, pp. 762–767.

[12] Hiestermann, J., 1990, Learning in Neural Nets by Genetic Algorithms, *Parallel Processing in Neural Systems and Computers*, North-Holland, pp. 165–168.

[13] Battiti, R., Tecchiolli, G., 1995, Training Neural Nets with Reactive Tabu Search, *IEEE Trans. on Neural Networks*, **6**(5), pp. 1185–1200.

[14] Whitley, D., Starkweather, T., Bogart, C., 1990, Genetic Algorithms and Neural Networks: Optimizing Connections and Connectivity, *Parallel Computing*, **14**, pp. 347–361.

[15] Reeves, C. R., Steele, N. C., 1992, Problem-solving by Simulated Genetic Processes: a Review and Application to Neural Networks, *Proceedings of the Tenth IASTED Symposium on Applied Informatics*, pp. 269–272.

[16] Stepniewski, S., Keane, A. J., Pruning back propagation Neural Networks using Modern Stochastic Optimization Techniques, to appear in *Neural Computing & Applications*.

[17] Voigt, H.M., Mühlenbein, H., Cvetković, D., 1995, Fuzzy Recombination for the Continuous Breeder Genetic Algorithm, *Proceedings of the Sixth International Conference on Genetic Algorithms*, Morgan Kauffmann.

[18] Wassermann, P. D., 1989, *Neural Computer Theory and Practice*, Van Nostrand Reihnold, New York.

[19] Yaw-Terng Su, Yuh-Tay Sheen, 1992, Neural Networks for System Identification, *Int. J. of Systems Sci.*, **23**(12), pp. 2171-2186.

[20] Mühlenbein, H., Schomish, M., Born, J., 1991, The Parallel Genetic Algorithm as Function Optimizer, *Parallel Computing*, **17**, pp. 619–632.

Fitness Causes Bloat

W. B. Langdon, R. Poli

School of Computer Science, The University of Birmingham, Birmingham B15 2TT, UK
{W.B.Langdon,R.Poli}@cs.bham.ac.uk http://www.cs.bham.ac.uk/˜wbl, ˜rmp
Tel: +44 (0) 121 414 4791, Fax: +44 (0) 121 414 4281

Keywords: genetic programming, structural complexity, introns, Occam's Razor, MDL, Price's Theorem

Abstract

The problem of evolving an artificial ant to follow the Santa Fe trail is used to study the well known genetic programming feature of growth in solution length. Known variously as "bloat", "fluff" and increasing "structural complexity", this is often described in terms of increasing "redundancy" in the code caused by "introns".

Comparison between runs with and without fitness selection pressure, backed by Price's Theorem, shows the tendency for solutions to grow in size is caused by fitness based selection. We argue that such growth is inherent in using a fixed evaluation function with a discrete but variable length representation. With simple static evaluation search converges to mainly finding trial solutions with the same fitness as existing trial solutions. In general variable length allows many more long representations of a given solution than short ones. Thus in search (without a length bias) we expect longer representations to occur more often and so representation length to tend to increase. That is fitness based selection leads to bloat.

1 Introduction

The tendency for programs in genetic programming (GP) populations to grow in length has been widely reported [1, 2, 3, 4, 5, 6, 7]. This tendency has gone under various names such as "bloat", "fluff" and increasing "structural complexity". The principal explanation advanced for bloat has been the growth of "introns" or "redundancy", i.e. code which has no effect on the operation of the program which contains it. ([8] contains a survey of recent research in biology on "introns"). Such introns are said to protect the program containing them from crossover [9, 10, 11]. [12] presents an analysis of some simple GP problems designed to investigate bloat. This shows that, with some function sets, longer programs can "replicate" more "accurately" when using crossover. That is offspring produced by crossover between longer programs are more likely to behave as their parents than children of shorter programs. [13] provides a detailed analysis of bloat using tree schemata specifically for GP.

In this paper we advance a more general explanation which should apply generally to any discrete variable length representation and generally to any progressive search technique. That is bloat is not specific to genetic programming applied to trees and tree based crossover but should also be found with other genetic operators and non-population based stochastic search techniques such as simulated annealing and stochastic iterated hill climbing.

In the next section we expand our argument that bloat is inherent in variable length representations such as GP. In Sections 3 and 4 we analyse a typical GP demonstration problem, showing that it suffers from bloat but also showing that bloat is not present in the absence of fitness based selection. Section 5 describes the results we have achieved and this is followed in Section 6 by a discussion of the potential advantages and disadvantages of bloat and potential responses to it. Finally Section 7 summarises our conclusions.

2 Bloat in Variable Length Representations

In general with variable length discrete representations there are multiple ways of representing a given behaviour. If the evaluation function is static and concerned only with the quality of a partial solution and not with its representation then all these representations have equal worth. If the search strategy were unbiased, each of these would be equally likely to be found. In general there are many more long ways to represent a specific behaviour than short representations of the same behaviour. Thus we would expect a predominance of long representations.

Table 1: Ant Problem

Objective:	Find an ant that follows the "Santa Fe trail"
Terminal set:	Left, Right, Move
Functions set:	IfFoodAhead, Prog2, Prog3
Fitness cases:	The Santa Fe trail
Fitness:	Food eaten
Selection:	Tournament group size of 7, non-elitist, generational
Wrapper:	Program repeatedly executed for 600 time steps.
Population Size:	500
Max program:	500
Initial population:	Created using "ramped half-and-half" with a maximum depth of 6
Parameters:	90% one child crossover, no mutation. 90% of crossover points selected at functions, remaining 10% selected uniformly between all nodes.
Termination:	Maximum number of generations G = 50

Practical search techniques are biased. There are two common forms of bias when using variable length representations. Firstly search techniques often commence with simple (i.e. short) representations, i.e. they have an in built bias in favour of short representations. Secondly they have a bias in favour of continuing the search from previously discovered high fitness representations and retaining them as points for future search. That is there is a bias in favour of representations that do at least as well as their initiating point(s).

On problems of interest, finding improved solutions is relatively easy initially but becomes increasingly more difficult. In these circumstances, especially with a discrete fitness function, there is little chance of finding a representation that does better than the representation(s) from which it was created (cf. "death of crossover" [14, page 222]). So the selection bias favours representations which have the same fitness as those from which they were created.

In general the easiest way to create one representation from another and retain the same fitness is for the new representation to represent identical behaviour. Thus, in the absence of improved solutions, the search may become a random search for new representations of the best solution found so far. As we said above, there are many more long representations than short ones for the same solution, so such a random search (other things being equal) will find more long representations than short ones. In GP this has become known as bloat.

3 The Artificial Ant Problem

The artificial ant problem is described in [15, pages 147–155]. It is a well studied problem and was chosen as it has a simple fitness function. Briefly the problem is to devise a program which can successfully navigate an artificial ant along a twisting trail on a square 32×32 toroidal grid. The program can use three operations, Move, Right and Left, to move the ant forward one square, turn to the right or turn to the left. Each of these operations take one time unit. The sensing function IfFoodAhead looks into the square the ant is currently facing and then executes one of its two arguments depending upon whether that square contains food or is empty. Two other functions, Prog2 and Prog3, are provided. These take two and three arguments respectively which are executed in sequence.

The artificial ant must follow the "Santa Fe trail", which consists of 144 squares with 21 turns. There are 89 food units distributed non-uniformly along it. Each time the ant enters a square containing food it eats it. The amount of food eaten is used as the fitness measure of the controlling program.

4 GP Parameters

Our GP system was set up to be the same as given in [15, pages 147–155] except the populations were allowed to continue to evolve even after an ant succeeded in traversing the whole trail, each crossover produces one child rather than two, tournament selection was used and the ants were allowed 600 operations (Move, Left, Right) to complete the trail. The details are given in Table 1, parameters not shown are as [16, page 655]. On each version of the problem 50 independent runs were conducted.

Since bloating is widespread in GP it is common to impose a limit on the size of programs. Commonly this is in

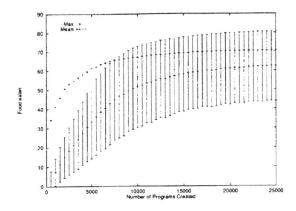
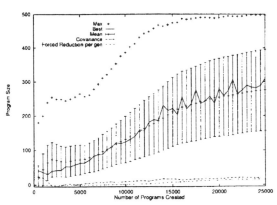

Figure 1: Evolution of maximum and population mean of food eaten. Error bars indicate one standard deviation. Means of 50 runs.

Figure 2: Evolution of maximum and population mean program length. Error bars indicate one standard deviation. Solid line is the length of the "best" program in the population. Means of 50 runs.

the region of 50, although Koza imposed a maximum depth restriction of 17 on his evolved programs. As we are studying program size the relatively large limit of 500 was used. As we shall see, this allows three separate stages of evolution to be studied: the initial population, bloat and bloat constrained by an upper bound. Note a limit of 500 allows the evolved programs to be far bigger than required to solve the problem. For example the 100% correct solution given in [15, page 154] takes about 543 time steps to traverse the Santa Fe trail but has a length of only 18 nodes and this is not the most compact solution possible.

5 Results

5.1 Standard Runs

In 50 independent runs 6 found ants that could eat all the food on the Santa Fe trail within 600 time steps. The evolution of maximum and mean fitness averaged across all 50 runs is given in Figure 1. (In all cases the average minimum fitness is near zero). These curves show the GP fitness behaving as expected with both the maximum and average fitness rising rapidly initially but then rising more slowly later in the runs. The GP population converges in the sense that the average fitness approaches the maximum fitness. However the spread of fitness values of the children produced in each generation remains large and children which eat either no food or only one food unit are still produced even in the last generation.

Figure 2 shows the evolution of maximum and mean program size averaged across all 50 runs. We see in the first four generations the average program length grows rapidly from an initial size of 23.5 to 73.5. Program size remains fairly static from generation 4 to generation 11, when it begins to grow progressively towards the maximum allowed program size of 500.

Surprisingly the mean program length grows faster than the size of the best program within the population and between generations 2 and 12 it exceeds it (cf. Section 5.1.1). In the later generations the mean program length and length of the best program on average lie close to each other. The size of the best program refers to that of a single individual. In generations with more than one individual program having the top score, one such program is chosen at random and labelled the "best". This random choice leads to random fluctuations in apparent program size which can be seen in Figure 2 after generation 27 despite averaging over 50 runs.

5.1.1 Fitness is Necessary for Bloat – Price's Theorem Applied to Representation Length

Price's Covariance and Selection Theorem [17] from population genetics relates the expected change in frequency of a gene Δq in a population from one generation to the next, to the covariance of the gene's frequency in the original population with the number of offspring z produced by individuals in that population (see Equation 1). We have used it to help explain the evolution of the number of copies of functions and terminals in GP populations [14, 18].

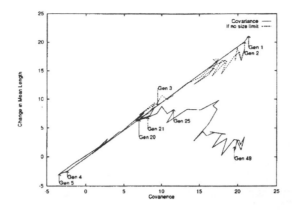

Figure 3: Covariance of program length and normalised rank based fitness v. change in mean length in next generation. Means of 50 runs.

$$\Delta q = \frac{\text{Cov}(z, q)}{\bar{z}} \tag{1}$$

In our GP the size of the population does not change so $\bar{z} = 1$ and the expected number of children is given by the parent's rank so in large populations the expected change is approximately $\text{Cov}(t(r/p)^{t-1}, q)$ as long as crossover is random. (t is the tournament size and r is each program's rank within the population of size p).

Where representation length is inherited, such as in GP and other search techniques, Equation 1 should hold for representation length. More formally Price's theorem applies (provided length and genetic operators are uncorrelated) since representation length is a *measurement function* of the genotype [19, page 28]. If it held exactly a plot of covariance vs. change in mean length would be a straight line (assuming $\bar{z}$ is constant). The solid line plot in Figure 3 shows good agreement between theory and measurement until generation 20. After Generation 20 the rise in program length is smaller than predicted by Equation 1. The bulk of the discrepancy is due to the restriction on program length, as can be seen from the dotted line in Figure 3.

Essentially Equation 1 gives a quantitative measurement of the way genetic algorithms (GAs) search. If some aspect of the genetic material is positively correlated with fitness then, other things being equal, the next generation's population will on average contain more of it. If it is negative, then the GA will tend to reduce it in the next generation.

For a given distribution of representation lengths Equation 1 says the change in mean representation length will be linearly related to the selection pressure. This provides some theoretical justification for the claim [2, page 112] that "average growth in size ... is proportional to selection pressure". Which is based upon experimental measurements with a small number of radically different selection and crossover operators on Tackett's "Royal Road" problem [2]. (Solutions to the Royal Road problem are required to have prespecified syntactic properties which mean "that the optimal solution has a fixed size and does not admit extraneous code selections". Such solutions are not executable programs.)

Referring back to Figure 2 and considering the length of the best solution and the covariance in the first four generations. We see the covariance correctly predicts the mean length of programs will increase. In contrast if we assumed the GA would converge towards the individual with the best fitness in the population, the fact that the mean length is higher than the mean length of the best individual would lead us to predict a fall in mean program length. That is Price's Theorem is the better predictor. This is because it considers the whole population rather than just one (albeit the best in the population).

Between generations 4 and 20 the mean program size can be reasonably described by a parabola, i.e. growth is quadratic. This corresponds with the covariance rising linearly (slope 0.76) with number of generations.

Where fitness selection is not used (as in the following sections), each individual in the population has the same fitness and so the covariance is always zero. Therefore Price's Theorem predicts on average there will be no change in length.

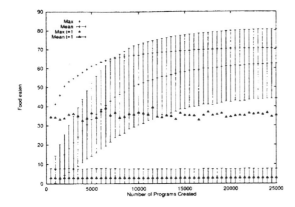

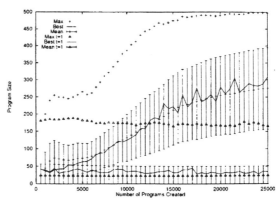

Figure 4: Evolution of maximum and population mean of food eaten. Error bars indicate one standard deviation. Means of 50 runs comparing tournament sizes of 7 and 1.

Figure 5: Evolution of maximum and population mean program length. Error bars indicate one standard deviation. Solid line is the length of the "best" program in the population. Means of 50 runs.

5.2 No Selection

A further 50 runs were conducted (using the same initial populations) and no fitness selection. Unsurprisingly no run found a solution and the maximum, mean and other fitness statistics fluctuate a little but are essentially unchanged (cf. Figure 4). Similarly program size statistics fluctuate but are essentially the same as those of the initial population (cf. Figure 5). This is in keeping with results with no selection reported in [2, page 112]. Given crossover produces random changes in length we might have expected the spread of lengths to gradually increase. This is not observed. The slow fall in maximum program size can be seen in Figure 5. There is also a small fall in mean standard deviation from 32.8 in the initial population to 27.3 at the end of the runs.

5.3 Removing Selection

A final 50 runs were conducted in which fitness selection was removed after generation 25 (i.e. these runs are identical to those in Section 5.1 up to generation 25). Of the six runs which found ants that could eat all the food on the Santa Fe trail within 600 movements, four where found before generation 26 and of these three retained such programs until generation 50 while in one case the maximum fitness dropped below 89. In another run a solution was found in generation 41 but no other generations of this run contained solutions.

The evolution of maximum and mean fitness averaged across all 50 runs is given in Figure 6. As expected Figure 6 shows in the absence of fitness selection the fitness of the population quickly falls.

After fitness selection is removed, if the crossover operator is unbiased, we expect the length of programs to fluctuate at random about the lengths of their parents. So the mean length should change little and the spread of lengths should increase. The length restriction means the crossover operator is biased because it must ensure the child it produces does not exceed the length limit. Where an unbiased choice of crossover points would cause the offspring to be too large, the roles of the two parents are reversed and a shorter (and legal) program is produced instead. This bias (which is more important as programs start to approach the length limit) explains the slow fall in program size seen in Figure 7. The spread of programs' lengths can also be seen. The average standard deviation rises from 92.3 in generation 26 to 107.6 at the end of the run.

5.4 Correlation of Fitness and Program Size

Figure 8 plots the correlation coefficient of program size and amount of food eaten by the ant it controls. (Correlation coefficients are equal to the covariance after it has been normalised to lie in the range $-1 \ldots +1$. By considering food eaten we avoid the intermediate stage of converting program score to expected number of children required when applying Price's Theorem). Figure 8 shows that in all cases there is a positive correlation (typically $+0.3$) between program performance and length of programs. Where selection is not used longer programs may be more fit simply because they are more likely to contain useful primitives (such as Move) than short

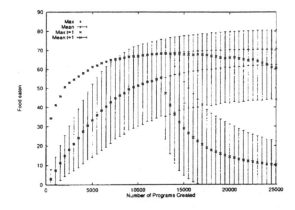

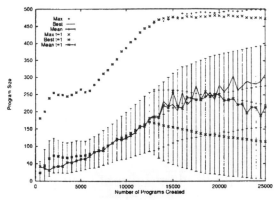

Figure 6: Evolution of maximum and population mean of food eaten. Error bars indicate one standard deviation. Means of 50 runs showing effect of removing fitness selection.

Figure 7: Evolution of maximum and population mean program length. Error bars indicate one standard deviation. Solid line is the length of the "best" program in the population. Means of 50 runs.

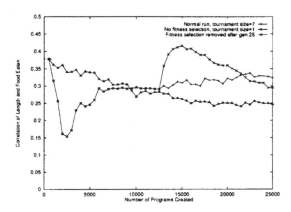

Figure 8: Correlation of program length and fitness, Normal runs, runs without selection and runs with selection removed halfway through. Means of 50 runs.

programs. When selection is used the correlation may be because when crossover reduces the length of a program it is more likely to disrupt the operation of the program and so reduce its fitness.

5.5 Effect of Crossover on Fitness

As expected in the initial generations crossover is highly disruptive (cf. Figure 9) with only 19.5% of crossovers producing a child with the same score as its first parent (i.e. the parent from which it inherits the root of its tree). However after generation 4 this proportion grows steady, by generation 18 more than half have the same fitness. At the end of the run this has risen to 69.8%.

The range of change of fitness is highly asymmetric; many more children are produced which are worse than their parent than those that are better. By the end of the run, no children are produced with a fitness greater than their parent. Similar behaviour has been reported on other problems [11] [20, page 183] [14, Chapter 7].

5.6 Non-Disruptive Crossover and Program Length

In Section 2 we argued that there are more long programs with a given performance than short ones and so a random search for programs with a given level of performance is more likely to find long programs. We would expect generally (and it has been reported on a number of problems) that crossover produces progressively fewer improved solutions as evolution proceeds and instead in many problems selection drives it to concentrate upon finding solutions with the same fitness. Apart from the minimum and maximum size restrictions crossover is

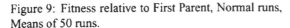

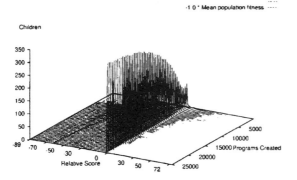

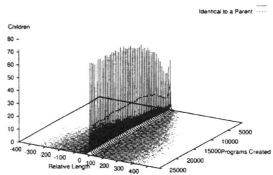

Figure 9: Fitness relative to First Parent, Normal runs, Means of 50 runs.

Figure 10: Change in length of offspring relative to first parent where score is the same. 50 Normal runs.

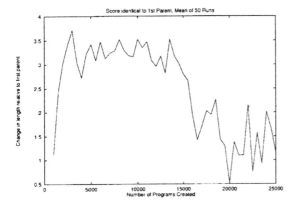

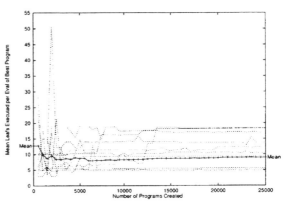

Figure 11: Mean change in length of offspring relative to first parent where score is the same. Normal runs.

Figure 12: Energy used per call of "best" program. 10 normal runs.

random so we would expect to see it finding more long solutions than short ones. The change in program length of programs produced by crossover which have the same fitness as their first parent is plotted in Figure 10. At first sight Figure 10 is disappointingly symmetric with a large central spike of programs with the same length and rapidly falling tails either side of zero. Initially 8% of crossovers produce a child which has the same length and fitness as its first parent. This proportion falls to a minimum of 6% in generation 4 and then rises progressively to 15% at the end of the run. The dashed line on Figure 10 plots the number of children in each generation produced by crossovers that are identical to one or other of their parents. On the ant problem 5% of crossovers produce such clones.

Closer examination of the data in Figure 10 reveals that on average (cf. Figure 11) there is a bias towards longer programs. On average programs with the same fitness as their first parent are always longer than it. From generation 4 to 23 they are on average about 3 program nodes longer. Presumably due to the average program length approaching the upper limit, the bias falls to about 1 after generation 33 but remains positive.

Part of the reason for the central peak of Figure 10 is a result of another aspect of GP crossover: almost all the genetic material is inherited from the first parent. Crossover points are typically drawn from close to the leafs of both parents simply because for large bushy trees most parts of a tree are close to its leafs. While the expected size of crossover fragments depends in detail upon the trees selected as parents and the relative weighting applied to functions and terminals, cf. Table 1, typically both the inserted subtree and the subtree it replaces consist of a function and its leafs. Since these subtrees are short together they produce a small change in total size. It seems likely that small changes are less likely to effect fitness than big ones, hence the spike in Figure 10. Inheriting principally from one parent is in sharp contrast to more traditional GAs (and indeed sexual reproduction in "higher organisms" in nature) where offspring receive about the same amount of genetic material from each parent.

Every program terminal uses one energy unit each time it is executed. Thus the amount of energy used when executing a program gives the actual number of terminals it has used. From Figure 12 we see typically only 5–10 terminals are used per execution (a move and four turns allows the ant to move forward whilst searching either side of it), While different parts of the program could be executed each time it is called, in the cases of those that have been analysed in detail, mostly code that has been used before is re-executed. That is large programs evolve which contain small functional parts.

6   Discussion

6.1   Do we Want to Prevent Bloat?

From a practical point of view the machine resources consumed by any system which suffers from bloat will prevent extended operation of that system. However in practice we may not wish to operate the system continually. For example it may quickly find a satisfactory solution or better performance may be achieved by cutting short its operation and running it repeatedly with different starting configurations [15, page 758].

In some data fitting problems growth in solution size may be indicative of "over fitting", i.e. better matching on the test data but at the expense of general performance. For example [1, page 309] suggests "parsimony may be an important factor not for 'aesthetic' reasons or ease of analysis, but because of a more direct relationship to fitness: there is a bound on the 'appropriate size' of solution tree for a given problem".

By providing a "defence against crossover" [11, page 118] bloat causes the production of many programs of identical performance. These can consume the bulk of the available machine resources and by "clogging up" the population may prevent GP from effectively searching for better programs.

On the other hand [3, page 84] quotes results from fixed length GAs in favour of representations which include introns, to argue we should "not ... impede this emergent property [i.e. introns] as it may be crucial to the successful development of genetic programs". Introns may be important as "hiding places" where genetic material can be protected from the current effects of selection and so retained in the population. This may be especially important where fitness criteria are dynamic. A change in circumstance may make it advantageous to execute genetic material which had previously been hidden in an intron.

In complex problems it may not be possible to test every solution on every aspect of the problem and some form of dynamic selection of test cases may be required [21]. For example in some cases co-evolution has been claimed to be beneficial to GP. If the fitness function is sufficiently dynamic, will there still be an advantage for a child in performing identically to its parents? If not, will we still see such explosive bloat?

6.2   Three Ways to Control Bloat

Three methods of controlling bloat have been suggested. Firstly, and most widely used (e.g. in these experiments), is to place a universal upper bound either on tree depth [15] or program length. ([22, 18] discuss unexpected problems with this approach).

The second (also commonly used) is to incorporate program size directly into the fitness measure (often called parsimony pressure) [15, 23, 24]. [13] gives an analysis of the effect of parsimony pressure which varies linearly with program length. Multi-objective fitness measures where one objective is compact or fast programs have also been used [25].

The third method is to tailor the genetic operations. [26, page 469] uses several mutation operators but adjusts their frequencies so a "decrease in complexity is slightly more probable than an increase". [10] suggests targeting genetic operations at redundant code. This is seldom used, perhaps due to the complexity of identifying redundant code. [7] showed bloat continuing despite their targeted genetic operations. Possibly this was because of the difficulty of reliably detecting introns. There was a route whereby the GP could evolve junk code which masqueraded as being useful and thereby protected itself from removal. While [13] propose a method where the likelihood of potentially disruptive genetic operations increases with parent size.

7   Conclusions

We have generalised existing explanations for the widely observed growth in GP program size with successive generations (*bloat*) to give a simple statistical argument which should be generally applicable both to GP and other systems using discrete variable length representations and static evaluation functions. Briefly, in general simple static evaluation functions quickly drive search to converge, in the sense of concentrating the search on trial solutions with the same fitness as previously found trial solutions. In general variable length allows many more long representations of a given solution than short ones of the same solution. Thus (in the absence of a parsimony bias) we expect longer representations to occur more often and so representation length to tend to increase. That is fitness based selection leads to bloat.

In Sections 3, 4 and 5 we have taken a typical GP problem and demonstrated with fitness selection it suffers from bloat whereas without selection it does not. We have demonstrated that if fitness selection is removed, there is a slight bias in common GP crossover (caused by the practical requirement to limit bloat) which causes a slow reduction in program size. NB fitness causes bloat in spite of a small crossover bias in favour of parsimony. Detailed measurement of crossover confirms after an extended period of evolution, most crossovers are not disruptive (i.e. most children have the same fitness as their parents). It also shows children with the same fitness as their parents are on average consistently longer than them.

In Section 5.1.1 we apply Price's Theorem for the first time to program lengths within GP populations. We confirm experimentally that it fits GP populations unless restrictions on program size have significant impact. We used Price's Theorem to prove fitness selection is required for a change in average representation length. In Section 6 we discussed the circumstances in which we need to control bloat and current mechanisms which do control it but suggest a way forward may be to consider more complex dynamic fitness functions.

Acknowledgements

This research was funded by the Defence Research Agency in Malvern and the British Council. We would like to thank Wolfgang Banzhaf, Terence Soule and Tom Haynes for helpful suggestions and critisim.

References

[1] Walter Alden Tackett, 1993, Genetic programming for feature discovery and image discrimination, *Proceedings of the 5th International Conference on Genetic Algorithms, ICGA-93*, University of Illinois at Urbana-Champaign, July, pp. 303–309.

[2] Walter Alden Tackett, 1994, *Recombination, Selection, and the Genetic Construction of Computer Programs*, PhD thesis, University of Southern California, USA.

[3] Peter John Angeline, 1994, Genetic programming and emergent intelligence, *Advances in Genetic Programming*, chapter 4, pp. 75–98, MIT Press, Cambridge, MA, USA.

[4] Walter Alden Tackett, 1995, Greedy recombination and genetic search on the space of computer programs, *Foundations of Genetic Algorithms 3*, Estes Park, Colorado, USA, 31 July–2 August, pp. 271–297.

[5] W. B. Langdon, 1995, Evolving data structures using genetic programming, *Genetic Algorithms: Proceedings of the Sixth International Conference (ICGA95)*, Pittsburgh, PA, USA, July, pp. 295–302.

[6] Peter Nordin and Wolfgang Banzhaf, 1995, Complexity compression and evolution, *Genetic Algorithms: Proceedings of the Sixth International Conference (ICGA95)*, Pittsburgh, PA, USA, July, pp. 310–317.

[7] Terence Soule, James A. Foster, and John Dickinson, 1996, Code growth in genetic programming, *Genetic Programming 1996: Proceedings of the First Annual Conference*, Stanford University, CA, USA, July, pp. 215–223.

[8] Annie S. Wu and Robert K. Lindsay, 1996, A survey of intron research in genetics, *Parallel Problem Solving From Nature IV. Proceedings of the International Conference on Evolutionary Computation*, Berlin, Germany, September, pp. 101–110.

[9] Tobias Blickle and Lothar Thiele, 1994, Genetic programming and redundancy, *Genetic Algorithms within the Framework of Evolutionary Computation (Workshop at KI-94, Saarbrücken)*, Im Stadtwald, Building 44, D-66123 Saarbrücken, Germany, pp. 33–38, Max-Planck-Institut für Informatik (MPI-I-94-241).

[10] Tobias Blickle, 1996, *Theory of Evolutionary Algorithms and Application to System Synthesis*, PhD thesis, Swiss Federal Institute of Technology, Zurich.

[11] Peter Nordin, Frank Francone, and Wolfgang Banzhaf, 1996, Explicitly defined introns and destructive crossover in genetic programming, *Advances in Genetic Programming 2*, chapter 6, pp. 111–134, MIT Press, Cambridge, MA, USA.

[12] Nicholas Freitag McPhee and Justin Darwin Miller, 1995, Accurate replication in genetic programming, *Genetic Algorithms: Proceedings of the Sixth International Conference (ICGA95)*, Pittsburgh, PA, USA, July, pp. 303–309.

[13] Justinian P. Rosca and Dana H. Ballard, 1996, Complexity drift in evolutionary computation with tree representations, Technical Report NRL5, University of Rochester, Computer Science Department, Rochester, NY, USA.

[14] W. B. Langdon, 1996, *Data Structures and Genetic Programming*, PhD thesis, University College, London.

[15] John R. Koza, 1992, *Genetic Programming: On the Programming of Computers by Natural Selection*, MIT Press, Cambridge, MA, USA.

[16] John R. Koza, 1994, *Genetic Programming II: Automatic Discovery of Reusable Programs*, MIT Press, Cambridge MA, USA.

[17] George R. Price, 1970, Selection and covariance, *Nature*, **227**, 1:520–521.

[18] W. B. Langdon and R. Poli, 1997, An analysis of the MAX problem in genetic programming, *Genetic Programming 1997: Proceedings of the Second Annual Conference*, Stanford University, CA, USA, July.

[19] Lee Altenberg, 1995, The Schema Theorem and Price's Theorem, *Foundations of Genetic Algorithms 3*, Estes Park, Colorado, USA, 31 July–2 August, pp. 23–49.

[20] Justinian P. Rosca and Dana H. Ballard, 1996, Discovery of subroutines in genetic programming, *Advances in Genetic Programming 2*, chapter 9, pp. 177–202, MIT Press, Cambridge, MA, USA.

[21] Chris Gathercole and Peter Ross, 1994, Dynamic training subset selection for supervised learning in genetic programming, *Parallel Problem Solving from Nature III*, Jerusalem, October, pp. 312–321.

[22] Chris Gathercole and Peter Ross, 1996, An adverse interaction between crossover and restricted tree depth in genetic programming, *Genetic Programming 1996: Proceedings of the First Annual Conference*, Stanford University, CA, USA, July, pp. 291–296.

[23] Byoung-Tak Zhang and Heinz Mühlenbein, 1993, Evolving optimal neural networks using genetic algorithms with Occam's razor, *Complex Systems*, 7, 199–220.

[24] Hitoshi Iba, Hugo de Garis, and Taisuke Sato, 1994, Genetic programming using a minimum description length principle, *Advances in Genetic Programming*, chapter 12, pp. 265–284, MIT Press, Cambridge, MA, USA.

[25] William B. Langdon, 1996, Data structures and genetic programming, *Advances in Genetic Programming 2*, chapter 20, pp. 395–414, MIT Press, Cambridge, MA, USA.

[26] K. Sims, 1993, Interactive evolution of equations for procedural models, *The Visual Computer*, **9**, 466–476.

TRIUNE–BRAIN INSPIRED UNIFYING VIEW OF INTELLIGENT COMPUTATION

Raju Surampudi Bapi

Centre for Neural and Adaptive Systems, School of Computing, University of Plymouth, Plymouth PL4 8AA United Kingdom, rajubapi@soc.plym.ac.uk

Keywords: dynamical systems, self-organisation, triune theory, soft computing, computational intelligence

Abstract

Paul MacLean, following the evolutionary scheme, proposed the idea of "triune–brain," in which the cortex is organised into three layers. He proposed that the three layers are responsible for instinctual behaviour, the motivational and emotional influences, and the rational influences on decision making, respectively. We borrow this metaphor of triune–brain to propose a unifying viewpoint for bringing together disparate themes of intelligent computation. In this framework, quantitative methods of statistics are the equivalent of behaviours at the instinctual layer, rule–based approaches of the symbol processing–kind, being rational and logical, sit at the top and methods of soft computing (neural networks, fuzzy logic, genetic algorithms, etc.) operate at the intermediate level. It is argued that just as all the levels are important for a functioning organism, this three-level interaction is crucial for theories of intelligent computation.

1. Introduction

Intelligence and computability are discussed in this section. In the next section various methods of intelligent computation are introduced. In the third section, a unified view of all these techniques is offered. The last section offers conclusions and pointers to future work.

1.1 Definitions of Intelligence

Defining intelligence is arguably one of the most difficult tasks. There has been little agreement as to what aspects need to be included in such a definition. In what follows, we summarise some of the definitions used by psychologists and then quote the definition of an artificial intelligence system theorist.

Wechsler, the designer of Wechsler Adult Intelligence Scale (WAIS) which is used extensively to assess human intelligence, gave his working definition of intelligence (in [1]): "Intelligence, as a hypothetical construct, is the aggregate or global capacity of the individual to act purposefully, to think rationally, and to deal effectively with his environment" (p. 79). Matarazzo [1] emphasised that functional intelligence is not identical with the mere sum of these abilities, however inclusive. The reasons, according to him, are, "(1) the ultimate products of intelligent behaviour are a function not only of the number of abilities or their quality but also of the way in which they are combined, that is, their configuration, (2) factors other than intellectual ability, for example, those of drive and incentive, are involved in intelligent behaviour, and (3) finally, whereas different orders of intelligent behaviour may require varying degrees of intellectual ability, an excess of any given ability may add relatively little to the effectiveness of the behaviour as a whole" (p. 79).

Howard Gardner [2], a Harvard Psychologist, defined intelligence "as the ability to solve problems, or to create products, that are valued within one or more cultural settings" (p. xiv), and he warned that his definition does not say anything about either the sources of these abilities or the proper means of testing them. Further he argued that "a human intellectual competence must entail a set of skills of problem solving — enabling the individual *to resolve genuine problems or difficulties* that he or she encounters and, when appropriate, to create an effective product — and must also entail the potential for *finding or creating problems* — thereby laying the groundwork for the acquisition of new knowledge" (p. 60). This definition emphasises the importance of the cultural context in which intelligent capabilities are acquired and assessed.

Earl Hunt [3] gave a lucid account of the recent controversies in the notions of intelligence and heritability. The details of this discussion are not relevant to the present exposition. However, it is worth looking at the differences he identifies between the notions of intelligence in psychometry and cognitive psychology.

According to psychometric views, human intellectual competence appears to divide along three dimensions, fluid intelligence, crystallised intelligence, and visual-spatial reasoning. *Fluid intelligence* is the ability to develop techniques for solving problems that are new and unusual, from the perspective of the problem solver. *Crystallised intelligence* is the ability to bring previously acquired, often culturally defined, problem-solving methods to bear on the current problem. *Visual-spatial reasoning* is the specialised ability to use visual images and visual relationships in problem solving. Cognitive psychology views intelligence in an information processing sense: Intelligence is the process of creating a mental representation of the current problem, retrieving relevant information from storage, and manipulating the representation in order to obtain an answer. Thus cognitive psychology views cognition as a process, whereas the psychometric view makes it a collection of abilities.

Allen Newell, an artificial intelligence system theorist and designer, takes a pragmatic view in defining intelligence. Newell [4] defined that "a system is intelligent to the degree that it approximates a knowledge-level system" (p. 90). A knowledge-level system is one that possesses a medium of knowledge and has a body of knowledge. This body of knowledge contains details of the external environment in which the system is embedded, its goals about how the environment should be, the sequence of actions it took over time, and the relations between them. Thus, as per this definition, if a system uses all of the knowledge that it has, it must be perfectly intelligent and if a system does not have some knowledge, failure to use it cannot be a failure of intelligence. If a system has some knowledge and fails to use it, then there is certainly a failure of some internal ability. This failure can be identified with a lack of intelligence.

Clearly, the common theme in all these definitions is that the system, artificial or natural, possesses general underlying abilities that are useful in perceiving the external world and acting on it to fulfill external or internal goals. Also most of these definitions distinguish between the abilities that are situation specific and those that are general and thus are applicable across situations. The other important distinction that Newell made is between possessing knowledge versus ability to bring all of it to bear on a given problem. These distinctions will be useful for any theory of intelligent computation. In the next section it will be shown how the metaphor of computability limits the space of possibilities in ways of thinking about natural intelligence (human cognition, mind) and artificial intelligence. This then would pave the way for the discussion of alternative perspectives.

1.2 Computability and Intelligence

Metaphors, the linguistic devices that provide insights into the nature of an unknown in terms of a known, have a significant heuristic value as guides to further investigation. In the realm of intelligent systems (artificial and natural), the computational metaphor has been a persistent one. It has been used to propose information processing view of the mind, i.e., mind as a computer and also the other way round, the computations in a computer as the analogue of the thinking aspect of mind. The computational metaphor also led to the formal systems view point of thinking and intelligence. For example, thinking has been equated to symbol manipulation and first order predicate logic has been used extensively as a representational framework for many AI systems (Newell, [4]). These formalisms have also been used to explain the workings of the mind (Fodor, [5]).

West & Travis [6] reflected on the computational metaphor and its use in artificial intelligence. They concluded that "the main consequence of the computational metaphor and formalist perspective is the need to recreate, inside the mind and inside the computer, a symbolic simulacrum of an extensive portion of the external world in such a manner that it is amenable to processing in pragmatically finite time. This need is not a consequence of the objective to emulate the mind but the perspective from which this problem was approached" (p.76). Thus they claimed that need for computability forced intelligence to be formalised. This formalist tradition has further restricted what mechanisms could count as intelligent processes. Although this computational metaphor is not in itself useful, it stands now as the declaration of the dualistic tradition in the design of intelligent systems: "(1) mind and nature are absolutely separate; (2) the mind manipulates abstract representations of the environment, never the environment directly, and it is this manipulation that constitutes thinking; and (3) the manipulations that make up mental functions can be expressed in a formal language; and (4) this formal language is basically mechanical and deterministic, at least sufficiently so that it can be embodied in a machine (computer) where it can function as a replicant of the functioning of the human mind" ([6], p.74). In a subsequent section, mechanisms that are non-deterministic and self-organising will be discussed as an alternative to these formalist systems.

The other shortcoming of the symbol manipulating approaches is the "grounding" problem. A discussion of how the "grounding" problem can be overcome using connectionist methods can be found in Honavar & Uhr [7] and Bapi et al. [8]. Harnad [9] argued that the woes of symbol manipulating machines (proposed in AI) relate to the "ungrounded" nature of the atomic symbols. He pointed out that the initial set of symbols, being arbitrary, is flexible enough to support compositionality (atomic symbols can be combined and molecular representations can be decomposed according to a formal syntax) and systematicity (the atomic and molecular symbols and the rules of syntax can be systematically assigned a meaning). He asserted however, that the very fact that these symbols are arbitrary, means they become "ungrounded", i.e., their meaning and interpretation reside in the subject's 'head' and are otherwise 'rootless.' His solution calls for grounding the categorisation in motor and sensory perception. By enabling the organism to interact with the environment, the resulting perceptions will have an action-based anchoring and the meaning does not depend on arbitrary interpretation by an external agent. The formalisms of self-organising dynamical systems and autopoiesis, to be discussed in a subsequent section, offer alternative ways of achieving the symbiosis of the organism (artificial or natural) with the environment.

2. Dynamical Systems Theory Viewpoints

In this section, two viewpoints based on dynamical systems approach will be discussed. Discussion of other viewpoints such as, the various soft computing methods (neural networks, fuzzy logic, genetic algorithms), artificial life, and genetic programming, is not attempted here as they are by now well known (for a discussion on AI perspectives see [10]). The primary aim of this article is to offer alternative and unifying views.

2.1 Self-Organisation

Erich Jantsch, a Viennese astrophysicist turned organisational theorist and futurist, proposed the notion of co-evolution (Jantsch, [11]). He tried to combine the dissipative system approach of Ilya Prigogine, who received a Nobel prize in chemistry in 1977, with the system's theory ideas of Ludwig von Bertalanffy. He synthesised the above ideas with the notion of *autopoiesis*, a process by which living organisms renew themselves by regulating processes that are responsible for the integrity of their structure. The idea of autopoiesis was first introduced by the Chilean biologists, Humberto Maturana and Francisco Varela (see Maturana & Varela, [12] for a detailed exposition of this idea). West & Travis [13] supported Maturana and Varela's autopoiesis idea as a viable formalism for intelligent system design.

Jantsch's goal was to propose a General Dynamic System Theory. Jantsch's attempt was to offer alternative paradigms for evolution. He thought that the Darwinian view of evolution as a process of adaptation through competition, was limited. His proposal was for evolution through mutual co-operation, and termed it *co-evolution*. The basic principle of co-evolution is that evolution at the micro and macro structural levels proceeds hand-in-hand as a whole. Co-evolution calls for a mutual influence of changes at the micro-level and the changes in the macro structural level on each other, without one being causal for the other. The causal connection is simultaneous and instantaneous. As a result of this complex feedback connection, the micro does not build the macro in gradual steps and also great shifts in macro-level are not necessarily reciprocated at the micro-level. He elaborated these co-evolution ideas using dissipative system framework. These co-evolution ideas are now being seriously pursued in the area of evolutionary computing [14].

A dissipative structure is one which is continuously dissipating entropy. Entropy represents the amount of unavailable (unavailable to the host system for any useful work) energy in the process of energy transformation from one form to another. Dissipative systems are self-organising and nondeterministic and operate far from equilibrium conditions. A vortex in a pool of water is an example of such a system. Vortex comes about due to the dynamic equilibrium between the inrush and outflow of water. Jantsch realised that increased autonomy means more instability and openness to change. Although this seems paradoxical, this is precisely what gives the system dynamic stability, that is openness to change than a state of static equilibrium. He arrived at this view finding that a dissipative structure is in a constant need of new input to maintain its form and hence is in intimate contact with the flow of the surrounding environment. In this view, John Doe, a generic human is pictured as a self-organising (autopoietic dissipative) process rather than merely as a distinct part of a whole. He is a collection of and is part of a collection of self-organising structures. He is simultaneously a macro structure feeding energy input into his internal processes to sustain himself as well as being the part of the other structures such as the corporation he works for, the city he lives in, the culture he belongs to, and the other external structures such as the ecosystem, solar system etc. There is a dynamic interaction among these various structures. One of the views this picture emphasises is the importance of the dynamic process between the so-called individual parts and the whole. Briggs & Peat [15] elegantly summarised this point:

> Co-evolution overthrows neo-Darwinism and asserts that life forms are not created piece by piece in small changes: they are dissipative structures arising spontaneously and holistically out of the flux and flow of macro and micro processes. Co-evolution explains the gentleness of the whale, the delicacy of the tropical fish, the gay markings of the butterfly, the curiosity of the human mind not as simply responses to the demands for survival but as the creative play and co-operative necessity of an entire evolving universe. (p. 214)

The language of nonlinear mathematics of the dynamical systems theory gives ways of capturing these insights in the design of systems and in understanding complex systems such as the Stock market, Global Economics, etc. He had a unifying vision of the relations between levels of planning, world views, guiding images, and system theoretic approaches. He demonstrated how planning may be organised as an evolutionary process. Jantsch's key idea in this hierarchical arrangement and also in his co-evolution theory, is that the top levels are open for entry and absorption of novelty and change. With the 'opening upward,' the manager of the higher conceptual levels (which are not necessarily the higher power levels) become the managers of change, whereas the lower conceptual levels need administrators, they serve as the dependable operators of established and only slowly changing processes. The three-level hierarchy will be discussed again when the triune brain theory is introduced in the next section. Jantsch also introduced the idea of mind and intelligence as the *quality* of the self-organising dynamics. These ideas of self-organisation and views of complex systems such as societies, markets are very useful in thinking about alternative ways of designing intelligent systems and the underlying algorithms for intelligent computation.

2.2 Dynamical Cognition

Jordan Pollack, an AI and Neural network researcher, in his book review of Allen Newell's *Unified theories of cognition*, offered a vivid picture of how the viewpoint of dynamical systems theory is very relevant to bringing in biological realism in artificial systems and in modelling complex systems [16]. He asserted that complex control systems such as — the insect and animal colonies, the immune system in the human body, the genetic control of fetal development, the evolutionary control of populations of species, co-operative control in social systems, autopoietic control system for maintaining the planet — all of these have some means of self-control while allowing for extreme creativity. He lamented that the current formalisms do not consider them as viable theories of computational intelligence. His research in neural networks had led him to believe that fractals and chaos have a role to play. Based on this research, he proposed a *dynamical cognition hypothesis*:

> The recursive representational and generative capacities required for cognition arise directly out of the complex behaviour of non-linear dynamical systems. (p. 367)

Although such dynamical systems formalisms offer qualitative advantages over the more logical formalisms, we are long way from using these in practical applications. However, the point we would like to make is that the field of computational intelligence, being still in its infancy, needs to explore various viewpoints and not get stuck in 'local minima.' In the next section, various viewpoints discussed in this article as well as the other softcomputing methods will be unified using the framework of triune–brain theory.

3. Triune–Brain Theory Based Unifying View

MacLean [17] hypothesised, based on extensive behavioural studies, that the human brain is divided into three 'layers' that arrived at different stages of evolution — the deepest layer is the "reptilian brain," responsible for automatic instinctive behaviour reflected in some basic maintenance patterns, and some habitual patterns. Above the reptilian brain is the "paleomammalian (old mammalian) brain," responsible for emotions such as fear, love, and anger, which focus on the needs for individual and species survival. Finally, at the top is the "neomammalian (new mammalian) brain," responsible for rational strategies and verbal capacities. This framework has been used by Bapi & Denham [18] to propose a unification approach for methods of computational intelligence. The underlying conceptual framework is due to Leven [19] who distinguished between three different problem solving styles, which he named after three well known mathematicians: DANTZIG problem solvers typically use direct-solving techniques — tackle one problem at a time; BAYESIAN solvers attempt to find the best possible solution; GODEL problem solvers seek meaning, look for causes, and adopt novel approaches. Also, Raghupathi, Levine, Bapi, & Schkade [20] used this framework to introduce three-level hierarchies in the process of legal reasoning.

This idea of triune–brain, that is that the brain can be viewed at three levels; instinctual, emotional and rational; can be used to unify various computational styles. One way to look at the computational theories is that the quantitative schemes such as the statistical and mathematical methods fit in with the instinctual level, as these schemes usually are rigid and depend on the nature of input exclusively. Usually there are no built in mechanisms in these techniques to adapt themselves to varying situations. The learning schemes of traditional AI, such as frames, scripts, case based reasoning systems, etc., on the other hand, fit in at the rational level where the behaviour is more logical and rational. Most of the soft-computing schemes, such as neural networks, fuzzy logic, genetic algorithms, classifier systems, etc., fit in at the emotional level or at the instinctual-emotional or emotional-rational borders. It is these schemes at the intermediate level that offer ways of studying how emotion colours perception, of deciding when to employ rational techniques and of promoting a shift between the levels of instinct, emotion and reason. The beginnings of how this can be achieved in neural networks has been indicated in Levine & Leven [21] and Bapi [22].

Table 1. Three-fold models in various domains

Evolutionary Brain		
Reptilian	Paleomammalian	Neomammalian
Forms of Rationality		
Calculative	Arational	Deliberative
Decision Making		
Dantzig	Godelian	Bayesian
Guiding Image for Manager (Intelligent Agent): Jantsch's Viewpoint		
Administrator	Co-ordinator	Catalyst
Theories of Intelligent Computation		
Quantitative Methods	Soft Computing	Symbolic AI
(Statistics, Mathematical methods)	(NN, Fuzzy Logic, GA, etc.)	(Frames, Scripts, CBR)

Note. From *Motivation, emotion, and goal direction in neural networks* (Chap. 8, p. 293) by D. S. Levine and S. J. Leven (Eds.), 1992, Hillsdale, NJ: Lawrence Erlbaum Associates. Copyright 1992 by Lawrence Erlbaum Associates, Inc. Adapted by permission.

Leven [23] culled three-fold models from various domains and discussed them in the light of the triune brain theory of MacLean [17]. Table 1 lists various three-fold models. Dreyfuss and Dreyfuss (see the discussion in [23]) argued, from their study of experts and nonexperts, that nonexperts employed *calculative rationality*, which involves looking for context-free cues and using precise rules. Competent thinkers developed hierarchical rule bases, organised information in semantic nets and use *deliberative rationality* to weigh options and make decisions. Whereas the experts, not so surprisingly, employ *arationality*. When things are proceeding normally, Dreyfusses claimed that experts do what normally works for them, they do not have to employ deliberative reasoning. This three-way organisation can also be discerned in Jantsch's view of the managerial decision process. Additionally, managers here are treated as metaphor for intelligent agents that are potentially performing intelligent computations. In Table 1, Jantsch's guiding image for managers is organised at three levels. A manager assuming responsibilities at the higher level needs to be open for change and needs to act as a *catalyst* for change. The middle-level manager acts as a *co-ordinator*, acting to convey the visions of the higher conceptual levels to the lower-level managers. The lower-level manager, can also be viewed as an *administrator*, the level at which the actual execution of plans and collection of information takes place. It needs to be emphasised here that, by virtue of their roles in an organisation, managers are placed at different levels of the hierarchy, their interactions, however, place them in a holistic heterarchy. This key point is applicable in all the other three-way organisations shown for various domains in the table.

4. Implications of Triune Models in Engineering Design and Manufacture Problems

Solution to engineering design and manufacture problems can also be attempted in a three-tier approach. Standard mathematical and engineering techniques sit at the lowest–tier, symbolic and rule–based methods lie at the highest–tier, and soft computing approaches such as neural networks, fuzzy logic, genetic algorithmic methods are situated at the middle–tier.

Given a design/manufacture problem, sub–problems that can be solved using traditional mathematical, statistical, and engineering methods need to be isolated from those sub–areas that need intelligent (heuristic) methods for their solution. This isolation stage is very crucial as heuristic techniques should not be applied when standard, well-understood techniques meet the desired objective. Development costs and times are reduced by using standard methods. There may be some rare exceptions when standard methods can solve the problem but are computationally inefficient. In this case the problem needs to be addressed by heuristic methods.

When it is determined that standard methods can not be utilised, then a solution using symbolic methods such as rule–based approaches, expert systems, case based reasoning, needs to be investigated first. Due to an explicit knowledge structure and a well–defined causal chain in the reasoning process, symbolic methods can trace an explanation of the results. Since such a facility is not always available with various soft computing approaches (NN, GA, etc.), it is recommended that symbolic methods are attempted first before resorting to soft computing methods. Symbolic approaches however assume that domain experts are available and appropriate symbolic representation of the problem is feasible. Additionally, it is assumed that the domain experts are willing and able to help develop a symbolic system.

When all else fails, problems can be attempted with soft computing approaches. Also a combination of approaches (hybrid methods) using standard mathematical, symbolic, and soft computing methods, can be utilised. Currently, one of the problems with methods at the middle–tier is that the solution process is not readily explainable and transparent to the user. For example, the input–output mapping learnt by a neural network from exemplars is not directly accessible. Although progress is being made in extracting rules from neural nets [24], the problem is still open.

There are two underlying themes in the three–tier problem solving process that is being proposed here —(i) to resolve the sub–problems at the middle–tier into either lower–tier or upper–tier problems so that the reasoning process becomes transparent and the solutions explicit and (ii) to organise the solution in an integrated way so that all the available methods are tried in a systematic way but taking care to isolate the problem sub–areas so that sub–problems are solved at an appropriate level. A similar three–layer architecture can be used for real–time control problems where the lower–layer has fast response times, the upper has slow response times and the middle–layer has medium response times. The control techniques that underlie each layer are organised in a similar way as proposed earlier —standard linear/nonlinear control techniques at the lower, neurofuzzy techniques at the middle, and symbolic methods at the upper–layer.

5. Conclusions

The framework of three-way organisation for the different theories of intelligent computation is offered as a guidance. It needs to be remembered that, just as an expert needs deliberative rationality when a decision scenario is out of the ordinary, just as in a functioning human being instinct, emotion and reason operate as a unified whole, various theories of intelligent computation would operate in a seamless and synergistic fashion in the designs of intelligent systems. It is unto us, the researchers in these areas, to build bridges and discern the unity in this trinity. The recent advances in the soft computing field herald some profound developments on the horizon of cutting edge theories of intelligent algorithms. It is hoped that the insights gained in this pursuit push the limits of our understanding of human intelligence and serve as guiding beacons to develop more intelligent machines.

Acknowledgments

The work presented here is supported by a grant from the Engineering and Physical Sciences Research Council (EPSRC), UK. The author benefitted from comments of Prof. Dan Levine, "RP" Raghupathi, on an earlier version. The author also appreciates the benifit from many informal discussions with Dr. Guido Bugmann, Prof. Jack Boitano, and Prof. Mike Denham.

References

1. Matarazzo, J. D., 1972, *Wechsler's measurement and appraisal of adult intelligence* (Fifth edition), Oxford University Press, New York, NY.
2. Gardner, H., 1993, *Frames of mind: The theory of multiple intelligences* (Second edition), Fontana Press, Harper Collins Publishers, London.
3. Hunt, E., 1995, The role of intelligence in modern society, *American Scientist*, **83**, 356-368.
4. Newell, A., 1990, *Unified theories of cognition*, Harvard University Press, Cambridge, MA.
5. Fodor, J. A., 1976, *The language of thought*, Harvester Press (Harvard University Press paperback), Sussex, UK.
6. West, D. M., and Travis, L. E., 1991, The computational metaphor and artificial intelligence: A reflective examination of a theoretical falsework, *AI Magazine*, **12** (1), 64-79.
7. Honavar, V., and Uhr, L., 1994, *Artificial intelligence and neural networks: Steps toward principled integration*, Academic Press, Boston, MA.
8. Bapi, R. S., McCabe, S. L., and Roy, R., 1996, Book Review of *Artificial intelligence and neural networks: Steps toward principled integration*, V. Honavar and L. Uhr (Eds.), 1994, Academic Press, Boston, MA , *Neural Networks*, **9**(3), 545–548.
9. Harnad, S., 1990, The symbol grounding problem, *Physica D*, **42**, 335-346.
10. Roy, R., and Bapi, R. S., 1996, AI at the cross-roads: Past achievements and future challenges, Book review of *Artificial Intelligence in Perspective* , D. Bobrow (Ed.), 1993, *SIGART Bulletin*, **7**(1), 10–13.
11. Jantsch, E., 1980, *The self-organizing universe: Scientific and human implications of the emerging paradigm of evolution*, Pergamon Press, Oxford.
12. Maturana, H. R., and Varela, F. J., 1992, *The tree of knowledge : The biological roots of human understanding* (Rev. ed.), Shambala.
13. West, D. M., and Travis, L. E., 1991, From society to landscape: Alternative metaphors for artificial intelligence, *AI Magazine*, **12** (2), 69-83.
14. Langton, C. G. et al., 1991, *Artificial Life II*, Addison-Wesley Publishing Co.
15. Briggs, J. P., and Peat, F. D., 1984, *Looking glass universe: The emerging science of wholeness*, Fontana Paperbacks, Harper Collins Publishers, London.
16. Pollock, J. B., 1993, On wings of knowledge: A review of Allen Newell's *Unified theories of cognition*, *Artificial Intelligence*, **59**, 355-369.
17. MacLean, P. D., 1990, *The triune brain in evolution*, Plenum Press, New York, NY.
18. Bapi, R. S. and Denham, R. S., 1996, Computational Intelligence: A Synergistic Viewpoint, *International Journal of Computational Intelligence and Organizations* , **1**, 2–9.
19. Leven, S. J., 1987, *Choice and Neural Process*, Unpublished doctoral dissertation, University of Texas at Arlington, Arlington, TX.
20. Raghupathi, W., Levine, D. S., Bapi, R. S., and Schkade, L. L., 1994, Toward connectionist representation of legal knowledge, In D. S. Levine and M. Aparicio IV (Eds.), *Neural networks for knowledge representation and inference*, Lawrence Erlbaum Associates, Hillsdale, NJ, pp. 269–282.
21. Levine, D. S., and Leven, S. J., 1992, *Motivation, emotion, and goal direction in neural networks* , Lawrence Erlbaum Associates, Hillsdale, NJ.
22. Bapi, R. S., 1996, Exploring the role of emotion in the design of autonomous systems, In M. van der Heyden, J. Mrsic-Flögel, and K. Weigl (Eds.), *Proceedings of the Helsinki Neural Network Group Workshop HELNET'95, Vacanciel, France, Oct 4-7, 1995* , VU University Press, Amsterdam, pp. 115–125.
23. Leven, S. J., 1992, Learned helplessness, memory, and the dynamics of hope, In D. S. Levine and S. J. Leven (Eds.), *Motivation, emotion, and goal direction in neural networks* , Lawrence Erlbaum Associates, Hillsdale, NJ, pp. 259–299.
24. Omlin, C. W., and Giles, C. L., 1996, Extraction of rules from discrete-time recurrent neural networks, *Neural Networks*, **9**(1), 41–52.

An Overview of Evolutionary Computing for Multimodal Function Optimisation

Rajkumar Roy[1] and Ian C. Parmee[2]

[1]The CIM Institute, Cranfield University, Cranfield, Bedford, MK43 0AL, UK;
Email: rroy@cim.cranfield.ac.uk or r.roy@ieee.org
[2]Plymouth Engineering Design Centre, University of Plymouth, Plymouth, PL4 8AA, UK;
Email: iparmee@plymouth.ac.uk

Keywords: evolutionary computing, multimodal genetic algorithms, multimodal function optimisation, search strategies, useful diversity.

Abstract

Evolutionary Computing (EC) is becoming popular among researchers in multimodal function optimisation (MFO). One of the major issues in MFO is maintaining the 'useful' diversity in the population. The 'useful' diversity is utilised in search either to achieve the global optimum or to maintain multiple sub-optima in the final population. In case of multimodal functions these two goals can be dependent on each other. The paper discusses the principle issues involved in MFO. All major EC techniques suitable for MFO are categorised, and their strengths and weaknesses are discussed. Application of EC techniques to solve real life MFO problems are difficult. Real life problems can pose some additional challenge than test functions. Real life problems are difficult mainly because of lack of prior knowledge. The paper mentions the challenges posed by such real life problems, and discusses how Adaptive Restricted Tournament Selection (ARTS) can address some of the challenges. From the discussion, direction for future research in MFO is identified.

1. Introduction

Evolutionary Computing (EC) like genetic algorithms (GA), simulated annealing and tabu search have been successfully applied to many optimisation problems where the aim is to identify the global optimum solution. Many real life problems require the identification of several good solutions (multimodal) in addition to the global optimum. This paper describes an overview of evolutionary computing techniques that are used in multimodal function optimisation.

In real life, an engineering problem often involves more than one objective. Thus, a true engineering solution is not necessarily the global optimum with respect to one criterion [1], [2], [3]. The solution is often selected by considering several criteria. For example, in case of multimodal design problems there may be quite different design solutions that perform similarly with respect to one criterion. On the other hand, the performance of the designs may vary considerably among each other considering other criteria. The other criteria can be quantitative and qualitative in nature, such as manufacturability, cost, maintainability, sensitivity and customer preference. Integrating all these evaluation criteria into one is difficult and can be misleading. In case of quantitative criteria only, multiobjective genetic algorithms [4] with pareto optimality are used to obtain a number of good solutions. The solutions are identified as the best possible compromises among different criteria. Roy et. al. [3] has developed a framework to handle real life problems with quantitative as well as qualitative criteria. The developed system, Adaptive Search Manager (ASM) [3], [5], [6], obtains multiple good design solutions based on the most important and computable criterion and then evaluates them qualitatively for other criteria. The optimisation technique developed for the multimodal problem (based on the most important criterion) is called 'Adaptive Restricted Tournament Selection (ARTS)' [7].

Identification of the best solution (the global optimum) or multiple 'good' solutions (multiple sub-optima) from a multimodal problem using GA falls in the realm of maintaining diversity in population. Maintaining the population diversity is a major issue in GA search. The diversity can be achieved randomly or by a systematic approach. The paper discusses 'diversity' versus 'useful diversity', and explains how useful diversity can be achieved. Different evolutionary computing techniques have been developed and applied to promote 'useful diversity' in a search process. The paper provides an overview of such techniques. It is observed that most of the techniques use prior information about the modality of the problem, and thus not suitable for real life problems. The paper mentions the challenges posed by real life problems. ARTS identifies multiple sub-optima in a multimodal problem, but does not require prior information about the modality of the problem. Key features of ARTS are discussed. Influence of one key parameter is tested on ARTS. This study finally points to the future research.

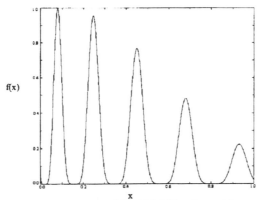

Figure 1: A multimodal test function, $f(x) = e^{-2\ln2((x-0.1)/0.8)^{**2}}\sin^6(5p(x^{3/4} - 0.05))$, where x is defined in [0, 1]. There are five local sub-optima in the function, but the left most peak represents the global optimum.

2. Diversity vs. useful diversity

The sub-optimum solutions are considered as "good" solutions in a multimodal function. A formal definition of a "good" solution is given below:

Let us assume a search space S and an objective function f (that assigns a real number to any member of S).

$$f : S \rightarrow R^n$$

Without loss of generality, let us assume that the goal is to maximise with respect to f. A neighbourhood of an element i of the search space S is defined by the resolution on each dimension. For any i ∈ S, N(i) ⊆ S is the neighbourhood of i in S. Where i can be considered a "good" solution or a sub-optimum member (or local sub-optimum) of the search space S if :

$$f(i) \geq f(j) \text{ for all } j \in N(i)$$

On the other hand, i can be considered the "best" solution or the global optimum in the search space S if:

$$f(i) > f(j) \text{ for all } j \in S$$

Figure 1 exhibits a multimodal function with five local sub-optima, and one of them is the global optimum. Maintaining the population diversity also means maintaining as many sub-optima as possible. In case of GA search early convergence can lead to a local sub-optimum, and thus attempts have been made in the past to stop quick convergence of the GA. A diverse search by the GA allows exploration of larger part of the search space in order to converge on a better, single solution. While doing a diverse search, the GA also explores different sub-optima.

The three main reasons for a quick convergence of the GA are: *selection pressure, selection noise* and *operator disruption*. In case of a finite population GA, use of the 'survival of the fittest' promotes high fitness individuals in the population. This introduces a *selection pressure* towards higher fitness individuals. In case of identically fit individuals the GA randomly selects one, thus there is a variance in the selection process. These variance results in *selection noise*, by which some fitter individuals are randomly thrown out of the population. The use of crossover, mutation and inversion can sometimes destroy the building blocks for higher fitness individuals this is known as *operator disruption*. One method of increasing the exploration by the GA is to reduce selection pressure and increase operator disruption. Operator disruption can be increased either by appropriate tuning or the introduction of more disruptive operations. This type of exploration is not necessarily useful, for example a very high mutation rate can lead to a random search. A useful diversity should explore the good building blocks [8]. An exploration can be called *useful* if it exploits the genotypic information present in the population to search through the interesting areas of the search space. The *useful exploration* should be goal directed. Diversity is utilised in search either to achieve the global optimum or to maintain multiple sub-optima in the final population. In case of multimodal functions these two goals can be dependent on each other. An exploratory EC search that tries to identify the global best in a multimodal function often encounters many local optima. Similarly, an EC search that tries to maintain many sub-optima is likely to do a useful exploration in the search space and thus also likely to find the global optimum in a multimodal function. The EC suitable for multimodal function optimisation (MFO) is called the *multimodal EC*. Techniques used to achieve the useful exploration for the multimodal EC are generally termed as the *niching methods*. The next section presents an overview of different EC techniques to-date used in MFO.

3. Multimodal evolutionary computing techniques

The earliest work reported on maintenance of population diversity is Cavicchio's dissertation [9]. As a method of preserving population diversity or variance he introduced a number of *preselection schemes*. The best selection scheme says: if a child is better (in terms of fitness) than the worse parent then replace the parent by the child for the next generation. Cavicchio assumed a parent as the closest member in the population to its child. This assumption may not be valid in case of many multimodal functions. Thus, the *preselection scheme* as described by Cavicchio suffers from high replacement error [10]. De Jong's dissertation [11] presented his model of multimodal function optimisation based on what is called the *crowding factor* or simply the *crowding model*. The crowding model was inspired from the ecological phenomenon that similar species compete with each other for survival whilst sharing a limited amount of resource. Different species live in different groups or niches, and thus dissimilar species do not compete among each other. The competition for survival to the next generation is local rather than global. This early work could maintain diversity of species present in the initial population; however it can not discover new species or niches. The model also suffers due to stochastic errors introduced in case of low crowding factor.

Evolutionary computing techniques that are generally used for multimodal function optimisation are Genetic Algorithms (GA), GA with clustering techniques, parallel GAs, GA that model naturally occurring niche and species formation, and GA with immune system model. The methodologies by which these techniques optimise a multimodal function or identify multiple local sub-optima in a multimodal function can be categorised as: *selection strategy*, *fitness sharing*, *parallel sub-populations*, and *sequential search*. The overview of multimodal EC techniques is divided into following four subsections as per the methodology used.

3.1 Selection strategy

Evolutionary computing techniques that perform MFO by modifying the selection and replacement procedure are generally placed under this category. Mahfoud [10] performed a detailed study on the different niching techniques used with GA, especially the crowding methods. Outcome of the study was an improved variant of the crowding technique called the *Deterministic Crowding* (DC) [12]. During his experiments with different crowding methods, Mahfoud found that by choosing members randomly for reproduction, and then providing the selection pressure by only replacing a parent with a fitter child better performance can be achieved. To determine which of the possible parent-child pairing should be used in comparing the parents to their children (that is either (parent1-child1 and parent2-child2) or (parent1-child2 and parent2-child1)), the total of the parent-child similarities (in terms of the Euclidean distance) for each of the two possible combinations are determined. The parents-children pairing that has the highest total similarity is used to determine if the child should replace the parent. The replacement is only possible if the child is fitter than the parent. Deterministic crowding has been applied on two-class and multi-class test problems. In case of multi-class problems it is apparent some peaks dominate over others, and eventually the dominated peaks are lost from the population. Although the method performs better than crowding, it is not clear if multiple solutions can be maintained for many generations using this method. The loss of some dominated peaks is a major limitation in case of real life multimodal problems, because there is always a possibility of losing some interesting peaks that are dominated by few others. Another limitation of DC is that it does not guarantee that the final population shall be distributed only among the peaks. This also limits the application of DC in real life problems, because in that case it is not clear whether what is returned from the algorithm is at least a sub-peak or not. Cedeno et. al. [13] developed the concept of multiniche crowding (MNC) in a genetic algorithm that permits one to simultaneously find several peaks of a multimodal function. In MNC both the selection and replacement steps are modified with a concept of crowding. The idea is to remove the selection pressure due to the fitness proportionate selection (FPR) whilst maintaining the diversity in the population. The method works with local mating and replacement strategy while allowing for some competition for population slots among the niches. In multiniche crowding the FPR is replaced by a *crowding selection*, where each member of the population has equal chance to mate in the next generation. First, an individual is selected either sequentially or at random. The partner for mating is selected from a random sample taken from the population (the size of the sample is defined by the *crowding selection group size* (C_s)). The MNC uses a replacement policy called *worst among the most similar*. In order to select an individual from the population for replacement by a child, crowding factor groups (the number of groups are defined by the *crowding factor* (C_f)) are defined by randomly selecting s (called as the *crowding factor group size*) number of individuals from the population per group. Next, one individual from each group is identified that is phenotypically the most similar to the child; and this constitutes a list of individuals ready for the replacement. The child replaces the lowest fit individual in the list. It is worth noting that the child could possibly have a lower fitness than the individual being replaced. The technique is applied on several test functions and also to determine the sequence of all nucleotide in a DNA molecule, from restriction-fragment data. The method works well for the test functions using the given set of crowding parameters. The paper does not comment concerning the quality of the solutions achieved. The parameters are set by trial and error and the paper also does not mention possible effects of the crowding parameters' values on the search.

Lin et. al. [14] have developed a cluster identification technique to identify clusters present in a GA search population. The technique defines a 'crowdness function' to identify the centre and radius of a cluster. The cluster identification technique in genetic algorithms helps the fast determination of local regions containing relative optima and provides good initial solutions (they are the centres of the clusters) for subsequent local searches. The technique uses many heuristics, and it is not clear how that affects the GA search. Parmee et. al. [15] describe a method of maintaining diversity and reinforcing the natural clustering (niching) tendencies of the GA by appropriate tuning of crossover and mutation probabilities. A shared near neighbour clustering algorithm is used after some pre set number of generations to further define the naturally occurring clusters present in the population. The clustering method does not impose any artificial shape on the niches present in a population. The method is suitable for rapid identification of 'good' regions in a problem space as opposed to the identification of individual optima. In this respect the technique is being developed to provide information to the engineer concerning high-performance regions of a complex, multidimensional search space. The technique requires no prior knowledge concerning the modality of the fitness landscape.

An improved tournament selection method for multimodal functions called the *Restricted Tournament Selection* (RTS) is developed by Harik [16]. The technique is based on the principle of local competition, that is a tournament among similar individuals (according to a distance metric). The method creates a new population as in a steady state GA [17]. Before an individual is allowed to the next generation it is placed into tournament with the closest (according to the distance metric) individual present within a random sample of the population. The size of the sample is kept fixed and is termed as the *window size*. This form of tournament selection should restrict an entering individual from competing with others, which are too different from it. For an individual, if the closest sub-optimum is selected in the random sample, the individual competes with the sub-optimum and fails to replace it. Thus, if the window size is big enough the replacement error is reduced. Therefore after the peaks are identified, the underlying distribution of the population is expected not to change for a long time. The procedure is dependent on the probability of a peak present in the sample taken from the population. This restricts the number of peaks the algorithm can maintain depending on the size of the window. That means the size of the window is determined using prior knowledge concerning the modality of the fitness landscape. RTS has been successfully applied to some multimodal test functions. It is observed that in a prolonged run some peaks start dominating others [7]. Thus RTS can not achieve a steady state of distribution and it carries the risk of losing some peaks. RTS can delay complete dominance of some peaks over others. But because of the presence of the dominance factor, distribution of individuals on several peaks changes. A steady distribution can be achieved by using a very large window size. The dominance factor becomes prominent when some dominating individuals start occupying a major part of the population. In case of real life problems, without any prior knowledge concerning the location and the number of peaks present, it becomes almost impossible to determine when to stop the GA so that the population is distributed among the peaks. Stopping early may mean converging to individuals which are not peaks. But delayed stopping can also lose some peaks because of the dominance factor.

3.2 Fitness sharing

Goldberg and Richardson [8] introduced what they called as the *sharing method*. In the sharing scheme, fitness is shared as a single resource among similar individuals. Fitness of an individual element of population is derated due to the presence of similar elements in the population. The concept of sharing is implemented by defining a sharing function, *share(d)* as shown below, where d is a measure of dissimilarity between two elements of the population :

$$\text{share(d)} = 1 - \left(\frac{d}{s_{share}} \right)^a \text{ , when d } \le s_{share}$$

$$= 0 \qquad\qquad d > s_{share}$$

where, s_{share} is defined as the dissimilarity threshold and a is a constant to determine the shape of the sharing function. An individual is compared with each member of the population to calculate the sharing function values. Summation of all the values due to individual members of the population defines the total sharing function value for the individual. The fitness of an individual is degraded by the total sharing function value, and the new fitness, F', can be described as follows :

$$F' = F \Big/ \sum_{i=1}^{N} share(d)_i \text{ , where N = population size}$$

Goldberg et. al. [18] have discussed the strengths and weaknesses of the above fitness sharing mechanism for optimisation of multimodal functions. Performance of the sharing scheme is very much dependent on the value of s_{share}. Determination of an appropriate value for s_{share} is a difficult task and is dependent on prior knowledge concerning the nature of the problem. Further work has been performed in the same direction by Oei et. al. [19], where they use tournament selection with a continuously updated sharing technique. The method updates the fitness (or calculates the shared fitness) with respect to the new population distribution as it is being developed. The technique claims to promote and maintain multiple sub-populations over many generations. But the technique is also dependent on prior knowledge regarding the fitness landscape. In an attempt to handle multimodal deceptive functions, Goldberg et. al. [18] used fitness scaling and the new fitness sharing scheme. Yin and Germay [20] presented their implementation of a faster genetic algorithm with the sharing scheme using a clustering technique. The clustering method is used to identify different niches present in the population. Niche count (that is the number of elements present in a niche) is used to degrade fitness of individuals present in the niche; thus sharing is local within one niche. Performance of the technique depends on the clustering method used. Setting of parameters for the clustering algorithm needs some trials and prior knowledge. The clustering algorithm also enforces an artificial shape (in this case spherical) to the niches, that may not necessarily be the natural shape for some niches. Jelasity and Dombi [21] described a niching technique called GAS. The technique dynamically creates a sub-population structure (they call it taxonomic chart) using a *radius function* instead of a single radius value, and a 'cooling' method similar to simulated annealing. The GAS algorithm uses a steady state GA and a high-level algorithm responsible for creating and maintaining the taxonomic chart. The technique allows the population to grow up to a limit and then to die off to reduce the population size to the starting level. The technique introduces a new function called *speed* of a species, that determines the *radius function*. It is not very clear how the technique would perform in case of multidimensional problems. The paper also does not elaborate on the computational complexity of the technique. In a recent work, Miller and Shaw [22] have introduced the *Dynamic Niche Sharing* for multimodal function optimisation. The technique is developed to be faster than the previous sharing method. The dynamic niching uses a greedy approach to identify peaks present in the population in every generation. Individuals are categorised according to the peak it belongs to (that is if within the s_{sh} radius of the peak). If an individual does not belong to any peak, it is categorised as 'non-peak'. Thus every individual belongs to a niche (or category), and the fitness of the individual is degraded by the size of its niche (*niche count*). Thus every individual within a dynamic niche has their raw fitness degraded equally. This means that there is no incentive to maintain distance between individuals within a dynamic niche. This allows the dynamic niching to explore the regions around the peaks of the niches more thoroughly than standard sharing. The overall performance of the technique is found to be better than the sharing technique and DC on a test function. It is not clear how efficient the technique would be for multidimensional problems. Setting a value for the s_{sh} would require prior knowledge about the problem, and that also restricts the use of the technique for real life problems.

Immune system model for pattern matching was first developed by Stadnyk [23]. The model could achieve niching by lowering the number of antigens used in computing the fitness of each population element. Smith et. al. [24] implemented an immune system model along with a GA in order to develop a GA which can search for diverse and co-operative populations. It is observed that the model exhibits an implicit fitness sharing which can be useful for multimodal function optimisation. The area of research is relatively new and needs further investigation before it can be useful for multidimensional multimodal real life problems. In a very recent work Darwen and Yao [25] compared the fitness sharing technique with the above mentioned implicit sharing. The authors used a realistic letter classification problem for the comparison. It is observed that the implicit fitness sharing searches the optima more comprehensively even when those optima have small basins of attraction, and also when the population is not large enough to form the species at each optima. In case of implicit sharing the individual closest to a peak is rewarded even if it's not particularly close to it and when another individual is almost as close. That means in case of implicit sharing there is greater relative selection pressure for the nearer individual and that helps in the better exploration. Whereas in case of fitness sharing the niching radius s_{sh} means the closest individual to a peak shares its payoff with all other individuals that are almost as close. But in case of small population the tendency of comprehensive peak coverage degrades the performance of the implicit sharing more than the fitness sharing.

3.3 Parallel sub-populations

Application of parallel sub-populations to evolve multiple solutions from a genetic algorithm was attempted by Grosso [26]. In his study he used some degree of communication between sub-populations to allow good building blocks to spread, but that caused reduced diversity and eventual convergence on one global peak. Without such communication the technique becomes equivalent to running a GA several times with a smaller population. Elo [27] presents a genetic algorithm with a dynamic division mechanism conceived on the Connection Machine-2 for multimodal function optimisation problems. The technique dynamically divides the population into an increasing number of sub-populations to allow specialisation on different maxima as discovered during the search process. This method allows the GA search to adapt to the topology of different multimodal optimisation problems. Without defining the control parameters explicitly, the dynamic nature of the algorithm enables divisions to occur appropriately when the maxima are discovered during the search process. Thus the method is flexible and requires very little knowledge about the fitness landscape. The use of parallel genetic algorithms to obtain multiple sub-optima from a multimodal function is a very promising area of research.

In an attempt to model naturally occurring Niche and Species formation, Davidor [28] developed a GA model called ECO GA, which uses a steady-state GA and is based on local and computationally inexpensive operators. In ECO-GA, the population of strings is held on a 2-D grid having its opposite edges connected together in such a way that each grid element has 8 adjacent elements. Initially individuals are placed at random, one on each grid point. ECO-GA randomly selects one grid element, and defines an 8-element sub-population around it, thus defining a sub-population of 9 elements. This definition implements implicitly parallel and overlapping sub-populations. A steady-state GA is applied with the population size of 9. Two individuals are selected probabilistically from the sub-population according to their relative fitnesses, and genetic operators are applied on them to produce two new individuals. The newly created individuals are probabilistically put back to the same grid positions depending on the relative fitnesses of the opponents (that is the already existing individuals at the two grid points). That means the children are more likely to stay in the vicinity of their parents. The smallness of the size of the sub-population helps the GA to converge very quickly. The technique works based on local convergence which is quick, and assumes that the global optimum can be obtained by the interaction of locally optimised individuals. It is not clear how the search is restricted due to the exploitation of only locally 'good' schema. The implicitly parallel overlapping sub-populations evolve locally but information migrates from one grid to adjacent grid elements because of the overlap. The technique intends to explore the search space in order to identify the global optimum in a multimodal function. The paper has presented some results with a standard one dimensional problem, but it is not clear how the technique would perform in higher dimensions. Further investigation is necessary for a better understanding of the strengths and weaknesses of the technique.

3.4 Sequential search

In real life problems, some time the model evaluation can be very expensive, and thus a smaller population size is used. All the techniques mentioned above try to maintain multiple peaks in one population. That means, in case of fixed sized population the identification of a number of peaks is restricted by the size of the population. An alternative approach called the *Sequential Niche Technique*, was proposed by Beasly et. al. [29] where peaks are identified one at a time. This generalised technique allows unimodal function optimisation methods to be extended to identify all optima and sub-optima of multimodal problems. The research implements the concept with a standard genetic algorithm. The method involves multiple runs of GA but uses knowledge obtained from previous runs to avoid re-searching the regions of the problem space where peaks (optima or sub-optima) have already been identified. Whenever one peak is located, in subsequent runs, region around the peak (defined by a *niche radius*) is depressed by applying a fitness derating function. That helps the search in concentrating in other interesting areas and thus identifying multiple peaks. The algorithm is dependent on the right selection of the niche radius. The use of the niche radius imposes a shape to the niches (in this case spherical). In case of problems where the maxima are not evenly distributed, the fixed size of the niche radius would underestimate the size of some niches whereas overestimating the size of others. An inappropriate selection of the niche radius can introduce false peaks, and that can misguide the search. Sequential niching can also offset a peak's location as a consequence of the fitness deration. The artificial shape may not match with the natural shapes of some niches. Prior knowledge concerning the problem would be helpful in determining a workable niche radius. This is a similar limitation as with the fitness sharing technique. In the fitness sharing method fitness landscape is modified every time an individual is evaluated, whereas in the sequential niche technique the fitness landscape remains static during one run. Thus the sequential niche technique overcomes the problem of exponential scaling of its fitness landscape. Another major limitation of the technique is that it does not allow transfer of the building block information to find one solution from another. This can restrict the GA's search capability in some applications. Mahfoud [30] compared other niching techniques with the sequential niching. The paper supports the above mentioned weaknesses of the sequential niching. It is also shown that, fitness sharing or DC performs better than the sequential niching over a wide range of functions.

4. Challenges posed by real life problems

Real life problems can pose some additional challenge than test functions. Test functions can be made very complex, but as a test function is developed with a goal in mind (say one wants to develop a multimodal two dimensional test function), it is easier to get some idea about the nature of the problem. Real life problems are difficult mainly because of the lack of prior knowledge. The techniques mentioned in the previous section are mostly tested on test functions. The main reason is that it is easier to visualise and measure the performance of an algorithm on test functions. Most of the techniques determine the search parameters assuming prior knowledge concerning the search space. Performance of the techniques is measured in terms of population distributions on known peaks. Only a few techniques are applied to real life problems, where the validation of the techniques is extremely difficult. A real life problem may be considered to have the following characteristics:

a) There is not much prior knowledge regarding the shape of the search space.

b) No prior knowledge regarding the performance and location of the optimum and sub-optimum points in the search space.

The lack of prior knowledge invites some difficulties for a multimodal GA search, such as:

a) The determination of search parameter values becomes extremely difficult in the absence of prior information regarding the modality of the search space.

b) It is very difficult to identify the state at which the GA distributes the population on the peaks.

c) The validation of the results obtained from the GA search becomes quite difficult because of the lack of knowledge concerning the quality and location of the peaks.

5. Adaptive restricted tournament selection

Adaptive Restricted Tournament Selection (ARTS) [7] identifies multiple sub-optima in a multimodal fitness landscape, where each sub-optimum represents a design option. The technique is an improvement over Restricted Tournament Selection (RTS) [18]. Without knowing how many peaks are present in the fitness landscape it is difficult in RTS to decide the size of the window (that is a fixed size sample). Thus RTS requires *prior* knowledge about the problem. In real life problems information about the modality of the fitness landscape is not available. In order to handle real life problems, ARTS uses a shared near neighbour clustering method [31] to define the closest point for a newly generated individual. For every generation this method identifies clusters of points present in the population. For each newly generated individual the closest point in the generation is determined by finding the closest point of the closest cluster present in the population. Thus the necessity for a fixed size window and prior knowledge about the problem (as in case of RTS) are eliminated in ARTS. A hybrid search technique using ARTS and a local search is used on a turbine blade design optimisation problem [3].

The principle behind ARTS is local competition while using the pool of building blocks present in the population. It is observed during the empirical trials with different multimodal test functions that ARTS exploits schema information at its initial stages of a run (i.e. the first few hundred generations). Once the population elements are distributed among the peaks a steady state is achieved where the competition is entirely local. During the initial stages of a run when the population is quite diverse the clustering algorithm tends to form wider clusters thus introducing some replacement errors in the ARTS search (*clustering error*). This causes a delayed convergence on the peaks. At the steady state of distribution, when the population is distributed among the peaks the clustering algorithm identifies the niches correctly. This helps to restrict the tournament within each niche and thus eliminates the dominance problem (that is discussed in section 3) as seen in the case of RTS. A simple genetic algorithm converges to a global optimum, whereas ARTS can maintain multiple peaks. ARTS also continues to search (even in later generations) a larger space by crossover between different niches present at the steady state of population distribution.

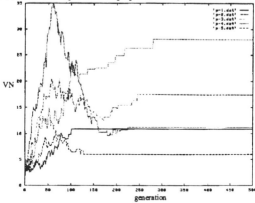

Figure 2: ARTS on F2 with KT = 15 and K = 15, where VN is the variance of the number of elements on each peak over ten random runs.

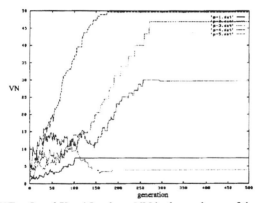

Figure 3: ARTS on F2 with KT = 8 and K = 15, where VN is the variance of the number of elements on each peak over ten random runs.

The shared near neighbour clustering technique is controlled by two parameters K and KT. It is important to understand the effect(s) of the two parameters on the ARTS based GA search. In order to study the effect, the ARTS based search is performed with different values of KT whilst keeping K constant. The value of K does not affect the clustering significantly, and generally K is fixed at 15 with a population size of 100. The runs use 100% crossover and no mutation or inversion. The value of KT is varied from 15 to 8. The study is performed on the multimodal function shown in figure 1. When K and KT are the same, that is when they are both 15, the tightest possible clusters are produced. Reducing the value of KT from 15 results in less tight clusters. Ten random runs are performed for each combination of K and KT. The average and variance of the number of elements present on each peak with KT equal to 15 and 8 only are plotted in figures 2 and 3. The experiments show that in all the cases ARTS is found to have achieved a steady state of population distribution and the performances are similar. ARTS could distribute the population on the peaks without using any prior knowledge concerning the modality of the problem space. Thus it is suitable for real life problems.

6. Conclusions and future research directions

This paper identifies the strengths and weaknesses of different EC techniques used in multimodal function optimisation. It is observed most of the techniques use some form of prior knowledge about the problem to set the search parameters. The recent techniques like DC and RTS are also observed to have 'dominance problem' where some peaks are lost due to the presence of dominant peaks. Optimisation of real life multimodal problems has some additional difficulty due to the lack of prior knowledge. ARTS is developed to optimise such real life problems. ARTS does not require prior knowledge about the modality of the problem to identify the peaks present in the problem space. One common difficulty with most of the techniques is the fixed size of the population. The size of the population also limits the number of peaks that can be represented in the population.

It is observed that in case of a real life problem it is almost impossible to guarantee that all the peaks have been visited. In some application, such as design, that may severely damage the confidence of the users on the results achieved from the search. There is a need for better understanding of the causes that can increase the confidence of the user. The search technique should address the issues to enhance users' confidence. One example can be developing a search algorithm that can produce results with confidence within an acceptable time limit. The use of penalty function for the constrained optimisation modifies the design fitness landscape. A multimodal GA algorithm that handles constraints without penalty functions would be very useful. Not much work is reported in parallel subpopulation search application to MFO. Parallel sub-population search techniques also have great potentials, needs further investigation. Research in the area of exploitation and exploration of good schemas present in the parallel sub-populations would result in better niching methods. The niching techniques also need further application in complex real life problems.

References

[1] Parmee, I. C., 1994, The implementation of adaptive search tools to promote global search in engineering design, *Proceedings of 3rd IFIP Working Conference on Optimisation-based Computer-aided Modelling and Design, 24-26 May, UTIA, Prague.*

[2] Parmee, I. C. and Denham, M. J., 1994, The integration of adaptive search techniques with current engineering design, *Proceedings of ACEDC'94 Conference, 21-22 September, Plymouth (UK)*, pp. 1-13.

[3] Roy, R., Parmee, I. C. and Purchase, G., 1996, Integrating the genetic algorithms with the preliminary design of gas turbine blade cooling systems, *Proceedings of ACEDC'96 Conference, 26-28 March, Plymouth (UK)*, pp. 228-235.

[4] Goldberg, D. E., 1989, *Genetic Algorithms in search, optimization, and machine learning*, Addison-Wesley, Reading, MA.

[5] Roy, R., Parmee, I.C. and Purchase, G., 1996, Sensitivity analysis of engineering designs using Taguchi's methodology, *CD-Rom Proceedings of the ASME DETC-Design Automation Conference, 18-22 August, Irvine, CA, 96-DETC/DAC-1455.*

[6] Roy, R., Parmee, I. C. and Purchase, G., 1996, Qualitative evaluation of engineering designs using fuzzy logic, *CD-Rom Proceedings of the ASME DETC-Design Automation Conference, 18-22 August, Irvine, CA, 96-DETC/DAC-1449.*

[7] Roy, R. and Parmee, I. C., 1996, Adaptive Restricted Tournament Selection for the Identification of Multiple Sub-Optima in a Multi-Modal Function', *Lecture Notes in Computer Science (Evolutionary Computing)*, **1143**, Springer - Verlag, 236-256.

[8] Goldberg, D. E. and Richardson, J. J., 1987, Genetic algorithms with sharing for multi-modal function optimization, *Genetic algorithms and their applications: Proceedings of the Second International Conference on Genetic Algorithms*, pp. 41-49.

[9] Cavicchio, D. J., Jr., 1970, *Adaptive search using simulated evolution*, Unpublished doctoral dissertation, University of Michigan, Ann Arbor, USA.

[10] Mahfoud, S. W., 1992, Crowding and preselection revisited, In R. Manner and B. Manderick (Eds.), *Parallel Problem Solving from Nature*, **2**, Elsevier Science Publishers B. V., 27-36.

[11] De Jong, K. A., 1975, *An analysis of the behavior of a class of genetic adaptive systems*, (Doctoral Dissertation, University of Michigan) Dissertation Abstracts International, 36(10), 5140B, University Microfilms No. 76-9381.

[12] Mahfoud, S. W., 1995, *Niching Methods for genetic Algorithms*; Doctoral thesis, University of Illinois, Urbana-Champaign, USA.

[13] Cedeno, W., Vemuri, V. R., and Slezak, T., 1995, Multiniche crowding in genetic algorithms and its application to the assembly of DNA restriction-fragments, *Evolutionary Computation*, **2**, no. 4, MIT Press, Cambridge, 321-345.

[14] Lin, C.-Y., Yang, Y.-J., and Liou, J.-Y., 1996, Genetic Search Based Hybrid Multimodal Optimization, *Emergent Computing Methods in Engineering Design*, NATO ASI Series, vol. 149, 74-81.

[15] Parmee, I. C., 1996, The maintenance of search diversity for effective design space decomposition using cluster-oriented genetic algorithms (COGAs) and multi-agent strategies (GAANT), *Proceedings of the ACEDC'96 Conference, University of Plymouth, UK*, pp. 128-138.

[16] Harik, G., 1995, Finding multimodal solutions using Restricted Tournament Selection, *Proceedings of the Sixth ICGA Conference, 15-20 July, Pittsburgh*, pp. 24-31.

[17] Syswerda, G., 1991, A study of reproduction in generational and steady-state genetic algorithms, In G. J. E. Rawlins (Ed.), *Foundations of Genetic Algorithms*, Morgan Kaufmann Pub., 94-101.

[18] Goldberg, D. E., Deb, K., and Horn, J., 1992, Massive multimodality, deception, and genetic algorithms, In R. Manner and B. Manderick (Eds.), *Parallel Problem Solving from Nature*, **2**; Elsevier Science Pub. B. V., 37-46.

[19] Oei, C. K., Goldberg, D. E., and Chang, S. J., 1991, *Tournament Selection, Niching, and the preservation of Diversity*, IlliGAL Report No. 91011, University of Illinois, Urbana-Champaign, USA.

[20] Yin, X. and Germay, N., 1993, A fast genetic algorithm with sharing scheme using cluster analysis methods in multimodal function optimization, In ; R. F. Albrecht, C. R. Reeves, and N. C. Steele (Eds.), *Proceedings of the International Conference on Artificial Neural Nets and Genetic Algorithms, Springer-Verlag, Berlin*, pp. 450 - 457.

[21] Jelasity, M. and Dombi, J., 1995, GAS, an approach to a solution of the niche radius problem, *Proceedings of GALISIA'95 Conference, Sheffield, UK, 12-14 September*, pp. 424-429.

[22] Miller, B. L. and Shaw, M. J., 1996, Genetic algorithms with dynamic niche sharing for multimodal function optimization, *Proceedings of 1996 IEEE International Conference on Evolutionary Computation, Nagoya, Japan, 20-22 May*, pp. 786-791.

[23] Stadnyk, I., 1987, Schema recombination in a pattern recognition problem, *Genetic Algorithms and their Applications: Proceedings of the Second International Conference on Genetic Algorithms*, pp. 27 - 35.

[24] Smith, R. E., Forrest, S. and Perelson, A. S., 1993, Searching for diverse, co-operative populations with genetic algorithms, *Evolutionary Computation*, **1**, no. 2, MIT Press, 127-149.

[25] Darwen, P. and Yao Xin, 1996, Every niching method has its niche: fitness sharing and implicit sharing compared, *Proceedings of PPSN IV Conference, Berlin, Germany, 22-27 September*, pp. 398 - 407.

[26] Grosso, P. B., 1985, *Computer simulation of genetic adaptation: Parallel subcomponent interaction in a multilocus model*, PhD Dissertation, University of Michigan, University Microfilms No. 8520908.

[27] Elo, S., 1994, A parallel genetic algorithm on the CM-2 for multi-modal optimization, *Proceedings of the First IEEE Conference on Evolutionary Computation, Florida, USA, 27-29 June*, pp. 818-822.

[28] Davidor, Y., 1991, A naturally occurring niche & species phenomenon: The model and first results, *Proceedings of the Fourth International Conference on Genetic Algorithms*, pp. 257 - 263.

[29] Beasley, D., Bull, D. R. and Martin, R. R., 1993, A sequential niche technique for multimodal function optimization, *Evolutionary Computation*, **1**, no. 2, MIT Press, 101-125.

[30] Mahfoud, S. W., 1995, A comparison of parallel and sequential niching methods, *Proceedings of the Sixth International Conference on Genetic Algorithms, Pittsburgh, Morgan Kaufmann Pub. Inc., 15-19 July*, pp. 136-143.

[31] Jarvis, R. A. and Patrick, E. A., 1973, Clustering using a similarity measure based on shared near neighbors, *IEEE Transactions on Computers*, **22**, no. 11, 388-397.

Evolution of Cellular-automaton-based Associative Memories

Marcin Chady and Riccardo Poli

School of Computer Science
The University of Birmingham (UK)
email: {mcc, rmp}@cs.bham.ac.uk

Keywords: Cellular Automata, Genetic Algorithms, Associative Memory, Emergent Computation

Abstract

Cellular Automata (CAs) are discrete dynamic systems composed of a large set of simple units organised into a regular one-, two- or multi-dimensional grid which update their state on the basis of their previous state and the state of a small number of neighbouring cells. CAs have been traditionally used for image processing and for hydrodynamics, thermodynamics and turbulence modelling. More recently CAs have also been used as mechanisms to study emergent computation, the phenomenon in which a large set of simple interacting elements with little information produce a complex coordinated information processing behaviour. In this paper, using the power of genetic algorithms, we study the ability of CAs to perform two very important forms of emergent computation: pattern association and associative memory.

1. Introduction

A cellular automaton is a discrete system which evolves in discrete space and time [1, 2]. It consists of a large number of cells which are organised in the form of an n-dimensional lattice. Each cell can be seen as a single processor which communicates only with the neighbouring cells and updates its state according to information received from them, as well as to its own present state. The state-update function is the same for all cells in the lattice. Although this is usually a very simple rule, on a global scale it can produce a very complex and often unpredictable behaviour.

Thanks to this complex behaviour CAs are an attractive platform for studying and implementing efficient forms of emergent computation, the phenomenon in which a large set of simple interacting elements with little information produce a complex, coordinated information processing behaviour. Interesting results with one-dimensional binary CAs have been obtained, for example, by Packard [3], Mitchell *et al.* [4, 5] and Andre *et al.* [6], who used evolutionary algorithms to discover CA rules which solve large-scale majority-classification and synchronisation problems.

In the work presented in this paper we also use binary CAs and evolutionary algorithms to discover state-update rules, but we concentrate on a considerably hard problem: the emergence of pattern-association and auto-association behaviour. In addition, differently from the previous work, we look at a particular class of CAs, which we term *feed-forward* CAs. Feed Forward CAs (FFCAs) are CAs with an update rule which yields a directional flow of information by using a neighbourhood which extends only in one direction, like in Figure 1.1.

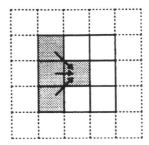

Figure 1.1 An example neighbourhood for a feed forward CA.

Two-dimensional FFCAs can be easily transformed into devices which perform global computation if we consider the left-most column as the input interface and the right-most column as the output interface, like in Figure 1.2.

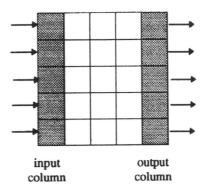

input   output
column   column

Figure 1.2 An example architecture for a CA-based pattern associator.

In our work we consider the upper and lower edges of the lattice as joined together to form a cylinder, so that information can pass freely between the top and bottom rows. The input and output columns, on the contrary, are not connected, so that information placed in the input column remains unchanged in the subsequent time steps, while information produced on the output column is discarded from one step to the next.

With this kind of CAs it is possible to map binary input vectors into corresponding output vectors, and, therefore, to implement pattern associators. In this paper we will mainly concentrate on a special class of pattern associators, namely the class of auto-associators or associative memories, where the CA is expected to retrieve incomplete binary vectors or rectify distorted ones, in a fashion similar to the operation of a Hopfield neural network [7].

In the next section we describe our approach to discovering CAs with emergent associative-memory properties, and in Section 3 we report the results obtained. Section 4 discusses these results, and Section 5 draws some conclusions and outlines some ideas for future research.

2. Evolving FFCAs

In order to make a CA perform a predefined global computation, i.e. to "program" it, one has to specify its initial configuration and the local rule that yields the required global behaviour. However, so far no method has been proposed to explain how to map the space of possible CA behaviours onto the corresponding local rules. In the literature on emergent computation with one-dimensional CAs, the rule table of binary CAs is usually encoded as a binary string and a genetic algorithm is used to evolve tables with the desired properties.

In our work we used the same strategy, but in addition to the rule table we also encoded in the chromosome the initial state of the CA. This may be very important in CAs for pattern association, as by modifying the behaviour of each cell in the automaton we allow the exploitation of local structure in the input/output patterns. A sample automaton and the corresponding chromosome are shown in Figure 2.1. The initial configuration of the input cells is not encoded, as these are always overwritten when an input is supplied.

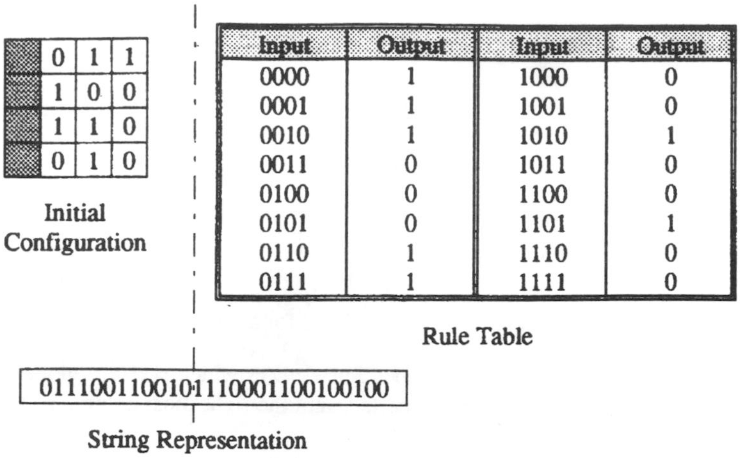

0	1	1
1	0	0
1	1	0
0	1	0

Initial
Configuration

Input	Output	Input	Output
0000	1	1000	0
0001	1	1001	0
0010	1	1010	1
0011	0	1011	0
0100	0	1100	0
0101	0	1101	1
0110	1	1110	0
0111	1	1111	0

Rule Table

0111001100101110001100100100

String Representation

Figure 2.1 An example CA encoded in a binary string.

In initial experiments aimed at tuning the GA and finding a good fitness function, we trained CAs to behave like hetero-associators, i.e. the input vectors were distinct from the corresponding output vectors. Some noise was applied to the input vectors, so that the CAs were trained to recognise distorted inputs. The input vectors in the training set were different enough from one another so that the noise could not make them indistinguishable.

The GA used one-point crossover, with both offspring preserved. The mutation rate was exactly one bit per chromosome. Each individual in the population was evaluated by running the corresponding automaton on a training set of input vectors and comparing the results obtained with the expected output vectors. The fitness of an individual was proportional to the number and severity of the errors made in each test. In the course of these experiments we found that the GA worked best with tournament selection and with the following fitness function:

$$f(s) = 1 - \frac{1}{n \cdot d_{max}^2} \sum_{i=1}^{n} |V_i - V_i'| (d_{max} - d_i),$$

where V_i is the output vector obtained when the i-th input vector of the training set is applied to the CA, $|V_i - V_i'|$ is the Hamming distance between the actual output vector and the desired output vector, i.e. the error produced by the CA, d_{max} is the maximum possible error, and d_i is the severity of input noise in test i (measured as the Hamming distance between the input vector and its distorted version fed into the input column of the CA). With this function CAs were less penalised for errors if their input was severely distorted.

Because the CAs were trained with noisy inputs, we decided to explore the effect of the generation gap, i.e. the percentage of population replaced by offspring at each generation, on the GA effectiveness. Experimentally we observed that the best results are obtained with small generation gaps. Indeed, the best generation gap was 30%. This can be explained as follows. Nondeterministic inputs introduce some amount of randomness and inaccuracy in the fitness estimation, because every individual is tested with a different set of inputs. It is likely that an individual's fitness estimation will be slightly different in each cycle of the genetic algorithm. Therefore, if an individual is estimated positively many times, it is more likely that it is indeed a good individual. A big generation gap shortens an individual's life and reduces its chances of being retested over many generations, thus reducing the generation gap gives greater stability of evolution and consequently leads to better results.

3. CA-based associative memory

The preliminary experiments mentioned above were quite promising and showed that genetic algorithms are an effective tool for producing FFCAs with pattern association behaviour. As a first step towards a rigorous study of the properties of FFCAs we decided to concentrate on their capability to act as associative memories. In the following subsections we report on our work towards an estimation of FFCA capacity to remember patterns.

3.1 Simple neighbourhood CA

In a first series of experiments on FFCA associative memories we used the neighbourhood shown in Figure 1.1. The task of rectifying a pattern may involve extracting global information that a single local rule using 4 cells cannot extract. It was hoped that the emergent behaviour of the CA would provide some means of global communication across the lattice, so that the whole pattern could be analysed. A helpful feature of FFCAs is that the input pattern placed in the input column stays constant throughout the computation, providing a steady reference for the rest of the automaton.

Two series of experiments were run, using

A. an 8×8 CA,

B. a 16×16 CA.

In each series the genetic algorithm was run repeatedly with an increasing number of pattern pairs to learn. In each pattern pair the input and output vectors were the same, but 20% noise was applied to the input vectors. The GAs were run with the parameters shown in Table 3.1.

Parameter	A	B	Input/Output
CA Width	8	16	0000000000000000
CA Height	8	16	1111111111111111
Minimum fitness	0	0	0101010101010101
Minimum generations	50	50	1010101010101010
Population size	400	400	0011001100110011
Chromosome length	72	256	1100110011001100
state encoding takes	56	240	1111000011110000
rule table takes	16	16	0000111100001111
Input noise level	20%	20%	0000000011111111
Number of CA steps per test	until relaxation	until relaxation	1111111100000000
Number of tests per pattern	100	100	
Number of patterns	2 - 8	2 - 10	

Table 3.1 GA/CA parameters for the CA-capacity experiments. For the 8×8 CA the input/output patterns were truncated after 8 bits.

Every individual was evaluated by feeding a distorted input vector into the input column and observing the CA's behaviour until it settled into a stable state. Individuals that did not settle within a given number of steps or fell into a periodic cycle were penalised by assuming the maximum error.

The GA ran for the minimum number of generations indicated in the table and then until no further progress was achieved, i.e. both the average and the best fitness in the population did not improve from one generation to another. For each experimental setting 3 independent runs of the GA were performed.

Table 3.2 shows typical rule tables obtained in the experiments.

Number of patterns	8×8 CA	16×16 CA
2	0001001100111111	0010001101111111
3	0001111100000111	0010111101001111
4	0000111100001111	0011111100000011
5	0000111100001111	0000111100001111
6	0000111100001111	0000110011001111
7	0000111100001111	0000111100001111
8	0000111101011111	0000111100001111
9	xxxxxxxxxxxxxxxx	0000111100001111
10	xxxxxxxxxxxxxxxx	0000111100001111

Table 3.2 Rule tables obtained in the simple-neighbourhood associative memory experiment.

It can be seen that, as the number of patterns grows, in both series the CA rule tables converge into one particular form, namely "0000111100001111," which is a simple copy rule that takes the state of the cell on the left and copies it into the state of the current cell.

The emergence of the copy rule is what we should expect to happen if the number of patterns stored in the CA exceeds its capacity, because the copy rule is the rule with the minimum error correction ability. This can be understood by considering the following equation

$$C = p \cdot \delta \implies \lim_{p \to \infty} \delta = \lim_{p \to \infty} \frac{C}{p} = 0,$$

where C is the total memory capacity of the CA, p the number of stored patterns and δ is the error correction capability. The formula simply shows that with a growing number of stored patterns the expected error correction capability approaches 0. Therefore, we can use the performance of CAs running a copy rule as the reference for measuring the CA error correction capability, i.e. we can assume that any CA having a higher fitness than a copy rule has a nonzero error correction capability. Furthermore, we can find the CA memory capacity by looking at the point where the evolved CA update rules converge to the copy rule.

Since $C = p \cdot \delta$, the maximum number of patterns stored in the CA will depend on the error correction requested from the CA, or in other words, on the level of noise applied to the input vectors. In these experiments, as well as the following ones, this level was set to 20%, which means the CAs were required to recover one fifth of the distorted input pattern.

The diagrams in Figure 3.1 show the performance of the simple-neighbourhood CA, compared with the performance of the copy rule.

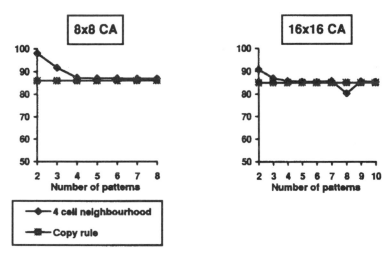

Figure 3.1 Performance of simple-neighbourhood CA compared with performance of the copy rule. The values denoted by diamonds correspond to averages of 3 runs of the GA.

From these results we can see that with up to 3 patterns the CA performance is above the copy rule. When the CA is required to store 4 patterns or more, the rules converge to the copy rule. This seems to indicate that the memory capacity is 3.

An interesting fact is that the 16×16 CA performance seems generally worse than 8×8 CA. A possible explanation for this effect is that the update rule with a small neighbourhood (i.e. 3+1 cells) is too simple to deal with 16-bit long patterns. Three new bits of information are processed by each cell in a single step, which is 37% of the total pattern length for the 8×8 CA, but only 19% for the 16×16 CA. Therefore, it might be relatively easy for a GA to discover rules that do better than the copy rule in a 8×8 CA, as the cells already have fairly large-scale information. On the contrary, the cells in a 16×16 CA really only have local information and therefore the GA has to solve the much harder task of discovering how to communicate information across the CA.

3.2 Increasing the CA capacity

The experiments described in the previous section were repeated with a neighbourhood including 6 cells, as shown in Figure 3.2.

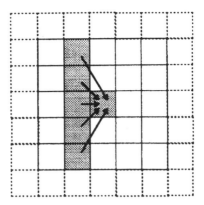

Figure 3.2 6-cell neighbourhood.

Since the rule tables for a 6-cell neighbourhood are quite large (they include $2^6 = 64$ bits) and difficult to understand, they are not shown in the report. However, from Figure 3.3 it can be seen that the experimental results follow a similar pattern to the results in the previous section, i.e. up to a certain number of patterns (in this case 4) both kinds of CA exhibit a positive error correction capability. After that point, the update rules converge to the copy rule, or never reach a performance comparable with it.

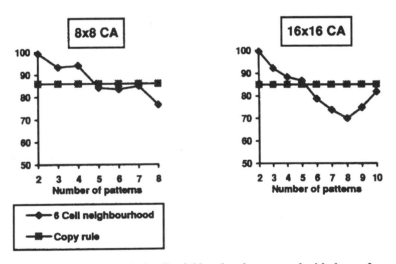

Figure 3.3 Performance of CAs with 6-cell neighbourhoods compared with the performance of the simple copy rule. The results are averages of 3 GA runs.

Comparing these results with those in the previous section one can see a clear increase in performance/capacity when using a bigger update rule neighbourhood. Both sizes of CA seem to have benefited from the bigger neighbourhood, although the 16×16 CA still does slightly worse than the 8×8 CA.

4. Discussion

It appears that increasing the size of CAs does not increase their memory capacity. In our experiments the capacity of the CA memory depends more on the update-rule neighbourhood size than on the size of the CA. Also, as it was pointed out in Section 3.1, the maximum number of patterns stored in the memory also depends on the level of noise applied to the input patterns. For this reasons, as well as because of the amount of time which would be required to carry out a more exhaustive investigation (e.g. to try different set of patterns, bigger neighbourhoods, different noise levels, etc.), it is impossible to give here a precise estimation of CA memory capacity in relation to its architectural parameters, as it was done by Hopfield in [7] for neural network-based associative memories. However, the results of the above experiments suggest that CAs can indeed have properties of associative memory.

The reduced capability of the 16×16 CAs to store patterns when using 4-cell neighbourhoods might be due to the fact that the proportion of the input pattern analysed by a single cell (19% of the total length) is smaller than the amount of noise present in the input (20%). It could also be related to more generic properties of CAs that affect their pattern matching capabilities. Some analysis of the pattern matching properties of CA has been done by Jen [8] who showed that there is a minimal radius of the update rule neighbourhood for which one-dimensional CA can "recognise" an arbitrary input string and retain it as invariant for the next time step. Although the processing done by feed forward CAs is much more complex than the processing of one-dimensional CAs, and the number of patterns being recognised in our experiments is greater than one, Jen's results can still be applied to the self-association problem in the FFCAs, since at every column of a FFCA a certain amount of pattern recognition takes place.

5. Conclusions

In this paper we have introduced feed-forward cellular automata models and presented a way of "programming" them by means of a genetic algorithm to perform a computationally useful function, namely pattern association. The investigation presented here showed the applicability of genetic algorithms to two-dimensional FFCA programming and indicated the potential of these CAs to act as associative memories.

In order to produce a full evaluation of CA associative memories and their practical applications, more work is required. These are some of the important issues that need to be addressed:

- The interdependence between the number of patterns stored in the memory and its error correction capability. The experiments in section 3 will have to be repeated with different noise levels to see how they affect the CA memory performance.

- Our "until relaxation" terminating condition was the simplest way for deciding when the CA computations have finished. However, other kinds of CA responses might favour the emergence of well-performing individuals.

- In the previous sections CAs were trained to remember very specific patterns, disregarding the fact that their capacity might increase by carefully selecting the patterns being stored. Also, the process of adapting the CA to a particular set of patterns is very lengthy, as the whole evolution process has to be repeated for every different set of memories. An alternative approach would be to look for CAs that remember well many patterns and try to understand how they do that, so as to find a more analytical approach to CA programming. We could look at the dynamic behaviour of CAs and pick up the individuals that have a large number of broad basins of attraction. This would allow storing many patterns (attractors) and being able to rectify them from a wide range of distorted patterns (basins of attraction).

References

[1] Toffoli, T., Margolus, N., 1987, *Cellular Automata Machines: A new Environment for Modelling*, MIT Press, London.

[2] Wolfram, S., 1994, *Cellular Automata and Complexity: Collected Papers*, Addison-Wesley.

[3] Packard, N.H., 1988, Adaptation toward the edge of chaos, in Kelso, J.A.S, Mandell, A.J., Shlesinger, M.F., *Dynamic Patterns in Complex Systems*, pp. 293-301, World Scientific, Singapore.

[4] Mitchell, M., Hraber, P.T, Crutchfield, J.P., 1993, Revisiting the Edge of Chaos: Evolving Cellular Automata to Perform Computations, *Complex Systems*, 7, pp. 89-130.

[5] Das, R., Mitchell, M., Crutchfield J.P., 1994, A Genetic Algorithm Discovers Particle-Based Computation in Cellular Automata, *Proceedings of the Third Parallel Problem-Solving From Nature Conference*.

[6]   Andre, D., Bennett III, F.H., Koza, J.R., 1996, Discovery by Genetic Programming of a Cellular Automata Rule that is Better than any Known Rule for the Majority Classification Problem, in Koza, J.R, Goldberg, D.E., Fogel, D.B., Riolo, R.L., *Genetic Programming 1996: Proceedings of the First Annual Conference*, MIT Press.

[7]   Hopfield, J.J., 1982, Neural Networks and Physical Systems with Emergent Collective Computational Abilities, *Proceedings of the National Academy of Sciences*, **79**, pp. 2554-2558.

[8]   Jen, E., 1986, "Invariant Strings and Pattern-Recognizing Properties of One-Dimensional Cellular Automata," *Journal of Statistical Physics*, **43**.

Simple Implementation of Genetic Programming
by Column Tables

Vladimír Kvasnièka, Jiøí Pospíchal

Department of Mathematics, Slovak Technical University, 81237 Bratislava, Slovakia
E-mail: kvasnic@cvt.stuba.sk, pospich@cvt.stuba.sk

Keywords: genetic programming, directed acyclic graphs, symbolic regression

Abstract

Simple implementation of genetic programming by making use of the column tables is discussed. Implementations of Koza's genetic programming in compiled languages are usually not most efficient when crossover is applied. If chromosomes are directed acyclic graphs, more efficient than rooted trees both in memory requirement as well as in evaluation time of chromosome, then crossover requires traversing the data structures and their preliminary analysis. Column tables inherently code directed acyclic graphs, the implementation of crossover is simple and needs neither traversing nor checking of integrity of resulting data structures and should be therefore more efficient. Stochastic transformation operation mutation is also easily defined. Column tables can represent graphs with several output nodes and may be used e.g. for optimization of feed-forward neural networks. Simple illustrative examples of symbolic regression based on the column tables are presented.

1. Introduction

Genetic programming is a name introduced by John Koza for an approach using genetic algorithms, where the chromosome is not a bit vector, but a rooted node-labeled tree with ordered branches. It is typically used for problems like symbolic regression, search for optimum strategy techniques in games, decision trees, or classification (representing in general any program). Let us have a directed tree, where some of the vertices are labeled by the same symbol and their subtrees are isomorphic (with equivalent evaluation of vertices and order of branches in case the operation is not commutative). These vertices can be merged, creating an evaluated directed acyclic graph. Such a graph can express a solution of a problem more efficiently both in memory requirement as well as in evaluation time of the solution [1,2]. Its representation in column tables contains also information about order of branches and moreover it remedies the problem of crossover. However, further described approach does not use minimal directed acyclic graphs like in [2], identical branches may appear coded in column tables. A crossover can be processed naturally in Lisp, which was the first programming language used for genetic programming, but it is slow, because interpreted. Crossover implementation is also not efficient in compiled languages like C or Pascal, when the tree structure usually expressed directly by pointers to nodes must be analyzed beforehand. In crossover operation two branches must be chosen, one from each of rooted (parse) tree corresponding to a chromosome, the branches are cut of and swapped. To achieve the root node of the branch usually means traverse through pointers starting from the root of the wlole tree. Mending of the problem by storing of some additional information about tree structure was neither consistent nor thoroughly theoretically founded [3-8]. Crossover for directed acyclic graphs is even more complicated. Recently, we employed graph theory to solve the problem by using Read's tree code [9-11]. The present communication shows that the so-called column tables with very simple structure allow a straightforward application of a recurrent algorithm for calculation of a functional value assigned to the graph, and a natural implementation of crossover.

2. Acyclic directed graphs and column tables

Let us study a *directed graph* [12] (its edges are directed) $G=(V,E)$, where $V=\{v_1,v_2,...,v_p\}$ is a nonempty vertex set and $E=\{e_1,e_2,...,e_q\}$ is an edge set, see Fig. 1. A *directed edge* $e \in E$ is determined as an ordered couple of vertices $v,v' \in E$, $e=(v,v')$. We say that an edge e starts at the vertex v and ends at the vertex v' (edge e is outgoing from the vertex v and incoming to the vertex v'). A directed graph is called *acyclic* if it does not contain directed cyclic path composed of a sequence of directed edges, where each outgoing vertex is also incoming vertex of the following edge.

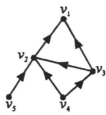

Figure 1. Directed graph with vertex set composed of five vertices, $V=\{v_1,v_2,v_3,v_4,v_5\}$, and edge set composed of six directed edges, $E=\{e_1,e_2,e_3,e_4,e_5,e_6\}$.

Indexing of vertices of a directed graph G is realized by a mapping φ that assigns unambiguously to each vertex an integer

$$\varphi : V \rightarrow \{1,2,...,p\} \tag{1}$$

The integer $\varphi(v)$ is called the *index* of vertex v. If a directed graph G is indexed by a mapping φ, then an *adjacency matrix* $A=(A_{ij})\in \{0,1\}^{p\times p}$ is determined as follows

$$A_{ij} = \begin{cases} 1 & \left(\text{for } \left(v_i,v_j\right) \in E\right) \\ \\ 0 & (\text{otherwise}) \end{cases} \tag{2}$$

This matrix for the directed graph in Fig. 1 has the form

$$\mathbf{A} = \begin{pmatrix} 0 & 0 & 0 & 0 & 0 \\ 1 & 0 & 0 & 0 & 0 \\ 1 & 1 & 0 & 0 & 0 \\ 0 & 1 & 1 & 0 & 0 \\ 0 & 1 & 0 & 0 & 0 \end{pmatrix} \tag{3}$$

According to this simple illustrative example we may conclude that properly indexed directed acyclic graphs are determined by adjacency matrices that contain nonzero entries only below the main diagonal, while on the other hand above the main diagonal (including the diagonal) all entries are zero.

Theorem 1 [12]. A directed graph $G=(V,E)$ is acyclic if and only if its vertices can be indexed in such a way that

$$\forall(v,v') \in E: \varphi(v)\rangle\varphi(v') \tag{4}$$

An indexing φ that satisfies the condition (4) is called the *canonical indexing*. According to Theorem 1, each directed acyclic graph can be canonically indexed. Simple illustrative example of the canonical indexing is displayed in Fig. 1, where vertices are already indexed so that the condition (4) is fulfilled. Vertices of canonically indexed graph may be divided into three disjoint subsets:

(1) *Input vertices*, these vertices are incident only with outgoing edges.
(2) *Intermediate vertices*, these vertices are incident simultaneously with incoming and outgoing edges.
(3) *Output vertices*, these vertices are incident only with incoming edges.

These three important notions may be illustrated by graph in Fig. 1, vertices v_4, v_5 are input, vertices v_2, v_3 are intermediate, and finally the vertex v_1 is output.

Canonically indexed directed acyclic graph with one output vertex will be called the *parse graph*. This type of graphs is of the great importance for simple implementation of the basic task of genetic programming (GP) [13,14] called the *symbolic regression*, where these graphs are used as an effective and simple representation of functions—subroutines. Let us consider, once again, the graph in Fig. 1. Each intermediate or output vertex is evaluated by a real number that is determined as a function value assigned to the vertex, its arguments are function values of vertices that are incident with edges outgoing from them and incoming to the given vertex, see Fig. 2. Then, after this convention, single vertices of graph in Fig. 1 are evaluated as follows

$$x_5 = c_5, \; x_4 = c_4, \; x_3 = f_3(x_4), \; x_2 = f_2(x_3, x_4, x_5), \text{ and } x_1 = f_1(x_2, x_3) \tag{5}$$

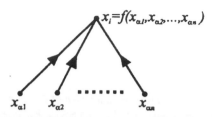

Figure 2. Each intermediate or output vertex of the parse graph is evaluated by function values with arguments corresponding to function values of vertices that are incident with edges outgoing from them and incoming to the given vertex. Input vertices are evaluated by constants.

For parse graphs, going successively from input vertices (that are evaluated by constants) through intermediate vertices to an output vertex, function values are recurrently calculated, final function value of the output vertex is considered as an evaluation of the whole parse graph. In other words, parse graph performs a mapping of input constants onto final function value assigned to the output vertex

$$x_1 = \mathbf{G}(c_4, c_5) \text{ or, in general, } x_1 = \mathbf{G}(c_i, c_j, \dots) \tag{6}$$

How to code parse graphs? As already noted in the previous part of this section, adjacency matrices expressed by lower-triangle matrices represent a universal tool for coding of parse graphs without ordered branches. From Theorem 1 and also from the above comment the following theorem may be formulated.

Theorem 2. A graph G is a *parse graph* if and only if its adjacency matrix A is lower-triangle so that each its row but the first one contains at least one element '1'.

A condition that each row but the first one contains at least one entry '1' corresponds to a property that a parse graph contains just one output vertex. If in a column below the matrix diagonal only zero entries appear, then the corresponding vertex is input (it does not have predecessors). A number of directed edges is determined by *input valences* (the number edges that are incoming to the vertex) or by *output valences* (the number of edges that are outgoing from the vertex) of all graph vertices

$$q = \sum_{v \in V} val_{in}(v) = \sum_{v \in V} val_{out}(v) \tag{7}$$

An application of adjacency matrices for coding of parse graphs is plagued by the following serious restrictions, these matrices are very sparse, i.e. adjacency matrices code the graph topology in a very "diluted" form, and a dominant part of entries is equal to zero. This is the main reason why we turn our attention to the so-called *column table* T_{column}, see Fig. 3.

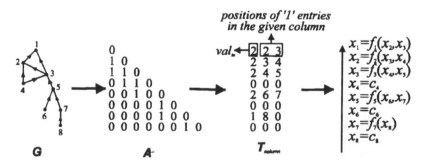

Figure 3. Illustration example of coding of parse graph by a column table. Parse graph G is initially coded by the adjacency matrix A, then this matrix is "condensed" to the form of a column matrix T_{column}. Going successively bottom-up through all table rows, function values of all vertices are recurrently constructed.

The column table T_{column} specifies positions of '1' entries in columns of the adjacency matrix A. The ith row of this table is composed of all positions of '1' entries in the corresponding ith column of adjacency matrix. In other words, the ith table row determines all predecessors of the ith graph vertex. Column tables express in a condensed form more information than adjacency matrices, because the vertices-predecessors in a row are in order, which is substantial for noncomutative functions. The column matrix $\mathbf{T}_{column}$ used for the determination of a parse graph G permits very simple recurrent calculation of the function value assigned to the output vertex, see Fig. 4 and Algorithm 1.

```
function Eval_Table(input : i) : real;
begin if T_{i1}=0 then
        begin {input vertex}
              Eval_Table:=c_i
        end else
        if T_{i1}=1 then
        begin {intermediate or output unary vertex}
              Eval_Table:=f_i(Eval_Table(T_{i2}))
        end else
        if T_{i1}=2 then
        begin {intermediate or output binary vertex}
              Eval_Table:=f_i(Eval_Table(T_{i2}),Eval_Table(T_{i3}))
        end;
end;
```

Algorithm 1. Pseudopascal implementation of evaluation the final function value assigned to the output vertex by making use a procedure - function with recursion, which acts on the column table $T'=(T_{ij})$. Function is initialized by the statement Eval_Table(1), its activation is finished on input vertices, i.e. $T_{i1}=0$. It is assumed that the parse graph is composed of vertices that have at most two predecessors ($val_{in}^{max}=2$), the table T has three columns and p rows. If this function is applied on the table displayed in Fig. 4, then we get the same final function value as the one presented in Fig. 4. Functions $f_1, f_2, ...$ are assigned to single vertices, they represent required function operations (plus, minus, times, division, sign change, ...). For input vertices, these functions correspond to given constants.

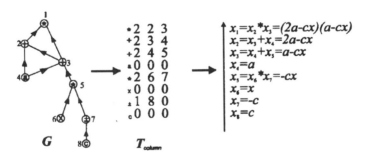

Figure 4. Illustrative example of calculation of function values of parse graph G presented in Fig. 3, its single vertices are now evaluated by algebraic operations. Column table T is enlarged by one column (the so-called 0-th column), which specifies functions assigned to vertices. Final function value is determined as a function value assigned to the output vertex, its numerical value is equal to $(2a-cv)(a-cx)$.

For purposes of symbolic regression, single vertices of a parse graph should be evaluated by functions in accordance with their input valences (arities), i.e. the ith vertex is evaluated by an function f_i. This requirement is simply realized so that the column table T is enlarged by a new the so-called 0-th column. This approach is illustrated in Fig. 4, where graph vertices are evaluated by simple algebraic operations.

Let us turn our attention to a random generation of column tables so that no repair process to these tables should be applied to get their semantically correct form. We shall postulate that the number of table rows is determined by the number p_{max}, and that the maximum input valence is determined by v_{in}^{max}. Then the column table is a matrix of the type $p_{max} \times \left(val_{in}^{max}+2\right)$, i.e. it contains p_{max} rows and $\left(val_{in}^{max}+2\right)$ columns (that are indexed by $0,1,..., (val_{in}^{max}+1)$. Since vertices of parse graph must be indexed canonically, entries of the ith row (T_{i1}, T_{i2}) must satisfy following conditions

$$i+1\le T_{i1}\le p_{max}, i+1\le T_{i2}\le p_{max}, T_{i1}\ne T_{i2} \qquad (8)$$

52

A graph determined by these conditions may not satisfy the basic requirement from the definition of parse graphs, i.e. though it is directed and acyclic it may contain more than one output vertex. Random generation of column tables so that conditions (8) are satisfied is outlined by Algorithm 2. Resulting column tables determine parse graphs, where only a part of table is active if we take into account recurrent solution initiated by output vertex indexed by '1', see Fig. 5.

```
procedure Gener_Parse_Graph(input : p_max, val_in^max; output T);
begin for i:=1 to p_max do
      begin T_i1:=random(max(val_in^max+1,p_max-i+1));
            case T_i1 of
              0 : begin {input vertex}
                        T_i0:=gener_type_function(0);
                  end;
              1 : begin {unary intermediate/output vertex}
                        T_i0:=gener_type_function(1);
                        T_i2:=i+1+random(p_max-i);
                  end;
              2 : begin {binary intermediate/output vertex}
                        T_i0:=gener_type_function(2);
                  repeat
                        T_i2:=i+1+random(p_max-i);
                        T_i3:=i+1+random(p_max-i);
                  until T_i1≠T_i2;
                  end;
            end {of case};
      end {of for};
end;
```

Algorithm 2. Pseudopascal implementation of random generation of column table composed of p_{max} rows and (val_{in}^{max}+2) columns, where $val_{in}^{max} \leq 2$. Function random(n) is random generator of integers from the closed interval [0,n-1] with uniform distribution of probability. Function genera_type_function(val_{in}) randomly generates the type of function assigned to ith vertex in a dependence on the input valence determined by $val_{in}=T_{i1}$; if $T_{i1}=0$, then the function determines a randomly generated "input constant".

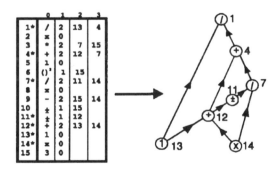

Figure 5. Illustrative example of randomly generated column table composed of 15 rows and four columns (p_{max}=15 and val_{in}^{max}=2). Vertices labeled by stars are *active* in the evaluation process corresponding to the output vertex index by '1'. Remaining vertices contribute in a construction of another parse graph with output vertex different than the vertex '1'. Drawn parse graph may be created by an activation of the function Eval_Table(1) from Algorithm 1. This means that each randomly generated column table correspond to a correct determination of parse function, though it is necessary to note that most part of the table may be inactive (redundant) in the process of its construction.

In this generalized approach to an interpretation of column tables the following situation may appear: Most part of table is not active in the construction of final evaluation assigned to the output vector indexed by '1'. In other words, the column table may contain some redundant information, see Fig. 5. Vertices that are used in construction of the final evaluation assigned to the output vertex are called *active* (these vertices are in Fig. 5 labeled by stars).

A *mutation* of table is understood as an operation that stochastically changes input table to a new output table

$$T' = O_{mut}(T) \tag{9}$$

where a stochasticity of this transformation is represented by a number P_{mut} and requires the following condition

$$\lim_{P_{mut} \to 0} O_{mut}(\mathbf{T}) = \mathbf{T} \tag{10}$$

A possible implementation of a mutation operation which satisfies all the above-required conditions is outlined in Algorithm 3 and in Fig. 6.

```
procedure Mutation_Table(input : T;
                         output : T');
begin for i:=1 to p_max do
      if random<P_mut then
      begin mutation_of_row(i); end else
      begin {copy of whole row}
            for j:=0 to T_i1+1 do
            T'_ij:=T_ij;
      end;
end;
```

Algorithm 3. A possible implementation of mutation of column tables $\mathbf{T}$ onto new table $\mathbf{T}'$. The parameter P_{mut} corresponds to a probability of a change of one table row. Single elementary transformations corresponding to the procedure `mutation_of_row(i)` are presented schematically in Fig. 6. The requirement (10) is automatically satisfied, i.e. if the probability asymptotically tends to zero, then a mutation of table T is inactive.

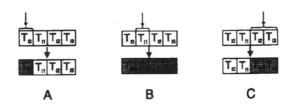

A   **B**   **C**

Figure 6. Schematic illustration of a mutation of the randomly selected table row (selected with probability P_{mut} see Algorithm 3). Three different cases are distinguished. In the case **A** only 0th row entry is mutated, it describes the function type and is generated by function `gener_type_function(T_i1)` from Algorithm 3. Other entries of the row remain unchanged. In the case **B** in an initial step the 1-st entry is mutated (it determines the input valence of the given vertex), then all remaining entries should be mutated. Finally, in the last case **C** are mutated those entries that correspond to predecessors of the given vertex.

A *crossover* of two tables T_1 and T_2 is a stochastic transformation that creates two tables T'_1 and T'_2 A stochasticity of this transformation consists in a fact that the so-called *crossing point* is randomly generated, starting from this point tables are exchanging their rows, see Fig. 7 and Algorithm 4. It is important to note that the crossover operator does not change a character of table, i.e. if input tables T_1 a T_2 are interpretable as parse graphs with order of branches, then also their "offspring" have this important property.

$$\left(\mathbf{T}'_1, \mathbf{T}'_2\right) = O_{cross}\left(\mathbf{T}_1, \mathbf{T}_2\right) \tag{11}$$

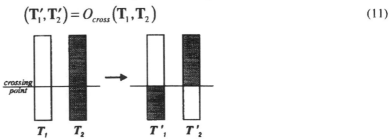

Figure 7. Illustration of crossover operation between two row tables. Crossing point is randomly generated, then rows below this point are mutually exchanged.

```
procedure Crossover_Table(input:T^(1),T^(2);
                          output:T'^(1),T'^(2));
begin crossing_point:=1+random(p_max-1);
      for i:=1 to crossing_point do
      begin for j:=0 to T^(1)_i1+1 do T'^(1)_ij:=T^(1)_ij;
            for j:=0 to T^(2)_i1+1 do T'^(2)_ij:=T^(2)_ij;
```

```
    end;
    for i:=crossing_point+1 to p_max do
    begin for j:=0 to T^(1)_11+1 do T'^(2)_ij:=T^(1)_ij;
          for j:=0 to T^(2)_11+1 do T'^(1)_ij:=T^(2)_ij;
    end;
end;
```

Algorithm 4. Simple implementation of crossover operation between two parental tables $\mathbf{T}^{(1)}$ and $\mathbf{T}^{(2)}$ that are transformed onto offspring tables $\mathbf{T}'^{(1)}$ and $\mathbf{T}'^{(2)}$.

3. Illustrative example

We have introduced two main operations over chromosomes - functions that are represented by column tables mutation and crossover. This means that some type of evolutionary algorithm (e.g. genetic algorithm [15], simulated annealing [16], evolutionary programming [17], etc.) may be used to solve the problem of symbolic regression. Let us have a regression set A composed of pairs x/y where x is a vector composed of input variables while y is a required function value. Each column table T is interpreted as a function (cf. eq. (6)) that assigns to input variables from the vector x a calculated functional value denoted by $T(x)$. Let as define the following objective function

$$E(\mathbf{T}) = \sum_{x/y \in A} |y - \mathbf{T}(\mathbf{x})| + \alpha |\mathbf{T}| \qquad (12)$$

where the last term on the r.h.s. corresponds to a penalization of the objective function (α is a small positive number) with respect to the number of active vertices $|T|$ in the parse graph T, i.e. the above objective function prefers parse graphs with smaller number of active vertices. In other words, let T_1 and T_2 be two parse graphs so that both correspond to the same first part of the objective function (12); if T_1 is composed of smaller number of active vertices than T_2, then $E(T_1) < E(T_2)$. The best (optimal) column table T_{opt} minimizes the above objective function over the "space" T of all possible functions representable by column tables

$$\mathbf{T}_{opt} = \arg \min_{T \in T} E(\mathbf{T}) \qquad (13)$$

The optimization problem (13) may be solved by any evolutionary method that could use the above defined mutation and/or crossover. In order to illustrate the present approach we used simple version of the genetic algorithm (population is composed of 200 chromosomes - column tables of the length $p_{max}=15$ and $v_{in}^{max}=2$, proportional selection realized by a roulette wheel, steady-state strategy with elitism). The regression set A was generated by a rational function $f(x)=(1+x-x^2)/(1+x+x^2)$ for 40 values of the variable $x \in [-10,10]$ that are equidistant with $\Delta x=0.5$. The terminal set was $\{x,1,2,...,9\}$, the function set $\{+,-,*,/,\pm\}$. The fitness was $1/E(T)$. Some parse graphs obtained in the course of history of genetic algorithm are displayed in Fig. 8. We see that the last graph corresponds to the function $((1/x+1)-x)/((1/x+1)+x)$, which after multiplication by x of both numerator and denominator exactly corresponds to the original function. The number of runs to achieve similar results was comparable with a genetic algorithm using tree structure instead of column tables, the generating function was probably too simple, so that "neutral changes" in unused parts of column tables did not have greater effect.

4. Conclusions

The data structure of column tables, introduced in this paper, allows to implement an operation of crossover for directed oriented graphs, without a need to use time-consuming algorithms of analyzing the directed oriented graphs. Directed oriented graphs are more efficient than rooted trees used by Koza, because substructures (branches), which appear more than once in a node-labeled rooted tree can be evaluated and stored only once, not redundantly several times.

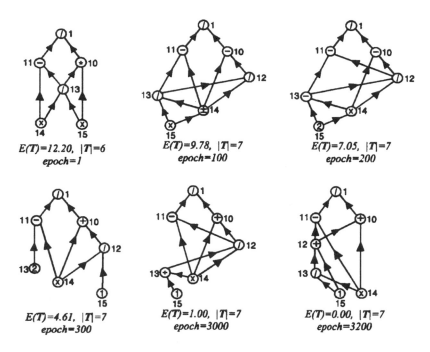

Figure 8. Illustrative parse graphs that appeared in the history of genetic algorithm, the last graph corresponds to a correct result precisely reproducing the regression set (i.e. its regression objective function $E(T)$ with the penalization term, see eq. (12)).

The use of directed oriented graphs in genetic programming was until now hindered by the fact, that implementation of crossover for these structures meant their preliminary analysis. Illustration of the problems with crossover applied for directed oriented graphs: the graph created by cutting an edge could be still connected and subgraphs of two directed oriented graphs could not be then swapped. If this happens, at least one more edge (from which there is no oriented path to the first cut edge) would have to be cut to divide the graph into two nonconnected subgraphs. The same number of edges satisfying the same conditions would have to be cut in the second directed oriented graph, so that the graph would be divided into disconnected components. Only then the two subgraphs could be swapped and new edges created. The application of crossover for column tables described in this paper makes unnecessary such a computational time demanding analysis. The aim of the presented approach is the similar to that one of Teller [18], which aim is to get an efficient crossover for graphs (not only trees) representing programs. However, instead of creating more complicated optimized programs performing crossover, the presented work simplified the representation of directed acyclic trees (general trees are not needed for problems in symbolic regression). This simple representation then allowed to use a simple crossover.

Moreover, the possibility of having more than one output vertex can be used for coding and simultaneous optimization of topology and parameters of feed-forward neural networks with several output nodes, or, in general, for coding of a net of functions.

The fact, that some of vertices with their corresponding connections then may not be taken into account, may be considered as a hindrance, because during reproduction an unnecessary information must be copied and more memory is used. However, this approach is in accordance with optimization of code in biology, because it can simulate evolution with punctuated equilibrium by accumulating of "neutral changes" into megamutation. The parts of code can correspond to biological introns, portions of the genotype (the DNA) that are not expressed in the phenotype [19]. Unlike introns defined in genetic programming as parts of code which must be evaluated, but do not have an effect (like adding a zero), introns in the present approach do not enter evaluation at all, unless a mutation or crossover brings them into play.

56

Acknowledgments

This work was supported in part by the grants # 1/2148/95 and # 1/4209/97 of the Scientific Grant Agency of Slovak Republic.

References

[1] Handley, S., 1994, On the use of a directed acyclic graph to represent a population of computer programs, *Proceedings of the 1994 IEEE World Congress on Computational Intelligence*, IEEE Press.

[2] Keijzer, M., 1996, Efficiently Representing Populations in Genetic Programming, *Advances in Genetic Programming II*, P.J. Angeline, K.E. Kinnear, eds., MIT Press. Cambridge, MA, pp. 259-278.

[3] Zongker, D., and Punch, B., 1996, *Program lil-gp*, Michigan State University, obtainable from http://isl.cps.msu.edu/GA/software/lil-gp/

[4] Hörner, H., 1996, *The Vienna University of Economics Genetic Programming Kernel*, Vienna University of Economics, obtainable from http://aif.wu-wien.ac.at/%7Egeyers/archive/gpk/

[5] Glowacki, D., 1994, *Geppetto: C Library for Writing Genetic Programming Applications*, The University of Texas at Austin, obtainable from http://www-cgi.cs.cmu.edu/afs/cs.cmu.edu/project/ai-repository/ai/areas/genetic/gp/systems/geppetto/

[6] Singleton, A., 1994, *GPQUICK: Simple GP system implemented in C++*, Creation Mechanics, Inc., PO Box 248, Peterborough, NH, obtainable from http://www-cgi.cs.cmu.edu/afs/cs.cmu.edu/project/ai-repository/ai/areas/genetic/gp/systems/gpquick/

[7] Tackett, W. A., and Carmi, A., 1993, *SGPC: Simple Genetic Programming in C*, obtainable from http://www.cs.cmu.edu/afs/cs/project/ai-repository/ai/areas/genetic/gp/systems/sgpc/0.html

[8] Fraser, A., 1994, *Program gpc++*, University of Salford, obtainable from http://www.salford.ac.uk/docs/depts/eee/genetic.html

[9] Kvasnièka, V., Pospíchal, J., and Pelikán, M., 1996, Read's linear codes and evolutionary computation over population of rooted trees, *Intelligent Technologies*, P. Sinèák, ed., vol. II, Her¾any, Slovakia, November, pp. 141-154.

[10] Pelikán, M., 1996, *Genetic programming*, student project (in Slovak), Faculty of Mathematics and Physics, Comenius University, Bratislava, Slovakia, April.

[11] Read, R.C., 1972, Coding of Unlabeled Trees, *Graph Theory and Computing*, R.C. Read, ed., Academic Press, New York, pp. 153-182.

[12] Harrary, F., 1969, *Graph Theory*, Addisson-Wesley, Reading, MA.

[13] Koza, J. R., 1992, *Genetic Programming: On the Programming of Computers by Means of Natural Selection*, MIT Press, Cambridge, MA.

[14] Koza, J.R., 1994, *Genetic Programming (II): Automatic Discovery of Reusable Programs*, MIT Press, Cambridge, MA.

[15] Goldberg, D.E., 1989, *Genetic Algorithms in Search, Optimization, Learning*, Addison-Wesley, Reading, MA.

[16] van Laarhoven, P.M.J., and Aarts, E.H.L., 1987, *Simulated Annealing: Theory and Applications*, Reidel, Dordrecht, The Netherlands.

[17] Fogel, D.B., 1995, *Evolutionary Computation: Toward a New Philosophy of Machine Intelligence*. IEEE Press, Piscataway, NJ.

[18] Teller, A., 1996, Evolving Programmers: The Co-evolution of Intelligent Recombination Operators, *Advances in Genetic Programming II*, P.J. Angeline and K.E. Kinnear, Jr, eds., MIT Press, Cambridge, MA, pp. 45-68.

[19] Watson, J.D., Hopkins, N.H., Roberts, J.W., Wiener, A.M., 1987, *Molecular Biology of the Gene*, Menlo Park, CA: The Benjamin/Cummings Publishing Company, Inc.

Part 2: Neural Networks

Papers:

XCS Classifier System Reliably Evolves Accurate, Complete, and Minimal Representations for Boolean Functions

Tim Kovacs

School of Computer Science, University of Birmingham
Birmingham U.K. B15 2TT
T.Kovacs@cs.bham.ac.uk
Phone: +44 121 414 3736
Fax: +44 121 414 4281

Keywords: XCS, Classifier Systems, Generalization Problem, Structural Credit Assignment Problem, Reinforcement Learning.

Abstract

Wilson's recent XCS classifier system forms complete mappings of the payoff environment in the reinforcement learning tradition thanks to its accuracy based fitness. According to Wilson's *Generalization Hypothesis*, XCS has a tendency towards generalization. With the *XCS Optimality Hypothesis*, I suggest that XCS systems can evolve *optimal populations* (representations); populations which accurately map all input/action pairs to payoff predictions using the smallest possible set of non-overlapping classifiers. The ability of XCS to evolve optimal populations for boolean multiplexer problems is demonstrated using *condensation*, a technique in which evolutionary search is suspended by setting the crossover and mutation rates to zero. Condensation is automatically triggered by self-monitoring of performance statistics, and the entire learning process is terminated by *autotermination*. Combined, these techniques allow a classifier system to evolve optimal representations of boolean functions without any form of supervision. A more complex but more robust and efficient technique for obtaining optimal populations called *subset extraction* is also presented and compared to condensation.

1 Introduction

This work is concerned with finding solutions to the *generalization* or *structural credit assignment* problem, which may be seen as the problem of generalizing accurately from experience. This is an important issue for machine learning systems as effective generalization is vital for systems to scale well to larger problems. Additionally, our folk concept of "understanding" learned material seems to imply an ability to generalize effectively. Certainly it is important for machine learning systems to be able to generalize from past experience in order to cope with novel situations.

Reinforcement learning (RL) systems attempt to learn complete maps of their environment from a limited input/action/payoff interaction[1] with it. A complete map is one which has an estimated payoff for each input/action pair. Many approaches to learning such mappings, and to producing accurate generalizations within them have been used by RL systems. For example, the popular tabular Q Learning technique [1] exhaustively enumerates condition/action pairs and maintains a payoff estimate for each. As a result it suffers from poor scalability, although modifications to allow generalization have been introduced (see for example [2, 3]).

Classifier systems (CS) are rule based learning systems which are able to generalize over their inputs, and thus have the potential to scale well, but they have not traditionally constructed complete payoff maps. The current work is based on Wilson's XCS system [4, 5], a new type of classifier system which, unlike traditional CS, does construct complete payoff maps thanks to its shift to accuracy based fitness.

---

[1] Or, using alternative terminology, a stimulus/response/reward interaction.

According to Wilson's *XCS Generalization Hypothesis*, XCS has a natural tendency towards accurate generalization.

Potential applications of XCS in design and manufacturing include those in which accurate generalization (rather than traditional point optimization) is desirable. XCS may also be suitable when a complete representation of the problem space is needed. One case where this is useful is when a variety of fitness functions need to be applied to a common structure. The current work enhances the ability of XCS to find useful solutions.

The focus of the current work is this: given that we have in XCS a reinforcement learning system with a tendency towards accurate generalization, how can we achieve optimal generalizations? In this paper I will present two techniques which allow XCS to evolve optimally general representations for boolean functions.

1.1 Optimal Populations

I will be referring to *optimal populations* or *optimal solutions* for given problems. For the purposes of this work, an optimal solution is defined as one which has three characteristics:

- It is complete, i.e. it describes all regions of the input/action space.

- It is non-overlapping, i.e. no part of this space is described more than once.[2]

- It is minimal, i.e. the minimum number of non-overlapping rules is used to described the problem space.[3]

The advantages of optimal solutions over those which are merely correct are:

- An optimal solution is more comprehensible than less succinct solutions. It is one thing for a system to find a solution to a problem and another for humans to understand this solution.

- Fewer resources are required to store and process the solution if it is minimal. The more concise the representation, the better the system will scale.

- An optimal population may be advantageous if the problem changes and further evolutionary search is necessary. In an optimal population concepts (in the form of segments of classifier conditions) are expressed most generally and may be more easily adapted by recombination to solve new aspects of the problem. At the same time, an optimal population is less diverse and may, depending on the problem, make a less useful base for further genetic search.

1.2 Overview of XCS

Classifier systems are a form of domain independent rule-based machine learning system introduced by John Holland (see [6, 7]). Classifier systems use a Genetic Algorithm (GA) to generate condition/action rules or *classifiers* which are evaluated during interaction with the problem environment. Classifiers typically use strings of characters composed from the ternary alphabet $\{0, 1, \#\}$ to represent conditions and actions. This is the binary alphabet of the standard GA augmented with a "don't care" symbol $\#$ which acts as a wildcard when matching environmental inputs to rule conditions. It is the use of the $\#$ which allows rules to generalize over inputs. Classifiers also have various parameters (statistics associated with them), such as estimates of the payoff the system will receive if it uses that rule, or the *fitness* of the rule (an estimate of its overall value to the system). Search proceeds in the GA by selecting parent classifiers (with fitter classifiers being more likely to be selected) and generating children from them using search operators. The two most common search operators used by the GA are *crossover* (in which

---

[2]Note that in a traditional classifier system we might consider classifiers to overlap if their conditions do, but in XCS both conditions and actions must overlap since we are dealing with a complete map of the input/action space.

[3]Allowing overlaps may reduce the population size depending on the problem and the representation used, but this issue will not be dealt with in the current work.

segments of two parent's strings are recombined to make new strings for their children) and *mutation* (in which a random change is made to a child's string).

XCS is a type of classifier system introduced in [4] and extended in [5] (additional detail and analysis of the system are available in [8]). The primary distinguishing feature of XCS is that classifier fitness is based on the accuracy of classifier payoff prediction rather than on payoff prediction (strength) itself. In addition, there are many more subtle differences between the traditional classifier system and XCS including the use of a niche GA, a Q-Learning like update mechanism and a deletion mechanism which seeks to balance classifier allocation between niches. XCS has many features in common with Wilson's earlier ZCS work [9].[4]

For an extended review and analysis of XCS the reader is referred to [8], which addresses many of the subjects of this paper in more detail, and includes overviews of classifier systems, reinforcement learning, payoff environments, multiplexer problems, representational difficulties with classifier systems and other subjects.

1.2.1 Accuracy-Based Fitness

In traditional CS, classifier strength plays a double role: it is used as a predictor of payoff in action-selection, and as a measure of fitness in the GA. In XCS, strength is replaced by three parameters: *(payoff) prediction, prediction error* and *fitness*. Prediction fills the role of strength in the action-selection component, and is also used to calculate prediction error, a measure of classifier accuracy. Fitness is an inverse function of the prediction error (put another way, it is a function of the accuracy of the prediction).

By relieving payoff prediction of its double role Wilson has significantly improved the classifier system architecture. In XCS, payoff prediction is only relevant in action selection which means XCS classifiers are free to map any region of the input/action/payoff space – as long as they do so accurately. As a result XCS tends to form "complete and accurate mappings X × A ⇒ P from inputs and actions to payoff predictions ... which can make payoff-maximising action-selection straightforward" [4]. The attempt to construct a complete mapping of the payoff environment is in the spirit of much reinforcement learning work, and indeed the function of the learning subsystem in XCS is related to the RL technique Q-learning.

1.2.2 Macroclassifiers

A second major difference of XCS is the use of *macroclassifiers*, the name given to a classifier with an additional *numerosity* parameter which indicates how many copies of that classifier are considered to be in the population. When a new classifier is created, the population is scanned to see if a classifier with the same condition and action exists. If so, the new classifier is discarded and the existing one has its numerosity incremented by one. An analogous process is used in deleting classifiers from the system. XCS treats a macroclassifier in all ways as the equivalent number of (micro)classifiers. Macroclassifiers increase the run-time speed of the system, make for a more compact and better scaling representation, offer interesting information on the distribution of resources within the system, and most importantly for the current work, make possible the use of techniques for obtaining optimal populations of classifiers which I will outline shortly. I have briefly evaluated macroclassifiers in [8].

1.2.3 Deletion Techniques

[4] describes two techniques for calculating the probability of a classifier being selected for deletion in the GA. I have compared these techniques and found that although the second technique results in smaller population sizes, it is highly detrimental to the development of members of the optimal population. I believe this is a result of excessive bias against newly evolved classifiers, and have found that it can be rectified by modifying this technique to only penalise low fitness classifiers if their experience parameter is greater than 20 (i.e. if they have had their parameters updated more than 20 times). I will refer herein to this modified form of the second technique as deletion technique 3.

1.2.4 Terminology and Notation

The following terminology and notation was drawn from [4] whenever possible.

---

[4]ZCS is intended as a minimalist classifier system whose mechanisms are more easily understood than those of the traditional CS. Wilson showed that learning in ZCS has strong similarities to the RL technique Q-Learning [1].

- Classifiers are written: <condition>:<action> ⇒ <payoff prediction>
  E.g. 111### : 0 ⇒ 100 would be interpreted as: if the input string begins with 111, then action 0 should be taken, and a payoff of 100 units will be expected.

- [M] The match set. This is the set of classifiers which match the current input to the system.

- [A] The action set. XCS forms [M], then selects an action from among those advocated by the classifiers in [M]. The classifiers in [M] which advocate the chosen action form [A].

- [P] The general classifier population (the classifier list).

- [O] The set of classifiers forming an optimal population for a given problem. An optimal population has three characteristics: it is accurate, complete and minimal (see section 1.1). [O] is sometimes referred to as existing within some larger population [P].

1.3   Overgeneral, Maximally General and Suboptimally General Classifiers

Classifiers express generalizations using the don't care symbol # in their conditions. For example, a classifier with condition 00# will match both 001 and 000 and treats these two inputs as equivalent. A classifier is either overgeneral, maximally (optimally) general, or suboptimally general in respect to the inputs it matches. Consider the following payoff landscape:

Input	Action	Payoff Rate
00	1	200
01	·1	200
10	1	100
11	1	100
##	0	0

Suppose an XCS system trained on this payoff landscape has a population consisting of the following classifiers:

Classifier	Condition	Action	Predicted Payoff	Prediction Error	Accuracy	Fitness
A	##	1	100	0.5	0.0	< low >
B	0 #	1	200	0.0	1.0	< high >
C	1 0	1	100	0.0	1.0	< high >
D	1 1	1	100	0.0	1.0	< high >
E	##	0	0	0.0	1.0	< high >

(The accuracy of A is 0.0 because its prediction error exceeds a threshold called the *accuracy criterion* [4, 8].) We can describe each classifier as one of the following:

- **Overgeneral** An overgeneral classifier matches too many inputs, specifically, some of the condition/action pairs it refers to pay off at different rates. Classifier A is overgeneral; its conceptualisation of the input/action space is inaccurate, and it should ideally be replaced by more specific classifiers whose conditions do not cross payoff level boundaries.

  If fitness is based on strength, overgeneral classifiers may survive as "guessers" which are sometimes correct and sometimes not. The prediction of A, 100, is an average of the payoffs it receives. To a system which bases fitness on strength (predicted payoff), A will look as valuable as C or D for use as a parent classifier. However, to a system which bases fitness on accuracy of payoff prediction, it is clear that A is not as valuable as the other classifiers.

- **Maximally General** A maximally general classifier is one which matches only inputs which pay off at the same rate, and which cannot become any more general (add any more #s) without becoming inaccurate (i.e. without matching inputs which would pay off at different rates). Classifiers B and E are maximally general.

- **Suboptimally General** Classifiers C and D are suboptimally general; each only matches inputs which pay off at the same rate, but there are other inputs which pay off at that rate which they could also match. Thus they could each be made more general without losing accuracy (notice that they are already perfectly accurate, i.e. their accuracy is 1.0). Ideally they would both be replaced by a single more general classifier with condition 1#.

1.4   The Generalization Hypothesis

It appears that evolutionary pressure to generate *maximally general classifiers* (i.e. classifiers which are as general as possible while remaining within some accuracy criterion) exists within XCS. Wilson's *Generalization Hypothesis* explains this process as follows:

"Consider two classifiers C1 and C2 having the same action, where C2's condition is a generalization of C1's. That is, C2's condition can be generated from C1's by changing one or more of C1's specified (1 or 0) alleles to don't cares (#). Suppose that C1 and C2 are equally accurate in that their values of ε are the same.[5] Whenever C1 and C2 occur in the same action set, their fitness values will be updated by the same amounts. However, since C2 is a generalization of C1, it will tend to occur in more match sets than C1. Since the GA occurs in match sets, C2 would have more reproductive opportunities and thus its number of exemplars would tend to grow with respect to C1's (or, in macroclassifier terms, the ratio of C2's numerosity to C1's would increase). Consequently, when C1 and C2 next meet in the same action set, a larger fraction of the constant fitness update amount would be "steered" toward exemplars of C2, resulting through the GA in yet more exemplars of C2 relative to C1. Eventually, it was hypothesised, C2 would displace C1 from the population." (Wilson 1995)

The logical conclusion of this preference for the more general of two equally accurate classifiers is that the *maximally general* classifier for a payoff level will tend to be evolved and tend to displace its competitors from the population. In the following I will make use of this tendency to obtain optimal populations.

1.5   Proposal of the Optimality Hypothesis

I propose a simple extension to the Generalization Hypothesis called the *Optimality Hypothesis*. It states that XCS will eventually evolve a maximally general classifier for each payoff level and that that classifier will have a greater numerosity than any other in its payoff level. The significance of the Optimality Hypothesis is that it holds that XCS will evolve a population of which a subset is an optimal solution and that we will be able to distinguish this subset from the general population based on the distribution of numerosity.

The predictions of the Optimality Hypothesis are naturally contingent upon such things as enough cycles and a large enough population size being used for the generalization process to succeed completely. Further, the generalization process is limited by the nature of the samples drawn from the environment (e.g. unrepresentative sampling of the problem environment may throw the generalization process off). However, such complications are inherent in all reinforcement learning systems.

1.6   The Multiplexer Function

The multiplexer function is defined for strings of length $L = k + 2^k$ where k is an integer > 0. In a 6 multiplexer problem ($k = 2$), the input to the system consists of a string of six binary digits, of which the first two (the address) represent an index into the remaining bits (the data), and the value of the function is the value of the indexed data bit. E.g., the value of of 101101 is 0 as the first two bits 10 represent the index 2 (in base 10) which is the third bit following the address. Similarly, the value of 001000 is 1 as the 0th bit after the address is indexed. The optimal population for this problem consists of the 8 classifier conditions of the form: 11###1 (i.e. where all digits in the data part are # apart from the indexed one), each advocating the 2 possible actions for a total of 16 classifiers.[6] The multiplexer is actually a rather

---

[5]ε represents the prediction error parameter of a classifier.

[6]See [4] or [8] for a longer introduction to the multiplexer function.

difficult problem to find an optimal population for, as there are a number of maximally general classifiers of the form 0#00## which are as general (both types have three #s) and as accurate as the members of [O]. However, maximally general classifiers of this second form tend to overlap and a population based on them requires more than 16 classifiers to completely map the input/action space. Because these two types of maximally general classifier are equally general and accurate, the generalization mechanism must distinguish the optimal population by considering the solution as a combination of interacting classifiers, not just a group of independent classifiers.

2 Obtaining Optimal Populations by Condensation

Wilson [4] discusses the use of *condensation* to reduce the size of the population once the system appears to have learned a given problem. Condensation consists of running the system with the mutation and crossover rates set to zero. This suspends the genetic search as no new classifier conditions can be generated, but allows the classifier selection/deletion dynamics in the GA to continue to operate. The result is a gradual shift in numerosity from less fit/less general classifiers to more fit/more general classifiers. Once a classifier reaches zero numerosity it is removed from the system. Over several thousand cycles the result is a condensation of the population to its fittest members.

Although condensation was applied to an XCS system in [4], optimal populations were not obtained as condensation was applied too soon. Because no new classifier conditions are generated once condensation begins, the optimal population must already exist within the general population at this point. Choosing when to begin condensation is thus of some importance, and, unfortunately, difficult to determine because of the variability in the time required for the completion of [O] and the indirectness of problem-independent techniques for detecting it. I investigated the use of several statistics for estimating the completion of [O] in [8] but none was very accurate, and all involved the use of a user-selected delay between the trigger condition being met and the commencement of condensation. I have triggered condensation for the 6 multiplexer problem when a moving average of the last 50 points of system error[7] has remained below 0.01 for 2,500 consecutive GA cycles (using deletion technique 1). With these settings the system was able to find optimal populations on 100 out of 100 runs. Unfortunately, this approach does not scale well, as a delay of 10,000 GA cycles is required when triggering condensation for the 11 multiplexer to again achieve 100% success on 100 runs.

2.1 Autotermination

The system can be left to run for a preset number of cycles, or alternatively can make use of autotermination to cease execution when an optimal population has been evolved. Autotermination works as follows: once condensation has begun, each time a classifier is removed from the system, each pair of classifiers which can be formed from those in [P] is compared to see if they overlap (i.e. to see if there is any input which both will match). Initially, there will be many overlaps in the population, but once condensation has reduced [P] to [O], no overlaps will remain (recall from section 1.1 that an optimal population has no overlaps). It should be noted that this only works if an optimal population is found, otherwise overlaps will likely remain and autotermination will not stop the system.

2.2 Accelerating Condensation

I have found that the rate of condensation can be greatly increased by the simple modification of truncating each action set to its single most numerous member once condensation has begun. In other words, each time the system forms an action set, all but the most numerous classifier in that set are removed completely from the entire system. When an optimal population is present in the general population, this has the same effect as the original condensation method because in an optimal population there are no overlaps and thus only a single classifier for each action set.

---

[7]System error is calculated on each cycle as (| sum of the fitness weighted payoff predictions of the classifiers in [A] - actual payoff received from the environment |) / (highest possible payoff - lowest possible payoff)

3 Obtaining Optimal Populations by Subset Extraction

Although optimal populations can be found by condensation, the approach is somewhat unsatisfactory in that the delay before commencing condensation must be set by the user and appears to be rather sensitive to problem complexity. Additionally, because of the considerable variability from run to run in the number of cycles required for the complete optimal population to evolve (before which condensation must not be started), a rather liberal delay must be used. A second approach to obtaining optimal populations called *subset extraction* improves on condensation by being less sensitive to problem complexity and yet finds optimal populations in fewer cycles. However, it has the disadvantage of being more complex than the straightforward condensation method.

Subset extraction attempts to find an optimal population by analysing the system at a particular instant – it does not rely on incremental calculations or a delay as condensation does. This means it can be applied at any interval, e.g. every 50th or 100th cycle in order to reduce processing resource use. Subset extraction is conceptually simple: if some well-evaluated subset of [P] with good accuracy ratings can be found which completely covers the input/action space and is non-overlapping, this subset is taken as a candidate optimal population. The third property of an optimal population, that it have as few members as possible, is then delt with by attempting to combine compatible classifiers. The following sections describe this process in detail.

3.1 A Complete, Non-Overlapping Covering of the Input/Action Space

The first problem which presents itself is that some classifiers in [P] may be inaccurate and should not be used in forming [O]. For example, ###### : 1 and ###### : 0 cover the input/action space completely and in a non-overlapping way, but for the 6 multiplexer problem clearly do not form an optimal population as they will often incorrectly classify the inputs they match. To avoid generating [O]s based on inaccurate classifiers we will simply ignore any classifiers which are not accurate (i.e. which do not have an accuracy of 1.0) or which are not experienced (i.e. which have been members of [A] and therefore had their parameters updated less than 20 times).

The second difficulty is dealing with the large number of possible subsets of [P]. In a typical 6 multiplexer experiment there might be a maximum of 400 microclassifiers allowed, although this might amount to only 100 macroclassifiers when we try to extract [O]. Given that some classifiers may be rejected out of hand due to a lack of accuracy or experience, we might be left with only 75 to consider. However, this leaves an impractical 2^n or $2^{75} = 3.777893e+22$ possible subsets to consider.

A simple technique was adopted to reduce the computational load, based on XCS's tendency to eventually allocate more numerosity to members of [O] than to any other classifiers in [P]. This technique requires the assumption that if a classifier x is an element of [O], then any more numerous classifiers will also be in [O]. This assumption is not always true, but as XCS redistributes numerosity it will tend to become true.

Given this assumption, we can sort the accurate, experienced members of [P] according to numerosity and use the following algorithm to identify elements of the optimal population:

P = List of all accurate, experienced classifiers sorted in order of decreasing numerosity.

O = Empty list.

1. If P is empty conclude optimal population does not exist in P, else remove head of P and add to O.

2. If classifiers in O form a complete non-overlapping map of the input/action space then conclude O is the optimal population. If classifiers in O overlap conclude an optimal population does not exist in P. If classifiers in O do not overlap but do not cover the complete input/action space return to 1.

With this approach we have at most 75 subsets of [P] to consider – making the assumption above changes the computational complexity of the algorithm from exponential to linear time. Although this algorithm can incorrectly conclude that a complete [O] does not exist within [P], this becomes less likely as numerosity is redistributed in the expected way. Because the algorithm is computationally inexpensive it can be applied at relatively frequent intervals, minimising the delay in extracting [O]. This assumption appears reasonable and justifiable given the Generalization Hypothesis. By this I mean that, although it is possible for some element in [P] not in [O] to be more numerous than some element in [O], this situation is counter to the trend of numerosity allocation in XCS. Consequently, although this algorithm may not extract [O] from [P] as soon as possible, it will not be long before the least numerous classifier in [O] becomes more numerous than the most numerous outside [O], at which point this algorithm will extract [O] from [P]. In other words, by making this assumption and attempting subset extraction relatively frequently we merely delay the extraction of [O] by some reasonably small number of cycles.

3.2 Achieving Minimality in [O] by Combining Classifiers

Once [O] has been extracted from [P] there is still the matter of minimality to contend with before [O] is truly optimal. The problem is that [O] may cover the input/action space completely, but may use more classifiers than necessary. Fortunately, it is always possible to detect and rectify this condition by combining classifiers. It may be possible for two or more classifiers to have their conditions combined such that a smaller number of conditions expresses the union of all the others. E.g. $000000 : 0 \Rightarrow 100$ and $000001 : 0 \Rightarrow 100$ can be combined into (and replaced by) $000000\# : 0 \Rightarrow 100$. This is not always possible due to limitations in the representational capacity of the ternary alphabet currently used in XCS classifier conditions (see [10] and [8] section 5.3. This is a difficulty with the ternary alphabet and thus only indirectly with XCS as it could employ some other representational scheme.) It is necessary that classifiers which are to be combined advocate the same action and have the same prediction, in order not to corrupt the representation of the input/action space. In principle, we might have a population in which many classifiers could be combined, e.g. 8 classifiers might be combined into a single one. In practice, the only combinations of classifiers which have been encountered in trying to minimise the candidate [O] in 6 and 11 multiplexer experiments are combining pairs of classifiers into a single classifier, and combining triplets into pairs. Thus it appears unnecessary to attempt all possible combinations of classifiers, thanks to XCS's tendency towards generalization which produces classifiers which are already maximally general, or close to it. Once all combinations have been made [O] has all three properties of an optimal population.

4 Evaluating Condensation and Subset Extraction

Condensation and subset extraction were tested on 6 and 11 multiplexer problems in order to compare their performance, although only 11 multiplexer problems are shown due to lack of space. Figure 1 shows condensation on an 11 multiplexer problem. In this test, the population size limit[8] was set to 1,000, condensation was triggered by 12,000 consecutive GA cycles of system error below 0.01, and deletion technique 3 from section 2 was used. [O] was first completed around 13,000 cycles and condensation began around 28,000 cycles. This excessive delay is the only cause for complaint with the condensation method as the accelerated form of condensation performed very well, reducing [P] to [O] within a few hundred cycles once it had actually begun. Unfortunately, the large delay before beginning condensation is necessary due to the considerable variance between runs between the trigger condition being met and the completion of [O]. If a more accurate estimate of the completion of [O] could be found it would greatly improve the efficiency and scalability of the condensation method.

Figure 2 shows the performance of subset extraction on the 11 multiplexer problem. Again, the population size limit was 1,000 microclassifiers and deletion technique 3 was used. [O] is first completed just after 16,000 cycles, but its final member is deleted, evolved again, and deleted again before reappearing around 18,000 cycles. [O] was finally extracted from [P] around 19,500 cycles. [O] was not extracted directly at 18,000 cycles because the numerosity allocation dynamics had not yet met the conditions of the numerosity distribution assumption made in section 3.1. However, the lag of 1,500 cycles in obtaining [O] from [P] is much better than that of 15,000 cycles for the condensation experiment. Note the difference between figures 1 and 2 in terms of when [O] was first completed (13,000 and 16,000 cycles) and when system error dropped below 0.01 (13,000 and 15,000 cycles). This sort of variability makes triggering condensation difficult and necessitates the use of long delays for reliable performance.

---

[8] Although **macroclassifiers** are used to implement the system, the population size limit is counted in terms of the equivalent number of **microclassifiers**. However, the population size curve represents the number of **macroclassifiers**, as graphs of the population size in microclassifiers quickly reach the size limit and remain there, and are thus of little interest.

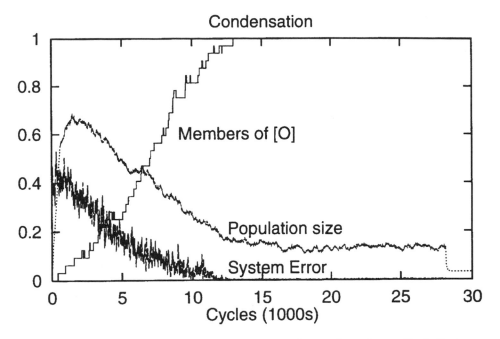

Figure 1: A single run of an 11 multiplexer experiment using the condensation method. System error is a measure of the difference between the system's prediction of payoff and the actual payoff received on each cycle. Members of [O] is the percentage of [O] present (generated using problem-dependent information). Population size is the number of macroclassifiers / 1000.

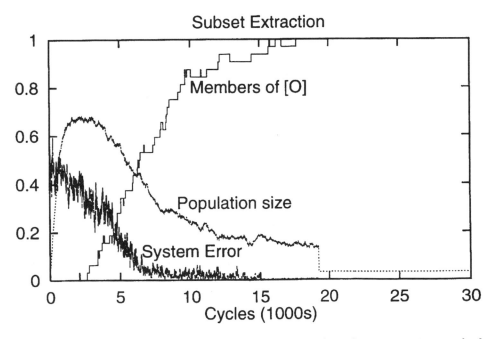

Figure 2: A single run of an 11 multiplexer experiment using the subset extraction method. [O] is extracted from [P] soon after it is complete, unlike in figure 1 where the condensation method shows a considerable lag between the completion of [O] and condensation of [O] to [P].

5 Conclusion

I have presented two methods of evolving optimal solutions for boolean functions using XCS. Condensation is the simpler of the two to implement, but requires the user to choose an appropriate delay for the trigger because our estimate (based on the system error statistic) of the time at which [O] is completed is imprecise. As a result of this delay more cycles are needed to finish and this approach scales poorly. Furthermore, if condensation begins too soon, [O] cannot be found, at least not without a costly restart of evolutionary search. In contrast, the subset extraction method works without modification on both the 6 and 11 multiplexer problems despite the latter problem being considerably more complex. It is computationally tractable thanks to XCS's tendency to allocate more numerosity to the members of [O] than to any other classifiers, which allows us to check only a small number of the possible subsets of [P] for completeness and a lack of overlaps. XCS also tends to evolve minimal solutions, and when this does not occur it is possible to combine classifiers to form a minimal population. Finally, subset extraction, unlike condensation, can be applied at any time without ill effects, as failure to extract [O] leaves the population unchanged.

Both methods rely on XCS's tendency towards accurate generalization and its ability to form complete mappings of the input/action space. These two characteristics are not found in other classifier systems, which consequently are incapable of evolving optimal populations. The success of these two methods demonstrates the effectiveness of XCS's tendency towards generalization and its ability to form complete mappings. Combined, these characteristics make XCS an RL system which scales well, and suggest that XCS will be a useful means of investigating RL based systems.

Acknowledgements

I would like to thank Stewart Wilson, Manfred Kerber, Ian Wright and Aaron Sloman for their help in all its forms. This work was funded by Tim Kovacs and the School of Computer Science at Birmingham.

References

[1] Watkins, C., 1989, *Learning from Delayed Rewards*, PhD Thesis, Cambridge University, U.K.

[2] Lin, L. J., 1992, Self-improving reactive agents based on reinforcement learning, planning and teaching. *Machine Learning*, **8**, 293-321.

[3] Munos, R., and Patinel, J., 1994, Reinforcement learning with dynamic covering of state-action space: Partitioning Q-Learning, in *From Animals to Animats 3: Proceedings of the Third International Conference on Simulation of Adaptive Behavior (SAB-94)*, pp. 354-363.

[4] Wilson, S. W., 1995, Classifier Fitness Based on Accuracy, *Evolutionary Computation*, **3**.

[5] Wilson, S. W., 1996, Generalization in XCS, Unpublished contribution to ICML '96 Workshop on Evolutionary Computing and Machine Learning.

[6] Holland, J., H., 1975, *Adaptation in Natural and Artificial Systems*, University of Michigan Press, Ann Arbor.

[7] Holland, J., H., 1986, *Machine Learning, an Artificial Intelligence Approach. Volume II*, chapter Escaping Brittleness: The possibilities of General-Purpose Learning Algorithms Applied to Parallel Rule-Based Systems, pp. 593-623, Morgan Kaufmann.

[8] Kovacs, T., 1996, Evolving Optimal Populations with XCS Classifier Systems, Technical Report CSR-96-17 and CSRP-96-17, School of Computer Science, University of Birmingham, United Kingdom, available from http://www.cs.bham.ac.uk/system/tech-reports/tr.html

[9] Wilson, S. W., 1994, ZCS: A zeroth level classifier system, *Evolutionary Computation*, **2**.

[10] Schuurmans, D., and Schaeffer, J., 1989, Representational Difficulties with Classifier Systems, in *Proceedings Third International Conference on Genetic Algorithms*, pp. 328-333.

Designing Neural Networks using a Genetic Rule-based System

Kwok Ching Tsui[1,3] and Mark Plumbley[2]

[1]*Department of Computer Science*
[2]*Department of Electrical and Electronic Engineering*
King's College London, Strand, London WC2R 2LS, UK
email:tsuikc@info.bt.co.uk and Mark.Plumbley@kcl.ac.uk

Keywords: neural network design, classifier system,
hill climbing, network performance estimation

Abstract

The application of neural networks is continuously increasing as the research interests in neural networks advance. Despite the experience accumulated, practitioners constantly face the obstacle of network design which is application dependent. There have been numerous attempts to automate the network design process. A popular approach uses evolutionary computing technique such as genetic algorithms (GAs) and has been successfully applied to design a wide range of network architecture. However, there are a few drawbacks such as high computational cost and incomprehensible network evolution. This paper presents a neural network design algorithm based on a classifier system which is enhanced with a hill climbing strategy to perform efficient search through the solution space. The proposed algorithm has been successfully used to design networks for two benchmark problems, namely XOR function learning and two spiral separation.

1. Introduction

Neural networks have been applied to many different areas as the research interests advance. The discovery of efficient learning algorithms, new network architecture and better understanding of their behaviour have contributed to the success. Despite these, practitioners constantly face the obstacle of network design which is application dependent. Although experience learned from one application may be used in another, there is no guarantee that the knowledge gained can be reused.

There have been numerous attempts to automate the network design process. A popular approach uses evolutionary computing technique such as genetic algorithms (GAs) [1] and has been successfully applied to design a wide range of network architecture, ranging from multilayer feedforward networks to probabilistic RAM networks [2, 3]. They relieve their users from the burden of repeating the laborious trial and error process. However, there are a few drawbacks. Firstly, due to the presence of a pool of candidate networks, the time for performing a GA cycle depends heavily on the speed of measuring the performance of the networks. Secondly, the use of binary digits for encoding the parameters of neural network in GAs make the development process almost incomprehensible. This renders experience gathering and reuse impossible.

This paper presents a neural network design algorithm based on an enhanced classifier system called *Maze Classifier System* (MaCS) [4]. The classifier system has a hill climbing strategy to perform efficient search through the solution space and has been applied to maze negotiation task. By recasting the neural network design task into one similar to maze negotiation, the benefits of *MaCS* can be reaped. The system for network design consists of a pool of rules which operate on a single neural network at any time. This reduces the computation time required for network performance evaluation.

The proposed algorithm has been used to design networks for two benchmark problems taken from the CMU

---

[3]current address: BT Laboratories, Martlesham Heath, Ipswich IP5 7RE, UK

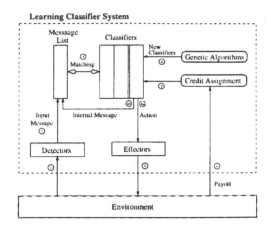

Figure 1: Block diagram of a CS processing cycle and the interactions between various components. The numbers in circle are the sequence number of the activities within a CS cycle.

Neural Network Benchmark Database[4]. The benchmark tasks are: learning the function XOR, and two spiral separation. These tasks have been widely used for comparing performance of various network architecture and learning algorithms. Details of them are given in the respective sections.

The primary objective of the experiments reported below is to evolve neural networks for the tasks with reasonable performance within a given period of time. Reasonable performance is defined as less than 10% difference between the expected output and the actual output of the networks. A time limit is imposed on every experiment up to which the experiments can run as a long evolution period would make the evolution process much less useful in practice.

Feedforward networks are used in all experiments where bypass connections (skipping layers) are allowed. BackProp [5] is used to train the neural networks. However, due to the demand for computation power, it is too expensive in terms of time to train every neural network from a random set of weight until perfection. A rough estimation of the performance is therefore adopted which takes the mean squared error of a neural network after being trained for a fixed number of epochs.

Section 2. is a brief introduction to classifier systems. The method used to encode a neural network and to approximate a neural network's performance are given in sections 3. and 4.. Sections 5. and 6. report the experimental results of neural network architecture optimisation using *MuCS*. Some observations regarding the rules used to construct and optimise a neural network are presented in section 7. and is followed by a summary of the results and possible future research directions.

2. Classifier Systems

Broadly speaking, a Classifier System (CS) [6] can be divided into several sections (Figure 1). There are rule-like structures called *classifiers* which are the storage of knowledge in the CS. The *system interfaces* work with the world external to the CS which is commonly known as *environment*. The *input interface* provides information of the current status of the problem. It also supplies feedback to the CS concerning the effect of the action as instructed by the selected classifier. The *output interface* carries out the action suggested by the CS. The most dynamic part of a CS is the *messages list*. In fact, we can say that a CS is a message driven machine learning system, as information from the environment is input to the system in the form of messages. Communication between classifiers is also achieved via posting messages to the message list.

Competition in various places in the processing cycle of a CS allows the system to select better rules. Conflicts arise naturally as a result of this competition and thus conflict resolution becomes a major component of a CS. In order to be useful for solving the problem in hand, a CS need to maintain a set of useful classifiers which can tackle different aspects of the problem. In conjunction with the feedback signal from the environment, a *credit assignment* scheme is used to reinforce the good and penalise the bad classifiers. Finally, injection

---

[4]ftp.cs.cmu.edu:/afs/cs/project/connect/bench

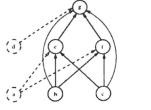
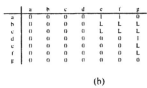
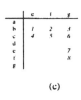

(a)

(b)

(c)

Figure 2: Internal Representation of the Connectivity of a Neural Network in *MaCS*. (a) A Feedforward Net with 5 Nodes and Bias Node (dotted circles). (b) Connection Matrix showing permanently connected '1', permanently disconnected '0' and learnable connections 'L'. The rows represent the source of a connection and the columns are the destinations of a connection. (c) A transformed connection matrix where locations of learnable connections are indexed.

of new rules is achieved through the periodic application of GAs. Figure 1 is a schematic description of the processing cycle of a CS with the major steps labelled sequentially.

MaCS is an enhanced version of the above basic CS by hill-climbing the solution space in search for optimal solutions. This is achieved by storing a solution and the classifiers that contributed to its discovery while searching for a new one. The newly found solution will not be adopted unless it is better than the current best in hand. This scheme has proved to be useful in some maze negotiation problems [4].

In the experiments conducted, there are two conditions and one action in every classifier. A set of input/output pairs is selected randomly from the training data and presented to *MaCS* in every generation as detector messages. Input and output training samples are prefixed by '0' and '1' respectively. Instructions to modify the neural network architecture are encoded into the actions of the classifiers.

Classifiers will bid using 10% of their strength in a bidding system where a Gaussian noise with variance 0.075 is added to each bid. For those classifiers which have bidden, a bid tax of 1% of its strength is charged, while those that do not bid will still suffer from 0.5% decay in strength. On the genetic algorithm front, the top 30% of classifiers of one generation are allowed to enter the next generation without competition. The crossover rate is 1 while the mutation rate is 0.02.

In the XOR learning task, there are 50 classifiers in the population. Evaluation of the neural network configuration is performed every 25 generations of classifier action and a genetic algorithm is applied every 50 generations. In the two-spiral separation task, 500 classifiers are used. Fitness of solution is evaluated every 50 generations and genetic algorithm is invoked every 100 generations.

3. Neural Network Representation Scheme

The way connections are made is encoded in a connection matrix [7] where the rows and columns represent the out-going and in-coming connections of the nodes respectively. Bias nodes are also included so as to learn the threshold values of the hidden and output nodes. The associated connections are made permanent while the others are learnable. The whole matrix is populated with '1', '0' and 'L' to represent permanently connected, permanently disconnected and learnable connections. Figure 2a is a sample network and Figure 2b is the corresponding connection matrix. For feedforward networks, all the learnable connections lie in the upper triangle of the connection matrix above the diagonal.

The action of the classifiers in *MaCS* for this study consists of two parts: operation and location. The operation can either be *connect* (1) or *disconnect* (0). The location identifier is implemented in two formats. For the XOR learning task, the location identifier is further divided into two parts which contains the source and destination of the operations. This is effective in case of small search space. It should be pointed out that there is a huge redundant space at the lower triangle of the connection matrix (including the diagonal). To avoid unnecessary search in the harder problem, the location identifier is modified to contain the index of the connection within

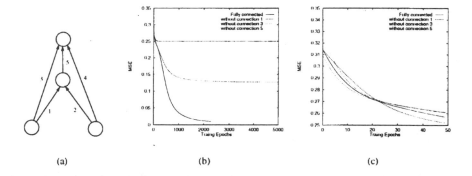

(a) (b) (c)

Figure 3: Learning Curve of a Fully Connected 2-1-1 Feedforward Net. (a) the network with all the connections labelled. (b) the learning curves with one of the odd numbered connections removed. Note that the learning curve for the network with connection 3 overlaps with the one of the network without connection 5. (c) magnification of learning curves in the first 50 epochs.

the learnable connections. In a fully connected feedforward network shown in Figure 2a, the search space in this seven node configuration (5 nodes and two bias nodes) is reduced from 49 dimensions to 8. The reduction is even more remarkable when bigger networks are involved.

4. Neural Network Performance Estimation

The neural networks to be evolved in the experiments are feedforward type networks with either one or two hidden layers. A naming convention of the form i-h_1-h_2-o where i and o are the number of nodes in the input and output layers respectively while h_1 and h_2 are the number of nodes in the first and second hidden layers respectively. In the event of a single hidden layer network, -h_2- is dropped for simplicity. Associated with every node in the hidden layers and output layer are bias values and they are learned in the same way as the connection weights. Moreover, these nodes use sigmoid as the transfer function.

A simple test was run using a fully connected 2-1-1 feedforward net (Figure 3a) to learn the mapping between a set of four input/output data. All the connections in the neural network are labelled from 1 to 5. As the network is symmetric along the central axis, one of the connections 1, 3 and 5 is taken away resulting in 3 different networks. They, together with the original fully connected network, are trained by BackProp for 5,000 epochs or until the mean squared error (MSE) falls below 0.01, whichever comes first. The same set of randomly generated connection weight in the range of ±1 is used.

Figure 3b shows the learning curve of the above four networks. It is clear that while weights can be tuned over time, one of the determining factors for network performance is the right mixture of connections. In fact, only the fully connected 2-1-1 network achieved the required performance using less training epochs (2,280) than the set limit. It seems there is not much chance for the others to match that performance even given more training epochs. It can be observed from the learning curves that the MSE decreases drastically at the beginning of a training session and levels off gradually when approaching convergence. It is quite easy to distinguish the good performer from the bad by looking at the learning curve at, say, 500 epochs. But it may be an expensive comparison exercise when a large number of neural networks are involved.

A closer look at the curves in the first 50 epochs (Figure 3c) reveals that there is already some hint regarding the relative performance of these networks with only one exception. Taking the symmetry into account, a 60% confidence can be obtained for the conjecture that performance of a neural network relative to the others can be determined after 50 BackProp training epochs. This approximation criteria will be used throughout this paper.

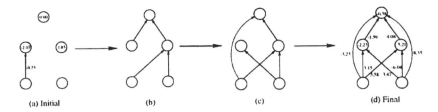

Figure 4: Evolution process of a feedforward network for learning the XOR function (seed A)

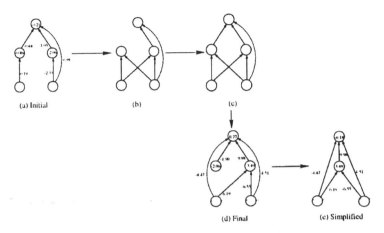

Figure 5: Evolution process of a feedforward network for learning the XOR function (seed B)

5. Learning XOR Function

This task is to train a 2-2-1 network to produce the boolean exclusive or (XOR) function. XOR has been the most popular benchmark problem in the neural networks field but is perhaps the simplest learning problem that is not linearly separable. This task serves as the litmus test of the effectiveness of the schemes devised above.

In the five independent runs reported in this section, three of the randomly generated networks do not learn the function at all (MSE=0.25). In contrast, all the five evolved networks achieved the goal (MSE≤0.01) well below the time limit. A comparison of the performance and details between the initial randomly generated network and the evolved network can be found in Table 1. They are labelled from A to E to denote their use of different seeds of the random number generator.

The stages of evolving the final networks in two cases are given in Figures 4 and 5. The initial and final networks in each case are shown with the connection weights and bias values obtained after the final training. The nodes in the neural networks are numbered sequentially from 1 to 5, from bottom (input) to top (output) and left to right within a layer.

Table 1: summary of results for learning the XOR function

Seed	Initial Network			Final Network		
	MSE	epochs	type	MSE	epochs	type
A	0.25	5,000	N/A	0.01	1447	2-2-1
B	0.25	5,000	2-2-1	0.01	1734	2-1-1
C	0.25	5,000	2-1-1	0.01	1779	2-1-1
D	0.14	5,000	2-2-1	0.01	1779	2-1-1
E	0.13	5,000	2-2-1	0.01	1779	2-1-1

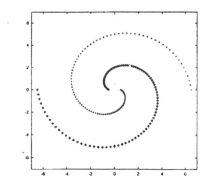

Figure 6: A Two-spiral with Single Turn

Among the five evolved networks, one of them (seed A) is a 2-2-1 fully connected network while the others are 2-1-1 fully connected networks; both are common solutions found in current publications [8]. It is observed that exactly the same network is evolved from three different randomly connected networks (seed C, D & E).

In the run using seed B, node 3 in the final network (Figure 5d) has no input connections but it is connected to the output node (node 5). As all the units are initialised at zero, this extra connection from node 3 to 5 becomes a constant input signal of strength $-1.9/(1 + \exp 0) = -0.95$. Therefore, the bias of node 5 can be adjusted to $-0.95 + 0.77 = -0.18$ when node 3 is eliminated.

6.   Two-Spiral Separation

The spiral separation task is to learn to discriminate between two sets of points on a two dimensional plane. These points are aligned in two different spirals coil around the origin and each other. The spirals used here are shown in Figure 6. There are all together 194 points equally divided between the spirals.

A few attempts to train a neural network to learn this task have been reported using different network architectures. Two examples are described here. Lang and Witbrock [9] used a 2-5-5-5-1 fully connected feedforward network with bypass connections to separate two spirals with 3 turns. Neural network learning algorithm has been used to train the network whose structure is pre-defined. An average learning time of 20,000 epochs has been reported but the spirals have three turns instead of one. Robbins et. al. [1] tested a single turn version with 32 points each using a single hidden layer feedforward network of 20 hidden nodes. A genetic algorithm has been used to optimise the structure of the network before a neural network learning algorithm is used to train the connection weights. The network which has learned the task has significantly less connections and fewer hidden nodes compared to its fully connected counterpart. Others have also used this task to test the performance of different network architectures and learning algorithms [10, 11, 12, 13, 14].

This section details an attempt to apply *MaCS* to this task. Two different network architectures with a single and two hidden layers are tested.

6.1.   Evolving a Single Hidden Layer Network

The neural network to be constructed consists of a single layer of up to 11 hidden nodes. There are two input nodes and one output node. Using the scheme described in section 3. and allowing bypass connections from the input nodes to the output node, there are 35 connections available to be modified.

On training data representation, inputs and outputs are normalised to lie within the range of $(-1, +1)$ in increments of $1/255$ and eight bits are used to encode this information. A one-bit prefix is added to differentiate the two types of data. Therefore, detector messages of nine bits long are presented to *MaCS*. One set of input/output is presented at a time. In this task where there are two inputs and one output, three detector messages are presented at a time. This applies to both spiral separation experiments using either single or double hidden layer networks.

As for the 7-bit classifier actions, there is a single bit operator code and a six-bit index pointing to the location

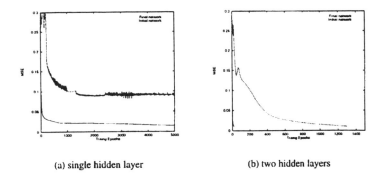

(a) single hidden layer   (b) two hidden layers

Figure 7: Training curves of initial and final networks for the two spiral identification task

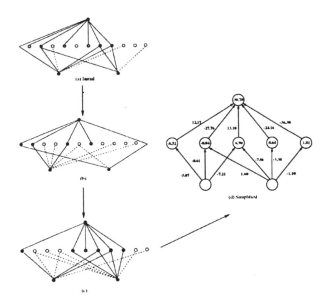

Figure 8: Two-spiral separation with single hidden layer feedforward net. Dark circles and solid lines are the nodes and connections which are contributing to the output of the network. Dotted lines and empty circles are not counted as valid because inputs via these paths do not reach the output. The shaded circles receive no input but they add extra bias to the output node.

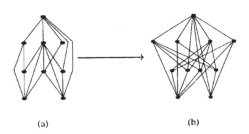

(a)   (b)

Figure 9: Two-Spiral Separation with Double Hidden Layer Feedforward Net

Table 2: The classifier actions for optimising a neural network for learning the XOR function

network configuration transition	configuration instruction	classifier conditions	interpretation
(a) → (b)	**Connect 2 and 4**	+10,+#1	if output=0 and input=1
	Disconnect 1 and 3	+11,+0#	if output=1
	Connect 4 and 5	+11,+0#	if output=1
	Connect 3 and 5	+01,+10	if input=1 and output=0
	Connect 1 and 4	+11,+0#	if output=1
(b) → (c)	**Connect 1 and 5**	+01,+10	if input=0 and output=1
	Connect 2 and 3	+11,+0#	if output=1
	Disconnect 3 and 5	+1#,+10	if output=0
(c) → (d)	**Connect 3 and 5**	+10,+10	if output=0
	Connect 1 and 3	+10,+10	if output=0
	Connect 2 and 5	+01,+11	if output=0 and input=1

to be changed. The six-bit index is enough to encode 64 different locations in the connection matrix in a similar way to that described in section 3. and shown in Figure 2. However, there is a total of only 35 learnable connections in the 2-11-1 network. *MaCS* has to learn to discard the unproductive actions. This is achieved by an operator which modifies the action at the spot whenever such a circumstance arises.

The initial network (Figure 8a) consists of three effective hidden units with 10 connections. The MSE is 0.28 after 50 training epochs. An intermediate network (Figure 8b) with six hidden units and 14 connections is found which has an MSE of 0.12. The final network (Figure 8c) found by *MaCS* is a 2-5-1 network with 12 connections. Unlike the previous two, there are no bypass connections. An MSE of 0.11 is achieved by this network. Common to all three networks is the presence of hidden nodes with no incoming connections but with a connection to the output node. These hidden nodes function as extra bias nodes. They are taken into account when the final network is simplified (Figure 8d). The learning curves of the initial and final network can be found in Figure 7a. Upon training for 5,000 epochs, the initial and final networks have achieved an MSE of 0.089 and slightly above 0.01 respectively.

6.2. Evolving a Double Hidden Layer Network

This experiment employs the same setup for *MaCS* as the task of evolving single layer feedforward net except that a 2-5-5-1 feedforward net is used instead. Out of the 57 learnable connections, the initial network (Figure 9a) has 17 connections in its 2-3-3-1 architecture. The MSE of the initial network is 0.14, which is significantly better than the single hidden layered counterpart. When the 1,000 generation limit is reached, a 2-4-4-1 network with 23 connections (Figure 9b) is found which has achieved 0.03 in MSE. Further training up to 5,000 epochs reviews the superiority of the evolved network as it achieved the target MSE limit of 0.1 after only 20 epochs while the initial network required 1,300 epochs. The learning curves are shown in Figure 7b. A similar 2-4-4-1 network with the same number of connections which is evolved using a genetic algorithm has been reported elsewhere [1] but *MaCS* evaluates only one neural network at any time and uses much fewer training epochs during the solution evaluation.

7. The Neural Network Optimisation Rules

An advantageous feature of classifier systems is its use of rule like structures. Although the rules (classifiers) are binary encoded, interpreting the meaning of the rules is still possible.

Consider the network which has learned the XOR function shown in Figure 4. The initial network has only one connection while the final network has all the 8 possible connections. The instructions as decided by the classifiers are listed in Table 2 referring to the networks in Figure 4. The net effect of the three network configuration transitions resulted in seven instructions, as highlighted in bold. Among those seven instructions, four of them solely depend on the setting of the output. A difference in value between the input and output triggers the remaining three.

Examining the instruction sets for the other 4 networks for the XOR function does not show a similar pattern. This is expected as all the networks are generated randomly. Thus the searches have started from different points. The networks therefore require different configuration instructions in order to arrive at an optimal configuration. A cohesive pattern might be observed were the searches to start from a fully connected network. Another point to consider is the optimality of the rule set. Since the classifiers are frozen once it is formed (section 2.) and only a limited number of generations were run, there may not be enough chance to refine the rule set so as to give a good description of the decision process. Future investigation along this line is definitely needed.

8. Summary

This paper has discussed the use of a classifier system to optimise neural network architectures in tasks of different complexity. For the XOR function learning task, networks with commonly recognised optimal architectures have been evolved. Actions by *MaCS* on four randomly generated network resulted in the same network architecture, although different instructions for architecture modification had been applied. For the two-spiral separation task, two neural networks using single or double hidden layers are evolved. Both of these compare favourably with results published by others.

The above results are made possible by enhancing the basic classifier system to perform hill climbing in the reward landscape. This is accompanied by the introduction of a neural network performance estimation method where only a limited number of training epochs are conducted before performance measurement is taken. It has been shown that this is not a fool-proof method but provides an effective way to cut down the computational cost. The stochastic nature of the search performed by *MaCS* and genetic algorithm provides some compensation for this deficiency. On the other hand, it is assumed that there is a mechanism to discard the bad solutions while searching for the good ones.

The hill climbing in *MaCS* provides a means for gradually improvement over time. Bad solutions are excluded from being taken up as the reference point for future search. Moreover, the basis for future search and network architecture modification, changes with every discovery of a better solution, albeit sub-optimal. The whole idea of gradual improvement can be found in other search algorithms such as simulated annealing. In fact, the acceptance of new solutions in a stochastic fashion in simulated annealing can be adopted as the neural network performance estimation, which would help to prevent the system from being stuck in a sub-optimal solution.

A minor point to mention is the rule-like structures in a classifier system such as *MaCS* has the potential of providing some kind of explanation tool for the task it has performed. However, *MaCS* has to work on the rule set for many more generations before a fine tuned rule set emerges.

In summary, *MaCS* provides an alternative neural network optimisation tool with the major advantage of low computational cost in terms of network evaluation.

Acknowledgments

K.C. Tsui is partly supported by the Rotary Foundation through the provision of 1994-1995 Rotary Ambassadorial Scholarship. Dr. Plumbley is partly supported by grant (GR/J383987) from the UK Science and Engineering Research Council.

References

[1] Robbins, G., Plumbley, M. D., Hughes, J. C., Fallside, F., and Prager, R., 1993, Generation and adaptation of neural networks by evolutionary techniques (GANNET), *Neural Computing and Applications*, 1(1):23–31.

[2] Schaffer, J. D., Whitley, D., and Eshelman, L. J., 1992, Combinations of genetic algorithms and neural networks: A survey of the state of the art, in Whitley, L. D. and Schaffer, J. D., editors, *Proceedings of International Workshop on Combinations of Genetic Algorithms and Neural Networks*, 1–37. IEEE Computer Society Press.

[3] Tsui, K. C. and Plumbley, M., 1995, An empirical study of two approaches to automated pRAM network design, in *Proceedings of the 1995 International Symposium on Artificial Neural Networks*, C1.17–24.

[4] Tsui, K. C. and Plumbley, M., 1997, A genetic-based adaptive hillclimber, To appear in the 5th European Congress on Intelligent Techniques and Soft Computing.

[5] Rumelhart, D. E., Hinton, G. E., and Williams, R. J., 1986, Learning internal representation by error propagation, in Rumelhart, D. E. and McClelland, J. L., editors, *Parallel Distributed Processing*, volume 1, chapter 8, 318–362, MIT Press.

[6] Holland, J. H., 1992, *Adaptation in Natural and Artificial Systems*, MIT Press, 2nd edition.

[7] Miller, G. F. and Todd, P. M., 1989, Designing neural networks using genetic algorithms, in Schaffer, J. D., editor, *Proceedings of the Third International Conference on Genetic Algorithms*, 379–384. Morgan Kaufmann: San Mateo, Calif.

[8] Beale, R. and Jackson, T., editors, 1990, *Neural Computing: An Introduction*, Adam Hilger.

[9] Lang, K. J. and Witbrock, M. J., 1988, Learning to tell two spirals apart, in Touretzky, D. S., Hinton, G., and Sejnowski, T., editors, *Proceedings of the 1988 Connectionist Models Summer School*. Morgan Kaufmann: San Mateo.

[10] Chua, H. C., Jia, J., Chen, L., and Gong, Y., 1995, Solving two-spiral problem through input data encoding, *Electronics Letters*, 31(10):813–814.

[11] Cios, K. J. and Liu, N., 1992, A machine learning method for generation of a neural network architecture: A continuous ID3 algorithm, *IEEE Trans. Neural Networks*, 3(2):280–291.

[12] Klagges, H. and Soegtrop, M., Limited fan-in random wired cascade-correlation, Anonymous ftp from ftp.cis.ohio-state.edu:/pub/neuroprose.

[13] Koza, J. R., 1992, A genetic approach to the truck backer upper problem and the inter-twined spiral problem, in *Proceedings of the International Joint Conference on Neural Networks*, 311–318.

[14] Waterhouse, S. R. and Robinson, A. J., 1994, Classification using hierarchical mixtures of experts, in *IEEE Workshop on Neural Networks for Signal Processing*, 177–186.

A Neural Network for Real-World Postal Address Recognition

Michael Blumenstein and Brijesh Verma

School of Information Technology
Faculty of Engineering and Applied Science
Griffith University, Gold Coast Campus
Parklands Drive, Gold Coast, Qld 4217 Australia
E-mail: {M.Blumenstein, B.Verma}@eas.gu.edu.au

Keywords: Artificial Neural Networks, Pattern Recognition, Classification, Postal Address Recognition.

Abstract

In this paper, we present a description of an implemented system for the recognition of printed and handwritten postal addresses, based on Artificial Neural Networks (ANNs). Two classification methods were compared for the task of character and address recognition. We compared two neural network techniques, measuring recognition rate and accuracy. The C programming language, a SUN workstation, and the SP2 Supercomputer were used for the experiments. The system has been successfully tested on real world printed and handwritten postal addresses. Some experimental results are presented in this paper.

1. Introduction

ANNs have been successfully used in such areas as pattern recognition [1], medical applications [2], fingerprint analysis [3] and signature verification [4] to name just a few. This paper attempts to take one step forward in producing another successful practical application: Recognition of handwritten postal addresses.

As outlined in [5] and more recently in [6], there has been an inordinate number of different methods employed for the task of handwritten character recognition. Some of the best results have been achieved with neural network based classifiers using an error backpropagation training algorithm. The recognition rates attained have ranged from 86 to 94%. However, all of these methods solely focused on handwritten numerals and in some cases highly constrained network structures were used. RBF neural networks have also been used for various classification tasks such as Hindi character recognition [7]. Recognition rates between 54 and 70% have been achieved using relatively small character databases for training and testing.

Although in some cases high recognition rates have been attained for character recognition, there has not been the same amount of success for the recognition of words and subsequently characters in postal address recognition. In some cases, the reason for low recognition rates in real world situations has been the difficulty in segmenting whole words accurately before using some method of classification [8]. Gilloux [9], also acknowledges segmentation as a critical aspect for accurate postal address recognition.

Some research has already demonstrated the success of ANN-based pattern recognition systems for the recognition of handwritten zip codes [10], [11]. Research in the field has been ongoing for a substantial amount of time. The ANN-based system for the recognition of handwritten zip codes is very significant due to the excellent recognition rates attained. The zip codes used for training and testing were sampled from an actual U.S. post office. The authors' work proved the accuracy and speed which could be attained using a backpropagation neural network. It was also demonstrated that their work could be extended for use with larger tasks.

This paper attempts to extend their research focusing on another difficult task which is handwritten postal address recognition. Sample handwritten addresses were collected and stored in a large database. The aim was to develop a system that could segment, preprocess, and classify the handwritten addresses acquired. An ANN was used for the actual classification of characters and addresses.

Another method for classification was tested: The Radial Basis Function (RBF) method. Both the RBF method and the backpropagation algorithm were first compared for the task of handwritten and printed character recognition. The technique producing the best classification rates was then applied to the task of postal address recognition.

The organisation of the remainder of the paper is as follows: section 2 discusses the proposed techniques, section 3 details the experimental results, section 4 discusses the results and finally a conclusion is drawn in section 5.

2. Proposed Techniques

There are many steps that need to be taken before handwritten words or characters on an envelope or a page can be recognised by an automated system. The techniques used in our experiments were as follows: 1. Scanning, 2. Binarisation, 3. Segmentation, 4. Preprocessing and 5. Classification. Figure 1 depicts the complete system.

2.1 Scanning

The handwritten characters were acquired from various students and faculty members around the university. Their handwriting was sampled on A4 sized paper. These handwritten characters were scanned using a flatbed Macintosh scanner and a Macintosh personal computer. The scanned images were saved in Tagged Image Format (TIF). Later the images were imported onto an IBM PC and converted into a Windows Bitmap format using Paint Shop Pro Version 3.11.

2.2 Binarisation

The Windows Bitmaps were then uploaded onto the SP2 Supercomputer. Using a program originally implemented for the recognition of Hindi characters [7], the Windows Bitmaps were converted into binary bitmap representations of the handwriting. The black pixels were represented as ones (1's) and the white background was represented by zeroes (0's). In this form, segmentation and preprocessing could take place more easily.

2.3 Segmentation

A segmentation program was implemented to separate the individual lines and characters of the handwritten postal addresses. The algorithm paid particular attention to separating hand-printed characters which were "touching" due to poor scanner resolution. The program first looked for characters which were not touching (separated by columns of zeros). If the search could not find a clear breakpoint, it tried to find a point where the density of pixels was sparse. An algorithm was devised especially for this purpose, which searched for a decrease of black pixel density. When a point like this was found the program monitored either side of the point to make sure that it was in between an area of high pixel density. The end result was a file containing characters of varying sizes. A simple preprocessing technique was then required to normalise the character matrices.

2.4 Preprocessing

A simple normalisation technique was employed to create a file of character matrices of the same height and width. First, the largest character in the set was found. The rest of the characters were then padded horizontally and vertically with "0's" to produce character matrices of equal size.

2.5 Classification

A feedforward, multi-layered neural network [7], was used for the recognition process. An Error Backpropagation (EBP) algorithm and the RBF technique were used for training the neural network. The characters were presented to the classifiers as multiple sets of 36 character training pairs (A-Z and 0-9).

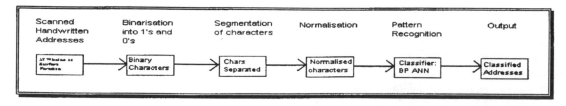

Figure 1. A system for postal address recognition.

3. Experimental Results

3.1 Character Database

The largest database of characters consisted of 4644 characters written by 11 different students and faculty members. The database was divided into a training set containing 4248 characters and a test set containing the remainder of the characters. Some samples of the characters used are shown in Figure 2.

Figure 2. Samples of training characters

Figure 3. Test samples for proposed system

3.2 Comparison of Classifiers

The experiments were executed on the SP2 Super Computer. As was previously mentioned, it was necessary to perform a comparison of two classifiers to select the most appropriate one for the task of address recognition. Both classifiers were ANN-based. The comparison was performed on the basis of their ability to recognise printed and difficult handwritten characters. Varying results were obtained by altering the parameters of the ANN-based methods. The main variables that were altered for the ANN-based techniques included the number of hidden units, learning rate and momentum. Two tables are presented which give the highest printed and handwritten character recognition rates. Table 1 presents results for the Backpropagation neural network and Table 2 presents results for the RBF neural network.

Table 1. Character classification rates using the EBP Neural Network

	Hidden Units	No. of Iterations	Learning Rate	Momentum	Classification Rate [%]	
					Training Set	Test Set
Printed	16	300	0.1	0.1	99.65	84.72
Handwritten	100	500	0.1	0.1	98.73	58.59
Printed & Handwritten	16	200	0.1	0.1	95.49	70.83

Table 2. Character classification rates using the RBF Neural Network

	Hidden Units	Classification Rate [%]	
		Training Set	Test Set
Printed	200	81.25	56.94
Handwritten	200	51.98	17.31
Printed & Handwritten	200	55.38	35.42

3.3 Experimentation with Real-World Postal Addresses

Further experimentation included the scanning of printed and handwritten postal addresses for testing the proposed system. Some samples are shown in Figure 3. We used the classification technique with the highest recognition rate for this task. The results presented in Section 3.2, clearly show the backpropagation ANN as the most successful classifier. Table 3 presents results for the recognition of 10 printed postal addresses using 10 different fonts. Table 4 presents seven of the best results for handwritten postal address recognition.

Table 3. Results for printed postal addresses

	Recognition Rate of Address (%)
Address 1 (Arial 9pt)	97.56
Address 2 (Book Antiqua 9 pt)	88.89
Address 3 (Bookman Old Style 9pt)	94.87
Address 4 (Calisto MT)	97.62
Address 5 (Century Gothic 9pt)	85.11
Address 6 (Century Schoolbook 9pt)	97.22
Address 7 (Comic Sans 9pt)	92.11
Address 8 (Courier 9pt)	83.33
Address 9 (News Gothic 9pt)	90
Address 10 (Times New Roman 9pt)	94

Table 4. Results for handwritten postal addresses

	Recognition Rate of Address (%)
Address 1	50
Address 2	60
Address 3	57.43
Address 4	51.28
Address 5	56.76
Address 6	68.75
Address 7	52.78

4. Discussion Of Results

As can be seen clearly from the tables presented in Sections 3.2 and 3.3, the results vary quite substantially. The best and most promising results for printed character recognition could be found when using the backpropagation ANN. The settings which provided the best results were 16 units in the hidden layer, very low values for η and α (0.1 and 0.1), and 300 iterations for training. For handwritten character recognition, the difficulty of the task could be reflected in the quality of results attained for the test set for both techniques. However, again the backpropagation ANN proved most successful. The best settings were found to be 100 units in the hidden layer, very low values for η and α (0.1 and 0.1), and 500 iterations for training.

Although on average the classification rates were quite high for the training sets, lower recognition rates were being continually received for the test sets. This could be attributed to the difficult nature of the handwritten samples used for experimentation. Out of the two methods tested, the backpropagation ANN proved its superior generalisation ability by providing the best results for both training and test sets.

Tables 3 and 4 show results for the classification of printed and handwritten postal addresses. The results in Section 3.2 clearly indicated the backpropagation ANN's superiority for classification. Therefore, it was the only method used for classification of the postal addresses. For printed addresses, the recognition rates were extremely high. Perfect classification rates were obtained when using fonts the ANN had been trained with. High results were also attained for the recognition of addresses using fonts the ANN had not been previously trained with. Finally, handwritten address recognition rates were presented. Although being substantially lower, the results were very impressive as the postal addresses were composed of difficult handwriting which the ANN had not been presented with previously. The results compared well with other postal address recognition systems [8], [9], however an in-depth comparison cannot be made due to the variation in postal address databases that were used for the testing of each system.

5. Conclusions

A preliminary intelligent system based on a multi-layered feed forward neural network, has been developed to recognise handwritten postal addresses. Two classification techniques were compared to give suggestions on the most successful classification method available for the task. The results presented, clearly indicated that the backpropagation ANN was the best classification method for the task. Over 4000 real world handwritten characters were used to train the ANN, varying the number of hidden units, the learning rate and the momentum. For printed characters, the results showed that by keeping the number of units in the hidden layer low, and using a low learning rate and momentum, good recognition rates could be achieved. It was also found that when using a large database of handwritten characters, better results could be achieved by increasing the number of hidden units. The system was tested on a number of real world printed and handwritten postal addresses, the results obtained were very promising. This work is still in progress, and methods are being devised to hopefully increase recognition rates. Further research shall use a more complex segmentation algorithm and feature extraction techniques. This shall ensure that problems with touching and very difficult handwriting are resolved.

References

[1]   Khotanzad, A., Lu, J., 1991, Shape and Texture Recognition by a Neural Network, *Artificial Neural Networks in Pattern Recognition*, Sethi, I. K., Jain, A. K., Elsevier Science Publishers B. V., Amsterdam, Netherlands. pp. 109-131.

[2]   Zheng, B., Qian, W., Clarke, L., 1994, Multistage Neural Network for Pattern Recognition in Mammogram Screening, *IEEE ICNN, Orlando*, pp. 3437-3448.

[3]   Kulkarni, A. D., 1994, *Artificial Neural Networks for image understanding*, Van Nostrand Reinhold, New York.

[4]   Han, K., Sethi, I. K., 1996, Handwritten Signature Retrieval and Identification, *Pattern Recognition Letters*, **17**, 83-90.

[5]   Suen, C. Y., Legault, R., Nadal, C., Cherier, M., and Lam, L., 1993, Building a New Generation of Handwriting Recognition Systems, *Pattern Recognition Letters*, **14**, 305-315.

[6]   Lee, S-W., 1996, Off-Line Recognition of Totally Unconstrained Handwritten Numerals Using Multilayer Cluster Neural Network, *IEEE Transactions on Pattern Analysis and Machine Intelligence*, **18**, 648-652.

[7]   Verma B., 1995, Handwritten Hindi Character Recognition Using RBF and MLP Neural Networks, *IEEE ICNN'95, Perth*, pp. 86-92.

[8]   Srihari, S. N., 1993, Recognition of Handwritten and Machine-printed Text for Postal Address Interpretation, *Pattern Recognition Letters*, **14**, 291-302.

[9]   Gilloux, M., 1993, Research into the New Generation of Character and Mailing Address Recognition Systems at the French Post Office Research Center, *Pattern Recognition Letters*, **14**, 267-276.

[10] Denker J. S. et al., 1989, Neural Network Recognizer for Hand-Written Zip Code Digits, *Neural Information Processing Systems*, **1**, 396-493.

[11] Le Cun, Y. et al., 1990, Handwritten Digit Recognition with a Back-Propagation Network, *Neural Processing Systems*, **2**, 323-331.

Neural Processing
in Semi-Rigid Connections of Steel Structures

G.E. Stavroulakis [1], A.V. Avdelas [2] *, P.D. Panagiotopoulos [3], K.M. Abdalla [4]

[1] *Dr.Ing., Institute of Applied Mechanics, Carolo Wilhelmina Technical University,
D-38023 Braunschweig, Germany, email: gs@r2.infam.bau.tu-bs.de*
[2] *Asst. Prof.,Institute of Steel Structures, Aristotle University,
GR-54006 Thessaloniki, Greece, email: aris@archytas.civil.auth.gr*
* *To whom all correspondence should be addressed*
[3] *Prof.,Institute of Steel Structures, Aristotle University,
GR-54006 Thessaloniki, Greece, email: pdpana@heron.civil.auth.gr*
[4] *Asst. Prof., Department of Civil Engineering, Jordan Institute of Science and Technology,
Irbid, Jordan, email: abdalla@just.edu.jo*

Keywords: semi–rigid connections, steel structures, nonlinear computational mechanics

Abstract

The problem of classifying measured one-dimensional relations (e.g. stress-strain laws, moment-rotation laws, etc.) into a predetermined family of laws is considered for semi-rigid steel connections. Experimentally measured data have been loaded in back-propagation neural networks and have been used in structural analysis tasks. A second application concerns the use of Hopfield neural networks for the solution of the nonlinear computational mechanics problem. The structural analysis problem of a structure with semi–rigid joints has been written as an inequality constrained optimization problem, or an equivalent linear complementarity problem. Both formulations are amenable to neural network computations by a Hopfield network.

Both above directions can be unified so that a complete neural network environment can be given from the experimental evaluation to the structural analysis of steel structures with semi-rigid joints.

1. Introduction

The highly nonlinear effects that have to be considered in the detailed modelling of semi-rigid steel structure connections (e.g. unilateral contact and friction-"prying effects", arising between adjacent parts of the connection [1]-[3], local plastification effects etc.) require the use of time-consuming and complicated software not always available to the practicing engineer. Another approach, permitted by modern design codes [4], is their treatment by means of simplified nonlinear relations. In the present paper, an alternative to the above procedures is proposed: The estimation of the mechanical behaviour of the steel structure connection by the use of learning algorithms in an appropriately defined neural network environment of experimental steel-connections data. An estimated simplified law results which will be used as a moment-rotation constitutive law for the joints in the structural analysis and design of the steel structure.

By the use of the error correcting back propagation algorithm [5], a multilayer feed-forward neural network can be trained to recognize and generalize the experimental data. In our case, the experimentally measured moment-rotation curves for various design parameters of the steel structure connection are used as training paradigms. Next, the neural network reproduces the moment rotation law for a given set of design variables and thus it can be used in every design or structural analysis procedure.

The neural network theory, which provides a solid basis for the construction of the model estimator, is considered highly suitable for the study of complex problems in mechanics and engineering. Some recent representative applications are mentioned next, without any aim on completeness, in the area of structural analysis and design parameter identification problems [6]-[9], material modelling [10],[11] structural analysis [12],[8],[13] and optimization problems [14].

The experimental data considered here concern the case of single angle beam-to-column, bolted steel structure connections. These experiments have been reported in [15] and are included in the steel connection data bases [16],[17]. The neural network analysis is given next (see ref. [14],[18],[19]). The experimentally measured moment-rotation laws for steel structure connections are first preprocessed in a form suitable for neural network treatment. These data are next used for training a multilayer feed-forward back propagation neural network. The trained network provides a model-free predictor, with a satisfactory accuracy.

It should be mentioned here that other artificial neural network models can be used as well instead of the back propagation neural method. It is also worth-mentioning that the methodology which is proposed in this paper could be generalized to include more complicated effects (for instance dynamic behaviour, fatigue and creep effects).

2. Back propagation and neural network theory

A neural network is defined by its node characteristics, the learning rules and the network topology. The learning rules control the improvement of the network performance through appropriate adaptive changes of the weights of the links. The basic factors which characterize the computational effectiveness of a neural network are the large connectivity degree of the neurons and the massive parallelism as well as their nonlinear analog response and their training or learning capabilities. Further, one of the most remarkable properties of neural networks is their considerable fault tolerance due to the increased numbers of locally connected processing nodes. Thus, some neurons or links out of order do not diminish considerably the whole performance of the network, as well as its learning capability.

Supervised learning is the case in which the desired output data, with respect to a given set of input data, are known. Then, the whole set of input-output learning paradigms can be used to adjust the values of the connection weights or some variables of the activation functions so as to be able to reconstruct the implicit highly nonlinear mapping between input and output variables. It must be noted that no specific model has been assumed for the mapping between input and output variables; thus, the trained neural network provides us with a model-free estimator which simulates, for instance, the mechanical behaviour of a structural component [9],[10] or a structure [10],[20].

Next, the algorithms expressing the supervised learning procedure of a back-propagation neural network and the generalization after the training of the network can be formulated as in [21].

The latter algorithm starts from an initial value of the synaptic weight w_{ij}, which is produced by taking variables in a reasonably small range (e.g. [-1,+1] according to [7], or even [-0.3,+0.3]). Nevertheless, we must note here that learning in a neural network is algorithmically a hard problem (in general it is proved to be a nonpolynomial difficult NP problem, see [22]) and therefore the size of both the network and the training examples must be kept to the minimum necessary. Thus, engineering experience and a set of good, representative experimental data must be used as learning examples for the neural network.Note here that learning algorithms are actually optimization algorithms and they can be easily adapted to the solution of minimum problems (cf. e.g. [14],[8],[23]-[25]).

It is well known that neural computers are not yet commercially available. Neural network computations are best suited for a parallel computer environment which even permit a fairly good hardware implementation [14]. Here all the computer simulations have been performed on a classical serial computer, on which the neural environment has been emulated.

3. Neural networks and elastoplastic analysis

One of the main applications of neural networks is the solution of optimization problems. This constitutes one of their major advantages. Therefore, in order to adapt the structural analysis methods to a neural computing framework, the structural analysis problem must be formulated as a minimization problem [12],[13],[26],[27],[8]. For the elastoplastic analysis, corresponding optimization problems have been derived in among others [28]-[36]. At this point, the more important advantages of a neural network environment concerning the calculations should be mentioned:

- In a neural network environment, the treatment of an inequality constrained Quadratic Programming Problem (Q.P.P.) is equally time consuming as an unconstrained Q.P.P. which is equivalent to a system of linear equations. The neural network approach does not need large computer memory.

- In a neural computing environment the parameter identification problem and the sensitivity analysis are treated directly in a natural way as supervised learning and unsupervised [19] learning problems respectively without the need for complicated algorithms.

In the following, the elastoplastic analysis problem will be formulated for a given $M - \phi$ law as a Q.P.P. solved again through a neural network model. Thus, we use two neural networks: the first one gives the $M - \phi$ law from the experimental results and the second one solves the resulting Q.P.P. The first neural network is based on the perceptron model [14],[19], the second on the Hopfield model [37],[38] which seems to be more direct for the calculation of the solution of the plasticity Q.P.P.

One of the most interesting applications of mathematical programming in mechanics is the holonomic and the incremental elastoplastic analysis of structures. The minimum propositions used in this paper, have been derived by Maier [28] by the use of the theory of the Linear Complementarity Problems (L.C.P.). Here, only the incremental elastoplastic analysis will be briefly presented. The holonomic analysis leads to Q.P.Ps having the same structure as the ones of the incremental theory. The relations of incremental elastoplastic analysis are given, formulated for the assembled structure. under the assumptions described by Maier in [28]-[34] etc:

$$\dot{e} = \dot{e}_0 + \dot{e}_E + \dot{e}_P \tag{1}$$

$$\dot{e}_E = F_0\dot{s}, \quad \dot{e}_P = W\dot{\lambda}, \quad \dot{F} = N^T\dot{s} - H\dot{\lambda} \tag{2}$$

$$\dot{\lambda} \geq 0, \ \dot{F} \leq 0, \ \dot{F}^T\dot{\lambda} = 0, \tag{3}$$

where $\dot{e}_0$, $\dot{e}_E$ and $\dot{e}_P$ are the initial, the elastic and the plastic strain vectors respectively, F_0 the natural flexibility matrix, $\dot{s}$ the stress vector, N the gradient of the yield functions which are zero, W the gradient of the corresponding plastic potentials, $\dot{\lambda}$ the plastic multipliers vector, $\dot{F}$ the yield functions vector and H the workhardening matrix depicting the moment-rotation relationship ($M-\phi$ curve) of the structural components [34]. Further, the equilibrium $G\dot{s} + K_G\dot{u} = \dot{p}$ and compatibility $\dot{e} = G^T\dot{u}$ equations hold, where G is the equilibrium matrix, K_G the geometric stiffness matrix, $\dot{u}$ the displacement vector and $\dot{p}$ the load vector. As it is well known, relations (3) form a L.C.P. [28],[33],[34]. By using the same notation, assumptions and substitutions as in [35],[13] and taking into account the above equilibrium and compatibility equations, it can be proved, either by the L.C.P. approach proposed by Maier or by the use of variational inequalities as in [35],[36], that relations (1)-(3) are equivalent to the Q.P.P. (primal problem)

$$\min\left\{ P(x) = \frac{1}{2}x^TMx + q^Tx | x \geq 0 \right\} \tag{4}$$

Matrix M is generally symmetric and positive semidefinite. Under the assumption that matrix K is nonsingular, the problem is formulated, expressed with respect to the plastic multipliers $\dot{\lambda}$. Through suitable substitutions and by the use again either of the L.C.P. approach or of variational inequalities and if matrix D is symmetric ($N = W, H = H^T$) and positive semidefinite, the minimization problem

$$\min\left\{ R(\lambda) = \frac{1}{2}\dot{\lambda}^TD\dot{\lambda} - (N^T\dot{s}^E)^T\dot{\lambda} | \dot{\lambda} \geq 0 \right\} \tag{5}$$

is obtained. Thus, the main problem may be put in the form:

Find $x \in \Re^n$ such as to solve the problem

$$\min\left\{ \frac{1}{2}x^TMx - q^Tx | x \geq 0 \right\} \tag{6}$$

Here $M = \{\mu_{ij}\}$ is a given matrix and $q = \{q_i\}$ is a given vector. For the solution of (6) the Hopfield neural network model is applied. It is a feedback or reccurent network which has parallel input and output channels and n-neurons which are fully interconnected, i.e. all neurons are connected to all the other neurons. These neurons are modeled as amplifiers with a general performance expressed through f_i and input resistors (resp. capacitors) ρ_i (resp. C_i), $i = 1, \ldots, n$. Then, V_i is the output voltage of the amplifier j and u_j is the input voltage. A *synapse* of the neurons i, j, is characterized by a conductance T_{ij}, connecting the output of the j-neuron with the input of the i-neuron. Both the case of the excitatory and of the recessional synaptic connections are taken into account by the assumption that each amplifier has two outputs, a positive and a negative one taking respectively the values [0,1] and [-1,0]. Then, if a synapse is excitatory (resp. recessional) and the synapse conductance is realized through a resistor R_{ij}, whose absolute value is equal to $1/|T_{ij}|$, we connect the resistor to the positive (resp. negative) output of the amplifier j. This can be written as

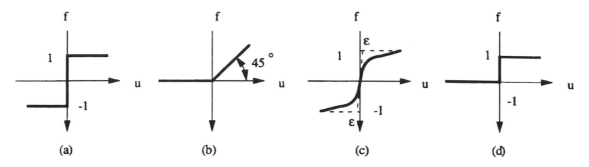

Figure 1: Certain common types of responses of neurons

$$\frac{1}{R_{ij}} = T_{ij} = \begin{cases} \frac{1}{R_{ij}^+} & \text{if } T_{ij} > 0 \text{ with } R_{ij}^- = \infty \\[2mm] \frac{1}{R_{ij}^-} & \text{if } T_{ij} < 0 \text{ with } R_{ij}^+ = \infty \end{cases} \qquad (7)$$

and

$$\frac{1}{R_{ij}} = T_{ij} = \frac{1}{R_{ij}^-} - \frac{1}{R_{ij}^-} \text{ if } R_{ij}^{\pm} > 0 \qquad (8)$$

Each neuron receives also an externally supplied input current I_i. The evolution with time t of the circuit, is described by equations derived by means of Kirchoff's law

$$C_i \left(\frac{du_i}{dt}\right) = \sum_{i=1}^{n} T_{ji} V_j - \frac{u_i}{R_i} + I_i, \quad V_j = f_j(u_j) \qquad (9)$$

Here, $R_i^{-1} = \rho_i^{-1} + \sum_{j=1}^{n} R_{ij}^{-1}$, that is a parallel connection of the input resistor ρ_i and the resistors R_{ij} of the synapse and f_j is a possibly multivalued monotone response function of the j-neuron. For given initial values of the neuron inputs u_i at t=0, the integration of (9) in a digital computer supplies the numerical results for the network under consideration. As it can be proved [37],[38], if in the fictitious network which we have introduced $T_{ij} = T_{ji}$ and obviously $T_{ii} = 0$ (the original Hopfield model), the solution of (9) converges to solutions with the outputs u_i of all neurons constant. Each such solution is called a *stable state* and makes stationary (here minimum) the quantity:

$$E = -\frac{1}{2} \sum_{i,j=1}^{n} T_{ij} V_i V_j + \sum_{i=1}^{n} \left(\frac{1}{R_i}\right) \int_0^{V_i} f_i^{-1}(V_i)\, dV_i - \sum_{i=1}^{n} I_i V_i \qquad (10)$$

which is also called the *Liapunov function* of the system. If f_i is the identity mapping, the above relation takes the following simpler form:

$$E = -\frac{1}{2} \sum_{i,j=1}^{n} T_{ij} V_i V_j + \sum_{i=1}^{n} \frac{V_i^2}{2R_i} - \sum_{i=1}^{n} I_i V_i \qquad (11)$$

This is valid for instance if f is given by Fig.(1b).

In this case, the minimum of the quantity E will be sought for $V_i \geq 0$. Indeed $V_i = f_i(u_i) = \{u_i$ if $u_i > 0$, or zero if $u_i \leq 0\}$. It must be noted that the second term in (11) influences the position of the minimum only for shallow sigmoidal curves. For curves tending to the dotted line of Fig. (1c), that is for narrow sigmoidal curves, the influence of the integral of the second term in the right hand side of (11) is negligible.

Thus, the network always fulfills the constraints $x_i \geq 0, i = 1, \ldots, n$ as it is obvious from the function f used for the calculation.

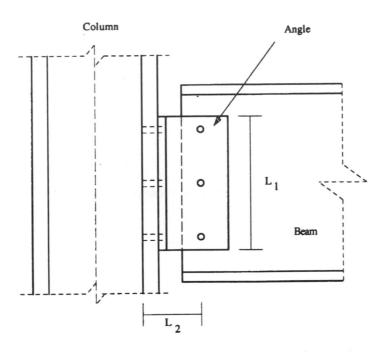

Figure 2: Single angle beam to column connection

4. Neural network configuration for steel structure connections

The experimental results of semi-rigid steel connections have been taken from the data bases described in References [16],[17]. In particular the case of single web-angle, beam-to-column connections with the angle bolted to both the beam and the column has been considered here (see Fig. 2).

A back-propagation artificial neural network as the one described previously is able to learn an input-output relation between appropriately preprocessed data. For each experiment, the design variables of the tested steel connections are taken as input data and the measured moment-rotation curve is considered to be the output data. For instance, the bolted steel structure connections used in the examples are determined by three design variables (Table 1). Thus, an input vector $Z = \{ z_1, z_2, z_3 \}$ is constructed in this case by an analogous preprocessing of the design variables. The experimental moment-rotation curve of a steel structure connection which is shown in Figure 3a is first considered. A number of points $\phi_i, M_i, i = 1, \ldots, m$ placed on the curve are used for discretization (Figure 3b). The output of the neural network model are the pairs $\{ \phi_i, M_i \}, i = 1, \ldots, m$ (or actually only the $\{ M_i \}, i = 1, \ldots, m$ values if the division of the $[0, max\phi]$ interval is assumed to be equidistant). The scaling factors α_1 and α_2 must be included additionally in the set or output variables. Thus, the set of variables $X = \{ \alpha_1, \alpha_2, M_1, \ldots, M_m \}$ uniquely determine the preprocessing transformation from the $M - \phi$ curve to the discretized $M\prime - \phi\prime$ one (see Figure 3).

Experiment	Nr of bolts	Angle thickness	Angle length
1	2	0.625	13.75
2	3	0.625	21.25
3	4	0.625	28.75
4	6	0.625	36.25
5	6	0.625	43.75
6	4	0.7813	28.75

Table 1: Experimental data for beam to column connection [cm]

The values of the design variables that describe the experimentally tested connections and which have been considered in the neural network model are summarized in Table 1. A detailed description of the experiments is given in [15].

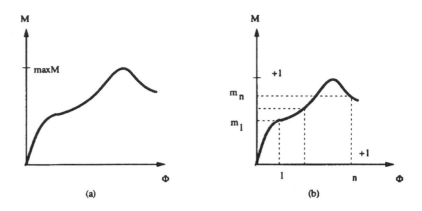

Figure 3: Moment-rotation experimental curve for use with the neural network model: (a) measured curve (schematic); (b) normalized curve

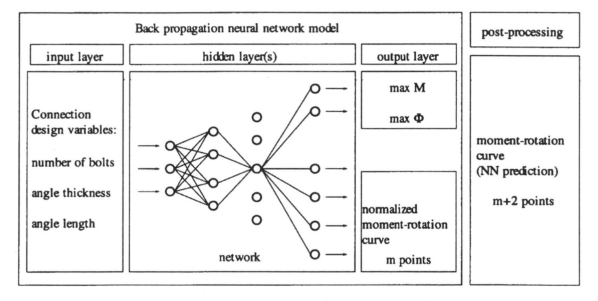

Figure 4: Neural network model for the connection problem; schematic representation

In the numerical example, $m = 20$ values have been used for the discretization of the experimental $M - \phi$ curve. Thus the neural network model will have three input variables (the design variables which identify the steel connection as it is given in Table 1) and 22 output variables. A fully connected feed-forward network has been chosen.

The experimental results (see Table 1) and the neural network predictions of the problem are shown in Figure 4. A $3 - 100 - 100 - 100 - 100 - 100 - 22$ network has been used. All available experiments have been used for both training and testing of the neural network. In Figure 4, the quality obtained in loading the experimentally gained information in the neural network, is depicted. The training phase of the network took approximately 9000 epochs, to reach an error less than 0.00001. A serial computer implementation has been used. The computation, on a $HP755$ workstation, has been completed in approximately 90 minutes CPU time. Figure 5 shows a schematic representation of the neural network model which has been developed. In ref. [21] the case of single plate beam-to-column connections with the plate bolted to the beam and welded to the column has been also treated in a similar way.

The above $M - \phi$ law can be used for the elastoplastic calculation of any desired steel structure. We formulate according to [34] the hardening matrix $\mathbf{H}$ and the corresponding Q.P.P., in the form of relation (5) for the holonomic case, characterizing the position of equilibrium. Then the Hopfield neural network is obtained permitting the

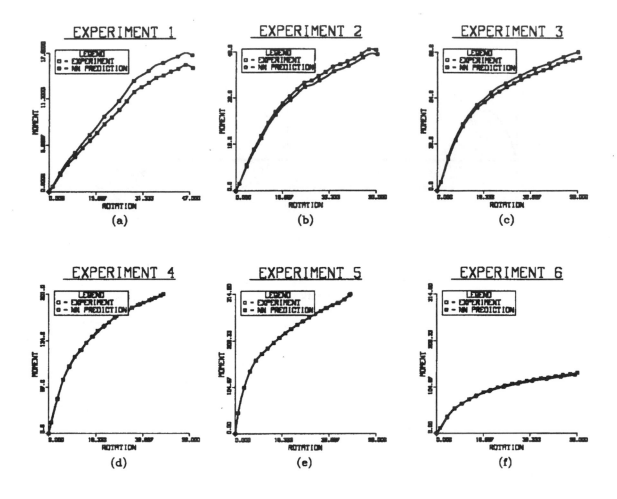

Figure 5: Neural network prediction versus experimental results

calculation of the solution of the minimum problem (5), i.e. the position of equilibrium of the elastoplastic structure. The reader is referred to [12],[13] for more details about numerical applications on the treatment of the elastoplastic analysis problems as Q.P.P. by the use of the Hopfield model. In Figure 6 the corresponding circuit realizing this model is given ([14] p. 43).

5. Conclusions

The observations derived from the numerical implementation of the method are briefly presented in the following:

- The performance of the neural network model may be seriously influenced by the optimal choice of the network configuration in which the variables must be chosen, for each case, after numerical experimentation.

- If the network uses interpolation from a given training set, the quality of the neural network prediction is better than in the case of extrapolation.

- In the method presented above, the neural network model treats all the parameters describing the experimental curves with the same accuracy not taking into account their different importance and contribution. In this subject, some research results will be presented by the authors in the near future.

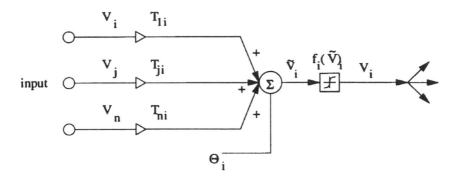

Figure 6: A typical artificial neuron; schematic representation

Acknowledgements

G.E.Stavroulakis is currently supported by the European Union (TMR Grant ERBFMBICT960987).

References

[1] Abdalla, K.M., Stavroulakis G.E., 1989, Zur rationalen Berechnung des "Prying actions" Phänomens in Schraubenverbindungen. *J. Stahlbau*, **58** (8), pp. 233-238.

[2] Baniotopoulos, C.C., Abdalla, K.M., 1993, Steel column-to-column connections under combined load: a quadratic programming approach. *Computers and Structures*, **46** (1), pp. 13-20.

[3] Abdalla, K.M., Alshegeir, A., Chen W.F., 1993, Strut-tie approach for the design of flat plates with drop panels and capitols. *Research Report CE-STR-93-19*, Purdue University, Struct. Engineering Department.

[4] *Eurocode 3. Design of steel structures-Part 1-1: General rules and rules for buildings*. ENV 1993-1-1:1992, CEN, 1992.

[5] Rumelhart, D.E., McClelland, J.L., 1986, *Parallel distributed processing*. Vols.I, II, III, The MIT Press, Cambridge MA.

[6] Adeli, H., Yeh, C., 1989, Perceptron learning in engineering design. *Microcomputers in Civil Engineering*, **4**, pp. 247-256.

[7] Vanluchene, R.D., Roufei, Sun, 1990, Neural networks in structural engineering. *Microcomputers in Civil Engineering*, **5**, pp. 207-215.

[8] Theocaris, P.S., Panagiotopoulos, P.D., 1993, Neural networks for computing in fracture mechanics. Methods and prospects of applications. *Computer Methods in Appl. Mechanics and Engineering*, **106**, pp. 213-228.

[9] Chassiakos, A.G., Masri, S.F., 1991, Identification of the internal forces of structural systems using feedforward multilayer networks. *Computing Systems in Engineering*, **2** (1), pp. 125-134.

[10] Ghaboussi, J., Garrett Jr, J.H., Wu, X., 1991, Knowledge-based modelling of material behaviour with neural networks. *Journal of Engineering Mechanics, ASCE*, **117** (1), pp. 132-153.

[11] Pidaparti, R.M.V., Palakal, M.J., 1993 Material model for composites using neural networks. *AIAA Journal*, **31** (8), pp. 1533-1535.

[12] Kortesis, S., Panagiotopoulos, P.D., 1993, Neural networks for computing in structural analysis: methods and prospects of applications. *Intern. Journal for Numerical Methods in Engineering*, **36**, pp. 2305-2318.

[13] Avdelas, A.V., Panagiotopoulos, P.D. and Kortesis, S., 1995, Neural networks for computing in elastoplastic analysis of structures. *Meccanica*, **30**, pp. 1-15.

[14] Cichocki, A., Unbehauen, R., 1993, *Neural networks for optimization and signal processing*. John Wiley and Sons Ltd..

[15] Lipson, S.L., 1968, Single-angle and single-plate beam framing connections. *Canadian Struct. Eng. Conf.*, Toronto, Ontario, pp. 141-162.

[16] Kishi, N., Chen,W.F., 1986, *Data base of steel beam-to-column connections*. Vol. I and II, Structural Engineering Report No. CE-STR-86- 20, School of Civil Engineering, Purdue University, West Lafayette IN.

[17] Abdalla, K.M., Chen, W.F., Kishi, W., 1993, Expanded database of semi-rigid steel connections. *Research Report CE-STR-93-14*, Purdue University, Structural Engineering Department.

[18] Kosko, B., 1992, *Neural networks and fuzzy systems: a dynamical systems approach to machine intelligence.* Prentice-Hall, Englewood Cliffs New Jersey.

[19] Beale, R. and Jackson, T., 1990, *Neural computing. An introduction.* Adam Hilger, Bristol.

[20] Berke, L., Hajela, P., 1992, Applications of artificial neural nets in structural mechanics. *Structural Optimization*, **4**, pp. 90-98.

[21] Abdalla, K.M., Stavroulakis G.E., 1995, A backpropagation neural network model for semi-rigid connections *Microcomputers in Civil Engineering*. **10**, pp. 77-87.

[22] Judd, J.S., 1990, *Neural network design and the complexity of learning.* The MIT Press, Cambridge MA.

[23] Theocaris, P.S., Panagiotopoulos, P.D., 1995, Plasticity including the Bauschinger effect, studied by a neural network approach. *Acta Mechanica*, **113**, pp. 63-75.

[24] Theocaris, P.S., Panagiotopoulos, P.D., 1995, Hardening plasticity approximated via anisotropic elasticity. The Fokker-Planck equation in a neural network environment. *ZAMM*, **75** (12), pp. 889-900.

[25] Brause, R., 1991, *Neuronale Netze.* B.G. Teubner, Stuttgart.

[26] Antes, H., Panagiotopoulos, P.D., 1992, *The boundary integral approach to static and dynamic contact problems. Equality and inequality methods.* Birkäuser Verlag, Basel, Boston.

[27] Panagiotopoulos, P.D., 1993, *Hemivariational inequalities. Application in mechanics and engineering.* Springer Verlag, Berlin, Heidelberg, New York.

[28] Maier, G., 1971, Incremental plastic analysis in the presence of large displacements and physical instabilizing effects. *Int. J. Solids Struct.*, **7**, pp. 345-372.

[29] Maier, G., 1968, A quadratic programming approach for certain classes of nonlinear structural problems. *Meccanica* , **3**, pp. 121-130.

[30] Maier, G., 1968, Quadratic programming and theory of elastic-perfectly plastic structures. *Meccanica*, **3**, pp. 265-273.

[31] Maier, G., 1969, Linear flow-laws of elastoplasticity, a unified general approach. *Rend. Acc. Naz. Lincei* , **VIII**, pp. 266-276.

[32] Maier, G., 1970, A matrix structural theory of piece-wise linear elastoplasticity with interacting yield planes. *Meccanica*, **8**, pp. 54-66.

[33] Maier, G., 1973, Mathematical programming methods in structural analysis. In: *Proc. Int. Conf. on Variational methods in engineering*, Vol. II, eds C.A. Brebbia and H. Tottenham, Southampton Univ. Press, Southampton, pp. 1-32.

[34] Cohn, M.J. and Maier, G. (eds), 1979, *Engineering Plasticity by Mathematical Programming.* Proc. NATO-ASI, Waterloo, Canada 1977, Pergamon Press, New York.

[35] Panagiotopoulos, P.D., 1985, *Inequality Problems in Mechanics and Applications,Convex and Nonconvex Energy functions.* Birkäuser Verlag, Basel, Boston (Russian Translation, MIR Publ., Moscow, 1989).

[36] Panagiotopoulos, P.D., Baniotopoulos, C.C. and Avdelas, A.V., 1984, Certain propositions on the activation of yield modes in elastoplasticity and their applications to deterministic and stochastic problems. *ZAMM*, **64**, pp. 491-501.

[37] Hopfield, J. J., 1982, Neural networks and physical systems with emergent collective computational abilities. *Proc. of the Nat. Acad. of Sciences*, **79**, pp. 2554-2558.

[38] Hopfield, J. J. and Tank, D. W., 1985, "Neural" computation of decisions in optimization problems. *Biol. Cybern.*, **52**, pp. 141-152.

Evolutionary Design of Neural Trees for Heart Rate Prediction

Byoung-Tak Zhang[1] and Je-Gun Joung[2]

[1] *Dept. of Computer Engineering, Seoul National University*
Seoul 151-742, Korea. E-mail: btzhang@comp.snu.ac.kr
[2] *Dept. of Computer Science and Engineering, Konkuk University*
Seoul 143-701, Korea. E-mail: jgjoung@ai.konkuk.ac.kr

Keywords: Neural network design, Evolutionary computation,
Neural tree induction, Time series prediction, Heart rate data.

Abstract

Some classes of neural networks are known as universal function approximators. However, their training efficiency and generalization performance depend highly on the structure which is usually determined by a human designer. In this paper we present an evolutionary computation method for automating the neural network design process. We represent networks as tree structures, called neural trees, in genotype and apply genetic operators to evolve problem-dependent network structures and their weights. Experimental results are provided on the prediction of a heart rate time-series by evolving sigma-pi neural trees.

1 Introduction

Recently, several evolutionary methods have been proposed for constructing and training neural networks. Existing representation methods for neural networks can be roughly divided into two categories [1]: direct and indirect encoding [1]. Direct encodings use a fixed structure, such as connection matrix or bitstrings that precisely specifies the architecture of the corresponding neural network. This encoding scheme requires little effort to decode. However, matrix structures have limited flexibility in expressing topologies of the network structure with variable layers. Bitstrings are not flexible enough to represent various partial connectivity without further annotation. Genetic operators need to be applied carefully to preserve the topological constraints of networks.

Indirect encoding schemes use rewrite rules to specify a set of construction rules that are recursively applied to yield the phenotype. Examples include graph generation grammars and cellular encoding. This approach is interesting in that it simulates in some sense the developmental process. Subtree crossover applies well to these representations. In addition, experimental evidence has shown that the cellular encoding scheme is effective in evolving modular structures consisting of similar substructures. However, the grammatical encoding does not seem appropriate for exploring a huge number of partial interaction possibilities as required in our application. In addition, grammatical encoding requires execution of rewrite-rules for every conversion from genotype to phenotype. This makes network training an expensive phase since training of neural networks requires a large number of evaluations and each evaluation needs a separate decoding.

In this paper we present an alternative representation scheme, called neural trees [9, 10]. A neural tree consists of a number of artificial neurons connected with weights in a tree structure. The leaves of the tree are elements of the terminal set X of n variables, $X = \{x_1, x_2, \cdots, x_n\}$. The root node of the tree is the output unit. All the nodes except the input units are non-terminal units. Each non-terminal node i is characterized by the unit type u_i, the squashing function f_i, the receptive field $R(i)$ and the weight vector $\mathbf{w}_i$. $R(i)$ is the index set of incoming units to unit i and can be different from unit to unit.

The neural tree representation combines the advantages of direct and indirect encoding schemes. It is powerful and flexible in expressing a broad class of neural architectures. The representation is decoding-efficient and convenient for genetic operations. The structure and size of the network is also automatically adapted during the evolutionary learning process. With sigma and pi units, for example, the method can build a higher-order functional structure of partially connected polynomial units. The explicit use of product neurons has been very useful for solving problems which are difficult for multilayer perceptrons [2].

The paper is organized as follows. Section 2 describes the neural tree encoding scheme in more detail. Section 3 is devoted to the description of the heart rate time-series data and the evolutionary computing method for designing neural trees for the prediction of the data. Section 4 reports the experimental results. Sectoin 5 contains conclusions.

2 Neural Trees

Let $\mathcal{NT}(d, b)$ denote the set of all possible trees of maximum depth d and maximum b branches for each node. The nonterminal nodes represent neural units and the neuron type is an element of the basis function set $\mathcal{F} = \{\text{neuron types}\}$. Each terminal node is labeled with an element from the terminal set $\mathcal{T} = \{x_1, x_2, ..., x_n\}$, where x_i is the ith component of the external input $\mathbf{x}$. Each link (j, i) represents a directed connection from node j to node i and is associated with a value w_{ij}, called the synaptic weight. The members of $\mathcal{NT}(d, b)$ are referred to as neural trees. In case of $\mathcal{F} = \{\Sigma, \Pi\}$, the trees are specifically called sigma-pi neural trees. The root node is also called the output unit and the terminal nodes are called input units. Nodes that are not input or output units are hidden units. The layer of a node is defined as the longest path length to any terminal node of its subtree.

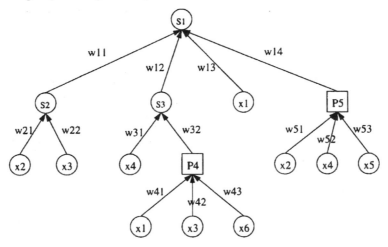

Figure 1: An example of neural tree structure.

Different neuron types are distinguished in the way of computing net inputs. For the evolution of higher-order networks, we consider two types of units. Sigma units compute the sum of weighted inputs from the lower layer:

$$net_i = \sum_j w_{ij} y_j \tag{1}$$

where y_j are the inputs to the ith neuron. Pi units compute the product of weighted inputs from the lower layer:

$$net_i = \prod_j w_{ij} y_j \tag{2}$$

where y_j are the inputs to i. The output of a neuron is computed either by the threshold response function

$$y_i = \sigma(net_i) = \left\{ \begin{array}{ccc} 1 & : & net_i \geq 0 \\ -1 & : & net_i < 0 \end{array} \right. \tag{3}$$

or the sigmoid transfer function

$$y_i = f(net_i) = \frac{1}{1 + e^{-net_i}} \tag{4}$$

where net_i is the net input to the unit computed by Eqn. 1 or Eqn. 2.

A higher-order network with m output units can be represented by m sigma-pi neural trees. That is, the genotype A_i of ith individual in our evolutionary framework consists of m neural trees:

The neural tree representation does not restrict the functionality since any feedforward network can be represented with a forest of neural trees:

$$A_i = (A_{i,1}, A_{i,2}, ..., A_{i,m}) \quad \forall k \in \{1, ..., m\}, \; A_{i,k} \in \mathcal{NT}(d, b) \tag{5}$$

The connections between input units to arbitrary units in the network is also possible since input units can appear more than once in the neural tree representation. The output of one unit can be used as input to more than one unit. The duplication does not necessarily mean more space requirements in trees than network representations since frequently-used fit submodules can be stored and multiply reused. This leads to the construction of modular structures and reduces memory requirements for representing the population [10].

Neural trees do not require decoding for their fitness evaluation. Training and evaluation of fitness can be performed directly on the genotype since both the genotype and phenotype are equivalent. Since subtree crossover used in genetic programming [3] applies without modification to this representation, we can use genetic programming as the main evolutionary engine.

3 Predicting Heart Rates by Evolving Neural Trees

Physiological systems are often considered to have machinelike properties, and a common objective in physiology is to discover how some system senses its current state and uses this information to respond automatically [6]. For example, it has long been known that heart rate is regulated by the baro-reflex, in which specialized nerves in the aorta and other blood vessels sense the blood pressure and convey this information to the brainstem, which in turn sends signals to the pacemaker region of the heart that determines the heart rate. The physiological signals are ordinarily nonstationary due, for example, to the difficulty of controlling voluntary behavior or to circadian influences. Furthermore, a variety of nonlinearities abound in the interaction between the components of physiological systems.

We used a multivariate physiological data recorded from a patient in the sleep laboratory. The signals were recorded from a 49-year-old male. He had been tentatively diagnosed as suffering from sleep apnea, a potentially life-threatening disorder in which the subject stops breathing during sleep. It was a part of the data used in the Santa Fe Institute Time Series Prediction and Analysis Competition [7]. This data set provides simultaneous measurements using the extra information to learn about how potentially variables interact. If one-variable time series is deterministic, there exists a scalar d (which is called the embedding dimension) and a function f such that for every $t > d$:

$$x_t = f(x_{t-1}, x_{t-2}, ..., x_{t-d}). \tag{6}$$

We have used neural trees to model and predict the heart rate time series. For the construction of neural models, we maintain a population $\mathcal{A}$ consisting of M individuals of variable size:

$$\mathcal{A}(g) = (A_1, A_2, ..., A_M). \tag{7}$$

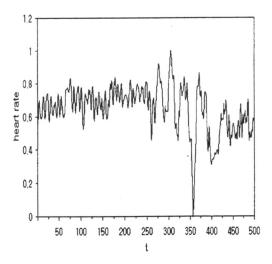 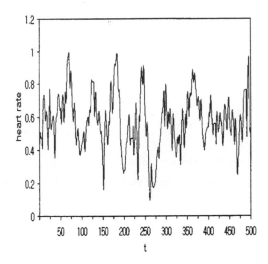

Figure 2: The heart rate time series used for training (left) and test (right).

Each individual A_i is a neural network represented as neural trees. The initial population $\mathcal{A}(0)$ is created at random. In each generation g, the fitness values $F_i(g)$ of networks are evaluated and the upper $\tau\%$ are selected to be in the mating pool $\mathcal{B}(g)$. The next generation $\mathcal{A}(g+1)$ of M individuals are then created by exchanging subtrees and thereby adapting the size and shape of the network. Mutation changes the node type and the index of incoming units. The best individual is always retained in the next generation so that the population performance does not decrease as generation goes on (elitist strategy).

Between generations the network weights are adapted by a stochastic hill-climbing search. This search method is based on the breeder genetic algorithm [4], in which the step size Δw is determined with a random value $\epsilon \in [0, 1]$:

$$\Delta w = R \cdot 2^{-\epsilon \cdot K}, \tag{8}$$

where R and K are constants specifying the range and slope of the exponential curve. In the experiments, the values were $R = 2$ and $K = 3$. This method proved very robust for a wide range of parameter-optimization problems. The fitness F_i of the individuals A_i is defined as

$$F_i = F(D|A_i) = \frac{E(D|A_i)}{m \cdot N} + \frac{C(A_i)}{N \cdot C_{best}}, \tag{9}$$

where m is the number of outputs, N is the number of training examples, and C_{best} is the complexity of the best individual at generation $g - 1$. The first term expresses the error penalty $E(D|A_i)$ for the training set. The second term penalizes the complexity $C(A_i)$ of the network, defined as the sum of the units and the number of layers in the tree. This evaluation measure prefers parsimonious networks to complex ones and turned out to be important for achieving good generalization [9].

4   Experimental Results

Since we do not know the embedding dimension for the time series, we varied the number of inputs for the prediction. In particular, we experimented with two different embedding dimensions, $d = 3, 5$:

- One-step ahead prediction from three inputs

$$x_t = f(x_{t-1}, x_{t-2}, x_{t-3}) \tag{10}$$

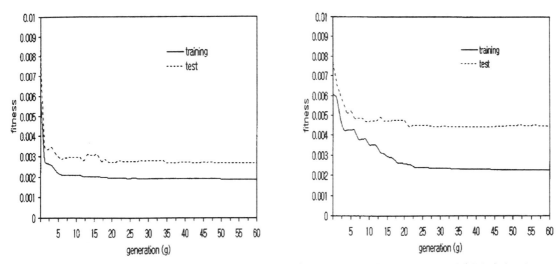

Figure 3: Fitness vs. generation plots for heart rate prediction: (left) three inputs and (right) five inputs.

- One-step ahead prediction from five inputs

$$x_t \;\; = \;\; f(x_{t-1}, x_{t-2}, x_{t-3}, x_{t-4}, x_{t-5}) \tag{11}$$

A time series of 500 points was used to generate the training set. The next 500 points of time series were used for generating the test set. The algorithm parameters used for the experiments are summerized in Table 1. Each run consists of 60 generations with a population size of 200.

Table 1: Parameter values used in the experiments.

Algorithm parameters	Values used
population size	200
max generation	60
crossover rate	0.95
mutation rate	0.1
no. of iterations	100
training set size	500
test set size	500

Figure 3 shows the change of fitness values of the best-of-generation neural trees. A fast reduction of errors can be observed during early generations. This seems attributed to the fact that the value x_{t-1} is usually important for the prediction of x_t and the evolutionary algorithm can find this variable in an early generation. Other authors, for example [5], also have found similar phenomenon in one-step ahead predictions of time series. The rest of the evolution attempts to find a better neural tree structure by both structural and parametric modification.

Figure 4 plots the evolution of tree size during the run. The graphs show a flexible change of network size. Close relationship between the change of the network structure and the change of the fitness value was observed. The use of complexity penalty in the fitness function was observed very useful; without it the network size usually grows without bound, resulting in poor prediction performance for unseen time series.

Figures 5 and 6 show the one-step ahead prediction results for the training and test data, respectively. For comparison the figures also contain plots for the difference between the measured and predicted values.

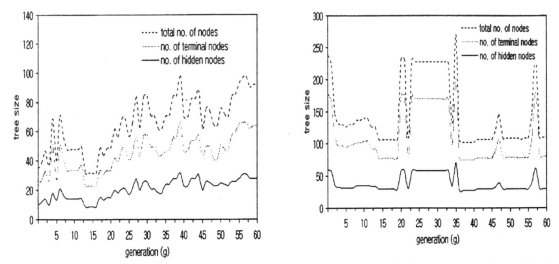

Figure 4: Tree size vs. generation plots for heart rate prediction: (left) three inputs and (right) five inputs.

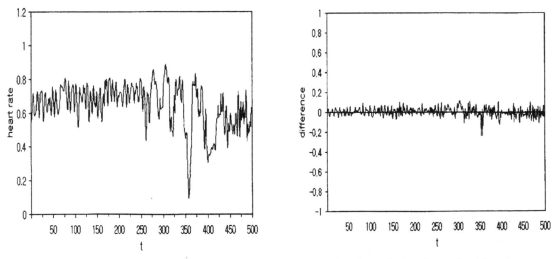

Figure 5: Performance for the training data: one-step ahead prediction from three inputs.

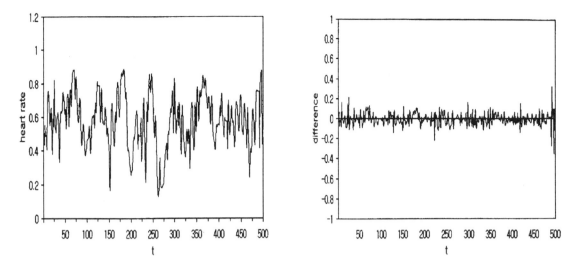

Figure 6: Performance for the test data: one-step ahead prediction from three inputs.

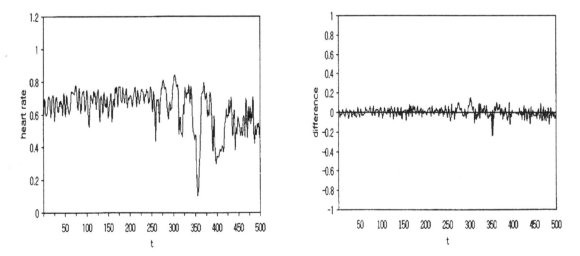

Figure 7: Performance for the training data: one-step ahead prediction from 5 inputs.

100

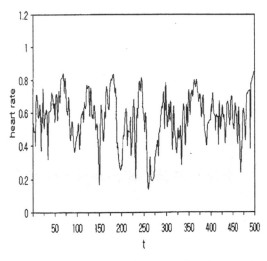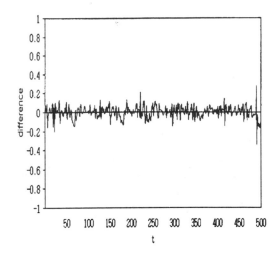

Figure 8: Performance for the test data: one-step ahead prediction from 5 inputs.

Considering the nonlinearity and nonstationarity of the heart rate, the method finds a reasonable model of the underlying structure within a few dozen generations.

Figures 7 and 8 show the results for the five inputs rather than three. Several runs have shown the tendency that using five points as input results in poorer performance than using three inputs. This suggests a possibility of the embedding dimension of this problem less than five. We did not try to analyze the underlying structure of this data.

5   Conclusions

We presented an evolutionary computation method for evolving neural trees and demonstrated its performance on a real-life physiological data. Unlike most conventional neural models, neural trees employ different types of neurons in a single network. The set of different types is defined by the application domain, and the specific type of each unit is determined during the evolutionary learning process.

Since evolutionary search does not require restrictive assumptions such as differentiability of neural activation functions, the method can be used to explore a wide range of novel neural architectures. The control of tree size was found important to make the approach practical faced with real-life data.

The physiological data seems to provide an interesting benchmark for the test of neural modeling techniques in several reasons. First, they contain much noise and information loss caused during the preprocessing of the signals. Second, the signals are ordinarily nonstationary due, for example, to the difficulty of controlling behavior (e.g., respirator or activity) or to circadian influences. Third, a variety of nonlinearities abound in the interaction between the components of physiological systems. Experiments are in progress to achieve a prediction accuracy in multi-step ahead prediction problems which is comparable to that of one-step ahead prediction.

Acknowledgements

This research was supported in part by the Korea Science and Engineering Foundation (KOSEF) under grant number 961-0901-001-2.

References

[1] Balakrishnan, K. and Honavar, V., 1995, *Evolutionary design of neural architectures*, CS-TR-95-01, AI Lab, Dept. of Computer Science, Iowa State University.

[2] Durbin, R. and Rumelhart, D. E., 1989, Product units: a computationally powerful and biologically plausible extension to backpropagation networks, *Neural Computation*, 1, 133–142.

[3] Koza, J. R. 1992, *Genetic Programming: On the Programming of Computers by Means of Natural Selection*, MIT Press, Cambridge, MA.

[4] Mühlenbein, H. and Schlierkamp-Voosen, D. 1994, The science of breeding and its application to the breeder genetic algorithm, *Evolutionary Computation*, 1(4), 335-360.

[5] Mulloy, B. S., Riolo, R. L., and Savit, R. S. 1996, Dynamics of genetic programming and chaotic time series prediction, *Proc. First Annual Conf. on Genetic Programming*, J. R. Koza, D. E. Goldberg, D. B. Fogel, and E. L. Riolo (eds.), Cambridge, MA: MIT Press, pp. 166-174.

[6] Rigney, D. R. et al., 1994, Multi-channel physiological data: description and analysis (data set B), *Time Series Prediction: Forecasting the Future and Understanding the Past*, Addison-Wesley.

[7] Weigend, A. S. and Gershenfeld, N. A. (Eds.), 1994, *Time Series Prediction: Forecasting the Future and Understanding the Past*, Addison-Wesley.

[8] Whitley, D. Starkweather, T. and Bogart, C. 1990, Genetic algorithms and neural networks: optimizing connections and connectivity, *Parallel Computing*, 14, 347-361.

[9] Zhang, B. T. and Mühlenbein, H. 1995, Balancing accuracy and parsimony in genetic programming, *Evolutionary Computation*, 3, 17–38.

[10] Zhang, B. T. Ohm, P. and Mühlenbein, H. 1997, Evolutionary induction of sparse neural trees, *Evolutionary Computation*, 5(3), to appear.

Part 3: Fuzzy Logic

___

Papers:

Hypertrapezoidal Fuzzy Membership Functions for Decision Aiding

Wallace E. Kelly[1], John H. Painter[2]

Texas A&M University, Department of Electrical Engineering
College Station, Texas 77843, [1]*wkelly@tamu.edu,* [2]*painter@tamu.edu*

Keywords: fuzzy sets, fuzzy membership functions, decision aiding, avionics.

Abstract

This paper reports research pushing forward the theory and application of fuzzy logic and fuzzy control. The push is in a theoretical area needed to apply fuzzy logic to the interpretation and management of systems of increasing complexity. Hypertrapezoidal fuzzy membership functions (HFMFs) are a new mechanism for designing multidimensional fuzzy sets. Unlike their one-dimensional counterparts traditionally used in fuzzy engineering, HFMFs model the correlation that exists between the variables of the state space. Additionally, the manner in which HFMFs are electronically stored makes them ideal for applications requiring training or on-line adaptation. This paper contains the background and practical application of HFMFs in an on-board pilot advisory system for general aviation aircraft. The pilot advisory system serves as an example of the usefulness of HFMFs for decision aiding. HFMFs have proved to be an essential element for flight mode analysis, and will likely find many other useful applications.

1. Introduction

The present theoretical work was motivated by an application in a high-technology development area known as *Smart-Cockpit Computing* [1]. This is an area similar to a 1986 USAF program known as *Pilot's Associate* [2]. The present work grew out of NASA-supported research over the period, 1989-1994, styled *Knowledge-Based Processing for Aircraft Flight Control* [3].

In the prior research, the basic idea was developed of using artificial intelligence to help an aircraft pilot process and manage information necessary to flying an airplane in today's increasingly complex Air Traffic Control System (ATC). In particular, the idea was to use computer technology and rule-based processing to make assessments of what flight-operational procedure was being executed, independent of pilot input to the computer. A companion idea was that of using the assessment, together with recorded flight-plan and ATC clearance information, to formulate automated control inputs to the aircraft, as in an automated flight-management system. By 1994, these ideas had matured to implementations for a *Flight Mode Interpreter* and a *Meta-Controller*. Both were exercised and tested by way of a computer simulation of a Boeing-737 aircraft.

In 1995, under new NASA commercialization funding, focus was shifted from automated flight-management of large transport aircraft to interpretation and advising for pilots of smaller general aviation aircraft. The Meta-Controller, which had previously driven the autopilot directly, now became a *Pilot Advisor*. A full-blown development project began, to create an engineering prototype of a commercializable system. The software development environment now included a fixed-base engineering flight simulator. Two hardware items were added, being a *Head-Down Display* (HDD), and a *Head-Up Display* (HUD). Three software modules were added, being the drivers for the HUD and HDD, and a *Navigation/Flight-Manager* module. The end-goal of the development project was a flight demonstration/test in a Rockwell Commander-700 light-twin research aircraft by the end of 1997. The present program of research and development is called Automated Safety and Training Avionics, or *ASTRA*.

2. A Fuzzy Logic Flight Mode Interpreter

The original research version of the Flight Mode Interpreter was implemented in standard fuzzy logic. As such, it was a decision system, to process flight variable data and identify the current flight operation, out of a predefined list of operations. Flight variables processed included airspeed, altitude, flight attitude, engine power setting, etc. Defined modes included Takeoff, Climb-out, Cruise, Hold, Initial Approach, Final Approach, and Landing. The flight operations were modeled in terms of trapezoidal fuzzy membership functions. For more background on the flight mode interpretation problem and the one-dimensional solution, see [4].

The choice of fuzzy logic was made for two reasons. The driving reason was to frame the decision problem using a decision-rule which directly models the uncertainty in the flight-mode definitions, with respect to the data variables. The second reason was that by using the soft fuzzy logical connectives [5] and by insuring that overlapping membership functions sum to unity, the fuzzy decision rule is isomorphic to the Bayes decision rule [6]. Moreover, the resulting decision algorithm is then the Minimum Probability of Decision Error algorithm, or Ideal Observer [7].

Fuzzy decision is implemented as follows. A set of measured data is input to the fuzzy processor and the membership functions of the various hypotheses are evaluated from the input data. The hypothesis which yields the greatest membership value for the data is then chosen as the decision. For flight interpretation, the hypotheses are the various possible flight operations, or modes.

Thinking of the various flight variables, such as airspeed, altitude, and engine power, as defining an orthogonal state space, a particular flight mode may be modeled as a subset of the state space. Furthermore, a sequence of flight modes, such as cruise, initial approach, and final approach, form contiguous, and in fact overlapping, subsets. Therefore, a particular point in the state-space may belong to more than one mode. The mode subsets are not disjoint. This modeling uncertainty at the subset boundaries motivates the use of fuzzy logic. The subsets are taken to be "fuzzy" at their boundaries, where they overlap. The membership functions represent this uncertainty in exactly the same way as probabilistic set functions.

In standard fuzzy logic, the membership functions are normalized to unity. They are one-dimensional set functions. To deal with subsets in an N-dimensional state-space, the one-dimensional functions are combined to obtain a membership function defined on the N-dimensional space. This combining is called "composition." The N-dimensional membership function is said to be composed from the N one-dimensional functions.

Using the soft fuzzy logical connectives, the composition is performed by multiplying together the N one-dimensional functions on the Cartesian product space formed from the data variables. For a two-dimensional function, composed from two one-dimensional functions, the fuzzy set's footprint in the plane will be rectangular. In the Bayes isomorphism, this corresponds to combining two probability functions as though the variables were independent, and hence uncorrelated. The problem is that for aircraft flight-modes, this is a very unrealistic assumption. In practice, it is found that the variables, such as airspeed and altitude are indeed correlated, and that the resulting subset shapes are anything but hypercubes. What is needed is true N-dimensional membership functions which can model a subset having any desired shape in the state-space.

It was to make this fundamental improvement in the state of the fuzzy art that the research reported herein was initiated. This particular improvement was financially supported by the State of Texas through its Advanced Technology Program [1]. Such an improvement is required, to yield the performance of the Fuzzy Flight-Mode Interpreter necessary to the Air Traffic Control problem. Who was it that said, "Necessity is the mother of invention?"

3. Hypertrapezoidal Fuzzy Membership Functions

Hypertrapezoidal fuzzy membership functions were proposed by the authors [8] as a convenient mechanism for representing and calculating multidimensional fuzzy sets. An important consideration in the development of this new technique was the requirement that the multidimensional fuzzy sets sum to unity over the entire state space, as shown in equation (1). Membership functions defined in such a manner are referred to as a *fuzzy partitioning* of the state space and are consistent with a Bayesian interpretation of fuzzy sets.

$$\sum_i \mu_i(\bar{x}) = 1 \quad \forall \bar{x} \tag{1}$$

All the previous work in the area of multidimensional membership functions encountered by the authors is based on the Gaussian probability density function. See, for example, [9] and [10]. The Gaussian PDF is easily extendible to N-dimensions, but does not satisfy the property of equation (1). Furthermore, membership functions based on the Gaussian PDF evaluate to unity at only a single point in the state space. Foster and Khambhampati address this last issue by extending the unity point of the Gaussian density along a vector in state space [11].

Another important consideration is that N-dimensional membership functions be specified with only a few parameters. While a one-dimensional fuzzy membership function in the shape of a trapezoid can be defined with four points, extending this point-by-point definition of membership functions becomes impractical in two dimensions. Trying to define the corners of an N-dimensional fuzzy set on three or more inputs becomes virtually impossible, especially if the sets must satisfy the requirement of equation (1).

As an alternative to trying to define all the corners of N-dimensional fuzzy sets, consider the use of a single point in the state-space as the defining parameter of an N-dimensional fuzzy set. Each fuzzy set in a fuzzy partitioning would then have an associated N-dimensional vector which is a typical value for that set. We chose to call such an N-dimensional vector the *prototype point*. The prototype point λ_i, for a fuzzy set S_i, with a membership function $\mu_i(x)$, satisfies the following equations.

$$\begin{aligned} \mu_i(\lambda_i) &= 1 \\ \mu_j(\lambda_i) &= 0 \quad j \neq i \end{aligned} \tag{2}$$

Figure 1 shows a simple example of a fuzzy partitioning in two dimensions using three prototype points to define three fuzzy sets.

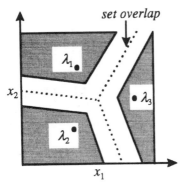

Figure 1. Prototype points defining a fuzzy partitioning.

A measured value, x, which is an N-dimensional point in the state space of a fuzzy partitioning, has a degree of membership in a fuzzy set based on its Euclidean distance from the prototype point for that set. For example, if $x = \lambda_1$, then $\mu_1(x) = 1$, $\mu_2(x) = 0$, and $\mu_3(x) = 0$. As another example, if x is equidistant from all three prototype points, then $\mu_1(x) = 0.333$, $\mu_2(x) = 0.333$, and $\mu_3(x) = 0.333$. This is the basis of *hypertrapezoidal fuzzy membership functions* and has proved to be quite useful in inferring operational modes of an aircraft.

One additional parameter is needed for defining an N-dimensional fuzzy partitioning. The *crispness factor* determines how much overlap exists between the prototype points of two adjacent fuzzy sets. We chose to define the range of the crispness factor, σ, to be [0, 1]. For $\sigma = 1$, no overlap exists between the sets, and the partitioning reduces to a minimum distance classifier. Figure 2 shows the resulting partitions of the above example for the two extremes $\sigma = 0$ and $\sigma = 1$.

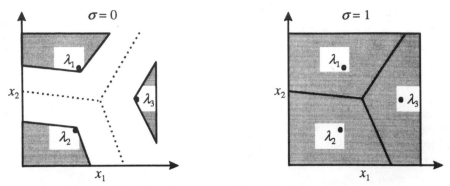

Figure 2. The effect of σ on a two-dimensional fuzzy partitioning.

Such a scheme for defining fuzzy sets can also be used to define standard one-dimensional fuzzy partitions. Figure 3 illustrates how varying σ in a one-dimensional partition evolves the membership functions from triangular fuzzy sets, through trapezoidal fuzzy sets, and finally to crisp, non-fuzzy sets.

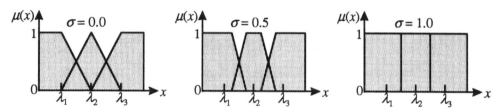

Figure 3. The effect of σ on one-dimensional fuzzy sets.

The authors chose to define the crispness factor according to equation (3) and Figure 4.

$$\sigma = \frac{2\alpha}{d} \qquad (3)$$

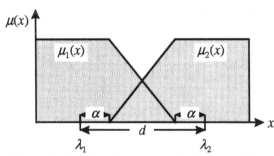

Figure 4. Defining the overlap of a fuzzy partition.

Multidimensional fuzzy membership functions defined by prototype points and a crispness factor are called *hypertrapezoidal fuzzy membership functions (HFMFs)*. Given a sensor measurement, x, the HFMFs can be calculated using standard trigonometry. First, a distance measure, $\rho_{i|j}$, is calculated for each pair of prototype points, as shown in equation 4. Here, $d(x,y)$ is the Euclidean distance between x and y.

$$\rho_{i|j}(x) = \frac{d^2(x,\lambda_i) - d^2(x,\lambda_j)}{d^2(\lambda_i,\lambda_j)} \qquad (4)$$

Then the *conditional membership functions* are calculated for each pair of prototype points, as shown in equation 5. Here, $\vec{v}_{ji}$ is a vector from λ_j to λ_i, $\vec{v}_{jx}$ is a vector from λ_j to x, and $\vec{v}_{ji} \cdot \vec{v}_{jx}$ is the dot product of the two vectors.

$$\mu_{i|j}(x) = \begin{cases} 0; & \rho_{i|j}(x) \geq 1-\sigma \\ 1; & \rho_{i|j}(x) \leq \sigma-1 \\ \dfrac{\vec{v}_{ji} \cdot \vec{v}_{jx} - \dfrac{\sigma}{2} \cdot d^2(\lambda_j, \lambda_i)}{(1-\sigma) \cdot d^2(\lambda_j, \lambda_i)}; & \text{otherwise} \end{cases} \tag{5}$$

Finally, the degree of membership $\mu_i(x)$, of measured input x, is given by equation 6, where M is the number of fuzzy sets in the partition.

$$\mu_i(x) = \frac{\displaystyle\prod_{j=1 \neq i}^{M} \mu_{i|j}(x)}{\displaystyle\sum_{k=1}^{M}\left(\prod_{j=1 \neq k}^{M} \mu_{k|j}(x)\right)} \tag{6}$$

Hypertrapezoidal fuzzy membership functions have proved to be a valuable asset for complex decision aiding problems. The generalized architecture adapted from the *ASTRA* system for monitoring and decision aiding systems is illustrated in Figure 5. HFMFs can play an invaluable role in determining the operating condition of a complex system, as shown in the following section.

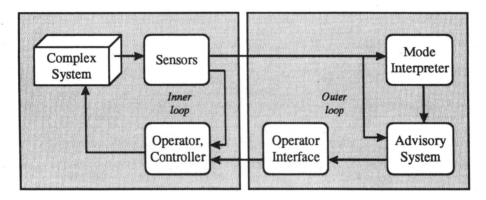

Figure 5. The generalized architecture for monitoring and decision aiding.

4. Examples in Flight Mode Interpretation

The development of hypertrapezoidal fuzzy membership functions was motivated by an on-going research effort in aircraft flight management. HFMFs are being used to infer the operational flight mode of an aircraft from sensor data. HFMFs play the role of the "Mode Interpreter" of Figure 5. This section details some of the results of this research.

Examples of hypertrapezoidal fuzzy membership functions designed for the *ASTRA* Flight Mode Interpreter are shown in Figure 6 through Figure 9. The HFMFs are three-dimensional fuzzy sets defined in the state space of altitude, indicated airspeed, and climb rate. The HFMFs partition the state space into operational flight modes. The flight modes modeled in this example are *takeoff, climbout, cruise, initapp* (initial approach), *finalapp* (final approach), and *landing*.

Unfortunately, fuzzy sets defined on three dimensions can not be illustrated on paper since the degree of membership, $\mu(x)$, must be plotted on the third axis of a three-dimensional plot. Therefore, only two inputs can be plotted at a time. In the following examples, the HFMFs are projected onto the axes of indicated airspeed (IAS) and climb rate (ROC). The third input, altitude above ground, is varied from 500 feet in Figure 6 to 4500 feet in Figure 9.

Notice in Figure 6 that at a lower altitude, given a positive rate of climb, the flight modes *takeoff* and *climbout* fill the state space. Similarly, a negative rate of climb would indicate that the current flight mode could be characterized by either *finalapp* or *landing*, depending on the indicated airspeed.

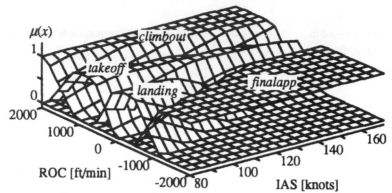

Figure 6. HFMFs for an altitude of 500 feet above ground.

As the altitude increases (see Figure 7), the flight modes *landing* and *takeoff* disappear from the state space and are replaced by *climbout*, *initapp*, and *finalapp*.

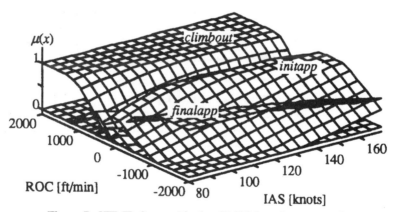

Figure 7. HFMFs for an altitude of 1500 feet above ground.

Figure 8 is a plot of the sample HFMFs at an altitude of 3000 feet. Notice that as the rate of climb approaches zero for a flight, the inferred mode becomes *cruise*.

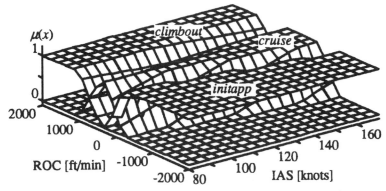

Figure 8. HFMFs for an altitude of 3000 feet above ground.

Figure 9 shows that as the altitude increases (in this case to 4500 feet), the fuzzy set modeling *cruise* expands to include more of the state space.

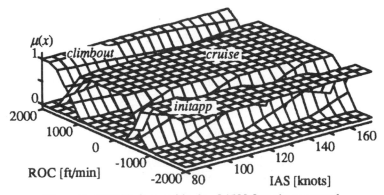

Figure 9. HFMFs for an altitude of 4500 feet above ground.

The important point to draw from Figure 6 through Figure 9 is that one set of prototype points can be used to build multidimensional fuzzy sets which model correlated regions in state space.

5. Multidimensional FMI Performance

The measure of the Flight Mode Interpreter's performance is how closely the inferred modes match the modes that a pilot would assign to each stage of a flight. Figure 10 is a plot of a test flight recorded in the Texas A&M Engineering Flight Simulator. The y-axis of the plot is the flight mode. The test flight of Figure 10 was flown by a U.S. Army test pilot.

During the test flight, the pilot informed the flight test engineer of the time of each transition into the predefined flight modes. The pilot-indicated flight mode is the standard by which the Flight Mode Interpreter's performance is measured. It is plotted as a solid line. The inferred flight mode is plotted as a dotted line and was obtained using the HFMFs plotted in Figure 6 through Figure 9.

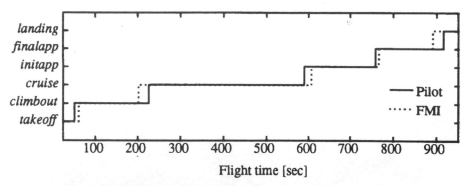

Figure 10. Comparison of FMI output with pilot-indicated flight mode.

The exact time of a transition from one mode to another is somewhat subjective. It is also dependent on pilot flying style, weather conditions, etc. Therefore, while the transitions between flight modes in Figure 10 do not match exactly, the Flight Mode Interpreter sufficiently provides a meaningful, real-time assessment of the aircraft's state. This real-time flight mode inference allows for dynamic, mode-driven cockpit displays and mode-dependent advice.

6. Conclusion

Hypertrapezoidal fuzzy membership functions have proved to be a valuable technique for inferring a qualitative assessment of an aircraft's flight-operational state. In fact, the described experiments have shown that using three-dimensional fuzzy membership functions, a state space can be partitioned into six operational flight modes. The ability of an on-board avionics computer to maintain an awareness of an aircraft's state enables dynamic, mode-driven cockpit displays and mode-dependent, on-board advice. HFMFs help meet the goal of an on-going research effort at Texas A&M University – reducing the possibility of information overload in the general aviation cockpit.

An extension to HFMF design allows for additional control over the shape of the membership functions. The authors have made use of *prototype vectors* for specifying the shape of the flight modes in multidimensional state space. In the flight mode interpretation problem, prototype vectors are a convenient method to directly incorporate "expert opinion" into the definition of the HFMFs.

While the development of HFMFs was motivated by the flight mode interpretation problem, the authors feel strongly that many other applications could benefit from this new technique. HFMFs are a theoretical advancement in the practical application of fuzzy logic to engineering problems. The fuzzy logic community understands the limitations inherent in designing systems based solely on one-dimensional fuzzy sets. HFMFs model the correlation that exists between the variables of a complex system.

While HFMFs are a convenient mechanism for specifying multidimensional fuzzy membership functions, determining the desired shape, and translating that into a set of prototype points is still quite a challenge, especially as dimensionality increases. Therefore, future papers on this subject will focus on the ease with which training and adaptation can be incorporated into systems based on HFMFs. Because each fuzzy set is parameterized by prototype points, training and adaptation will become one of the most useful aspects of HFMFs. The design of the HFMFs used in this paper was accomplished through some preliminary attempts at automatic training of multidimensional fuzzy sets.

References

[1]   Ward, D. T., Painter, J. H., 1995, *Smart Cockpit Computing Technology*, Task #-999903-108, Proposal to the State of Texas Advanced Technology Program, 1996-1997 .

[2]   Anonymous, 1993, *Final Report of the Pilot's Associate Program*, Technical Report WL-TR-93-3090, Lockheed Aeronautical Systems Co., Wright Patterson Air Force Base, Ohio, (export restrictions- limited distribution).

[3] Painter, J. H., Glass, E., Economides, G., Russell, P., 1994, *Knowledge-Based Processing for Aircraft Flight Control*, NASA Contractor Report 194976, Langley Research Center, Hampton, VA.

[4] Kelly, W. E., Painter, J. H., 1996, Soft Computing in the General Aviation Cockpit, *Proceedings of the First On-line Workshop on Soft Computing*, On the Internet, served by Nagoya University, Japan, pp. 151-156.

[5] Bellman, R. E., Zadeh, L. A., 1970, Decision-Making in a Fuzzy Environment, *Management Science*, No. 4, pp. 141-164.

[6] Kelly, W. E., 1997, *Dimensionality in Fuzzy Systems*, Ph.D. Dissertation, Texas A&M University, Department of Electrical Engineering, College Station, Texas.

[7] Hancock, J. C., Wintz, P. A., 1966, *Signal Detection Theory*, McGraw-Hill Book Co., p. 35.

[8] Kelly, W. E., Painter, J. H., 1996, Hypertrapezoidal Fuzzy Membership Functions, *Proceedings of the 5th IEEE International Conference on Fuzzy Systems*, New Orleans, Louisiana, pp. 1279-1284.

[9] Berenji, H. R., Khedkar, P. S., 1993, "Clustering in Product Space for Fuzzy Inference," *Second IEEE International Conference on Fuzzy Systems*, San Fransisco, CA, pp. 1402 - 1407.

[10] Wang, L. X., 1994, *Adaptive Fuzzy Systems and Control*. Englewood, NJ: PTR Prentice Hall.

[11] Foster, G. T., Khambhampati, C., 1994, "Optimal Set Placement and Multi-dimensional Fuzzy Sets for Fuzzy Logic Controllers," *Proceedings of the International Conference on Control'94*, pp. 658 - 663.

An Iterative Two-Phase Approach to Fuzzy System Design

F. Abbattista[1], G. Castellano[2] and A. M. Fanelli[1]

1 Universita' degli Studi di Bari, Dipartimento di Informatica
Via E. Orabona, 4 - 70126 Bari - ITALY
2 Istituto Elaborazione Segnali ed Immagini - C.N.R.
Via Arnendola, 166/5 - 70126 Bari - ITALY
E-Mail:fabio@gauss.uniba.it, casta@iesi.ba.cnr.it, fanelli@gauss.uniba.it

Keywords: Fuzzy Systems, Genetic Algorithms, Neural Networks, Neuro-fuzzy systems.

Abstract

This paper proposes an integrated approach to rule structure and parameter identification for fuzzy systems. The rule structure problem is formulated as a structure reduction process of the neuro-fuzzy network used to model a fuzzy system and is solved through an iterative algorithm aiming at selecting the minimal number of rules for the problem at hand. The parameter identification problem is solved, as an optimization problem, by means of a genetic algorithm. The integrated algorithm allows manipulation of a fuzzy system to minimize its complexity and to preserve its level of accuracy. Experimental results demonstrate the algorithm's effectiveness in identifying reduced fuzzy systems with equivalent performance to the original one.

1 Introduction

In the design of a fuzzy system, the problems of structure and parameter identification have to be solved. The first one essentially concerns the determination of the structure of the fuzzy rules so as to minimize both the output error and the complexity of the underlying model. The second problem refers to the tuning of the parameters associated with membership functions and is merely an optimization problem with an objective function. Though several method, such as gradient-based algorithms [I], [2], [3] and genetic algorithms [4], [5], [6] have been proposed to solve the latter problem, the identification of rule structure is still an open question that usually turns out to be a trade-off between the complexity of description (number of rules) and the resulting accuracy of the fuzzy model.

This paper deals with both rule structure problem and parameters tuning, which are solved under the requirement of simplifying a fuzzy system once a satisfactory structure is available.
Specifically, we transform the fuzzy model into a neural network form as usual for many neuro-fuzzy approaches [7], [8], [9], [10]. In the first phase, we attempt to reduce the structure of the neural network to obtain a simplified fuzzy system with unchanged performance. This is accomplished by adapting the complexity reduction method developed for neural networks in [11], [12], [13], so that it can handle the rule structure optimization problem for fuzzy systems. The second phase, based on genetic algorithms (GA) [14], [15] aims to optimize the membership function parameters. The two phases are iterated until a stopping condition is met. The resulting method is able to select the minimal number of necessary rules in a fuzzy system with a satisfactory level of accuracy.

The outline of the paper is as follows: section II introduces the adopted neuro-fuzzy model; in section III we presents the approach; section IV shows an example to demonstrate how the proposed method succeedes in simplifying the fuzzy-neural model while keeping unchanged the performance over the training data. Conclusions are summarized in section V.

2 The Neuro-Fuzzy Network

Let us consider a fuzzy system with n inputs $x_1, ..., x_n$ and and m outputs $y_1, ..., y_m$.
We assume to adopt singleton fuzzification, product inference rule and center average defuzzification. Under these assumptions, the output fuzzy set of each fuzzy rule is a singleton.
The network adopted to model this fuzzy system is a four-layer feed-forward network (fig. 1) also used in [16].

The first layer, denoted with L_I is the input layer, that consists of n units representing the linguistic input variables. No computation is done by these nodes.

Units in the second layer L_F, called the fuzzification layer, compute the membership degree for an input vector $\mathbf{x}$. This layer is composed of n groups, each including R neurons. Each unit $i_j \in L_F$ computes the membership value of the ith input variable x_i for the ith fuzzy term in the jth rule. With a gaussian membership function, the output of neuron $i_j \in L_F$ is:

$$\mu_{ij}(x_i) = \exp(-net_{ij})$$

where $net_{ij} = \left(\dfrac{x_i - w_{ij}}{\sigma_{ij}} \right)^2$ is its net input. The weight w_{ij} between unit $i \in L_I$ and unit $i_j \in L_F$, and the parameter σ_{ij} represent respectively the mean and the width of the membership function μ_{ij} for the ith input variable in rule j.

The third layer L_R performs the inference of rules by means of product operator. The number of nodes in this layer is equal to the number of rules in the fuzzy system. The output of neuron $j \in L_R$ is

$$\mu_j(x) = \prod_{i \in L_I} \mu_j(x_i)$$

that corresponds to the activation strength of the jth rule for input $\mathbf{x}$. Note that neurons in layer L_R are fixed, as no modifiable parameter is associated with them.

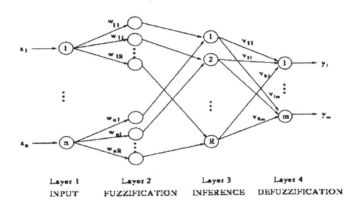

Layer 1   Layer 2   Layer 3   Layer 4
INPUT   FUZZIFICATION   INFERENCE   DEFUZZIFICATION

Figure 1. Architecture of the neuro-fuzzy network

The fourth layer L_D performs the defuzzification phase by means of the Center of Area method. Given an input vector $\mathbf{x}$, the output of unit $k \in L_D$ is:

$$y_k(x) = \frac{net_k}{\displaystyle\sum_{j \in L_R} \mu_j(x)}$$

where

$$net_k = \sum_{j \in L_R} \mu_j(x) v_{jk}$$

is the net input of unit $k \in L_D$. Weight v_{jk} connecting unit $j \in L_R$ to unit $k \in L_D$ represents the fuzzy singleton of the kth fuzzy output variable in the jth rule.

The jth fuzzy rule can be be written as:

$$\text{IF } (x_1 \text{ is } w_{1j}) \wedge ... \wedge (x_n \text{ is } w_{nj})$$

$$\text{THEN } (Y_1 \text{ is } v_{j1}) \wedge ... \wedge (Y_m \text{ is } v_{jm})$$

Thus fuzzy rules are defined once the premise parameters w_{ij}, σ_{ij} and the consequence parameters v_{ik} are determined. This can be accomplished by training the network through a back-propagation technique like in [16].

3 The designing approach

The proposed approach integrates a rule structure definition phase, in which the smallest set of rule is found, with a parameter definition phase aiming at the parameter optimization for the reduced network, in order to safe the accuracy of the resulting network.
In this section we first describe the two phases separately and then we briefly present the combined approach.

3.1 The rule selection phase

We solve the rule selection problem by means of a systematic procedure that sequentially removes rules and investigates the fuzzy models with fewer rules. When a system with a satisfactory rule set is given, a rule is identified to be removed and the remaining ones are updated so as the performance of the reduced system remains approximately unchanged.

Our rule selection algorithm starts from a neuro-fuzzy network trained for satisfactory performance, and iteratively produces networks with fewer rule nodes, by means of a step by step rule elimination.
Figure 2 shows the effect of removing a rule node in a single step of the algorithm. It can be noted that the elimination of a rule node leads to the elimination of the associated term nodes, therefore the size of the L_F layer is also reduced.

The selection algorithm works as follows:
$t := 0$;
Repeat
  1. identify the unit $h \in$ to be removed
  2. remove associated term nodes $i_h \in L_F$ with relative connections $\{ih\}_{i=1..n}$
  3. remove connections $\{hk\}_{k \in L_D}$
  4. update remaining weights $\{v_{jk}\}_{k \in L_D}$ according to

$$v_{jk}^{(t+1)} := v_{jk}^{(t)} + \delta_{jk} \quad \text{for } j \neq k$$

where δ_{ij}'s are appropriate adjusting factors
5. $t := t + 1$;
until *stopping condition* is met.

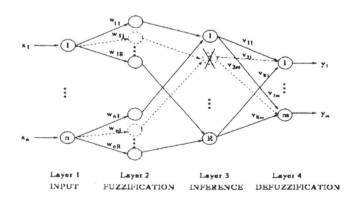

Figure 2. Elimination of a rule during a step of the selection algorithm

The update quantities δ_{ij}'s are derived by imposing that, after the elimination of unit $h \in L_R$, the net input net_k of each output node $k \in L_D$ remains approximately unchanged.
This amounts to requiring that, for each training pattern $\mathbf{x}_p$ and for each node $k \in L_D$, the following relation holds:

$$\sum_{j \in L_R} v_{jk} \mu_j(x_p) = \sum_{j \in L_R - \{h\}} (v_{jk} + \delta_{jk}) \mu_j(x_p) \tag{1}$$

Simple algebraic manipulations yield the following linear system:

$$\sum_{j \in L_R - \{h\}} \delta_{jk} \mu_j(x_p) = v_{hk} \mu_h(x_p) \tag{2}$$

The quantities δ_{ij}'s are then computed by solving the linear system (2) in the least-squares sense throught an efficient preconditioned Conjugate-Gradient method [18].

The criterion for identifying the rule to be removed at each step has been suggested by the adopted least-squares method. Such a method provides a better solution with faster convergence if the system being solved has a small known term vector (in terms of Euclidean norm). Since in system (2) the known terms depend essentially on the unit h being removed, our idea is to choose the unit for which the norm of the known term vector is minimum.

The *stopping condition* has been left pourposely undefined in the algorithm since it can be defined in different ways, according to the user requirements. For example, if the application requires keeping the original behavior of the system over the training data, the algorithm will be stopped as soon as the performance of the reduced model worsens significantly over the training set. As well, if a good generalization ability is requested to the simplified neuro-fuzzy system, a stopping contition that takes into account the performance over the test set can be used, regardless of the behavior over the training data. Also, the algorithm could be stopped after a predetermined number of rules have been removed.

3.2 The parameter tuning phase

The parameters optimization phase consists of identify the better distribution of weights between the two layers L_I and L_F of the reduced neuro-fuzzy network produced in the first phase. This is solved by means of a GA. Genetic algorithms are parallel search procedures inspired from the mechanisms of evolution in natural systems [14], [15]. Genetic algorithms work with a constant-size population of points, called individual or chromosomes. Every individual is associated with a fitness value that indicates its probability of surviving in the next generation; the higher the fitness, the higher the probability of survival. In all the optimization problems the fitness of individual must be related to the corresponding value of the function to be optimized. In our application, each individual represents the weights between the layers L_I and L_F of the reduced neuro-fuzzy network, being each weight a real number. The GA starts with an initial population, randomly chosen, and evolves it by means of three simple operator: reproduction, cross-over and mutation. In the present case, we inserts in the initial population a copy of the reduced network, produced in the rule selection phase. The reproduction operator, in its simplest form, is implemented to select individual to be copied in the next generation according to their fitness value. Specifically, the probability of an individual i to be reproduced is given by:

$$p_i = F_i / \sum_k F_k$$

where F_k represents the fitness value of the kth individual. After reproduction, the cross-over operator is applied between pairs of selected individual in order to produce new offspring individual. The simplest form of cross-over, single point cross-over, operates in two steps. First, two member are randomly chosen; next a cut point is determined randomly and the corresponding right-hand parts are swapped. The cross-over operator is applied with a frequency controlled by a fixed parameter p_c. Doing so, the cross-over does not create new weights, but simply exchanges them between individuals. The production of new weights is the task of the mutation operator, which randomly change the value of every gene of the chromosome, with fixed probability p_m, adding (algebraic sum) a small real value to the weight.

In their basic form, GAs are maximization procedures as they favor high fitness value. However, the goal of the parameters tuning phase is to minimize the error of the neuro-fuzzy network. To apply GAs to this problem we defined the fitness of individual i as:

$$F_i = E \max - E_i$$

where Emax is the maximum error allowed for a network, and E_i represents the error of the individual i. It has to be noticed that, as an individual represents only the weights between the first two layers of the neuro-fuzzy network, to evaluate the error of every individual, the connections among the other layers of the network are the same for all the individuals, and they are taken from the reduced network obtained in the first phase.

3.3 The two phase approach

The combination of the two phases is performed in an iterative manner. Starting with an initial network N, the rule selection phase attempts to reduce the number of rules of the network. As described in section 3.1, the first phase produces reduced networks from which to choose the smaller one with the same accuracy of the original network N.

In order to further optimize the network design, we select, from the reduced networks obtained, the first one (W) with an error value worse than the original one for more than a fiexd tolerance value. This network is inserted in the initial population of the genetic parameters optimizer. At the end of the GA run, the final result represents the network W optimized in order to have the same accuracy of the original network N.

The two phases can be iterated, substituting the original network N in the first phase with the network W produced by the second phase, until a stop criterion is met.

4 Experimental results

To verify the algorithm's effectiveness in identifying a fuzzy system with simplest structure, we considered a problem of non linear plan identification, also used in [16] and [17].

The plant to be identified is described by the following second-order difference equation:

$$y(k+1) = f[y(k), y(k-1)] + u(k) \tag{3}$$

where the form of the unknown function f is

$$f[y(k), y(k-1)] = \frac{y(k)y(k-1)[y(k)+2.5]}{1+y^2(k)+y^2(k-1)} \tag{4}$$

and $u(k)$ is a random signal uniformly distributed in $[-2,2]$.

To learn the output (3) of the plant, a neuro-fuzzy network with two inputs and one output was considered and a training set of 100 points was generated with $u(k) = \sin(2\pi k/25)$.

We applied our algorithm to a network with 40 rule nodes trained for 200 epochs, as reported in [17], where a similarity based technique is used to reduce the rule structure. Fig. 3 shows the output of the plant as well as the output of the trained network for $u(k) = \sin(2\pi k/25)$.

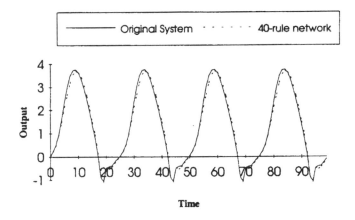

Figure 3. Output of the original system and the 40-rule neuro-fuzzy network

We adopted the following error measure:

$$E = \frac{\sqrt{\sum_{k=1}^{m} \sum_{p=1}^{P} (t_k^p - y_k(x^P))^2}}{\sum_{k=1}^{m} \sum_{p=1}^{P} |t_k^p|}$$

where t_k^p is the kth target component in the pth training sample.

In the reported experiment, the GA was run for 500 generations, using 50 individuals in the population, with a cross-over probability equal to 0.7 and a mutation probability of 0.01.

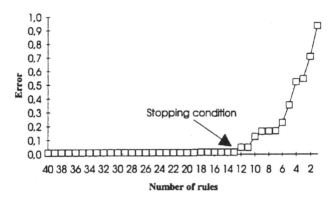

Figure 4. Performance of the neuro-fuzzy network during the rule reduction process

Figure 4 shows the error E of the reduced networks obtained during the rule selection phase with no stopping criteria. It can be noted that the procedure, if stopped when an increment of 0.01 or more is observed in the error E, is able to reduce the model from 40 to 13 rules with no degradation in accuracy. Stopping here, results in a great complexity reduction since the number of adjustable parameters of the neuro-fuzzy model is reduced from 200 (=40 x 5) of the original 40-rule model to 65 (=13 x 5).

Next, we selected the 12-rule network (that is the fitrst network found with an error greater than 0.01 in respect of the error of the original 40-rule network) and insert it in the GA population. We run the genetic optimizer and, at the end, we selected the better individual found in all the generations. The error value for this individual was 0,01 (cmp. fig. 4). This network has been used in the rule selection phase to further reduce it. After 2 iteration of the two phases, the procedure was able to find a 3-rule network with the same accuracy of the original 40-rule network, see figure 5, and an error value of 0.01.

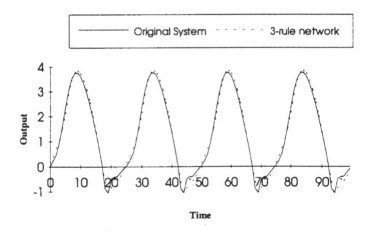

Figure 5. Output of the original system and the 3-rule neuro-fuzzy network

Finally, we compared our method and the method proposed in [17] in terms of complexity reduction. As it can be seen from table 1, our final model is much simpler. Moreover, our method provides a reduced well performing system with no need of retraining (fig. 5). Conversely, 100 epochs of learning are further required in [17] to have a simple system (22 rules) with a desiderable output.

Network	#rules	#parameters
original	40	200
reduced by method [17]	22	84
reduced by our method	3	15

Table 1. Comparison of the neuro-fuzzy models in terms of complexity

120

5 Conclusions

The approach to fuzzy system design developed in, combines the problems of structure and parameter identification of fuzzy models.

The main advantage of our selection algorithm is the iterative nature: at each step the designer can evaluate the reduced model in terms of tradeoff between accuracy and complexity and retain it as a potential final model.
The iterative nature of the algorithm allows the network's designer to evaluate a number of potential simpler models and to define a stopping condition according to specific requirements (i.e. accuracy of solution or structure complexity).

Experimental results have shown that the proposed approach is able to produce simpler models with very efficient results. Further work is in progress to test the approach on different practical problems.

References

[1] H.R. Berenji and P Khedkar, 1992, Learning and tuning fuzzy logic controllers through reinforcement, *IEEE Transactions on Neural Networks*, 3(5):724-740.

[2] Y. Shi, M. Mizumoto, N. Yubazaki, and M. Otani, 1996, A learning algorithm for tuning fuzzy rules based on gradient descent method, *Proceedings of the Fifth IEEE International Conference on Fuzzy Systems,*New Orleans, Louisiana, September, pp. 55-61.

[3] E. Blanzieri and A. Giordana, 1996, An incremental algorithm for learning radial basis function networks, *Proceedings of the Fifth IEEE International Confrrence on Fuzzy Systems*, New Orleans, Louisiana, September, pp. 667-673.

[4] R. J. Stonier and M. Mohammadian, 1994, Tuning and optimization of membership functions of fuzzy logic controllers by genetic algorithms, *IEEE International Workshop on Robot and Human Communication (RO-MAN'94)*, Nagoya, Japan, July , pp. 356-361.

[5] C.C. Wong and S.M. Feng, 1995, Switching-type fuzzy controller design by genetic algorithms, *Fuzzy Sets and Systems*, 74(2): *175-185.*

[6] C. Perneel, J.M. Themlin, J.M. Renders, and M. Acheroy, 1995, Optimization of fuzzy expert systems using genetic algorithms and neural networks, *IEEE Transactions on Fuzzy Systems*, 3(3):300-312.

[7] J.S.R. Jang and C.T. Sun, 1995, Neuro- fuzzy modeling and control, *Proceedings of the IEEE*, 83(3):378-406.

[8] C-T. Lin and C.S.G. Lee, 1991, Neural-network-based fuzzy logic control and decision system, *IEEE Transactions on Computers*, 40(12): 1320-1 336.

[9] S. Horikawa, T. Furuhashi, and Y. Uchikawa, 1992, On fuzzy modeling using fuzzy neural networks with the back propagation algorithm, *IEEE Transactions on Neural Networks*, 3(5)801-806.

[10]H. Takagi, N. Suzuki, T. Koda, and Y Kojima, 1992, Neural networks designed on approximate reasoning architecture and their applications, *IEEE Transactions on Neural Networks*, 3(5):752-760.

[11]G. Castellano, A.M. Fanelli, and M. Pelillo, 1997, An iterative pruning algorithm for feedforward neural networks, *IEEE Transactions on Neural Networks*, to appear.

[12]G. Castellano, A.M. Fanelli, and M. Pelillo, 1994, Pruning in recurrent neural networks, *Proceedings of International Conference on Artificial Neural Networks (ICANN'94)*, Sorrento, Italy, May, pp. 451-454.

[13]G. Castellano, AM. Fanelli, and M. Pelillo, 1995, Iterative pruning in second-order recurrent neural networks, *Neural Processing Letters*, 2(3): 5-8.

[14] J.H. Holland, 1992, Adaptation in Natural and Artificial Systems, MIT Press, USA..

[15] D.E. Goldberg, 1989, Genetic Algorithms in Search, Optimization, and Machine Learning, Addison-Wesley.

[16] K.S. Narenda and K. Parthasarathy, 1990, Identification and Control of Dynamical Systems using Neural Networks, *IEEE Trans. on Neural Networks*, 1(1):4-27.

[17] C.T. Chao, Y.J. Chen, and C.C. Teng., 1996, Simplification of fuzzy-neural systems using similarity analysis, *IEEE Transactions on Systems, Man and Cybernetics*, 26(2):345-354.

[18] A. Bjiorck and T.Elfving, 1979, Accelerated projection methods for computing pseudoinverse solutions of systems of linear equations, *BIT*, 19:145-163.

A Fuzzy Control Course on the Internet

Jan Jantzen[1], Mariagrazia Dotoli[2]

[1] *Technical University of Denmark, Dept. of Automation, bldg. 326, DK-2800 Lyngby, Denmark, Tel +45 4525 3561, fax +45 4588 1295, e-mail jj@iau.dtu.dk*

[2] *Politecnico di Bari, Dipartimento di Elettrotecnica ed Elettronica, Via Re David, 200, I-70125 Bari, Italy, Tel +39 80 5460 312, fax +39 80 5460 410, e-mail dotoli@poliba.it*

Keywords: Education, distance learning, automatic control, pendulum, Matlab

Abstract

A course in fuzzy control has been offered on the Internet (http://www.iau.dtu.dk/~jj/learn) since late 1995. Now that about 100 students have taken the course it is possible to report some results and experiences. The course concerns fuzzy logic for automatic control. The objectives are to teach the basics of fuzzy logic, to show how to use fuzzy logic, and to teach how to design a fuzzy controller. A ball balancer, an inverted pendulum problem, acts as a case study implemented in a software simulator in Matlab. The paper aims at teachers who might be interested in running a similar course, and it presents aspects of the development, the delivery, and the use of the course. The course differs from other distance learning courses in the close teacher-student interaction based on e-mail. It has been set up and run by one person.

1. Introduction

For years, people in Greenland, the Faroe Islands, and at sea have participated in Danish long-distance courses. These were until the late 1980s correspondence courses on paper. It was convenient to be able to study in remote, rural areas, but the response times were discouraging, depending on weather conditions and the mail service. With the event of the datanets *Earn* and *Bitnet*, the courses could be e-mail based, and the *Royal Danish School of Educational Studies* together with *IBM Denmark* started to build an educational system in 1987 [6]. Now distance learning can be offered to more people on the *Internet* and the *World Wide Web*.

Distance learning is rather widespread, most notably in the UK, USA, and Australia. Universities are offering courses for distance learners and open university courses, for example the *University of Southampton* which offers entire degree programmes [22]. The *Open University* in the UK has existed for more than 25 years, and they have developed a range of courses in engineering subjects [1]. There are also multi-institutional efforts, for example the *Collaboration for Interactive Distance Visual Learning* (CIDVL), which is built on a two-way video conferencing system. CIDVL consists of 11 universities and corporations in the USA seeking to create a virtual university environment in the field of engineering [7]. Over ten thousand courses and degree programmes are listed in the Globewide Network Academy [16].

There are several organisations in Europe for distance learning [2], [8]. The *International Council for Distance Education* was created as early as 1938. Among other things, its objectives are to set standards of practice, and to provide information through *Open Praxis*, their bulletin. *EUROSTEP* is a non-profit organisation of institutions and companies using multimedia and satellites for education and training across Europe. In *EUROPACE* 2000 members will have PhD satellite seminars, PhD ISDN seminars, computer conferencing, and distribution by mail/satellite. EADTU stands for the *European Association of Distance Teaching Universities*. They wish to establish a clearing house function, courses and programmes, and a qualifications framework. The last item should guarantee quality standards in European open and distance learning with a view to assessment and certification. In France there is the *Centre national d'enseignement a distance* (CNED), the *Federation interuniversitaire des enseignements a distance* (FIED) [17], and *Conservatoire national des arts et metiers* (CNAM). Canada has the *Teleuniversite du Quebec*, one of the oldest in distance learning. The major motivation for an online course is flexibility, but there are other key issues as well in the development, delivery and use of such a course. These include costs, technology, and credit unit transfer.

2. Key issues

The initial development of the course should be inexpensive, or virtually cost free for the university. It was deemed necessary to develop and test a prototype on one or two students, before launching it on the Internet, and to adjust the course on the fly. With higher visibility it is easier to apply for funding, for instance from the European Commission. From the initial developments in the *DELTA* exploratory action to the *Telematics* applications programme, there have been many experimental projects and networks [9]. The current *SOCRATES* programme has a sub-programme for *Open and Distance Learning* (ODL) [19]. Some of its objectives are to develop pedagogical frameworks, to improve the quality of organisational ODL frameworks and of the pedagogical materials, to enhance the ODL skills of teachers, and to encourage the recognition of qualifications obtained through ODL. Projects can be found in the *CORDIS* database [20].

The delivery of the course should be based on fairly simple technology. Main requirements are short response times and dependable communication links. The technology used for Internet courses may include animation, audio, inline video, remote control via Internet, and Java [3]. For instance, the Open University complements printed texts with BBC radio and television broadcasts, audio- and video-cassettes, home experiment equipment, computing packages and laboratory-based summer schools, and they have created an Interactive Media Group, that produces multimedia tools such as CD ROMs. The *Lotus Notes* software provides — apart from standard facilities — group communications, database management, and applications development [7].

The main issue is to get students to use it. This depends among other things on the possibility of credit unit transfer. A way to measure and transfer units from one institution to another is the *European Credit Transfer System* (ECTS), developed in order to guarantee academic recognition of studies abroad [18]. ECTS was initially established under the Erasmus programme (1988-1995), and it has been tested over a period of 6 years in a pilot scheme involving 145 higher education institutions. A total of 38 new universities and 36 non-university institutions implements ECTS during 1996-1997.

3. Methodology

The course is based on the framework developed at the Royal Danish School of Educational Studies. There are three main components of the course:

- a textbook on paper,
- a simulator (software), and
- online lessons

The textbook material existed beforehand, but the process of building the course forced much more modularity and organisation into the material. The course had to be broken down into lessons, and each lesson had to be linked to a certain part of the textbook. Incidentally, transparencies for lectures or external presentations follow the same organisation.

The simulator is written in *Matlab*. To make it as independent as possible, the simulator only requires the student edition of Matlab [10] and no extra toolboxes. If the student does not have access to Matlab, it is necessary to buy the student edition. The students use a variety of platforms, but the simulator runs on all without problems (after some initial adjustments to the code). The case study is an inverted pendulum problem, a ball balancer, because this type of problem is widely used for education and benchmarking. It is guaranteed free of computer virus, since all programmes are ascii files. The simulator has limits, however, since it is a home-written piece of software lacking all the facilities of a commercial package. It runs slowly on a PC, and in the long run it seems rather monotonous.

Each lesson is a text file to be downloaded from the Web. The student writes the answers straight into the file, and sends it back to the instructor inside an e-mail note. There are 12 short files to download. It is recommended not to download all files at once, because the teacher may change or correct the contents during the course as a result of feedback from the students. The course has one Web page, which describes the course objectives, contents, and work mode. The Web page has links to the lesson files. Once a lesson is downloaded, the student does not depend on the Web anymore, and thus only needs an Internet access to do the lesson. The teacher does not need Web access either, and this makes it simpler to teach away from the home base.

For each lesson, students

1. download the lesson as a text file;

   2. include it inside an e-mail;

   3. work with it and make notes about results and comments they want to pass to the instructor; and

   4. write their answers directly into the file and e-mail it to the instructor.

The instructor will comment on their solution and return the e-mail note to the student. It will be in the mail box the day after, or within 48 hours at the latest. The instructor may urge the student to proceed with new questions related to the topics in the lesson. A lesson may therefore continue after they submit their solution; sometimes the first solution will spawn a suite of activities afterwards. They must communicate with the instructor at least once every fortnight; either by sending solutions or by e-mail explaining their situation. Drawbacks in this kind of procedure are delays caused by traffic on the Internet, break downs of connections, as well as misunderstandings, often caused by having to communicate at a distance being limited to simple text messages.

The delivery of the course promotes a deeper learning compared to lectures. An online lesson forces the student to be active and alert. Student responses are written and not oral, so they have to be precise and well thought through. The teaching style is like tutoring, because the communication sessions are iterative. Compared to a three days crash course, the learning is deeper, simply because it is extended over a period of, say, six weeks to six months.

During this period the teacher and the student communicate by e-mail: messages are sent back and forth containing questions, remarks, ideas and answers regarding the lessons. Graphics and national characters are a problem since the electronic messages are ascii files; for example the special Danish characters (æ, ø, å) created so many problems, that it was a relief to translate the course into English. Graphics will have to be transmitted separately or by fax. This simplifies the course for people with no experience in handling e-mail advanced options.

The lessons reflect the three step learning model that the course uses,

   1. read some textbook material,

   2. do an exercise, and

   3. solve some problems.

The steps are increasingly activating for the student. In the first step the student reads in the textbook, offline during spare hours at home, or perhaps in the bus or train. The exercises are related to the simulator and must be done on a PC or workstation. The student does not answer any questions in this step, he/she only has to follow certain steps and observe and explore. The student is of course allowed to ask questions during this step if necessary. The third step is a matter of solving four to ten specific problems related to the text and the simulator together.

4. Outline of the course

The course has twelve lessons, which have been changed and reorganised as a result of student feedback. Most changes came in the prototyping phase and just after the initial launch. The text is stable now, and the list below contains the titles of the lessons.

   0. Course introduction

   1. Introduction to fuzzy control

   2. Analysis: operator's knowledge

   3. Fuzzy basics

   4. Design: defuzzification, inference

   5. Design: choices, control surface, gains

   6. Implementation: Matlab, Pendulum, cart controller

   7. Test: tune ball controller

   8. Test: tune cart + ball

   9. Self-organizing controller

  10. FUZZY II, Neuro-Fuzzy

  11. Course evaluation

Each lesson consists of a preparation, requiring the reading of the course book and/or the Matlab user's guide, some exercises, usually dealing with simulations of the case study (Figure 3), and an assignment, the most stimulating part and the one requiring more active participation from the student. Lessons end with management questions, such as time spent per person on the lesson, time spent on hardware or software problems and so forth. For details, see the web address http://www.iau.dtu.dk/~jj/learn.

The first step is to answer lesson 0, which requires providing personal data together with information on the student's control and fuzzy logic background. The official start is when DTU receives the Guest Student Form filled in by the student, who by that moment on is officially a foreign DTU student. The next step is to install the simulator.

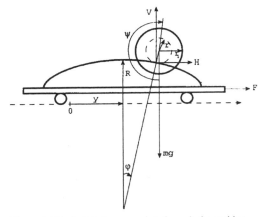

$$\begin{cases} \ddot{y} = a\varphi + bF \\ \ddot{\varphi} = c\varphi + dF \end{cases}$$

$$a = -m^2 r^2 g / (Mmr^2 + (M+m)I)$$

$$b = (I + mr^2) / (Mmr^2 + (M+m)I)$$

$$c = (M+m)mr^2 g / ((R+r)(Mmr^2 + (M+m)I))$$

$$d = -mr^2 / ((R+r)(Mmr^2 + (M+m)I))$$

(I moment of inertia of the ball, M mass of the carriage excluding ball, m mass of the ball)

Figure 1. The ball balancer, related symbols, and its equations after linearisation. (Adapted from [5])

Students receive the textbook material [4], a paper on the course case study [5] and the software simulator PENDULUM by ordinary mail. The simulator may, more easily, also be received by electronic mail as an attachment of its self-unzipping executable version. The programs are Matlab m-files written for the Matlab Student Edition v 4.0 for Windows; but they will also run under the professional edition and under Macintosh, DOS and UNIX. The simulator emulates the well-known case study of the ball and beam system (Figure 1, Figure 2), which is of course an advantage for those who might have come across this plant in laboratory tests. Moreover, one can compare results obtained during the course with those already available in literature when using intelligent or classical techniques to control this system. This can be easily accomplished by studying graphics such as the time and phase plane plots (Figure 3) that are available in the simulator. A cart-ball laboratory model is physically located at the Technical University of Denmark. Another advantage in this simulator is its use of the powerful and widespread calculus software Matlab. In particular, a student edition is available at low cost, but on the other hand the limitation to this edition prevents the use of a more flexible simulator, for instance Simulink based and/or compatible with the Matlab Fuzzy Logic Toolbox. Moreover, the use of such a simulator would be more suitable if coupled with the real laboratory process, which of course depends on the availability of a control laboratory in the student's university. The student edition user's guide [11] is the last part of the paper material necessary for the course.

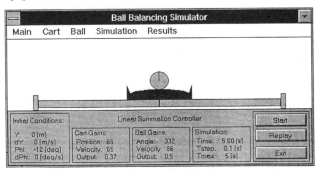

Figure 2. The top level user interface to the simulator.

The course ends when the student submits lesson 11, a lesson of comments and evaluation of the whole course. The *Technical University of Denmark* (DTU) will automatically issue a certificate, rated at 2.5 credit units in the ECTS.

The overall costs of the course are rather low for students, null in most cases. The main cost is the Internet account, for those who do not have free access to the web at the university. Another cost is the Matlab student's

126

edition for those who do not have it at the university. The rest of the course material is all free or paid by DTU, including the simulator.

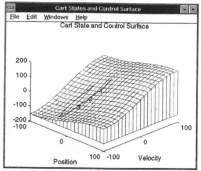

Figure 3. Example of a plot in the simulator.

5. Results and experience

Players in this project are the university, the teacher (first author), and the student (second author); the next sections present some experiences seen from their viewpoints.

5.1 University views on the course

The *Technical University of Denmark* (DTU) aims at involving both national and international aspects in the studies, and therefore emphasises international collaboration. This is partly done by attracting foreign students and teachers, partly by promoting trips abroad for Danish students for one or two semesters. The flow is somewhat unbalanced according to the records, however [24]; for example the outflow two years ago was 242, whereas the intake of foreign students was 136. Distance courses are an alternative way to attract foreign students. The fuzzy control course will probably have 100 students in 1997, of which 75 are foreigners.

The students who have taken or are taking the fuzzy course are from various countries (Figure 4). Denmark is high on the list, because the online lessons are part of a full, local course in fuzzy control. Sweden is high on the list, because of an agreement with the Lund Institute of Technology (Malmo branch) to deliver 40 hours of teaching as a part of a more general control course. Columbia is high because a class of students from Ponteficia Universidad Javeriana have registered for the course due to one student's marketing towards his fellow students.

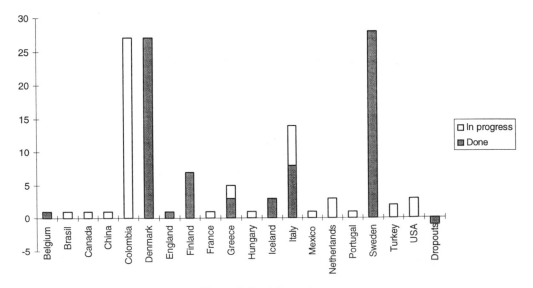

Figure 4. Participants by country.

The university requires now individual assessment of the students according to a relatively new regulation. The fuzzy course can be taken by groups of students, in fact this is recommended for pedagogical reasons, but it is difficult to devise a system for individual testing of the students when they are remote. One solution would be to have an oral examination over the telephone; this would then require an external examiner on the line as well. This is technically feasible, but not convenient when there is a large time zone difference. The best solution is probably to devise a multiple choice examination system, where each student gets a different test, which can be corrected by a program to minimise the work load.

There is no course fee for students, yet they receive a photocopied and soft-bound textbook as well as some software. The direct, running costs comprise printing, binding, postage, diskettes, and telephone charges, which amount to about 25 US dollars per student. The running costs are not high, but the department may not wish to pay for it indefinitely. If not, there will be a lot of administration around billing and accounting which should be allowed for when setting a fee.

5.2 Teacher views on the course

The level of the course is difficult to get right. DTU is just about to classify the course as a PhD course for PhD students; it has turned out that maybe two thirds of the participants are PhD students. This does not mean, however, that MSc students cannot take it. Presently it seems that it fits MSc students best, while PhD students may find the material a little too easy; some of them would like a more interesting case study, a larger one that really shows the strengths of fuzzy control, and more theoretical material. Some BSc students find the course time consuming, not well structured, and lacking entertainment; they expect multimedia in an Internet course.

The major teaching tool, apart from the textbook, is the simulator with its graphical user interface. It has turned out to be appropriate, except perhaps for its slow execution on medium sized PCs. Other tools such as dynamic generation of equations, use of Java applets, interactive tutorials are not part of this course. The focus is on the dialogue with the student, and over time a database of questions, answers, and viewpoints has been built. This has turned out to be an important resource, in the sense that certain problems tend to reoccur, and then it is easy for the teacher to look into the database and pick a paragraph that fits the discussion. Such a database makes it easier to train other teachers to run the course.

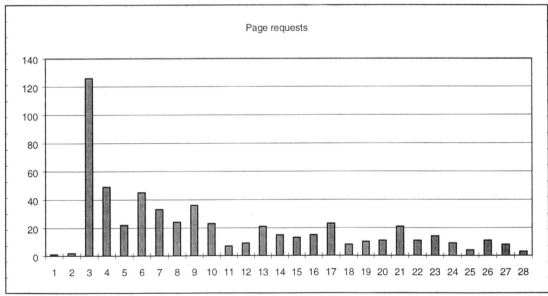

Figure 5. Web page requests around February 1997.

It is interesting to measure the Web activity around the course (Figure 5). The single web page is visited perhaps five times per day. The course has not been advertised anywhere, except for a link from a *MathWorks* page [15], and a link from a list of fuzzy and neuro-fuzzy resources [14]. Students find it themselves, mostly by means of search engines. One day in February the number of requests went up drastically. That was due to an e-mail note that mentioned the course in the fuzzy newsgroup on Internet (comp.ai.fuzzy). The figure shows how the note triggered some interest which levelled off after a month, almost like an impulse response. The steady level of five page requests per day generates roughly one serious inquiry for the course every 14 days. That activity level is almost too high for one teacher, which is why the course has not been advertised in magazines, newsletters, newsgroups, indexes, etc.

The teacher load is concentrated around three activities. The initial development and writing of the lessons is 60 - 200 man hours, if the online course is based on an existing course. The software development is unpredictable, but Matlab provides good facilities for designing a user interface and for plotting. The teaching load when running the course is as a rule of thumb one tenth of the student load; the course is a 60 hours course, so it requires 6 hours of an experienced teacher. There is not, however, an economy of scale; the teaching load is proportional to the number of student teams.

Other distance learning courses in engineering seem to use more advanced technology, while at the same time relying more on self-study. Students get a video, a program, or online textbook material (HTML), and they are asked to solve assignments and perhaps submit them via e-mail. The teacher corrects the assignments, but there is not an on-going dialogue. An exception is video-conferences, but this is more like classroom teaching instead of the tutoring style in the present course. Many courses start and end on fixed dates, opposed to an all year round admission in this course. It is difficult to set up success measures other than the number of students passed, not passed, or dropped out. The final lesson has a student evaluation form, but it is not a suitable basis for a systematic validation of the course.

5.3 Student views on the course

Students' motivation for taking the course is obviously the interest in fuzzy control; in particular the course proves to be helpful for newcomers, who are not able to find a way for their first approach to fuzzy control in the multitude of papers, seminars, conferences and tutorials related to the field. The course is not only helpful for newcomers in fuzzy control, but also for learners who already have an experience in the field. In particular, attention is drawn to two aspects of fuzzy control that textbooks rarely consider from a practical point of view:

1. the tuning of the fuzzy controller control surfaces;
2. the equivalence of a fuzzy and a linear controller.

Another motivation is the desire of a personalised course, with flexible work load and schedule, not requiring the student to move. The motivation issue is rather important, since if teaching in a classroom to unmotivated students is difficult, doing it at a distance is quite impossible.

The main objectives of a student taking this distance course are the achievement of personalised teaching, together with a certain flexibility during the development of the course itself, not forgetting the possibility of having access to knowledge that is physically outside the student's own university.

Firstly, teaching is personalised: there is no class attending the course, just a single student (sometimes a team of up to three students), that decides, together with the teacher, which subjects to focus on, and plans a personalised schedule of activities.

This naturally leads to a certain flexibility in the course development: the student by-passes all academic filters, directly interacts with the teacher and may change the course planned activities and schedule at any time.

Another major objective for the student obviously consists in having access to knowledge outside his university: he does not simply interact with a teacher, but with a foreign one, with a different background and points of view from those the student is accustomed to.

As for the communication tools used in the course, electronic mail is both low-cost and powerful, having the advantages of distance learning such as overcoming physical distance and different working hours, without its drawbacks such as dead times and slowness: even if the Internet can sometimes be a bottleneck, the course is not on line, and while waiting for an answer from the teacher the student can employ that time as he likes. On the other hand, communicating with a foreign teacher, at a distance, might cause misunderstandings, which is the reason why sometimes an e-mail with a lesson may go from student to teacher and then back more than once.

The use of such a communication tool allows a close connection between teacher and student, personalising the course itself and tailoring it to a single student (or team of students). This is made possible by a flexible schedule of activities as well as work load, the latter depending on the student's ability and desire to go deeply into a matter.

The appraised average work load is of 60 hours. Naturally, because of lack of uniformity in the students' background, the work load may be judged low or excessive by different students. In our opinion, this limit could be overcome by splitting the course into two parts: a basic one for students approaching fuzzy control for the first time, and an upper level for people already acquainted with fuzzy control. The latter could be more flexible than the former, depending on the students' background, and dealing with soft control techniques: neuro-control, evolutionary control, fuzzy adaptive control, and all hybrid methodologies. This might be a way to overcome the lack of uniformity in the students' background.

In any case, independently from the work load, the course is an easy means to enable knowledge transfer, in a way proportional to the student's interaction with the teacher. The course is ideal for students wanting access to knowledge outside their own university but unwilling or unable to move: it often happens that students having interests in fuzzy control do not find any courses focusing on this subject at the university they are in.

Discipline is another main point of the course. The student must be self-motivated: the course does not require the traditional class attendance, but relies highly on the student's self-discipline. Sometimes students do not seriously study the textbook or do not try and make an effort to understand the material themselves, expecting the instructor to explain it, which is obviously not feasible, especially at a distance. Team work might help in this respect, even if it might produce the opposite effect: firstly, sometimes it is difficult to co-ordinate the different activities of people in a team, who might happen to be abroad for their studies or simply busy with other projects; secondly, the team work strategy is winning if students in the team first work by themselves and then discuss together ideas, problems and solutions; it is not if it leads to unbalances in the group work load.

This leads us to considering the prerequisites for people taking the course: students come from different nations, from different education systems, so that it is difficult to define standards; a minimum threshold is generally a degree. To this respect a good inquiry by means of lesson 0 might be important, especially when students underestimate the background needed in such a course: as students learn at a distance, it is very important for them to have a strong background in the field.

7. Conclusions and further developments

The fuzzy control course is based on three components: textbook, simulator, and lessons. Each lesson is built around a learning model which prompts the student to enter three levels of increasing self-activity. The technology is simple; the course has only one World Wide Web page, the rest is exclusively based on e-mail. In our opinion the technology is not the main problem (anymore), it is the pedagogical side that deserves most attention.

Further developments of the course involve an application for funding from the European Commission in the framework of the Esprit programme. *ERUDIT* is an Esprit Network of Excellence around fuzzy and uncertainty techniques. The network has 250 members with 40 percent industry and 60 percent university members. We wish to make use of the network to promote open and distance learning (ODL) within the fuzzy logic and neural network community. Therefore an application has been submitted to the SOCRATES/ODL programme. The idea is to have universities join industrial companies in order to build distance learning courses around industrial software tools. The expected result is a system for placing courses on Internet. Each course will consist of text-book material, lessons to download, and a software simulator. The project aims to give learners access to a selection of courses in the fuzzy and neural network area, and to enable teachers to pool their educational resources.

References

[1]   Crecraft, D.I., 1995, Engineering Studies at the Open University. *European Journal of Engineering Education*, 20 (2), 201-210.

[2]   Croft, M., 1995, The International Council for Distance Education, *European Journal of Engineering Education*, 20 (2), 235-236.

[3]   Ghasemi, M., 1996, *Distance Learning and the World Wide Web*. Technical University of Denmark: Dept. of Automation, (project report). Also as Web document [25].

[4]   Jantzen, J., 1994, *Fuzzy Control*, Technical University of Denmark: Department of Electric Power Engineering, Fourth ed. (First ed. 1991), 122 pages.

130

[5] Jørgensen, V., 1974, A ball-balancing system for demonstration of basic concepts in the state-space control theory. *Int. J. Elect. Engineering Educ.*, 11, 367-376.

[6] Malmberg, A.C.: *LEARN - Education Via Data Network.* Royal Danish School of Educational Studies: INFA, Emdrupvej 115B, DK-2400 Copenhagen NV, 4 pp.

[7] Minoli, D., 1996, *Distance Learning Technology and Applications*, Artech House, 685 Canton Street, Norwood, MA 02062, USA.

[8] Ramalhoto, M.F., 1995, A Selected Complementary Information List of Flexible and Distance Learning Institutions Worldwide, *European Journal of Engineering Education,* 20 (2), 247-253.

[9] Rodriguez-Rosello, L., 1995, Telematics for Education and Training in the European Union: Experience from Research, *European Journal of Engineering Education,* 20 (2), 145-153.

[10] The MathWorks Inc, 1995, *The Student Edition Of Matlab - v 4 for MS Windows.* Englewood Cliffs, NJ: Prentice Hall, 833 pages + 5 diskettes (equivalent versions for MacIntosh/PowerMac).

World Wide Web links, overviews and resources

[11] The Comprehensive Distance Education List of Resources,
http://www.dacc.cc.il.us/~ramage/disted.html

[12] Internet Resources for Technical Communicators, http://www.rpi.edu:80/~perezc2/tc/

[13] The World Wide Web Virtual Library: Distance Education,
http://www.cisnet.com/~cattales/Deducation.html

[14] Fuzzy Logic and Neurofuzzy (Neuro-fuzzy) Courses, http://www-isis.ecs.soton.ac.uk/research/nfinfo/fzcourse.html

[15] Teaching With MATLAB,
http://education.mathworks.com/teaching/courseguide.html

[16] The Globewide Network Academy, http://www.gnacademy.org/

[17] Fédération Interuniversitaire de l'Enseignement à Distance, http://newsup.univ-mrs.fr/~wctes/

Europe related

[18] European Credit Transfer System,
http://europa.eu.int/en/comm/dg22/socrates/ects.html

[19] Download SOCRATES Vademecum and Guidelines for Applicants,
http://europa.eu.int/en/comm/dg22/socrates/download.html

[20] CORDIS homepage, http://www.cordis.lu/cordis/cord4400.html

[21] ERUDIT contents page, http://www.mitgmbh.de/erudit

Others

[22] The modular Msc at Southampton University,
http://modmsc.ecs.soton.ac.uk/adtech.htm

[23] DTU, Technical University of Denmark, http://www.dtu.dk

[24] Graph of student exchanges at DTU,
http://www.adm.dtu.dk/velkomst/giffiler/95tal/intnat95.gif

[25] Distance Learning and the World Wide Web,
http://www.iau.dtu.dk/~jj/erudit/odl.html

FUZOS -- Fuzzy Operating System Support
For Information Technology

A.B. Patki[1], G.V. Raghunathan[2] and Azar Khurshid[3]

[1]*Department of Electronics, Government of India New Delhi, India*
[2]*Department of Electronics, Government of India, New Delhi, India*
[3]*Department of Electronics Engineering, Jamia Millia Islamia University, New Delhi.*
E-mail: patki@xm.doe.ernet.in, raghu@xm.doe.ernet.in, azark@giasdla.vsnl.net.in

Keywords: Corporate Computing, Fuzzy Operating System, Soft Computing, Information Technology, File Handling Systems.

Abstract

At present fuzzy system is being viewed and applied more as a control system tool for decision making purposes. A pioneering work is being reported for taking fuzzy systems decision making capabilities of handling imprecise and vague concepts to the Operating System level, in terms of fuzzy file handling system commands. **Shell level commands** have been modified to incorporate fuzzy decision making ability in a context sensitive fashion. FUZOS - a retrofit solution is implemented using object oriented methodology and fuzzy class library. The system is able to handle **dir, list, del, copy and move** fuzzy DOS commands.

1. Introduction

Currently available marketing literature of the companies describe fuzzy logic products as rule based fuzzy controllers incorporating fuzzy inference engine. By and large these applications have only scratched the surface of the revenue raising potential of fuzzy systems. Authors are of the view that comparing and contrasting impact of fuzzy logic in limited domain of consumer and industrial sector of process controllers can be viewed as a temporary revenue raising effort, which had yielded only a little revenue, since these efforts have introduced limited value added products. Recent advances in the area of fuzzy hardware and software research & development have now paved way for Operating System and microprocessor environment which will bring in the application of fuzzy logic technology in the Information Technology (IT) area in a large way with special reference to Corporate Computing. Lately, the world of Corporate Computing has been overwhelmed with large amounts of information, at the same time facing problems of a) total lack of decision making support at the Operating System level and b) limited support at the application level. To provide information for the masses by the turn of the century, it would be essential to shift to fuzzy logic based information system which will need support at the Operating System level [1], [2], [3].

Till such time fuzzy computer system becomes available to the common user, we have designed a retrofit solution - FUZOS, which shall add fuzzy computing capacity to Operating System commands, with special reference to MS-DOS. The FUZOS seeks to supply a retrofit solution in the sense that all present DOS commands are maintained as they are, and additional features of intelligent querying added to it. It is designed with a view to keep it forward integrable by establishing a direction of approach towards the fuzzy logic based OOOS [4].

A complete fuzzification of the required commands demands certain kernel level support. Such support requires FUZOS module to be integrated with the OS shell layer, there, it will handle the demands of the users and running processes which require some kind of decision support regarding files.

Basic shell level commands like **dir, list, copy, move and del** are chosen to include fuzzy conditions to demonstrate the power of fuzzy logic based decision making for Corporate Computing at the Operating System level. For example, if a manager wants to look at only the **recent and small** mails, there are no means by which he can avoid unwanted or stale mail, without some selection procedure. In such a scenario of information overload, a context sensitive decision making support at the Operating System level is needed. The choice of these commands was made by considering that only those Operating System commands required for decision making by Corporate Computing users should be fuzzified. For example commands like format, label etc., are not used by managers to arrive at any kind of decision. Thus it is to be clearly understood that the above requirements are not analogous to the filter approach adopted in the File Manager in the existing Windows systems for size criteria, which are based on Boolean decision. Essentially the proposed FUZOS incorporates fuzzy dynamic context sensitive decision making feature as against a Boolean filter operation. As an example it is not introducing a filtering option over the feature in the MS Windows 95: **View- Arrange Icons - by Size I I by Dates** in the form of sorting.

The design requirements for FUZOS are that it should be able to form fuzzy logical temporary groups of files and perform the desired operation on them. Any other Operating System level shell command not covered in its domain is simply passed to the Operating System for implementation. The principles governing the fuzzy relational query structure are used as the basis of design for the command structure to incorporate fuzzy decision making ability.

The existing public domain research and related literature in Fuzzy System and/or Operating System software does not address the issues about Fuzzy Operating System as brought out in this section. This paper discusses issues of semantics, structure of fuzzy Operating System commands incorporating the fuzzy termsets/ attributes, fuzzy decision support, design and framework, software issues, implementation and performance of the these set of commands. In subsequent sections authors have adopted the style of text oriented description for various DOS command structure unlike the popular BNF style of writing such things in computer science. This approach has been adopted to cater to wider range of researchers in fuzzy logic domain, such that in future fuzzy logic applications even for control purposes a limited embedded Operating System level support will be provided by control system designers and developers.

2. Semantics, Fuzzy Command Structure and Algorithm

A DOS command like dir gives the information which can be viewed as a relation - **files** which is defined by the set of attributes **<filename, size, date, time>** in the form of tupels: <t1, t2, ..., tn>.

Any fuzzy decision making query incorporated with the file handling commands will operate on the relation - **files.** Hence it is considered to incorporate the features that exist in the fuzzy relational query structure in the file handling commands while retaining the basic DOS command structure intact. This would help application and other shell level programs to use these commands to do Operating System parameter based information processing and decision making.

A typical fuzzy query is given below:

> **select < attributes > from < relations > where <fuzzy conditions>**

This can be considered to do fuzzy processing if we consider a relation R defined over a set of attributes { A_1, A_2, A_3, ...A_P }. A set of selected tuples { t_1, t_2, t_3, ...t_n } is determined on the basis of the extent to which a tuple satisfies < fuzzy condition > [5].

The existing DOS command structure for a command like **dir** is as follows.

> **Dir [d:] [path] [filename] [ext] [switches]**

This structure is modified to include fuzzy conditions to provide decision making ability, while maintaining the existing DOS command structure. We have arrived at the following command structure.

> **[DOS Command] [fuzzy condition in a domain of attributes] [and/or] [fuzzy conditions in a domain of attributes] [and/or] .....**

The fuzzy conditions could be viewed as operations on the relation **files**, and the results are given in the form of tuples <t1, t2, t3, ...tn>, covering the attributes <filename, size, time and date>, along with membership functions satisfying the conditions after executing the decision making using the connectives 'and/or' for the entire operation appended to the DOS command. We would need to retain the feature of redirecting the output to any i/o devices. Example: the feature of >prn or >filename for sending to the printer device or the storage device needs to be retained.

The time and size are the two fuzzy domains on which these fuzzy conditions might be based [2]. The authors had discussions with several users, to arrive at the typical termset to be used for the domains of size and time, hedges, modifiers and connectives as well as the overall command structure. These are given below:

Termset:-

> Size : [large, average, small]
> Time : [old, medium, recent].

The domain of the fuzzy condition is determined by the directory and/or extension specifications. The **dir** is the basic command which brings in the domain of files on which fuzzy conditions apply. The DOS commands like **dir, list, copy, move & del** are the operations that are to be performed on the group of files qualifying the fuzzy condition.

Various possible types of queries/ modifications to the commands that the users might like to use were analyzed during discussions with various DOS users desirous of seeking fuzzy decision making support. A few examples of different types of possible queries/ commands are given below.

i) dir \programs*.txt large files

ii) copy \programs*.* old files to a:\

iii) move *.* very small and moreorless recent files to \junk

iv) del small and very old files

v) list not very large files

The domain of the fuzzy set large in first example is text files contained in directory \programs. The word **large** is interpreted in the context of the domain size. Therefore, suitable membership functions will be generated with reference to the minimum and maximum file size of the domain as the boundaries. All files that qualify to have a non zero membership of this fuzzy set are therefore displayed on the terminal screen, thus providing the dynamic nature to such commands.

While analyzing the typical commands, the following considerations were observed

i) The command could pertain to a single domain, say size for the file or a combination of domains, say size and time.

ii) In the context of file, it is essential to interpret the termset and compute each fuzzy set for the termset of each domain dynamically i.e., the behavior of the command will depend on the context of domain attribute range, the largest and the smallest file sizes or times defining the boundary of the range, leading to membership function generation.

iii) The membership functions for the terms large, medium etc. should be as close to the human thinking i.e., the curves representing the membership functions should be continuous and smoothly varying. Thus, unlike control system applications of fuzzy logic where triangular and trapezoidal curves/ shapes of membership functions are widely used, for Operating System level support for Corporate Computing, applications of s, pi and inverse s curves for membership function representation were adopted. On this basis, structural ambiguities are allowed in the fuzzy conditions to provide functional freedom to the user, but with certain logical restrictions to maintain consistency for processing the DOS command. These are,

i) the first word should always be a DOS command,

ii) files, to, from, size, etc. are terms which might be used optionally but do not result in any processing,

iii) any domain may be represented in a command/ query,

iv) not more than two modifiers are allowed for each term in the command (example 'very very' and 'not very' are allowed but 'very very very' large is not permitted.)

v) there is no precedence of any domain, i.e. time domain or size domain are allowed interchangeable positions in a complex query. Even two terms from the same domain are allowed in the case of complex conditions.

The methodology adopted for interpreting and processing the fuzzy termset dynamically with reference to time and size in context of domain are given in the succeeding paragraphs.

Once the domain of operation is identified from the input command, all the files in the domain form a universal set. Similarly, based on the domain, universal sets of times and sizes are determined. Then all operations are performed through domain mappings.

An expression in fuzzy condition like **large** can be seen as an operation on these sets, and during the implementation, the size of each file in S - F plane need to be mapped from the μ - S plane to the μ - F plane.

The termset for size and time need to be interpreted/ computed dynamically, in the context of domain of the files which are being handled. For the purpose of illustration, let us define a domain of files defined by the set,

$$F = \{ f_1, f_2, f_3, \ldots\ldots\ldots, f_n \}$$

whose values for size and time are drawn from the domain defined by the sets,

$$Size \ S = \{ s_1, s_2, s_3, \ldots\ldots\ldots, s_m \}$$

$$Time \ T = \{ t_1, t_2, t_3, \ldots\ldots\ldots, t_k \}$$

size defined in bytes/ Kbytes, and time as date/time of creation/modification of the files.

Each file in the set F can be defined by its tuples say,

$< s_i, t_i >$ for file f_i, where $s_i \in S$ and $t_i \in T$, where S and T are orthogonal.

The termset for size, [**large,   average,   small**]

are defined as fuzzy sets over the set **S**. As an example,

$$large = \sum^m (\mu_i / s_i), \quad \mu_i \rightarrow [0, 1], \text{ which is defined over the plane } \mu - S.$$

Similarly, the termset for time, [**old**, **medium**, **recent**] can be defined as fuzzy sets over the set **T**. As an example,

$$old = \sum^{k} (\mu_i` / t_i), \quad \mu \rightarrow [0, 1],$$ which is defined over the plane μ - **T**.

The size of the files can be defined on the plane **S** - **F** and the time of creation/ modification of files can be defined on the plane **T** - **F**. In order to carry out Tnorm - S-norm operations (such as 'and' and 'or'), the terms are to be mapped on the μ - **F** plane to obtain membership for each set in the termset for time and size dynamically. For example,

large files $= \sum^{n} (\mu_i / f_i) : \sum^{n}(s_i/f_i) \rightarrow \sum^{m} (\mu_i/S_i)$

This is shown graphically in figure 1.

Let the largest size and the smallest size of files be s_1 and s_2 respectively. Then average is defined in the context of s_1 and s_2 such that $(s_1 + s_2)/2$ is the mid point and $(s_1+s_2)/4$ is the starting point and $3*(s_1 + s_2)/4$ is the end point for the linguistic term average as pi curve. Similarly for every member of termset, the fuzzy set definable in the μ - **F** plane is computed dynamically. For example,

fa = **average file** $= \{ \mu_i``/f_1, \mu``_2/f_2....\}$

fo = **old files** $= \{ \xi_1/f_1, \xi_2/f_2..........\}$

then, **average and old** is computed as $\quad f_a \ominus f_o.$ Similarly, the hedge or modifier operations are also performed on the μ - **F** plane.

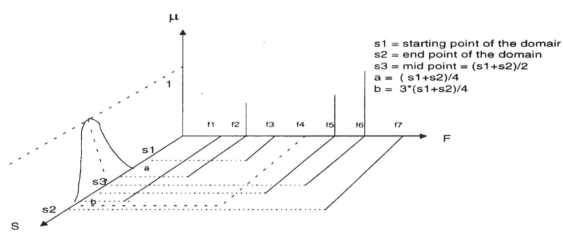

Figure 1. Domain Mapping from S-F to μ-F planes

2.1 Operator for Interpreting the Connective for decision making

The computer queries for single domain say size only or from orthogonal domains like size and time needs to be handled with decision making ability thereby suitably interpreting the connectives '**and/or**' very close to the corporate user requirements. In addition, another desired feature would be to incorporate the learning ability into the operator, so that as the system progresses, the operator parameters and functions could be characterized. The following are considered.

a) The connective 'and' could be interpreted as a pessimistic decision on the meaning of interpretation 'large size' in the **files** domain and average size on the **files** domain, i.e. 'large size' and 'average size' files could be interpreted as a pessimistic decision, and can be interpreted by 'min' operator and any of the other T-norms which would suit the context. The connective 'or' can be interpreted as an optimistic decision, and one of the S-norms could be used [6], [7].

M(large size files and average size files) = M(large size files) ^ M(average size file)

M(large size files or average size files) = M(large size files)V M (average size files)

For final decision, to select the files for display, could be based on the threshold membership value. The files which have μ above the threshold 'μ' could be selected. In this case it is proposed to use the center of gravity method for selection of files for final decision, as this would be closer to interpretation instead of selection , where all files are selected, indicating their membership values [7].

b) In case of complex query involving orthogonal domains, like size and time e.g. large and recent files, where the two domains have nothing in common, 'or' operation could be viewed as an optimistic decision on the meaning and 'and' operation as pessimistic decision on the meaning of the statements. In any case of and/or needs to be based on the given dependencies. The choices include T-norm, T-Conorm or a complex loaded operator.

An alternative approach could be to replace and/or with Θ which is a compensated operator [7].

$$C = A \; \Theta_\gamma B$$
$$\mu = (1 - \gamma).(A_x \wedge B_x) + \gamma (A_x \vee B_x) \qquad \gamma = \text{degree of compensation.}$$
If $\gamma = 1$ then $\Theta_\gamma = \vee$. and if $\gamma = 0$, then $\Theta_\gamma = \wedge$.

The decision could thus be modeled on the meaning of the query in the file domain [7].

C) Since the capability of complex query processing is attempted to be incorporated in the DOS command structure, the fuzzy connectives need to be interpreted in the lines of information retrieval capabilities,

It would be desirable to have an operator representing product through drastic sum that has learning function for each individual user, the probable candidate would be the fuzzy connective with learning capabilities as proposed in [8], which is given below.

The operator Θ has the requisite properties, and is defined as,

$$\Theta = m.S + (1 - m).T$$

where,

$$m = p_1 - (p_1 - p_2)X_1 - (p_2 - p_3)X_2$$
$$p_1 < p_2, p_3, \qquad 0 <= p_1, p_2, p_3 <= 1, \qquad 0 <= -p_1 + p_2 + p_3 <= 1$$

and p_1, p_2, p_3 are parameters.

The fuzzy operator Θ, as expressed above, as the connective T-norm and T-conorm are linearly defined by using variable m, which can be derived from values of X_1 and X_2 as indicated by the equation of m.

This provides a feature for adjusting the weights between T-norm and T-conorm in accordance with values of X_1 and X_2.

d) Yet another probable candidate for the operator would be through the rough set approach, with its ability to deduce rules and incorporate learning. It would be possible to define elementary sets and map the decisions over the relations as objects which are roughly definable. It would be possible to deduce the relevant rules with reference to the decision by partitioning the decision table. This method can be used to incorporate learning over the operator and as the system progresses over time, the operator would be trained to operate in that environment. [9].

3. System Design and Architecture

Object oriented methodology has been adopted in the design of FUZOS. The guidelines for the design being the future goal of merging it as a module into the fuzzy OOOS and to maintain the existing DOS command structure, to provide a retrofit solution for compatibility with the existing DOS.

The main part of designing of the system was the identification of various modules and objects that would be needed, their interface with each other and the overall systems' interface with the Operating System kernel.

File handling is a typical Operating System level job and to access information regarding files, system calls are used to design an interface to extract information regarding files, their location and other attributes. Also, any unhandeled or conventional system commands need to be simply passed on to the Operating System shell to be acted upon. Thus the system works as one of the modules of the Operating System, working in close conjunction with the kernel to provide Operating System level support needed for Corporate Computing..

The schematic block diagram of the system architecture is shown in figure 2. The module **parser** forms the main interface between the user, the decision making system and the Operating System shell. Depending upon the input Operating System command it triggers a process that prepares other modules (objects of classes fuzset, fileset, etc.) needed for computation work. These modules include two basic modules of fuzzy sets and a list of qualifying files and their attributes. Once the computation job is over, the information stored in these modules forms the data on which the command operates.

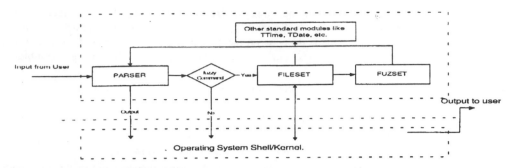

Figure 2. Schematic diagram of FUZOS implementation and architecture

4. Implementation and Results

The class fuzset, which forms the core class, is used to implement the datatype for the fuzzy set and is used to do all the fuzzy processing and supply modifier and connective support. It forms a fuzset object by considering the various files and their attributes. It implements the datatypes developed in [3] in the form of a class library, thus leading to code reusability. An example of membership function defining large files is illustrated in Appendix 1.

The parser class forms the user interface. It further communicates and controls the behavior of classes 'fileset' and 'fuzset' with the help of certain other helper classes, which are drawn from the standard C++ library like TTime, TDate, TISListIterator, etc..

Abstraction and inheritance properties of object oriented methodology are used to the advantage, such that the Operating System command is analyzed by just one call from the main program. Of the two other main classes used, one is designed for parsing and providing user interface, while the other is used to implement and handle file lists and their attributes. The parser class forms the interface object. It parses the input command as per the set of rules brought out in the earlier section and either implements it or passes it to the Operating System. The fileset class is used to provide and store a list of files in the domain and supply useful information when needed.

For decision making purpose, as proposed in 2.1, the various operators would also be required to be implemented in the class form. At this stage, several operators are under consideration. As the research work progresses, such a class structure which would take file sets, fuzzy sets and rough sets for methods used for decision making is expected to be developed. Such a class would typically contain functions for supporting various compensated operators, operators with learning capability, operators incorporating rough set methodologies, etc.

At the present stage, the first approach (a) as indicated in section 2.1 has been implemented. Some of the sample test results are in tables 1-3. For simple queries from same domain for decision making for selecting files the Centre of Gravity method for threshold μ has been implemented. For complex queries from the same domain the 'and' is interpreted as pessimistic decision and implemented as min operation; and the optimistic decision 'or' is implemented as max operation. In this case too the threshold μ is implemented using the Centre of Gravity method. In the case of complex queries involving orthogonal domains like size and time for testing the concept, the pessimistic and optimistic interpretations of 'and' and 'or' have been implemented and a fixed value of 0.5 as threshold has been used.

Table 1. Recent files

C:\MARCH\TEST\>dir recent files				
MASTRPLN.IDE	7544	April 3, 1997 3:44:58 pm	1	
INVNTORY.WRK	27500	March 27, 1997 4:53:40 pm	1	
RESOURCE.EXE	40448	January 12, 1996 9:26:10 am	0.82	
DIRECTRY.EXE	52917	November 6, 1996 3:15:34 pm	0.86	
LAYOUT.BIT	6051	April 3, 1997 3:40:08 pm	1	
FINANCE.LOG	4830	April 1, 1997 2:23:44 pm	1	
BUDGET97.PM5	32832	July 18, 1996 5:30:00 pm	0.84	
Total Files =	7			
Centre of Gravity = 0.7				

Table 2. Average files

```
C:\MARCH\TEST\>dir average files
FACTORY1.WRI        48512      June 17, 1993 9:42:06 pm    .88
DIRECTRY.EXE        52917      November 6, 1996 3:15:34 pm    .99


Total Files =        2
Centre of Gravity = 0.45
```

Table 3. large and average files

```
C:\MARCH\TEST\>dir large and average files
PARTNERS.WRK        73131                              December 4, 1994 12:13:16 pm         1


Total Files =        1
Centre of Gravity = 0.99
```

4.1 FILESET AND PARSE Classes

The typical class representation for classes **fileset** and **parse** used in the FUZOS are given below.

Typical listing of **class fileset:**

```
class fileset {
public:

        TTime createTime;
        TTime modifyTime;
        long size;
        uint8 attribute;
        char fullName[_MAX_PATH];

        //constructors

        fileset(){ }
        fileset( const char*, TTime&, long,
        uint8);
        fileset( const fileset&);
        ~fileset(){ delete [] fullName;}

        //operators

        fileset& operator=(const fileset& fs);
        int operator==(const fileset& fs)    const;
        int operator<(const fileset& fs) const;
        friend ostream& operator <<
        (ostream&, fileset);
};
```

Typical listing of **class parse:**

```
class parse: public fuzset
{
public:
        struct command *ComArr;
        struct command MainCom;
        fuzset fuzzyset1;
        fuzset fuzzyset2;
        container dir;
        fileset *file;
        //constructor
        parse(){ }
        parse(char*);
        ~parse(){ }
        //functions
        short impliment(int);
        fuzset getfuzset(const char*);
        friend short carryout();
private:
        float valtime(TTime &);
        short buildFileset(char*);
        char CommandList[60];
        char ConnectiveList[60];
        char QualifierList[60];
        char HedgeList[60];
        char info[20];
        friend class fuzset;
};
```

The constructor of parse class parses input command and decides if it can be handled, i.e. it checks for its consistency. It then stores each part of the command as a token in an array of a special structure called command. Based on each command, and information regarding the domain, the fileset and fuzset objects are created. The function implement( ) iteratively implements the command array and modifies the fuzset object. The getfuzset(const char*) function is a special function which assigns membership values to files based on the context and the fuzzy terms passed to it. The function carryout( ) executes the command once the fuzset and fileset objects are ready, or passes it back to the Operating System. The carryout function is called from the main program ascertaining the success of the fileset and fuzset building process only. Based on the input command, it also implements any switch functions and also gives redirection support. The function valtime(TTime&) is a function which returns the complete 'time word' based on the time as well as date into a single number. The function errormsg(char*) implements the exception handling and error message and checks the program flow in case of error. Container dir implements a list of fileset objects in an indirect pointer format.

The program permanently resides in the main memory and takes control of all system commands from the user. It simply passes those commands which are not implemented in the program.

4.2 Features Incorporated in FUZOS

For toggling between FUZOS and DOS a simple provision is implemented. If a double enter is given the system switches back to the Operating System shell i.e. DOS command mode. Typing **exit** in this mode loads the FUZOS back into the main memory.

For **del** command, verify feature has been added in the form of a switch in order to prevent accidental erasure, i.e. the system will ask for authorization with the message " Are you sure you can delete the file **filename.ext**", and waits for the user reply before carrying out the deletion.

The redirection feature has been maintained for fuzzy commands also. The ordinary switches like /p , /w are also maintained.

The FUZOS was tested on Intel based Pentiums running WINDOWS 95 / DOS as their Operating System. FUZOS when executed, shows the prompt ' A:\\>', the double slash indicating that the control to all DOS commands has been taken over by the FUZOS and it now resides in the memory. For the test machine, the memory locations occupied by the FUZOS are as given in table 4.

Table 4. Memory location

```
A:\>fuzos
Welcome to the Fuzzy Operating System Support - FUZOS
A.B. Patki, G.V.Raghunathan, Azar Khurshid.
Department of Electronics, Govt. of India, New Delhi.
FUZOS, (C)Copyright 1997.
.............................................................
A:\\>mem/m:fuzos
Error : Invalid parameter after /
 Standardization of  various switches is being considered for the FUZOS
FUZOS is using the following memory:


Segment Region    Total     Type
-------  ------  ----------------  --------
 009D6            352  (0K) Environment
 009EC        134,160 (131K) Program
                ----------------
Total Size:    134,512 (131K)
```

Unlike format, share, and other similar commands which also operate at Operating System level but are usually considered more as utilities since they are seldom used compared to shell level commands, FUZOS will be more on the lines of shell / kernel commands. In MS Windows 95 version, command.com file is of 92.8 Kb size, whereas fuzos.exe occupies 132Kb. Although at the outset it may appear as large to be treated at par with utility commands, the potential freedom it will provide in corporate computing merits its inclusion as a shell level program [10].

5. Performance Testing

The software was included in the 'autoexec.bat' file to be executed at the time of bootstrapping the system. Succeeding paragraphs present the observations from testing and user feedback.

It was observed that FUZOS resides over the Operating System shell. Any attempt to call the FUZOS commands from any other program or running process are routed to the Operating System shell and are not directed towards the FUZOS. Though solutions like modifying 'dos.h' or including another library to C or C++ could be possible, it is essential to implement the fuzzy commands at the Operating System shell level itself in order to realize its potential for a new programming environment.

During the test runs and user feedback phase, problems involved in providing fuzzy logic based computing were faced and attempts made to solve them.

When a fuzzy term such as **old** is given in context of time, it will be of different use and therefore interpreted differently by managers and software developers or system analysts. While a manager looks at just the dates, the system analyst will try to get version information from a single days work. This poses a resolution problem. If only dates are considered, the system analyst cannot use the FUZOS with full satisfaction. If both date and time are include, and files on the domain are all created on a single day, and cannot resolve further in time and gives resolution problem.

The nature of the membership function curves for these kinds of problems is still an issue of active debate. Definition of linguistic fuzzy term set could be subjective (personal bias), but a fuzzy Operating System should be generic. Some of the users associated during the performance feedback exercise were of the view that complex commands from the same domain should not be allowed as they serve no useful purpose. For the testing and observation purposes however, we have used s, pi and inverse s curves due to close resemblance with human perception as continuous and smoothly varying representation. Upon feedback from the users, **to** has been made a reserved word in **copy** and **move** commands. With a view to incorporate user defined membership function, the approach of setting up the same at the time of system generation time is thought of as a provision in FUZOS future versions.

It was felt that the current implementation of FUZOS does not exhibit the full capabilities of the fuzzy file handling systems since it has to use the file organization structure of a conventional system. An Operating System level approach to the file system will greatly alleviate the performance and utility of the system. This involves standardization issues requiring formation of a joint working group of fuzzy system researchers, computer software specialists, IT industry representatives etc. With a view to play a proactive role, authors would consider providing the software for use and feedback, through an ftp site if sufficient response is received.

In respect of the other proposed operators in section 2.1, work is in progress for implementation. As the research work progresses a clear cut structure for decision making class which could support variety of operators will emerge. Also, future research in converting the FUZOS DOS commands under Windows 3.1 environment is in progress.

6. Conclusion

Commercialization of Fuzzy Logic Technology has been primarily in the area of Control systems, in consumer electronics appliances area, etc. This paper is a major step towards extending the scope of fuzzy logic for Information Technology -- Operating System level support needed for successful use in Corporate Computing for 21st century wherein Soft Computing will play a major role. Authors are of the view that commercialization of the reported research work will lead to design of Fuzzy Operating System resulting in increased revenue in IT area as well as increased productivity in the corporate sector. Initiatives need to be taken by academic forums like IEEE, IEE to bring in various groups and researchers in the areas of Fuzzy Systems, Soft Computing, Operating Systems, Computer Science, IT industry, etc. to arrive at Fuzzy Operating System standards. Efforts by industry associations like NASSCOM, INDIA, CSSA, UK, and ITAA, USA would help in bringing out a commercial fuzzy operating system.

Acknowledgments

The authors wish to acknowledge the fruitful discussions with various management executives, IT experts and users and other professionals. The authors also acknowledge the useful discussions with Mr. R. Bandyopadhyay, Director, DoE, and his valuable comments. The authors would also like to thank Dr. A.K. Chakrawarty, Adviser, and Dr. U.P. Phadke, Senior Director, DoE, for encouraging interdisciplinary activities. Special thanks are due to Mr. W.R Deshpande, Director, DoE, and Mr. S.Sivasubramanian, Joint Director, DoE, for testing the software and organizing user feedback. Authors would like to thank Mr. A.Q. Ansari, Reader, Dept. of Electrical Engineering, Jamia Millia Islamia University, for permitting Mr. Azar Khurshid to carry out investigations in the DoE as a student trainee.

References

[1] Bandyopadhyay, R., 1996,Multimedia, Multilingual, Information Services Network Operating System, *Software, Electronics Information and Planning*, **23**, No.4. 205-230.

[2] Patki, A.B., and Raghunathan, G.V., 1996, Trends in Fuzzy Logic Hardware, *WSC1 - Proceedings of the First On-line Workshop on Soft Computing, Nagoya, Japan, Aug.19-30.* pp 180-185.

140

[3] Patki, A.B., Raghunathan, G.V., and Narayanan, N, 1996, On Datatypes for Object Oriented Methodology for Fuzzy Software Development. *WSCI - Proceedings of the First On-line Workshop on Soft Computing, Nagoya, Japan. Aug. 19-30*, pp 163-167.

[4] Patki, A.B., Raghunathan, G.V., and Khurshid Azar, Soft Computing Based Operating System - A Fuzzy Approach for Design and Development Considerations, Technical Report, No. MDD-SDD/4/97, Department of Electronics, New Delhi, India.

[5] Bosc, P., and Pivert, O, 1992, About Equivalence in SQLf, A Relational Language Supporting Imprecise Querying, *Fuzzy Engineering: Towards Human Friendly Systems*, (Edited by Terano, T., Sugeno, M., Mukardono, M., and Shigemasu, K.),Ohmsha, Tokyo, Japan., pp 309-320.

[6] Bellman, R.E., and Zadeh, L.A., 1970, Decision Making in Fuzzy Environment, *Management Science*, **17**, No. 4, 141- 164.

[7] Novac, V., 1986, *Fuzzy sets and Their Applications*, Adam Hilger, Bristol and Philadelphia (English translation 1989).

[8] Hayashi, I., Naito, E., and Wakami, N.,1992, A Proposal of Fuzzy Connective with learning Function and Its Application to Fuzzy Information Retrieval., *Fuzzy Engineering: Towards Human Friendly Systems*, (Edited by Terano,T, Sugeno,M., Mukardono, M, and Shigemasu, K.), Ohmsha, Tokyo, Japan.

[9] Pawlak, Z., Rough Sets, 1982, *International Journal of Computer & Information Sciences*. **11**, No.5, 342-356.

[10] 1997, User Manual for FUZOS, Department of Electronics, New Delhi, India.

Appendix 1

The membership functions are dynamically generated in the context of extreme file size/time limits of the files as brought out in section 2. As an example, membership function of '**large**' files are given in tables below.

Table A-1. File Sample space

```
Volume in drive C is MS-DOS_6
 Volume Serial Number is 1F4D-6392
 Directory of C:\march\test
.         <DIR>      04-01-97  3:40p .
..        <DIR>      04-01-97  3:40p ..
DIARY   WRK       104,115 11-01-93  3:11a Diary.wrk
EMPLOYEE FON       22,512  06-11-93  9:51a Employee.fon
MANAGERS DOC       89,248  09-15-92  1:00a Managers.doc
FACTORY1 WRI       48,512  06-17-93  9:42p Factory1.wri
MEDICAL  PRO       82,815  05-01-95  9:27p Medical.pro
MASTRPLN IDE       27,544  04-03-97  3:44p Mastrpln.ide
INVNTORY WRK       27,500  03-27-97  4:53p invntory.wrk
INTERNET DOC       12,120  12-13-90  4:24p Internet.doc
RESOURCE EXE       40,448  01-12-96  9:26a Resource.exe
DIRECTRY EXE       52,917  11-06-96  3:15p directry.exe
LAYOUT   BIT        6,051  04-03-97  3:40p layout.bit
PARTNERS WRK       73,131  12-04-94 12:13p Partners.wrk
ANUALPLN DOC       39,462  10-19-93 10:45a AnualPln.doc
FINANCE  LOG        4,830  04-01-97  2:23p finance.log
BUDGET97 PM5       32,832  07-18-96  5:30p Budget97.pm5
   15 file(s)    664,037 bytes
```

Table A-2. μ-F for large

C:\MARCH\TEST\>dir large files			
Filename	Size	Time	μ
DIARY.WRK	104115	November 1, 1993 3:11:00 am	1
MANAGERS.DOC	89248	September 15, 1992 1:00:00 am	0.82
Total Files =	2		
Centre of Gravity = 0.71			

Industrial Application of Chaos Engineering

Tadashi Iokibe

System Engineering Division, Meidensha Corporation
36-2, Hakozaki-cho, Nihonbashi, Chuo-ku, Tokyo, 103 Japan
Tel: +81-3-5641-7507   Fax: +81-3-5641-9303
E-mail: iokibe-t @honsha.meidensha.co.jp

Keywords: Chaos engineering, Deterministic non-linear short-term prediction, Fault diagnosis, Deterministic system, Stochastic process

Abstract

Recently, the study of chaos is attracting attention, and a wide range of academic fields is actively involved. On the other hand, Aihara proposed the term "chaos engineering" to describe the application of chaos theory for engineering purposes, and its possibilities have been demonstrated. Examples of applications reported so far include "Oil Fan Heaters (Sanyo Electric Co., Ltd.)," "Air-conditioners and Dish Washing Dryers (Matsushita Electric Industrial Co., Ltd.)," "Washing Machines (Goldstar Co., Ltd.; Korea)" and other home appliances and "Application to Health Care (Computer Convenience)." However, industrially, there has been only one application which is the "Tap Water Demand Prediction (Meidensha Corporation)." This paper first reviews the history of chaos research. Next, deterministic chaos is described. Time series forecasting and fault diagnosis are discussed as prospective industrial applications, and the related methodology is explained using practical examples.

1. Introduction

The history of chaos research can be traced back to the end of the 19th century to the research done by H. Poincaré and J. Hadamard [1][2]. At the beginning of the 20th century, H. Poincaré predicted the presence of chaos in his research on the 3-body problem [3]. In the 1960s, E. N. Lorenz discovered one property of chaos when carrying out a simulation of air circulation. He found that the solutions of his equations diverged exponentially with time, if very small differences were made in the initial values [4]. This was later called the butterfly effect, and it illustrates the sensitive dependence on initial condition, one of the most prominent features of chaos. Also S. Smale discovered chaos motion in a dynamic system that can be differentiated [5]. From Poincaré to Lorenz, more than 40 years had passed. This was because computers were not available in the age of Poincaré, so all computations were done by hand. In the 1970s, D. Ruelle and F. Takens discovered the strange attractor when studying the turbulent flow of a fluid [6]. In addition, T. Y. Li and J. A. Yorke found the conditions for generating chaos in a one-dimensional discrete dynamical system [7]. These conditions included "mapping shall have 3 periodic points," "topological entropy is positive" and "maximum Lyapunov's exponent is positive." From about this time onwards, the use of the word "chaos" became popularized. In the latter half of the 1980s, the study of chaos applications was launched, and in the 1990s, many papers began to be published.

In Japan, in 1991, the Bio-information Application System Study Committee (Chaired by Kazuyuki Aihara) was organized by the Japan Electronic Industry Development Association. The committee surveyed possible engineering applications of chaos, and a report titled "Questionnaire Survey on the Prospects of Chaos Engineering" was published in 1992 [8]. This report used the word "chaos engineering" for the first time in Japan, and since then, R&D on chaos applications has been more widely implemented.

Academic sectors which have possible relations to chaos include basic fields such as mathematics, physics, chemistry and biology, the engineering fields of electricity, electronics, mechanical engineering, control engineering, information engineering and civil engineering, and medical and economics fields. In other words, the prospective subjects to which chaos can be applied can be found all around us.

The status of recent studies of chaos is described below, with particular emphasis on industrial applications of chaos.

142

2. Present Status of Chaos Research

According to the report "Questionnaire Survey on the Prospects of Chaos Engineering" mentioned above, chaos engineering includes the following subjects.

- Chaos computing
- Deterministic non-linear short-term prediction
- Identifying and modeling chaos
- Bio-chaos
- Chaos memory

- Coding chaos

- Recognizing chaos patterns
- Chaos art
- Using chaos fluctuations
- Mounting chaos (circuit)
- Generating and using chaos in non-linear engineering systems
- And other topics

The author surveyed papers published mainly in Japan at academic meetings during the last several years, using the above classification. As a result, it was found that more than 200 papers were published from 1991 to 1996 as shown in Figure 1. In particular, it can be clearly seen that research work is concentrated on chaos computing, deterministic non-linear short-term prediction and mounting chaos (circuit). However, one reason why so many papers have been published about mounting chaos (circuit) might be that such experiments can be easily done in the laboratory of a university etc.

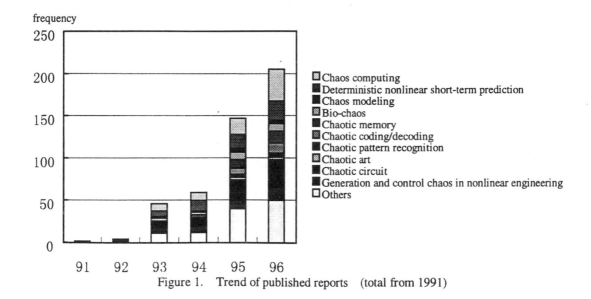

Figure 1.   Trend of published reports   (total from 1991)

3. Examples of Industrial Applications

The approaches to the engineering applications of chaos are generally classified, into two groups: (1) creative methods and (2) analytical methods. In the former, chaos conditions, consistent with the particular application, are produced in any case, and this behavior is used in a positive manner. In the latter, a phenomenon is examined to see whether it is chaotic or not, and if it is a deterministic chaos, the latent rules involved are extracted and used for a practical purpose. Applications based on the creative methods include chaos computing, chaotic memory, chaos cipher communications, chaos art, applications using chaos fluctuations, chaos circuits, non-linear engineering systems, etc. Applications using the analytical methods include deterministic non-linear short-term forecasting, identifying chaos, modeling, bio-chaos, etc.

Practical applications of chaos have most frequently been in the field of consumer appliances. However, there are only three examples in Japan. But, the situation is different from the time when fuzzy logic applications were booming. Manufacturers give only very simple explanations in small print in their catalogs, and they do not positively emphasize "chaos." In Korea, Gold Star applies chaos to washing machines.

(1) Oil fan heaters
These heaters were developed by Sanyo Electric, and this application of chaos to the fan heaters was the first time in the world, that chaos had been used in consumer goods. Here, an attempt was made to improve comfort

by providing a "fluctuation" made in which the temperature setting was varied by a fluctuation of $1/f^{\alpha}$ produced by intermittent chaos.

(2) Air-conditioners
Matsushita Electric Industries has developed this application in which Lorentz chaos is used to control the flaps of air-conditioners.

(3) Dish washers
Matsushita Electric Industries also developed this application. As in the air-conditioners, water sprayed from the cleaning nozzle of a dish washer, is controlled with a non-periodic motion (chaotic motion) in addition to the conventional cyclic operation, to improve the cleaning capability.

The following paragraphs describe "deterministic nonlinear short-term prediction" and "system diagnosis" which have been acknowledged as being successful industrial applications of chaos.

3.1 Deterministic non-linear Short-term Prediction

Short-term prediction is one of the most important methodologies in many fields such as forecasting tap water demand, electric power demand, local weather conditions, stock prices, and foreign currency exchange rates, and so many forecasting tools have been developed to improve the accuracy of forecasting. Although many attempts have been made to formulate forecasting models, a typical model is ARIMA (Auto Regressive Integrated Moving Average) which can be applied to data which has seasonal variations. According to this method, a forecastable structure is derived from observed time series data and the data is broken down into characteristic elements, using a summation filter, an auto regressive filter and a moving average filter. Using the results, a tentative model is produced, and the parameters of the model are estimated, together with an estimate of the validity of the model for diagnosis and for forecasting the future status. Conventional linear forecasting using this method often runs into difficulties in ascertaining the parameters after doing the modeling.

On the other hand, modern chaos applications include the problem of non-linear short-term prediction of time series data that behave chaotically. With this method, an attempt is made to find something similar to deterministic rules for phenomena that have been conventionally considered as noise or irregular sequences, in order to forecast the status in the near future. In practice, observed time series data are reconstructed in a multi-dimensional state space using Takens' embedding theorem [9], and using the proximity vectors of data vectors including recently observed data, the original data are locally reconstructed. Thus, data in the near future can be predicted from data observed at a particular time, for the short period before the deterministic causal relations are lost due to chaos having a sensitive dependency on initial conditions. Available local reconstructing methods include the tessellation method [10], the Gram-Schmitt's orthogonal system method [11] and the local fuzzy reconstructing method [12].

The following paragraphs describe "embedding" and "local reconstruction," important elements of deterministic non-linear short-term prediction methods, using the logistic map. In particular, we present the local fuzzy reconstruction method as local reconstruction method in detail. And the results of application for the tap water demand by the local fuzzy reconstruction method is introduced.

For more detailed information of the tessellation method and the Gram-Schmitt's orthogonal system method, see the references.

Embedding

A typical logistic map that generates chaos is the difference equation shown in Equation (1). The time series data and the actual numerical time series are shown in Figure 2 and Table 1, respectively. However, the entity that we can observe in the real world, is not the deterministic rule of Equation (1) but the numerical time series shown in Table 1 or Figure 3.

$$y_{t+1} = ay_t(1 - y_t) \tag{1}$$

By plotting the vector $\mathbf{x}_t = \{y_t, y_{t+1}\}_{t=0}^{n-1}$ in two-dimensional state space, Figure 3 is obtained. Obviously, the original deterministic law is reproduced. In other words, by embedding observed time series data, the deterministic law that lies behind the data can be recovered.

Table 1.　Numerical series of the logistic map

y_t	Value	y_t	Value	y_t	Value	y_t	Value
y_0	0.200000	y_4	0.823973	y_8	0.514181	y_{12}	0.880864
y_1	0.624000	y_5	0.565661	y_9	0.974216	y_{13}	0.409276
y_2	0.915034	y_6	0.958185	y_{10}	0.097966	y_{14}	0.942900
y_3	0.303214	y_7	0.156258	y_{11}	0.344638	y_{15}	0.209975

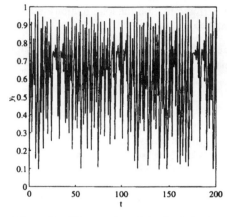

Figure 2.　Time series of the logistic map

Figure 3.　Estimation of determinism

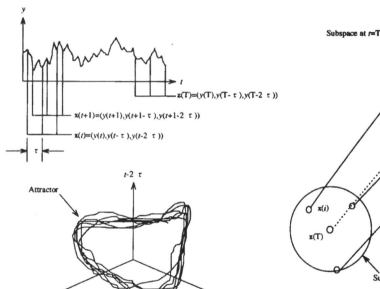

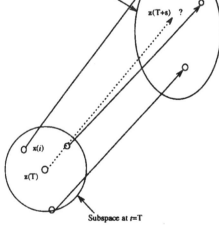

Figure 4. Embedding time series

Figure 5. Short-term prediction using local reconstruction

Generally, referring to Figure 4, a vector $\mathbf{x}(t) = (y(t), y(t-\tau), \ldots, y(t-(d-1)\tau))$ can be produced from some observed time series data $y(t)$ (τ is a delay time). This vector defines a point in a d-dimensional reconstructed state space R^d Therefore, by changing t, a trajectory can be drawn in this d-dimensional reconstructed state space. Assume that the original system is a deterministic dynamical system and that the observed time series data have been obtained by means of an observation system which corresponds to a C^1 continuous mapping from the state space of this dynamical system to a one-dimensional Euclidean space R, then the reconstructed trajectory provides an embedding of the original deterministic dynamical system, provided d is large enough. That is, if any attractor appears in the original dynamical system, another attractor that preserves the topological structure of the attractor will be generated in the reconstructed state space. d is called the "embedding dimension." F. Takens proved that the following equation (2) provides a sufficient condition for a reconstructing to be an "embedding."

$$d \geq 2m+1 \tag{2}$$

Where m is the dimension in the original dynamical system. However, this is a sufficient condition, so depending on the data, a smaller value of d can possibly give an embedding.

Local Reconstruction

Local reconstructing is defined as follows, for a state space reconstructed by an embedding operation and providing the trajectories of attractors. Using the data vector $z(T)$ which includes recently observed time series data, and the trajectories of its proximity data vectors $x(i)$ to $x(i+s)$, the near future trajectories of data vectors which include the most recently observed time series data are estimated, and the predicted values $\hat{z}(T+s)$ of data vectors $z(T+s)$ for "s" steps from the present are obtained. The procedure is shown in Figure 5. Then, the data vector $\hat{z}(T+s)$ is decomposed into components in the reconstructed state space (that is, into components corresponding to the embedding dimension), and predicted values $\hat{y}(t+s)$ of the data $y(t+s)$ for "s" steps from the most recently observed data $y(t)$ in the original time series are obtained.

Local Fuzzy Reconstruction Method

This method is a local non-linear reconstruction method, unlike Gram-Schmitt's orthogonal system method or the tessellation method. The space occupied by the proximity data vector $x(i)$ in the proximity space of $z(T)$, is fuzzily divided, and when $x(i)$ is moved to $x(i+s)$, s steps later, the trajectory dynamics are represented by a linguistic expression as a fuzzy rule with an IF-THEN- format. Then, the predicted value $\hat{z}(T+s)$ of $z(T+s)$ is estimated by a fuzzy inference in which a fuzzy rule is applied to $z(T)$ as an input data. In practice, data projected onto each axis of the reconstructed state space that constitutes $x(i)$ are used.

We plot the data vector $z(T)$ resulting from the latest observation in the d-dimensional reconstructed state space and replace the neighboring data vector with $x(i)$. And, we replace the state of data $x(i)$ $x(i)$ at "s" steps ahead with $x(i+s)$ and similarly the predicted value of data vector $z(T+s)$ at s steps ahead with $\hat{z}(T+s)$. If the behavior of the observed time series corresponds to deterministic chaos, the transition from state $x(i)$ to state $x(i+s)$ after "s" steps can be assumed to be dependent on the dynamics subjected to determinism. This dynamics can be described in a linguistic form as shown below.

$$IF\ z(T)\ is\ x(i)\ THEN\ z(T+s)\ is\ x(i+s) \tag{3}$$

$x(i)$ is the data vector neighboring to $x(i)$ $z(T)$. Therefore, before step "s" loses deterministic causality, it can be assumed that the dynamics from transition from state $z(T)$ to state $z(T+s)$ is approximately equivalent to that from state $x(i)$ to state $x(i+s)$. When the attractor embedded in the d-dimensional reconstructed state space is smooth manifold, the trajectory from $z(T)$ to $z(T+s)$ is influenced by the Euclidean distance from $z(T)$ to $x(i)$.

Now let us remember the following relations.

$$x(i) = (y(i),\ y(i-\tau),\ \ldots,\ y(i-(d-1)\tau))$$
$$x(i+s) = (y(i+s),\ y(i+s-\tau),\ \cdots,\ y(i+s-(d-1)\tau)) \tag{4}$$

This formula can be rewritten as follows when focusing attention on the "j" axis in the d-dimensional reconstructed state space.

$$IF\ ax_j(T)\ is\ y_j(i)\ THEN\ ax_j(T+s)\ is\ y_j(i+s)\quad (j=1,\ldots,d) \tag{5}$$

Where, $ax_j(T)$ is j-axis component of $x(i)$ value neighboring to $x(i)$ in d-dimensional reconstructed state space, $ax_j(T+s)$ is j-axis component of $x(i+s)$ in d-dimensional reconstructed state space.

Also, the trajectory from $z(T)$ to $z(T+s)$ is influenced by Euclidean distance from $z(T)$ to $x(i)$. This

146

influence becomes nonlinear because of a smooth manifold. Hence, for rendering a nonlinear characteristic, Expression (5) can be expressed by fuzzy function as follows.

$IF\ ax_j(T)\ is\ \tilde{y}_j(i)\ THEN\ ax_j(T+s)\ is\ \tilde{y}_j(i+s)\quad(j=1,...,d)$ (6)

Now let us remember the following .

$z(T) = (y(T), y(T-\tau),..., y(T-(d-1)\tau))$

Therefore, the j-axis component of $z(T)$ becomes equal to $y_j(T)$.

Accordingly, the j-axis component of the predicted value $\hat{z}(T+s)$ of data vector $z(T+s)$ after "s" steps of $z(T)$ is obtainable as $ax_j(T+s)$ by a fuzzy inference with $y_j(T)$ substituted into $ax_j(T)$ of Expression (6). This method is called "Local Fuzzy Reconstruction method".

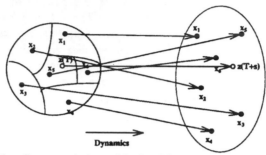

Figure 6.   Conceptual view of the local fuzzy reconstruction method

Explanation is given below on a concrete example where the dimension of embedding = 3, delay time = 4 and the number of neighboring data vectors = 2. Let us assume each data vector as follows.

$z(T) = (y_1(T), y_2(T-4), y_3(T-8))$ $z(T+s) = (y_1(T+s), y_2(T+s-4), y_3(T+s-8))$

$x(t_0) = (y_1(t_0), y_2(t_0-4), y_3(t_0-8))$ $x(t_0+s) = (y_1(t_0+s), y_2(t_0+s-4), y_3(t_0+s-8))$

$x(t_1) = (y_1(t_1), y_2(t_1-4), y_3(t_1-8))$ $x(t_1+s) = (y_1(t_1+s), y_2(t_1+s-4), y_3(t_1+s-8))$

On this assumption, the fuzzy rule given in Expression (6) can be represented by Expressions (7), (8) and (9) regarding 1st, 2nd and 3rd axis respectively.

$1st\ axis:\ IF\ ax_1(T)\ is\ \tilde{y}_1(t_0)\ THEN\ ax_1(T+s)\ is\ \tilde{y}_1(t_0+s)$

$IF\ ax_1(T)\ is\ \tilde{y}_1(t_1)\ THEN\ ax_1(T+s)\ is\ \tilde{y}_1(t_1+s)$ (7)

$2nd\ axis:\ IF\ ax_2(T)\ is\ \tilde{y}_2(t_0-4)\ THEN\ ax_2(T+s)\ is\ \tilde{y}_2(t_0+s-4)$

$IF\ ax_2(T)\ is\ \tilde{y}_2(t_1-4)\ THEN\ ax_2(T+s)\ is\ \tilde{y}_2(t_1+s-4)$ (8)

$3rd\ axis:\ IF\ ax_3(T)\ is\ \tilde{y}_3(t_0-8)\ THEN\ ax_3(T+s)\ is\ \tilde{y}_3(t_0+s-8)$

$IF\ ax_3(T)\ is\ \tilde{y}_3(t_1-8)\ THEN\ ax_3(T+s)\ is\ \tilde{y}_3(t_1+s-8)$ (9)

Also, since $x(t_0)$ and $x(t_1)$ are neighboring data vectors around $z(T)$ each axis of the reconstructed state space in the statement of fuzzy rules (7), (8) and (9) has the membership function shown in Figure 7. Note that the membership functions in the subsequent statement are all of a crisp expression for speeding up computation.

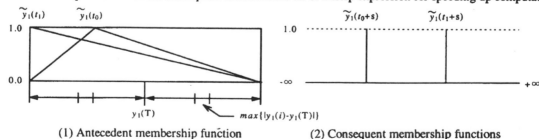

(1) Antecedent membership function   (2) Consequent membership functions

Figure 7.   Membership functions

For the dynamics expressed by the above fuzzy rules and membership functions, we now conduct a fuzzy inference with the following taken as input data; $ax_1(T) = y_1(T), ax_2(T) = y_2(T-4)$

In consequence, we obtain: $\tilde{y}_1(T+s) = ax_1(T+s),\ \tilde{y}_2(T+s) = ax_2(T+s-4)$ (10)

Thus, the predicted value $y_1(T+s)$ after the original time series data $y_1(T)$ has advanced "s" steps is available as $ax_1(T+s)$. According to Equation (10), fuzzy inference rule set described in Expression (7) should be only computed.

Application to Tap Water Demand

Waterworks facilities belong to a system that is composed of facilities for water storage, water intake, water delivery, water purification, water conveyance and water distribution and equipment for water supply.

The demand rate for the water created through these processes is varying by periodic elements with the life of the people reflected and also by other elements including weather. The water purification rate, on the other hand, is fixed regardless of the time, because of the characteristics of the water purification plants. The facility which is requested to provide the function to execute coordination between the demand rate, which varies in time, and the water purification rate, which is fixed. The service reservoir tank, therefore, executes water conveyance while changing the water storage rate because of variation of the demand rate. The service reservoir tank is requested to upkeep specified water level and water pressure even on occurrence of an accident on the upstream side of the service reservoir tank. The service reservoir tank is requested to keep a high water storage rate to fully exhibit this function.

When practical conduct of a service reservoir tank is considered, it is desirable that these two required functions are satisfied with a good balance and the service reservoir is run with minimum cost using equipment, which are not excessive.

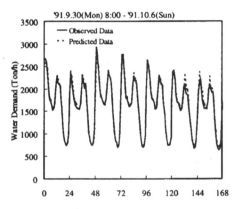
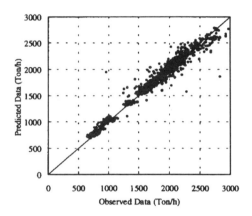

Figure 8.   Prediction Result (time series)   Figure.9.   Prediction Result (correlation)

In order to satisfy these requirements, prediction of tap water demand with high accuracy is requested. The authors applied the local fuzzy reconstruction method to the tap water demand. The data used is observed every hour from 1991.8.1 8:00 to 11.1 7:00. For this prediction, embedding dimension, delay time and nos. of neighboring data vectors are 7,3, and 7 respectively.

When forecasting using a linear probability process such as by a conventional auto regressive model (ARX model), descriptive variables, such as the order number of the model, air temperature, humidity, weather, time, day of the week and season, must be used. However, when the deterministic non-linear short-term prediction method described in this paper is applied, absolutely no data other than one line of observed time series data is needed, and yet a good forecasted result can be obtained.

Of the various local reconstruction methods mentioned above, the local fuzzy reconstruction method is most excellent from the points of view of forecasting performance and the time required for processing.

3.2 Application to Diagnosis

The next way of applying chaos to a process which seems to be hopeful is for diagnosis. Most process equipment is designed and manufactured so that it will work with the highest efficiency. However, the efficiency decreases as the equipment deteriorates. That is, the energy loss increases. Energy losses include radiated energy (for example, noise and vibration) and stored energy (for instance, heat and distortion). By measuring such energy losses, the extent to which the equipment has deteriorated can be measured indirectly.

In particular with rotating or reciprocating equipment, more than 70% of the deterioration might show up as vibration, including abnormal vibration, friction, abnormal noise and other failure modes, so measuring vibration is an effective way of diagnosing the equipment. Equipment diagnosis by measuring vibration has developed from simply listening to noise by an expert using a vibration rod, diagnosis by just measuring vibration levels to check for abnormalities, to modern precision vibration diagnosis using Fast Fourier Transformation (FFT) to check for failure causes.

Normally when rotating or reciprocating equipment fails, there may be changes in the general trends of vibration time-series data measured on the equipment. For example, a frequency component or a vibration amplitude may change. We may understand from this, that the dynamics of observed time-series data will change regardless of the cause.

But, depending on the cause of an abnormality, the phenomena described before, i.e. "A particular frequency appears or the vibration level changes" might not be observed. An approach using the amount of mixed random noise can be used effectively in such cases, and the procedure, chaotic diagnosis, is described below.

When time series data is embedded in a d-dimensional state space, the direction of a free data vector is substantially identical to the direction of a tangential unit vector for the trajectory of the data vector in a proximity space, provided that the time series is based on an ideal deterministic theory as shown in Figure 10 (1). With a stochastic system, however, the directions are random shown in Figure 10 (2). Using this characteristic, time series based on a deterministic theory and a stochastic process can be discriminated from each other or identified. Various algorithms have been proposed based on these principles, including the "Trajectory Parallel Measure Method [13]" proposed by the authors, which has the advantages of a high computation efficiency and rather a small number of parameters that have to be set. Details of this method are given below.

According to the method, checks are made of the variations between the directions of trajectory vectors for vectors randomly sampled from inside of a reconstructed topological state space and their proximity vectors. Therefore, it is sufficient to set a suitable number of proximities and appropriate number of samples, in addition to setting the embedding parameters (number of dimensions and delays). Practical methods are as follows.

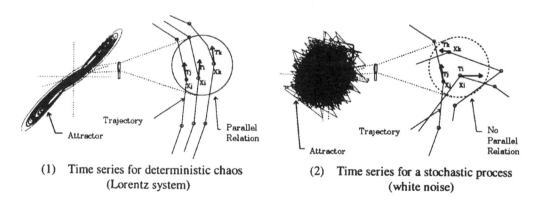

(1) Time series for deterministic chaos (2) Time series for a stochastic process
         (Lorentz system)   (white noise)

Figure 10.   Trajectories in local space, depending on the type of time series

When part of the trajectories of an embedded deterministic chaos time series is enlarged, it can be seen that over a very limited range, the trajectories flow in substantially the same direction. The directions of the trajectories are examined locally, and the attractors are processed in the same way in each local space, and their mean values are determined.

A free vector Xi of an embedded trajectory is selected, and a number m, of proximity vectors Xj : ($j=1,2,....,m$), close to Xi in terms of their Euclidean distances, are selected. Next, the tangential unit vectors Ti and Tj, for the trajectories of the data vector Xi and the proximity vector Xj, are derived. Using the tangential unit vector Ti thus obtained as a reference, variations in the directions of the tangential unit vectors Tj of the proximity vectors are calculated by Equation (12).

$$\gamma_i = \frac{1}{4m} \times \sum_j^m \|Ti - Tj\|^2 \tag{12}$$

This processing is applied to the local proximity space from which k samples were randomly selected from the total number of attractors. Next, to judge the condition of the trajectories of all the attractors statistically, a mean value is calculated by Equation (13).

$$\Gamma = \frac{1}{k} \sum_i^k \gamma_i \tag{13}$$

The closer the value of Γ is to 0, the more identical, statistically, are the directions of the vectors in the proximity space. That is, the space is a deterministic topological space. Conversely, the closer the result is to 0.5, the more orthogonal are the trajectory vectors in the proximity space. The closer to 1, the more the trajectory vectors face each other in the opposite direction. In particular, when the result is close to 0.5, the topological space is that of a stochastic process. When the method is applied practically, the computations are iterated to reduce the statistical errors.

For a deterministic system, the directions of the trajectory vectors are identical in a local region of the topological space.

Application to Diagnosis of Automatic Transmission Unit

The following paragraphs describe the results of applying this method to diagnosing failure in a rotating mechanical system consisting of planetary gears. The causes of the abnormalities, which we are concerned with here, are not "decentering," "faulty bearings" and "rough teeth surfaces" that can be identified by FFT analysis, but are "dents on teeth surfaces and shaving defects" that cannot be detected by conventional methods. With either of the latter causes, random noise with infinite harmonics is included in the time series data that is observed, so these defects cannot be diagnosed by FFT analysis. Figure 11 shows observed acoustic time-series data, Figure 12 shows the results, of an FFT analysis and Figure 13 gives the analysis results using the trajectory parallel measure method. Conditions for the analysis using the trajectory parallel measure method included embedding dimensions: 2-10, delay: 1, number of selected proximity data vectors: 4, number of samples: 400, and number of computations: 20. As can be seen in Figure 12, the FFT analysis, a typical, widely-used, equipment diagnosis method, cannot discriminate harmonics which have different frequency components. On the contrary, as shown in Figure 13, the trajectory parallel measure method can clearly identify a normal unit from an abnormal unit.

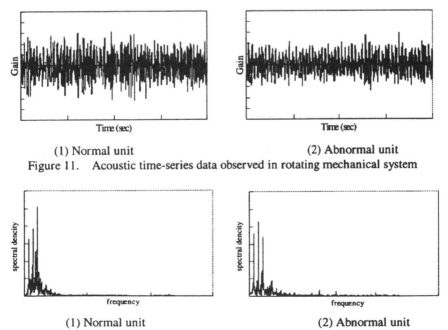

(1) Normal unit   (2) Abnormal unit

Figure 11.   Acoustic time-series data observed in rotating mechanical system

(1) Normal unit   (2) Abnormal unit

Figure 12.   Results of FFT analysis

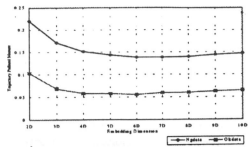

Figure 13.   Results of analysis using the trajectory parallel measure method

4. Conclusions

Although research into engineering applications of chaos has been actively promoted recently, regretfully there have, so far, been few practical applications in industrial fields, particularly in industrial processes. However, a considerable number of reports have been presented, and practical applications will be realized sooner or later. The authors are convinced that the application of chaos to processes will spread quickly in the next several years.

References

[1]   Poincaré, H., 1893, *Les Méthodes Nouvelles de la Mécanique Céleste2*, Gauthiers-Villar, Paris

[2]   Hadamard, J., 1898, Les Surfaces à Courbures Opposées et Leurs Ligns Géodésiques. *J. Math. Pure* Appl. 4,27.

[3]   Poincaré, H., 1913, *Sciencr et Méthode"*, Paris: Flammarion, 1912. English translation: *Science and Method.* Science Press Lancaster, P.A

[4]   Lorenz, E. N., 1963, Deterministic Nonperiodic Flow. *J.Atoms, Sci.*, **20**, 130-141

[5]   Smale, S., 1967, Differentiable Dynamical Systems. *Bull. Amer. Math. Soc.*, **73**, 747-817

[6]   Ruell, D., and Takens, F., 1971, On the Nature of Turbulence., *Commun. Math. Phys.*, **20**, 167-192

[7]   Li, T. Y., and Yorke, J. A., 1975, Period Three Implies Chaos. *Amer. Math. Monthly.* **82**, 985-982

[8]   Aihara, K., 1992, Chaos Engineering. *Scientific American (Japanese edition)*, 22.3.26, Nikkei Science, Inc., Tokyo

[9]   Takens, F., 1981,*In Dynamical Systems and Turbulence*, (eds. Rand and Young), Springer, Berlin,   366-381

[10]   Mees, A. I., 1991, Dynamical Systems and Tessellations : Detecting Determinism in Data. *Int. Journal of Bifurcation and Chaos*, Vol.1, No.4, 777-794

[11]   Jimenez, J., Moreno J. A., and Ruggeri, G. J., 1992, Forecasting on Chaotic Time Series : A Local Optimal Linear-reconstruction Method. *Physical Review* A, Vol. 45, No.6, .3553-3558

[12]   Iokibe, T., 1995, *Fusion of Chaos and Fuzzy Logic, and its Applications, Industrial Applications of Fuzzy Technology in the World* (eds. K. Hirota and M. Sugeno), World Scientific, 271-296

[13]   Fujimoto, Y., Iokibe, T., and Tanimura, T., 1996, Trajectory Parallel Measure Method for Discriminating between Determinism and Stochastic Process Property in Time Series. *Journal of Japan Society for Fuzzy Theory and Systems*,   (printing)

Part 4: Genetic Algorithms

Papers:

A Genetic Method for Evolutionary Agents in a Competitive Environment

Kousuke Moriwaki, Nobuhiro Inuzuka, Masashi Yamada,
Hirohisa Seki, Hidenori Itoh

Department of Intelligence and Computer Science
Nagoya Institute of Technology
Gokiso, Showa-ku, Nagoya 466, Japan
Tel: +81-52-735-5475 Fax: +81-52-735-5477
E-mail: moriw@juno.ics.nitech.ac.jp

Keywords: Genetic Algorithm, Co-evolution, Food-chain, Quasi Ecosystem

Abstract

A gene expression n-BDD is proposed in order to investigate co-evolutional environment. The gene expression is suitable for agent models to decide agent's behavior using information from environment. We can apply genetic algorithm to evolve behavior of agents. In a quasi ecosystem, a herbivore, a carnivore and plants make a food chain, which is a typical competitive environment and causes co-evolution. A herbivore using an n-BDD more rapidly acquire its ability to survive from a carnivore than using finite state automaton. Moreover, in the environment which carnivores and herbivores are co-evolved, we have seen a food chain relation in an experimental field.

1 Introduction

Co-evolution is attractive because it gives a method to evolve two kinds of competitive entities in an environment. D. Hillis presented that co-evolution is effective to solve a kind of sorting problem[1]. The relation between carnivores and herbivores makes a typical co-evolutional competition in the natural world. In the world animals compete and evolve each other through an interactive relationship of the food chain. The food chain or the flow of resources is also an important aspects of ecological organization[2], and it is necessary to study active environments of agents. We simulate a food chain model which consists of three kinds of artificial agents, herbivores, carnivores and plants, and give them ability to adapt to the environment by genetic methods.

The efficiency of genetic algorithms depends on the way how genes are expressed. A gene expression using finite state automata(FSA) is proposed in [3] and the expression is used in order to simulate to a quasi ecosystem[4, 5]. Agents in the system have finite state automata as their genes. The genes, automata, are used to decide behavior of agents using information from the environment. This gene expression is, however, not necessarily suitable for genetic operations, because a small operation to a gene of an agent with an automaton may cause a large change in their behavior as we will see later. We propose an gene expression using an n-BDD, which is introduced in the following section. The n-BDD gene expression expresses the behavior of agents and lets genetic operations work more efficiently in a co-evolutional environment.

We take up a simulation of quasi ecosystem as an experimental model to confirm the efficiency of our method compared with a system using finite state automata. We also compare it with a system using classifier systems, which is a well-known gene expression[6]. A herbivore or a carnivore has a gene expressed by an n-BDD, a finite state automata or a classifier system.

We propose two models for experiment, mini-model and food-chain-model. The former model is used to investigate basic ability of the gene expressions, and the latter is used to investigate co-evolutional ability.

In the following section the gene expression n-BDD is proposed and genetic operations, mutation, insertion and deletion, for n-BDD are also proposed. Section 3 reviews other gene expressions. Section 4 explains our model of quasi ecosystem. Information which are used to decide actions of an animal and a set of animal's actions are also defined. Section 5 gives experiments which compare the n-BDD gene expression with finite state automata and classifier systems. Section 6 discusses the results of experiments. We also discuss on time and space complexity of genetic expressions.

2 A gene expression n-BDD

2.1 n-BDD

BDD (Binary Decision Diagram) is a method to express logical functions, and is originally proposed by Akers in 1978[8]. Bryant developed methods to operate BDDs[9]. Because of its storage efficiency and processing speed, BDD has been applied to various fields such as CAD system of LSI design. Moreover, it has been used also for combinatorial optimization problems and plays a large part to solve the problems, which had not been solved in realistic time until recently.

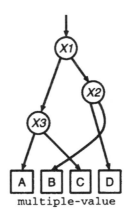

Figure 1: n-BDD(n-output binary decision diagram)

Figure 1 shows an n-BDD, which expresses a function that takes three-bit values $(X1, X2, X3)$ and outputs a value from a set $\{A, B, C, D\}$. In the figure a circle denotes an input bit and is called a decision node. Each decision node has three directional edges, one of which comes from outside and two of which go outside. An output value is denoted by a square which is called a terminal node. With input bits $(X1, X2, X3)$ an output values are calculated as follows: First look at the top decision node $X1$ and take a left edge if $X1$ is 0 or a right edge if $X1$ is 1. We call the left edge and the right edge a 0-edge and a 1-edge, respectively. Iterate this for the decision node indexed by the 0-edge or 1-edge until an edge indexes a terminal node. If the edge indexes a terminal node the value in it is the output.

While a BDD gives true or false with an input, an n-BDD gives a value from any set of values. That is, a BDD has only two terminal nodes labeled true and false, but an n-BDD can have more than two. Hence, functions expressed by n-BDDs are not restricted to logical functions. We can expect to use n-BDDs in many fields and so they have a wide-applicability. Because normalization algorithm of BDD can be applied also to n-BDD, it is possible to express a function using n-BDD compactly and uniquely. Moreover, many techniques developed for BDD such as graph-theoretic operations can be applied to n-BDD.

One-dimensional expressions like bit strings are used in many GA systems as their gene expressions, but structural expressions like n-BDD are sometimes suitable.

2.2   Genetic operation for n-BDD

Genetic operations, mutation, insertion and deletion, are defined as follows.

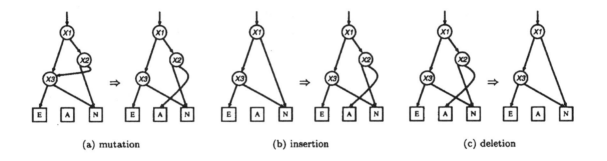

(a) mutation   (b) insertion   (c) deletion

Figure 2: Genetic operation for n-BDD

Mutation

Mutation changes a direction of an edge to a randomly selected new node. This new node must be subordinated by the node which the edge rises from(Figure 2(a)). Because of this restriction a loop and a cycle are never caused as a result. The node which the new edge points to may be an decision node or a terminal node.

Insertion

Insertion inserts a new decision node on a randomly selected edge. In Figure 2(b), the edge which pointed to the node C becomes to point to the decision node $X2$ which is newly added. Either of 0-edge or 1-edge of the new decision node is randomly selected to point to the node C. The other edge becomes to point to a subordinate node randomly selected.

Deletion

Deletion deletes a randomly selected decision node. In Figure 2(c), a node $X2$ is being deleted. The edge pointing to the node $X2$ becomes to point to one of the nodes B or C to which the 0-edge and 1-edge of the node $X2$ pointed.

3 Other gene expressions

In order to evaluate effectiveness of the gene expression of n-BDD proposed in the paper, we use other gene expressions, finite automata and the classifier systems.

3.1 Finite State Automata

A finite state automata repeats state transition by using input bits until all bits are consumed. A label of state in which all bits are consumed is an output of the automata with the input.

Genetic operation for finite state automata is based on [3]. Three kinds of genetic operation, mutation, insertion and deletion, are defined for finite state automata, which are illustrated in Figure 3. Mutation changes the direction of an arrow from a state chosen randomly or changes an output value at random. Insertion adds a state to a position chosen randomly. Conversely, deletion deletes a state chosen randomly.

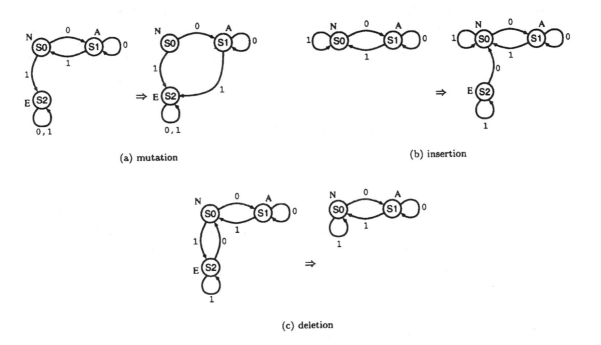

(a) mutation   (b) insertion

(c) deletion

Figure 3: Genetic operation for Finite State Automata

3.2 Classifier System

There are two ways to implement a classifier system. One is called Michigan approach which regards a classifier as a genetic individual and the other one is called Pitts approach which regards a classifier system as an individual. We take the Pitts approach in our experiment.

When a condition part of a classifier matches with an input message, an action part of the classifier is outputted. When there are two or more classifiers matching with an input message, conflict is resolved by using the roulette strategy with strength of the Bucket-Brigade algorithm. Because a conflict resolution is probabilistic, classifier systems are also probabilistic, i.e. the output is not uniquely decided, while n-BDD and finite state automata uniquely provide an output for an input.

Two kinds of genetic operation, crossover which exchanges the same number of classifiers and mutation which rewrite a part of information on classifier, is used. Figure 4 illustrate the operations.

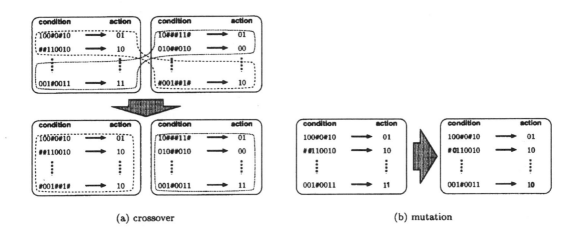

(a) crossover   (b) mutation

Figure 4: Genetic operation for Classifier System

About the Michigan approach, we confirmed the failure in the evolution of the action strategy though we used the Bucket-Brigade algorithm[7] called to be able to evolve even if a sub-goal is not specified.

4 Models of animals

A gene, i.e. an n-BDD, a finite state automaton, or a classifier system, takes a bit string as an input, the bit string which gives perceptional information from the environment shown in Table 1. The information includes states of an animal, hungry or repletion, and visual perception of a carnivore, a herbivore and plants. A gene outputs a value from values walk, runaway, eat and do-nothing. A gene of carnivore does not output a value runaway. An animal acts like the output value. Figure 5 shows the detail of the actions that correspond to the values.

Table 1: Assignment of bit string

$X0 = E$	hungry
$X1 = F$	repletion
$X2 = C_f$	carnivore is visible far
$X3 = C_n$	carnivore is visible near
$X4 = H_f$	herbivore is visible far
$X5 = H_n$	herbivore is visible near
$X6 = G_f$	plant is visible far
$X7 = G_n$	plant is visible near

$$\underset{\text{hungry}}{E} , \underset{\text{repletion}}{F} , \overbrace{C_f, C_n}^{\text{carnivore}}, \overbrace{H_f, H_n}^{\text{herbivore}}, \overbrace{G_f, G_n}^{\text{plant}}$$

An input bit string is assigned to the decision nodes of an n-BDD or is given to a finite state automaton as an input string. The output values are assigned to terminal nodes of the n-BDD or final states of the automata are labeled by the output values.

o walk

Animal moves to 8 neighborhoods at random.

o runaway

Animal moves in the opposite direction to the carnivore.
Animal moves in a random direction if the carnivore in not in view.

o eat

Animal moves in the direction of the food, and eat it if reached.
Animal moves in a random direction if the food is not in view.

o do nothing

Animal does not move.

o approach

Animal approaches the same kind of animal.
Animal moves in a random direction if the food is not in view.

O carnivore
● herbivore
⌂ plant

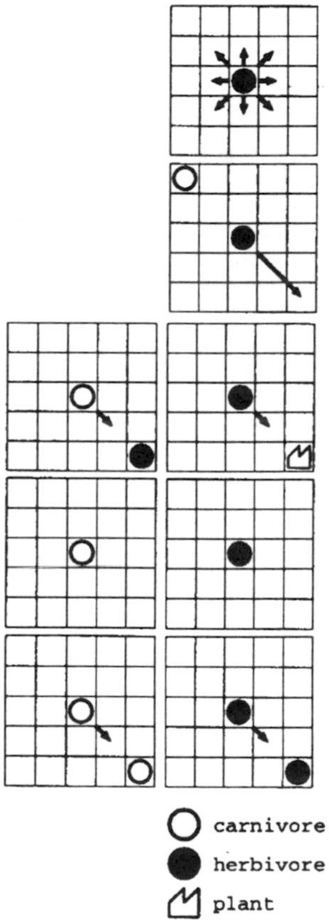

Figure 5: Actions of Herbivore

5 Experiments in a quasi ecosystem

To compare ability of three gene expressions in an environment of a quasi ecosystem we implemented a simulator. The simulator has two models, a small model and a larger mode. A small model is called mini-model, where a carnivore with a fixed rule and an evolvable herbivore act on a field. This model is used to investigate basic ability of each gene expression. The other model is called food-chain-model, whose objective is to investigate the phenomenon of food chain and co-evolution. In the following paragraphs, we give results in two models.

5.1 An experiment with mini-model

The field of mini-model is a two-dimensional 20×20 array. The herbivore and the carnivore act according with actions decided by their genes, n-BDDs, finite state automata or classifier systems. According to perceptional information, a gene selects an action from the four actions shown in Figure 5. Several plants are appeared in the simulator. When a plant is eaten by a herbivore, another plant are appeared in a randomly selected position.

In our experiment a gene of carnivore is fixed, and so the carnivore always acts with a fixed rule. On the other hand, herbivore's gene is evolved by GA. A stage in our simulator of quasi ecosystem starts with a herbivore, a carnivore and some plants and it terminates when the herbivore dies of hunger or being eaten.

To execute GA with our simulator, a group in a generation is composed of 30 individuals of herbivore. Each individual is let in the field and a stage is simulated with the herbivore. The number of steps for which the herbivore survived in the field is called fitness, which is used to select the next generation of herbivores. Five herbivores with the best fitness are included in the next generation. Five herbivores with the worst fitness are thrown away. Twenty-five herbivores are generated by the genetic operations from other twenty herbivores and the best five herbivores. The best herbivores are used twice.

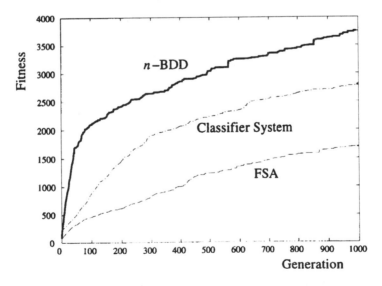

Figure 6: Fitness of herbivores

In the experiment we observed the speed of convergence with each gene expression. Figure 6 shows the average values of the maximum fitness of herbivores using n-BDDs, finite state automata and classifier system at each generation when experimenting 100 times. It is understood that the rise of the fitness of n-BDDs is faster than that of finite state automata and classifier system.

5.2 An experiment in food-chain-model

Here we explain another experiment in food-chain-model, where many herbivores and many carnivores exist and both of them are evolvable. This model aims to observe co-evolution among competitive animals through the food-chain.

In the simulator fields is a 150×150 two dimensional array. Both of animals have an input perceptual eight bit string shown in Table 1. Carnivores selects an action from eat, approach and do-nothing. Herbivore may select runaway, as well. A herbivore (a carnivore) acquires energy by eating plants (herbivores, respectively). When its energy reaches a threshold, it makes a child and the energy is distributed equally between it and its child. Its energy decreases in a constant step by every action until it becomes zero. When it becomes zero the animal will die and plants will grow around the body. Figure 7 illustrates the relation of food chain.

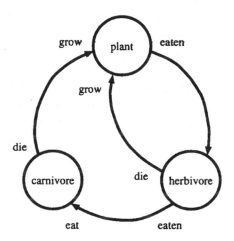

Figure 7: Food chain relation

Figure 8 shows a result with food-chain-model. We can observe many peaks of populations. Between 2000th step and 9000th step, every peak of herbivores' population follows a peak of carnivores' population. This shows that a food chain is caused during the period, and animals are evolved efficiently.

6 Discussions

In this paper, we proposed n-BDD as a gene expression with their genetic operations to use an genetic algorithm effectively. We confirmed the effectiveness by the experiments. The further study is to limit the search space using the normalization algorithm on n-BDDs. To compare with classifier system as a gene expression is also a research topic.-

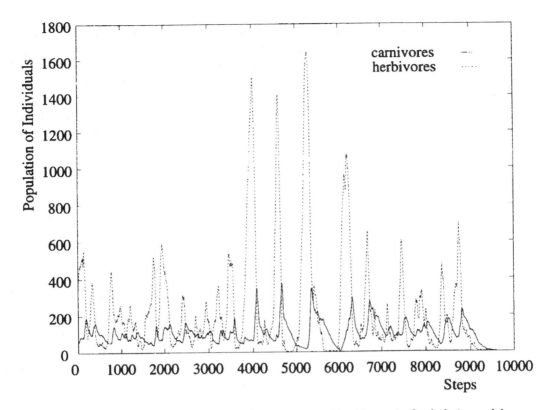

Figure 8: Transition of the number of carnivores and herbivores in food-chain-model

From experimental results, we can find that n-BDD expression is more superior to other expressions using finite state automata and classifier systems. In the section we compare n-BDDs with other two expressions. First we discuss on reasons of the superiority of the n-BDD gene expression from the aspect of local search ability and global search ability.

A decision node of n-BDD corresponds to a perceptual information bit. Hence a genetic operation to a decision node affects only the corresponding perceptual bit. This means that n-BDD gene expression gives local search ability. On the other hand, a state of a finite state automaton does not correspond to any individual information bit, because loops and cycles may exist in a transition rule. Because a genetic operation is done to the state which may accept many cases of input strings and the operation changes transitions of the state, the operation causes a large change of the function expressed by the automaton. We can say that it is inferior in the local search ability.

The disadvantage of finite state automata in local search is an advantage in a global search. The global search ability for n-BDDs is, however, obtained by doing an genetic operation to the near part to the root in an n-BDD. We can conclude that n-BDD is superior in the search ability to finite state automata.

Comparing n-BDDs with classifier systems, we should note that classifier system has a probabilistic factor. It output action by selecting a rule from a set of competitive rules probabilistically. On the other hands an n-BDD outputs actions deterministically. It prevents classifier systems from increasing their fitness.

162

We should also discuss on the time and space cost of the methods. The number of nodes of an n-BDD is at most 2^k, which is happen if the n-BDD is a complete binary tree, where k is the length of perceptual bit string and this value can be considered a constant value. Hence, necessary space to store an n-BDD is bounded by $O(2^k)$. On the other hands neither the number of states of a finite state automaton nor the number of classifiers of a classifier system are not bounded by any constant. Because of the bound for the size of an n-BDD, genetic operations, mutation, insertion and deletion, with a constant time can be implemented. Mutation, insertion and deletion for finite state automata and classifier systems and crossover for classifier systems take a proportional time to the size of genes, however.

References

[1] W. Daniel Hillis, 1990, *Co-Evolving Parasites Improve Simulated Evolution as an Optimization Procedure*, ARTIFICIAL LIFE II, Santa Fe Institute, Addison-Wesley.

[2] Lindgren Kristian and Mats G. Nordahl, 1994, *Artificial Food Webs*, ARTIFICIAL LIFE III, Santa Fe Institute, Addison-Wesley.

[3] L. J. Fogel, A. J. Owens, and M. J. Walsh, 1967, *Artificial Intelligence Through Simulated Evolution*, John Wiley & Sons.

[4] T.Takashina and S.Watanabe, 1994, *Study of self adaptive behavior in quasi-ecosystem*, Proceedings of the Third Parallel Computing Workshop, Kawasaki Japan, November, P2-U.

[5] T. Takashina and S. Watanabe, 1995, *Simulation model of self adaptive behavior in quasi-ecosystem*, IEICE Trans. on Fundamentals of Electronics, Communications and Computer Sciences, Vol. E78-A, No. 5.

[6] Holland,J.H, 1975, *Adaptation in natural and artificial systems*, University of Michigan Press.

[7] Holland,J.H, 1985, *Properties of the Bucket Brigade*, Proceedings of an International Conference on Genetic Algorithms and their Applications, 1-7. John J.Grefenstette (Ed.). Carnegie-Mellon University, Pittsburg.

[8] S. B. Akers, 1978, *Binary Decision Diagrams*, IEEE Trans. Comput., pp. 509-516.

[9] R. E. Bryant, 1986, *Graph-Based Algorithms for Boolean Function Manipulation*, IEEE Trans. Comput., pp. 677-691.

[10] Gregory M.Werner and Michael G. Dyer, 1993, *Evolution of herding behavior in artificial animals*, From Animals to Animats 2 Proceedings of the Second International Conference on Simulation of Adaptive Behavior, The MIT Press.

[11] R.Haberman, 1977, *Mathematical Models*, PRENTICE HALL.

Empirically-Derived Population Size and Mutation Rate Guidelines for a Genetic Algorithm with Uniform Crossover

Edwin A. Williams[1], William A. Crossley[2]

[1]*School of Aeronautics and Astronautics, Purdue University, West Lafayette, IN 47907-1282*
williame@ecn.purdue.edu
[2]*School of Aeronautics and Astronautics, Purdue University, West Lafayette, IN 47907-1282*
crossley@ecn.purdue.edu

Keywords: genetic algorithm, population size, mutation rate, uniform crossover

Abstract

The Genetic Algorithm (GA) is employed by different users to solve many problems; however, various challenges and issues surround the appropriate form and parameter settings of the GA. One of these issues is the conflict between theory and experiment regarding the crossover operator. Experimental results suggest that the uniform crossover can provide better results for optimization, so many users wish to employ this approach. Unlike for the single-point crossover GA, no established set of guidelines exists to assist in choosing appropriate population sizes and mutation rates when using the uniform crossover. This paper presents the results of an empirical study to determine such guidelines by examining several parameter combinations on four mathematical functions and one engineering design problem. The resulting guidelines appear to be valid over these test problems. They are presented and discussed, with the intent that they may provide assistance to users of GAs with uniform crossover.

1. Introduction

The Genetic Algorithm (GA) [1] has found increasing applications in many problem solving areas, including function optimization and optimal engineering design. The ability to find near globally-optimal solutions in a non-convex, multimodal and/or discontinuous design space without requiring an initial design point has made these algorithms appealing. While, as a whole, genetic algorithms work very well on a wide range of problems, different versions of the GA display different performance when solving different problems. This problem dependence has been noted in many instances, including in the use of crossover schemes.

Different crossover approaches provide an opportunity to improve the designs generated by a GA. But, a conflict between theory and empirical evidence exists regarding the effectiveness of crossover operators. Theory supports the single-point crossover [1], while the empirical evidence suggests that uniform crossover [2] is better for GAs applied to function optimization and engineering design. While much published work continues this debate [3,4], there are GA users who wish to employ the uniform crossover strategy but do not have adequate guidelines for appropriate population sizes and mutation rates.

The work described in this paper used several mathematical test problems in an empirical study of the effects that population size and mutation rate have on GA performance when binary coding and uniform crossover are used. As a result of this work, a simple set of guidelines for these parameters has been derived to provide good GA performance, in both "optimality" of the results and computational expense. These guidelines were then checked by solving an engineering design problem via a GA with uniform crossover.

1.1 Crossover as an Operator

As an operator in the genetic algorithm, crossover provides the real engine for search through the design space. The GA can so quickly focus on the most promising parts of a design space because the crossover operator combines parts of good solutions to form new potential solutions [1]. Information contained in one string combines with information contained in another string, and the resulting child strings will either have a good fitness and survive to exchange this information again or will have a poor fitness and will not survive, taking the poor information with them when they perish.

Crossover also plays an important role in what is termed *schema disruption* [1]. Schemata are patterns of 1's and 0's in the binary code that are associated with certain behavior of the design. The schemata associated with highly-fit individuals are propagated rapidly through the population. For example, the string "100110" contains the schema "10****" (where the * signifies a location not specified as a one or a zero), the schema "***110", and several other schemata. This example string also contains the schema represented by "1****0", which has a long length between the bits of concern. If the example string "100110" exchanges information with the string "011011" in a single point crossover (Fig. 1), the schema of "1****0" is disrupted. In this contrived example, if "1****0" is associated with higher fitness than the resulting "0****0" and "1****1" schemata, the offspring will likely have lower fitness because the desirable schema no longer exists. In the same crossover, short length schemata, like "10****", are not as likely to be disrupted during crossover.

$$100|110 \longrightarrow 100011$$
$$011|011 \longrightarrow 011110$$

Figure 1. Single point crossover with disruption of schema "1****0".

The notion of schema disruption led most early GA researchers to use single-point crossover techniques. Holland's Fundamental Theorem of Genetic Algorithms [5] deals explicitly with the number of schemata that are processed by the GA, leading to the "implicit parallelism" label given to genetic algorithms. De Jong's work used a generalized crossover model and suggested that single-point crossover performed better than multi-point schemes over his test function suite, because fewer schemata were disrupted by the single-point crossover [6].

Recently, researchers experimenting with GA have seen contrary results. The notion of the uniform crossover approach, in which each bit in the parent strings is evaluated for exchange with a probability of 0.5, was shown to have promising capability [2]. This leads to essentially $\ell/2$ crossover points on strings with length ℓ [3]. From a schema processing standpoint, the uniform crossover is a poor method; however, empirical evidence seems to suggest that uniform crossover is a more *exploratory* approach to crossover than the traditional *exploitative* approach that maintains longer schemata. This apparently results in a more complete search of the design space while still maintaining the exchange of good information. For example, a two bit-order schema remains intact with a probability of 0.5. Eshelman and Schaffer [4,7] compared the uniform crossover to more traditional approaches, with the uniform crossover appearing to have better performance. In light of this, Spears and De Jong [3] attempted to uncover the reasons for the observed performance advantages of uniform crossover over the more traditional single-point method. Unfortunately, this debate still lingers, as no satisfactory theory exists to explain the discrepancies. Still, many GA users wish to take advantage of the observed benefits of uniform crossover.

1.2 Population Size

Choosing the appropriate population size, N, presents a challenge for the genetic algorithm user. A large N may provide good coverage of the design space, yet may be so diverse that it is statistically difficult to combine two exceptional individuals to generate nearly-optimal individuals. Conversely, a small N may not provide enough diversity for the GA to discover designs close to the global optimum and will prematurely converge. "Rules of thumb" have been used to select the initial population sizes for the GA, which are often loosely based on the length of the binary strings used to represent individuals in a population.

Goldberg [8,9] has derived the optimal population size for binary coded strings, when the GA uses single-point crossover. The optimal population size for serial GAs grows exponentially with the string length used. Using these guidelines for a binary-coded GA with uniform crossover appears to give unacceptable performance. This motivated the work discussed herein.

1.3 Mutation Rates

The appropriate choice for the probability of mutation, P_m, is also an issue of debate in GAs, and evolutionary programming in general. Some researchers claim that mutation is a primary operator in a genetic search, so the mutation rate should be high. Others argue that crossover provides the majority of the GA's search abilities, so mutation rates should be low.

While this work does not attempt to solve this argument, it does assume that the crossover operator provides most of the search power of the GA, even though uniform crossover disrupts the theoretically important "long-length" and "high-order" schema. Probability of mutation examined in this effort varied from a low of $1/2N\ell$, to a high of $(3\ell-1)/2N\ell$ which effectively encompassed the commonly suggested ranges for mutation as cited in both Smith [10] and in Bäck[11] ($1/N\ell \leq P_m \leq 1/N$). This range also allowed exploration of the effects of probabilities lower and higher than previous guidelines. The four highest rates represent linearly increasing values of P_m, while the lowest breaks from this trend to ensure a positive value.

2. Runs and Results

To investigate the various population sizes and mutation rates, a serial genetic algorithm was written in Fortran 90 and compiled on a Sun Sparc Ultra-2 server. The GA incorporated tournament selection, uniform crossover, and a stopping criterion to halt the GA after five consecutive generations with no improvement in the best individual. This GA was used to solve four mathematical optimization test problems and an engineering design problem to determine appropriate guidelines for population size and mutation rate.

The four mathematical problems were chosen to present increasingly challenging cases to the genetic algorithm. These problems start with a simple, two-dimensional, uni-modal problem and increase in complexity by presenting multi-modal functions and additional numbers of design variables. Finally, a stiffened panel constructed from composite materials was designed for minimum weight subject to several constraints. This final problem is intended to represent a "real-world" engineering problem for which a GA is a suitable solution technique.

Each problem was solved by 100 runs of the GA for a given set of parameters to reduce the effect of probabilistic "noise". Multiples of the string length (ℓ, 2ℓ, 4ℓ, and 8ℓ) provided the population sizes for this investigation. The investigated mutation rates were also functions of the string length and of the population size. As an example, Table 1 displays the matrix of parameters used for the engineering test problem with a string length of 34 bits.

Table 1. Population sizes and associated mutation rates studied for $\ell = 34$.

population sizes		mutation rates				
	N	$(3\ell - 1)/2N\ell$	$1/N$	$(\ell + 1)/2N\ell$	$1/N\ell$	$1/2N\ell$
ℓ	34	0.04369	0.02941	0.01514	0.00087	0.00043
2ℓ	68	0.02184	0.01471	0.00757	0.00043	0.00022
4ℓ	136	0.01092	0.00735	0.00378	0.00022	0.00011
8ℓ	272	0.00546	0.00368	0.00189	0.00011	0.00005

For each run, the best fitness value and the number of generations were recorded as measures of the GA's effectiveness. These values were averaged over the 100 runs for each set of N and P_m to provide representative performance of a general GA run. For illustration, the average performance values for each parameter set were then normalized to those values from runs with the population size of ℓ and the mutation rate of $(3\ell - 1)/2N\ell$.

2.1 Test Problem 1

The first problem examined in this study was the "banana" function of Vanderplaats [12]; it is so named, because a contour plot of the function has a banana-like shape. This two-dimensional problem is smooth, convex and has continuous derivatives. Although this can generally be solved by other methods, it provides a slightly challenging starting point to test the GA's behavior. The "banana function" problem is stated as

$$\text{minimize } f(\mathbf{x}) = 10x_1^4 - 20x_1^2 x_2 + 10x_2^2 + x_1^2 - 2x_1 + 5 \tag{1}$$

For representation in the binary-coded GA, the variables x_1 and x_2 were each coded to 10-bit binary strings, providing a minimum value of -4 and a maximum value of 12 with a resolution of 1/64. The total chromosome length for this problem was 20 bits. Figure 2, below, shows a plot of this function. The known optimum of this function occurs at $\mathbf{x}^* = [1\ 1]^T$.

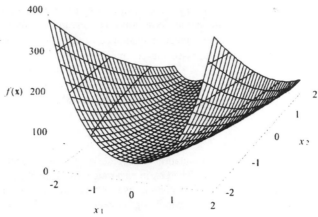

Figure 2. Vanderplaats' "banana" function.

Figure 3 shows a graph of the normalized best fitness data and normalized function evaluations obtained for this first test problem, using the data reduction methods detailed above.

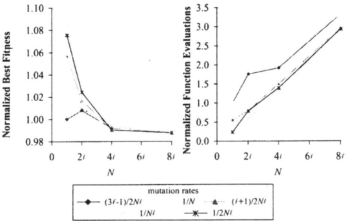

Figure 3. Normalized best fitness values (left) and normalized function evaluations (right) versus population size and mutation rates for the "banana" function.

The plots above show that a larger N usually provided a more accurate solution (nearer the global optimum), but very little improvement was observed for population sizes above 4ℓ. The computational cost, for which the general trend is a linear increase of computational effort for an increase in population size, must also be considered. While there was little improvement in the best fitness for populations larger than 4ℓ, the computational effort continued to increase. These results suggest that $N = 4\ell$ is an appropriate compromise for best fitness and reasonable computational effort for the banana function.

This also shows a similar trade-off with respect to the mutation rate. For this problem, the rate of $1/N$ appeared to yield the most accurate results, yet this was one of the more "expensive" rates. Since the relative scale of improvement for the mutation rate was smaller than the population size, it seems prudent to choose as best the rate that provided good fitness for a minimal cost increase. This was the $(\ell+1)/2N\ell$ mutation rate.

2.2 Test Problem 2

The two-dimensional, multimodal "egg-crate" problem was used as the second case to examine N and P_m. This would be challenging to solve with a calculus-based method but is well suited to the GA. The objective function is

$$\text{minimize } f(\mathbf{x}) = x_1^{\ 2} + x_2^{\ 2} + 25\left(\sin^2 x_1 + \sin^2 x_2\right) \tag{2}$$

The parameters, x_1 and x_2, were coded to binary strings of 10 bits each, ranging from $-\pi$ to $+\pi$, with a resolution of $\pi/512$; this made the total chromosome length 20 bits. Figure 4 shows a surface map of the "egg-crate" function, whose known global optimum is at $\mathbf{x}^* = [0\ 0]^T$.

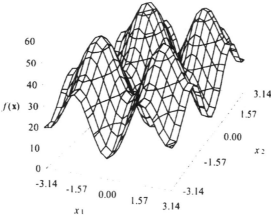

Figure 4. Surface plot of the "egg-crate" function.

As with the first test problem, performance data was collected for 100 runs at each combination of parameter settings, then was averaged and normalized. Figure 5 shows a graph of the best fitness and function evaluation data obtained for this problem. The scale of the normalized best fitness plot was "cut-off" at 1.4. High fitness values are not desirable for minimization, and the runs with a population size of ℓ and mutation rates of $1/N\ell$ and $1/2N\ell$ have normalized fitness values of 22 and 36, respectively.

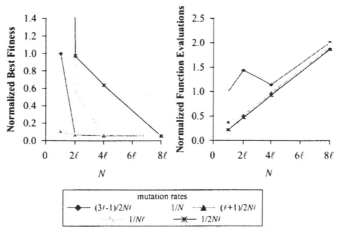

Figure 5. Normalized best fitness values (left) and normalized function evaluations (right) versus population size and mutation rates for the "egg-crate" function.

For mutation rates other than $1/N\ell$ and $1/2N\ell$, there was very minimal improvement in best fitness for population sizes greater than 2ℓ, and essentially no improvement above 4ℓ. These best-fitness results must be weighed against the computational cost. The trend for this plot is a generally linear progression of computational time with increasing N, as previously observed in Fig. 3.

Mutation rates of $(\ell+1)/2N\ell$ and $1/N$ provide good performance for all of the population sizes used on the egg-crate problem. For these mutation rates, there was little difference in the best fitness over the whole range of population sizes. However, the mutation rate of $(\ell+1)/2N\ell$ consistently required fewer function evaluations.

2.3 Test Problem 3

The third problem examined was a six-dimensional version of Rosenbrock's function. This function was chosen since the greater number of variables would add more complexity to a problem similar to the "banana" function. The objective function is

$$\text{minimize } f(\mathbf{x}) = \sum_{i=1}^{5}\left[100\left(x_{i+1} - x_i^2\right)^2 + \left(1 - x_i\right)^2\right] \tag{3}$$

The six parameters were coded to binary strings of 12 bits each, ranging from -20 to 20, with a resolution of 5/512. The total chromosome length for this problem was 72 bits. In two-dimensional space, Rosenbrock's function is very similar to the "banana" function, as displayed by Figure 6.

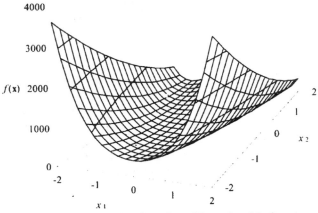

Figure 6. Two-dimensional version of Rosenbrock's function.

Figure 7 shows a graph of the normalized best fitness and function evaluation data obtained from the analysis of this function for each combination of N and P_m. As before, the best fitness graph was "cut-off" at 1.4; the runs with a population size of ℓ and mutation rates of $1/N\ell$ and $1/2N\ell$ had normalized fitness values of 33 and 6, respectively, and lower values were easily obtained for other combinations.

The fitness plot shows that for population sizes greater than 2ℓ, there was little improvement in the fitness for P_m of $1/N$, $(\ell+1)/2N\ell$, and $(3\ell-1)/2N\ell$. For all mutation rates, there was even less improvement for N above 4ℓ. Computational expense was again generally increasing linearly with population size.

The normalized function evaluations graph also shows that the optimization cost increases with an increasing mutation rate. For the three mutation rates providing good fitness performance, there was little improvement in the average fitness after 2ℓ. Of these, the mutation rate with the lowest cost was $(\ell+1)/2N\ell$.

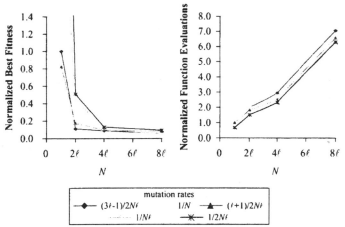

Figure 7. Normalized best fitness values (left) and normalized function evaluations (right) versus population size and mutation rates for the six-dimensional Rosenbrock's function.

2.4 Test Problem 4

As a last mathematical optimization problem, a four-dimensional version of Griewank's function was used. This function would be challenging to conventional methods since it is non-linear and multimodal. The greater number of variables adds to the complexity of the problem. This function was expressed as

$$\text{minimize } f(\mathbf{x}) = \sum_{i=1}^{4} \frac{x_i^2}{200} - \prod_{i=1}^{4} \cos\left(\frac{x_i}{\sqrt{i}}\right) + 1 \tag{4}$$

The four parameters were coded to binary strings of 12 bits each. Each of these parameters were given ranges from -10 to 10. This gave a resolution of 5/1024 and a total chromosome length of 48 bits. Figure 8 shows a surface plot of the two-dimensional Griewank's function.

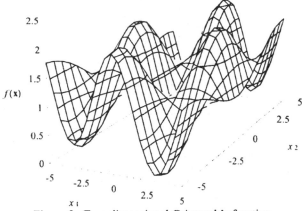

Figure 8. Two-dimensional Griewank's function.

Figure 9 shows a graph of the normalized best fitness and function evaluation data obtained from the solutions of this function. Once again, there was minimal improvement in the fitness above a population of 4ℓ and a generally linear increase in function evaluations with increasing N. As before, this figure shows that the computational cost increased with an increasing mutation rate, while higher mutation rates give better fitness values. The graphs show that a population size of approximately 4ℓ was adequate, and that the mutation rate of $(\ell+1)/2N\ell$ gives "good" fitness results and a computational cost that was (relatively) inexpensive.

170

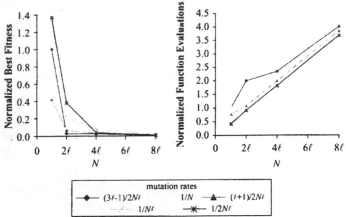

Figure 9. Normalized best fitness values (left) and normalized function evaluations (right) versus population size and mutation rates for the four-dimensional Griewank's function.

2.5 Stiffened Composite Panel Design

As a final problem for investigating the population size and mutation rate trends, a stiffened composite panel design problem was used as a realistic engineering problem. For this problem, the objective was to minimize the weight of a panel 8 in. deep and 16 in. wide which much support an evenly distributed 10,000 lb load without buckling, using a 150% safety factor. Figure 10 displays a representative panel.

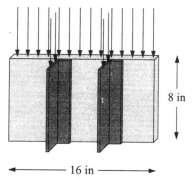

Figure 10. Representative stiffened composite panel

This panel design problem is well suited for application of the GA, because a combination of continuous, discrete and integer variables form the necessary design variable vector. For this problem, the objective function was the panel weight, and constraints were imposed to prevent buckling, to ensure a balanced plate laminate, and to maintain a minimum contact dimension between the stiffeners and the plate. These were enforced via an exterior "step-linear" penalty [13]. Eight design variables described the possible panel designs. The panel material, stiffener shape and the plate ply orientation angles were discrete design variables and were coded to binary strings. The integer variables, number of stiffeners and number of plate plies were coded with a resolution of one, while the plate and stiffener dimensions (continuous variables), were also discretized in order to map them to binary strings. Total length of the chromosome was 34 bits. A full description of this problem can be found in Ref. 13.

Figure 11 shows graphs of the normalized best fitness and normalized function evaluations performance data from this problem.

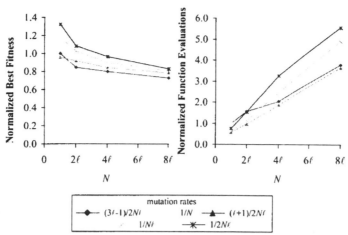

Figure 11. Normalized best fitness values (left) and normalized function evaluations (right) versus population size and mutation rates for the stiffened composite panel design problem.

This data shows similar trends as the mathematical problems. Compared to the previous test problems, the best fitness performance improved with increasing population size with a more noticeable, but still only slight, improvement beyond $N = 4\ell$. The computational cost again increased almost linearly with population size. However, the highest mutation rate of $(3\ell-1)/2N\ell$ appears to have outperformed the other rates on the best fitness measure. When evaluating the computational cost, the "best" choice for minimum number of evaluations is $(\ell+1)/2N\ell$. For this problem, it is suggested to choose a population size of 4ℓ and a mutation rate of $(\ell+1)/2N\ell$. This gives good fitness performance for low computational cost.

3. Summary and Conclusions

For all of the problems examined in this effort, similar trends of best fitness and computational expense performance versus population were observed. For the small population size, $N = \ell$, fitness performance was generally poor, but improved rapidly with increasing N. This continued until the population size began to exceed 4ℓ. Increases in N beyond this value generally gave little or no improvement in fitness performance. The number of function evaluations to reach the GA stopping criterion for all of the problems increased in an essentially linear relationship with population size. In view of this increasing trend in cost, the minimal improvement in fitness beyond $N = 4\ell$ does not appear to be worth the related expense.

At small population sizes, the mutation rates appeared to have a marked impact on best fitness performance, but as population sizes increased to $N = 4\ell$, this effect became minimal. Additionally, at larger N, the improvement gained in fitness from one mutation rate to another was much smaller than the effect of population size. A similar statement can be made about the impact of P_m on the number of fitness evaluations. As the complexity increased in moving from the 2-D to the 4-D, 6-D and mixed variable type problems, higher mutation rates provided slightly better fitness at a slightly increased cost but not with complete regularity. Because of the secondary effect of mutation at prudent population sizes, the value $P_m = (\ell+1)/2N\ell$, which provides an excellent compromise between fitness performance and computational cost performance, makes a good guideline for mutation rate.

For each of the five problems, the overall performance trends are similar and support the above claims, but there are some specific differences for each problem. The issue of problem dependence could be at work here, in that some problems may be better suited to larger populations. The empirically derived guidelines of $N = 4\ell$ (four times the string length) and $P_m = (\ell+1)/2N\ell$ should be viewed as such, but may be used knowing that they should provide good performance for a serial GA with uniform crossover.

Because the population size of $N = 4\ell$ appears to work well, this suggests a computational advantage over the serial GA with single-point crossover for problems using long-length chromosomes. For example, using Goldberg's guidelines [8, 9] for optimal population size, $N = 8$ for $\ell = 10$. With the uniform crossover, $N = 40$

for $\ell = 10$. However, for a string length of $\ell = 50$, the single-point crossover would require $N = 1952$; the uniform crossover, $N = 200$.

These empirically derived guidelines should provide beneficial information to GA users who wish to take advantage of the experimentally observed advantages of uniform crossover, but have no good starting point from which to choose crucial GA parameters.

References

[1] Goldberg, David E., 1989, *Genetic algorithms in search, optimization, and machine learning*, Addison Wesley, Reading MA.

[2] Syswerda, G., 1989, Uniform crossover in genetic algorithms, *Proceedings of the third international conference on genetic algorithms*, George Mason University, Fairfax, VA, pp. 2-9.

[3] Spears, W. M. and De Jong, K. A., 1991, On the virtues of parameterized uniform crossover, *Proceedings of the fourth international conference on genetic algorithms*, University of California at San Diego, San Diego CA, pp. 230-236.

[4] Eshelman, L. J. and Schaffer, J. D., 1993, Crossover's niche, *Proceedings of the fifth international conference on genetic algorithms*, University of Illinois at Urbana-Champaign, Urbana, IL, pp. 9-14.

[5] Holland, J. H., 1992, *Adaptation in natural and artificial systems*, MIT Press, Cambridge, MA.

[6] De Jong, K. A., 1975, *An analysis of the behavior of a class of genetic adaptive systems*, PhD Dissertation, University of Michigan, Ann Arbor, MI.

[7] Eshelman, L. J. and Schaffer, J. D., 1991, Preventing premature convergence in genetic algorithms by preventing incest, *Proceedings of the fourth international conference on genetic algorithms*, University of California at San Diego, San Diego, CA, pp. 115-22.

[8] Goldberg, D. E., 1985, Optimal initial population size for binary-coded genetic algorithms, TCGA Report No. 85001, The Clearinghouse for Genetic Algorithms, The University of Alabama, Tuscaloosa, AL.

[9] Goldberg, D. E., 1989, Sizing populations for serial and parallel genetic algorithms, *Proceedings of the third international conference on genetic algorithms*, George Mason University, Fairfax, VA, pp. 70-79.

[10] Smith, R. E., 1993, Adaptively resizing populations: an algorithm and analysis, TCGA Report No. 93001, The Clearinghouse for Genetic Algorithms, The University of Alabama, Tuscaloosa, AL.

[11] Bäck, T., 1993, Optimal mutation rates in genetic search, *Proceedings of the fifth international conference on genetic algorithms*, University of Illinois at Urbana-Champaign, Urbana, IL, pp. 2-8.

[12] Vanderplaats. G. N., 1992, *Numerical optimization techniques for engineering design*, McGraw-Hill, New York.

[13] Crossley, W. A., and Williams, E. A., 1997, A study of adaptive penalty functions for constrained genetic algorithm based optimization, 35th Aerospace Sciences Meeting and Exhibit, Reno, NV, AIAA Paper 97-0083.

Analysis of Various Evolutionary Algorithms and the Classical Dumped Least Squares in the Optimization of the Doublet

Darko Vasiljević, Janez Golobič

Department of Optics and Optoelectronics, VTI VJ

Avalska 17, Belgrade, Yugoslavia

darko@afrodita.rcub.bg.ac.yu

Keywords: genetic algorithm, evolutionary strategies, optical design, optimization, doublet

Abstract

Doublet is a very simple optical system with only two lenses. Optimization of the doublet with various algorithms both classical (dumped least squares) and evolutionary (adaptive steady state genetic algorithm without duplicates, (1+1) evolutionary strategy and (μ, λ) evolutionary strategy with or without recombination) is given. All the algorithms are described and their advantages and shortcomings are presented. Simulation results of the doublet optimization for each algorithm are presented and discussed.

1. Introduction

Problem of the automatic lens design and optimization of the optical system is very old. Many researchers proposed various methods or their improvement in order to solve this problem, which belongs to a class of highly nonlinear optimization problems. All optimization problems can be classified in two broad groups:

- classical optimization methods which are based on the computations of the first and the second partial derivatives. In the optimization of the optical systems some kind of the dumped least squares (DLS) is mostly used.
- modern optimization methods which are based on analogies in nature like simulated annealing (SA) and evolutionary algorithms like genetic algorithms (GA), evolutionary strategies (ES) and evolutionary programming

Currently many automatic lens design programs are commercially available or developed for proprietary use and most of them employ some variant of the dumped least squares. Our intention is to try to implement modern optimization techniques in the optimization of the optical systems. In this paper we present following evolutionary algorithms applied in the optimization of the optical systems:

- adaptive steady state genetic algorithm without duplicates - (ASSGA)
- (1+1) evolutionary strategy - (1+1)ES
- (μ, λ) evolutionary strategy without recombination - (μ, λ)ES
- (μ, λ) evolutionary strategy with recombination - (μ, λ)ESR

All those algorithms are compared among themselves and with the classical dumped least squares optimization and the best evolutionary algorithm is selected.

Simple optical system like doublet which consist of only two lenses, one positive and one negative, is chosen for comparison of the optimization algorithms. All optimizations are done on proprietary optical design program called ADOS (Automatic Design of Optical Systems) described in the Master of Science thesis by Vasiljević [1] in serbian language.

This paper is continuation of the research presented in the paper by the same authors on the First Workshop on Soft Computing [2].

2. Classical dumped least squares optimization

DLS optimization is a modification of the Newton - Raphson method first developed by Levenberg [3]. It was introduced into optics by Rosen and Eldert [4], Merion [5,6], Wynne [7] and others. Our implementation of

DLS optimization is based upon works of researchers from Imperial College in London (Wynne, Wormell and Kidger [8,9,10,11]). Mathematical theory of the DLS optimization is well known and will not be presented here.

DLS is very much dependent of the starting point of the optimization. If the starting point is badly chosen this may prevent DLS optimization in finding sufficiently good optical system which will satisfy all design criteria. Finding good starting point for optimization is highly dependent on the skill and experience of the designer and he may have difficulty in searching out satisfactory solution for "state of art" design requirements.

3. Evolutionary optimization

Evolutionary optimization is one of possible ways to improve classical optimization. All classical optimization methods belongs to the local optimization methods since it guaranties finding only local minimum nearest to the starting point. They don't take no explicit account of the fact that there may be many local minima of the merit function in the search space. Evolutionary optimization is looking for as many local minima as it can find and chooses best among them. It can be said that evolutionary optimization belongs to global optimization methods.

Evolutionary optimization is based on analogy in the nature. All life in our planet evolved according Darwinian theory of the evolution. So we try to apply simplified Darwinian theory of the evolution in the technical systems optimization. Genetic algorithms emphasize selection process and crossover, while evolutionary strategies emphasize normally distributed mutations.

3.1. Genetic algorithm

We chose adaptive steady state genetic algorithm without duplicates which is described in Handbook of Genetic Algorithms by Davis [12] and applied it to the optical design, because it works with real numbers and according to Davis has substantial advantages over other genetic algorithms like simple genetic algorithm or elitist genetic algorithm which usually works with bit strings. For us it is very important that genetic algorithm works with real numbers because it is incorporated in our proprietary program for complete design of optical systems. All important information about an optical system is placed in one record and that record represent one individual of the population.

The population of optical systems is initialized randomly. Because of the fact that good starting point for the optimization is not known in advance, best way is to start with randomly chosen starting points, and let the GA do the rest.

One of the most important things in defining the GA optimization is a proper determination of the evaluation function. The merit function used in the classical DLS, which consists of the sum of the squares of weighted aberrations, is also used the GA.

Because of randomly chosen starting points for the optimization, the merit functions of the optical systems usually differ a lot. In order to be possible to compare them the linear normalization, to interpolate the merit function from one starting point to the end point, is introduced. Values of the starting point, the end point and the step are parameters of our implementation of the GA optimization and can be chosen differently for each optimization run by the optical designer.

The parent selection technique is an roulette wheel parent selection, which gives more reproductive chances to the optical systems that have better merit functions (smaller aberrations).

GA optimization uses steady state without duplicates for reproduction technique because every member of the population will be different and the best individuals from all generations will be kept together so no variable genetic information will be lost.

When two different parents are selected only one genetic operator is applied, chosen by roulette wheel operator selection from:

- uniform crossover;
- average crossover;
- real number mutation;
- big number mutation;
- small number mutation.

The new offspring competes with all members of population for place in it. If it is better than the worst member in population, the offspring is accepted, and the worst member is deleted from population.

3.2. Evolutionary strategies

We implemented three evolutionary strategies which are described in Evolution and optimum seeking by Schwefel [13]:

- (1+1) ES - in this evolutionary strategy there is only one parent and offspring, which is made by mutating parent. Mutations are done according to the Gaussian normal distribution law. Better optical system, i.e. with smaller merit function and aberrations eider parent or offspring, is chosen for next generation.

- (μ, λ) ES without recombination - in this evolutionary strategy there are μ parents and they produce λ offspring's $(\lambda > \mu)$. Only μ best offspring are kept for next generation. Offsprings are made from parents only by mutation according to the Gaussian normal distribution. In our implementation number of parents are 10 and number of offsprings are 100 which is in accordance with theory that number of offspring must be $\lambda \geq 6\mu$.

- (μ, λ) ES with recombination - this is same evolutionary strategy as previous with following addition. New offsprings are made from parents with two genetic operators:

  - mutation which is same as in earlier implementations of ES,

  - recombination which combines two randomly chosen optical systems into one that is used as starting point for mutation.

All implementations of the evolutionary strategies use starting point which designer of the optical system must specify. Here isn't critical to specify good starting point because ES has mechanism for escaping local minima. (μ, λ) ES form initial population by mutating starting optical system. All mutations are done according to the Gaussian normal distribution law.

4. Simulation results and discussion

We chose one relatively simple optical system like a doublet to be starting point for optimization. This is because the evolutionary algorithms are rather time consuming and if there are lot of variable design parameters they may take time before it come with results. They also have stochastic nature so they ought to be run several times. In our case each evolutionary algorithm is run five times which is minimum value for calculating necessary average values.

Optimization by each algorithm is done when:

- only radiuses are variable,
- radiuses and separations are variable.

We don't varied glasses because they are discrete variables that are taken from glass database. Radiuses and separations are continuous variables and the evolutionary strategies are optimization algorithms that use only continuous variables for the optimization.

Principal optical data for the chosen doublet are given in table 1:

Table 1: Principal optical data for doublet

focal length [mm]	f = 125
relative aperture (f-number)	f / 5
aperture stop	at the first surface
field angle [°]	$2\varpi = 10°$

Results of the optimization is characterized by the merit function of the optical system. Results from the optimization when only radiuses were variable are presented in table 2:

Table 2: Optimization of the doublet with only radiuses variable

	ASSGA			(1+1) ES	(μ, λ) ES	(μ, λ) ESR
	pop=100	pop=200	pop=300			
I st run	23.295	23.274	23.272	23.270	23.270	23.270
II nd run	23.548	23.271	23.272	23.270	23.270	23.270
III rd run	22.299	23.273	23.271	23.270	23.270	23.270
IV th run	23.273	23.271	23.272	23.270	23.270	23.270
V th run	23.279	23.273	23.272	23.270	23.270	23.270
average	23.339	23.272	23.272	23.270	23.270	23.270
standard deviation	0.117	0.001	0.000	0.000	0.000	0.000

One can see very small variation of the average merit function with the change of the optimization algorithm. ASSGA with population of the 200 and 300 optical systems found optical systems with same average merit function 23.272. Three types of the evolutionary strategies (1+1) ES, (μ, λ) ES and (μ, λ) ESR found in all runs optical system with merit function 23.270. This optical system can be considered as a global optimum for the optimization of the doublet with only radiuses variable. Optical systems found by the ASSGA with the population of the 200 and 300 optical systems are very close to the global optimum. Optical systems found by the ASSGA with the population of the 100 optical systems are little more different. This can show that population size of the 100 optical systems is not big enough to find global optimum.

Results from the optimization when radiuses and distances are variable are presented in table 3:

Table 3: Optimization of the doublet with radiuses and distances variable

	ASSGA						(1+1)ES	(μ, λ)ES	(μ, λ)ESR
	max coef = 5			max coef=50					
population	100	200	300	100	200	300			
I st run	17.810	9.414	9.166	5.028	10.138	8.848	2.423	2.446	2.423
II nd run	15.636	11.764	10.494	14.401	12.130	11.891	2.427	2.427	2.423
III rd run	16.948	10.662	12.481	12.653	4.945	6.989	2.773	2.437	2.423
IV th run	11.718	13.881	10.293	6.322	13.935	4.408	2.638	2.427	2.409
V th run	17.603	11.214	12.323	5.183	9.910	4.345	2.698	2.433	2.427
average	15.943	11.387	10.951	8.717	10.212	7.296	2.592	2.434	2.421
standard deviation	2.510	1.644	1.419	4.462	3.368	3.188	0.160	0.008	0.007

One can see large variations of the average merit function with change of the optimization algorithm. All evolutionary strategies found optical systems with very similar merit function. Each optical system found by ASSGA has different merit function. Explanation of this fact is complex:

- In the evolutionary strategies variable radiuses and distances are continuous variables and can take every possible value, while variable radiuses and distances in the ASSGA can take values only from predefined interval of values. This interval is calculated for each variable radius and distance separately by taking value of the starting optical system variable radius or distance and multiplying it with the maxcoef (maximum coefficient) that is parameter of the ASSGA. Optimization was done for two values of maxcoef 5 and 50 because when maxcoef was 5 optical system that represent global minimum was out of reach of the allowed distances from the interval 0 to 5*variable distance. One can clearly see that much better optical systems, i.e. optical systems with smaller merit function, are found but they are still far from global minimum.

- Main reason why ASSGA is not so successful as ES lies in the foundations of the both algorithms:

  - ASSGA generates initial population randomly. Each variable parameter is randomly chosen from the interval 0 to maxcocf*variable parameter of the starting optical system. It is obvious that when the maxcoef is smaller randomly generated optical systems are closer to the starting optical system and have better merit function, then when the maxcoef is large. In order to be possible to reach global minimum maxcoef must be large and that implies randomly generating large number of the optical systems with very bad merit function.

- Evolutionary strategies use initial optical system as a starting point and form population by mutating, according to the Gaussian normal distribution law, starting optical system. There most of the population is very similar to the starting optical system and have good merit functions in comparison to the optical systems generated by ASSGA.
- In the evolutionary strategies mutation is main and most important genetic operator, while in the ASSGA mutation is auxiliary operator and selection and crossover are main operators.
- (μ, λ) ES and (μ, λ) ESR are typical representatives of generational optimization algorithms i.e. optical system can live only one generation, while in the ASSGA optical system can live for several generations if it has good merit function. In fact only merit function of the given optical system determines whether it will live or die.

Comparison of the optimization results for each evolutionary algorithm (ASSGA, (1+1) ES, (μ, λ) ES and (μ, λ) ESR) and classical DLS is presented in the table 4.

From the table 4 is clear that the all optimization methods found very similar results when the radiuses of the doublet are only variable. For ASSGA are presented average results. One can see that population of 100 optical systems is too small and that average results are worse then results obtained from ES and classical DLS. When population grow to 200 and 300 optical systems situation changed and average results obtained by the ASSGA is same as the classical DLS. All types of the ES found little better results (23.270 for ES and 23.272 for DLS). It is interesting to note that all runs of the each ES variant (1+1) ES, (μ, λ) ES and (μ, λ) ESR have found same optical system with the merit function 23.270, which is global minimum.

Table 4: Comparison of the results for the optimization of the doublet

Optimization type	variable parameter		merit function
ASSGA	r	pop = 100	23.339
		pop = 200	23.272
		pop = 300	23.272
ES	r	(1+1) ES	23.270
		(μ, λ) ES	23.270
		(μ, λ) ESR	23.270
DLS	r		23.272
ASGGA	rd	pop = 100	8.717
		pop = 200	10.212
		pop = 300	7.296
ES	rd	(1+1) ES	2.592
		(μ, λ) ES	2.434
		(μ, λ) ESR	2.421
DLS	rd		2.427

When the radiuses and distances of the doublet are variable situation is changed. ASSGA failed to find optical system with merit function comparable with other two types of the optimization. Merit functions of the optical systems found by the ASSGA are three to four times grater then merit functions of the optical systems found by the (1+1) ES, (μ, λ) ES and (μ, λ) ESR and DLS.

Evolutionary strategies in fifteen executions found nine different optical systems with merit functions from 2.773 to 2.409. Each of the two optical systems with merit functions 2.423 and 2.427 are found four times and represent dominating local minimums. Classical DLS also found optical system with merit function 2.427 as a local minimum. Here one can compare various evolutionary strategies among themselves. (1+1) ES is the worst evolutionary strategy with average merit function 2.592. In five executions (1+1) ES found following optical systems:

- two dominating local minimums are found i.e. optical systems with merit functions 2.423 and 2.427,
- three optical systems with the merit function worse then average.

(1+1) ES is very simple simulation of evolutionary process with only one parent and one offspring and only one genetic operator - normally distributed mutation. For obtaining better results simulation of the evolution must be complicated i.e. must be closer to the reality. This means that population of the optical systems must be greater then one and there must be various genetic operators like mutation and recombination. All this necessary conditions are incorporated in (μ, λ) ESR. One can clearly see that (μ, λ) ESR found optical systems with best merit functions. In five executions (μ, λ) ESR found:

- three times optical system with the merit function 2.423 which is one of the dominating local minimums,
- one optical system with merit function 2.427 which is other dominating local minimum (also found by classical DLS),
- one optical system with merit function 2.409 which is the best optical system found in those optimizations.

In four of five executions (μ, λ) ESR has managed to find optical system with merit function that is better then merit function of the optical system found by classical DLS.

5. Conclusions

In this paper simple optical system, doublet, is optimized by various evolutionary algorithms which belongs to the two main branches: genetic algorithm (ASSGA) and evolutionary strategies ((1+!) ES, (μ, λ)ES, (μ, λ)ESR. All optimized optical systems are compared with optical system optimization by classical DLS.

When the radiuses are only variable parameters of the optical system all optimization algorithms are found similar results. ASSGA with population 200 optical systems and greater and classical DLS found optical systems with same merit function. Evolutionary strategies managed to found optical systems with little better merit function.

When the radiuses and distances are variable parameter of the optical system all evolutionary strategies and classical DLS found similar optical systems, while the ASSGA failed. (1+1) ES and (μ, λ) ES found optical systems that have merit functions little worse then optical system found by classical DLS. (μ, λ) ESR found best optical systems that have merit function little better than optical system found by classical DLS.

From presented results of the doublet optimization it is clear that it is better to use evolutionary strategies then genetic algorithms in the optimization of the optical systems.

The best results are obtained by using (μ, λ) evolutionary strategy with recombination.

References

[1] Vasiljevi• D., 1989, Contribution to the lens design optimization with personal computers, MSc Thesis, University of Belgrade, Belgrade, Yugoslavia (in serbian)

[2] Vasiljevi• D., Golobi• J., 1996, Comparison of the clssical dumped least squares and genetic algorithm in the optimization of the doublet, Proceedings of the First Worksshop on soft computing, Nagoya, Japan, August, pp. 200 – 204

[3] Levenberg K., 1944, A method for the solution of certain nonlinear problems in least squares, Q. J. Appl. Math., vol. 2, pp. 164-168

[4] Rosen S., Eldret C., 1954, Least squares method for optical correction, JOSA, vol. 44, pp. 250-252

[5] Meiron J., 1959, Automatic lens design by the least squares method, JOSA, vol. 49, pp. 293-298

[6] Meiron J., 1965, Damped least squares method for automatic lens design, JOSA, vol. 55, pp. 1105-1109

[7] Wynne C., 1959, Lens designing by electronic digital computer, Proc. Phys. Soc. vol. 73, pp. 777-783

[8]   Wynne C., Wormell P., 1963, Lens design by computer, Appl. Opt., vol. 2, no.12, pp. 1233-1238

[9]   Kidger M., Wynne C., 1967, The design of double Gauss systems using digital computers, Appl. Opt., vol. 6, no.3, pp. 553-563

[10]   Wormell P., 1978, Version 14, a program for the optimization of lens designs, Opt. Acta, vol. 25, no.8, pp. 637-654

[11]   Kidger M., 1971, The application of electronic computers to the design of optical systems, including aspheric lenses, Ph. D. Thesis, University of London, London, UK

[12]   Davis L., ed., 1991, Handbook of genetic algorithms, Van Nostrand Reinhold, New York

[13]   Schwefel H-P., 1995, Evolution and optimum seeking, John Wiley , New York

Genetic Programming with One-Point Crossover

Riccardo Poli and W.B. Langdon

School of Computer Science
The University of Birmingham
Birmingham B15 2TT, UK
E-mail: {R.Poli, W.B.Langdon} @cs.bham.ac.uk

Keywords: genetic programming, one-point crossover, parity problems, schema theorem

Abstract

In recent theoretical and experimental work on schemata in genetic programming we have proposed a new simpler form of crossover in which *the same* crossover point is selected in both parent programs. We call this operator *one-point crossover* because of its similarity with the corresponding operator in genetic algorithms. One-point crossover presents very interesting properties from the theory point of view. In this paper we describe this form of crossover as well as a new variant called *strict one-point crossover* highlighting their useful theoretical and practical features. We also present experimental evidence which shows that one-point crossover compares favourably with standard crossover.

1 Introduction

Genetic Programming (GP) has been applied successfully to a large number of difficult problems like automatic design, pattern recognition, robotic control, synthesis of neural architectures, symbolic regression, image analysis, natural language processing, etc. [1, 2, 3, 4, 5, 6, 7, 8, 9]. However, only a relatively small number of theoretical results are available which try and explain why and how it works (see [10, pages 517–519] for a list of references).

Holland's schema theorem (see [11] and [12]) is often used to explain why genetic algorithms (GAs) work. For binary GAs, a schema is a string of symbols taken from the alphabet $\{0,1,\#\}$. The character # is a "don't care" symbol, so that a schema can represent several bit strings. One way of creating a theory for GP is to define a concept of schema for parse trees and to extend the GA schema theorem.

Unfortunately, until very recently the efforts in this direction have given limited results. In the last few years alternative definitions of schema have been proposed [1, 13, 14]. All these definitions are based on the idea that a schema is composed of one or more trees or fragments of trees and that each schema represents all the programs in which such trees or tree fragments are present. These notions of schema have led to some theoretical results which, however, have a limited explanatory power. There is a simple reason for this.

A schema is a subspace of the space of possible solutions, ideally represented using some concise notation (rather than enumerating all the solutions it contains). A population of strings or programs samples many subspaces in parallel. A schema theorem is an attempt to explain which subspaces will be sampled at the next generation. So, the crucial feature for schemata to be useful in explaining how GP searches is that their definition must make the effects of selection, crossover and mutation comprehensible and relatively easy calculate. The problem with the definitions of schema for GP mentioned above is not that they are not clear or concise, it is that they make the effects on schemata of the genetic operators used in GP too difficult to evaluate mathematically.

In recent work [15] we have reconsidered all this and proposed a new definition of schema for GP which is very close to the original concept of schema in GAs. We define a *schema* as a tree composed of functions from the set $\mathcal{F}\cup\{=\}$ and terminals from the set $\mathcal{T}\cup\{=\}$, where $\mathcal{F}$ and $\mathcal{T}$ are the function set and the terminal set used in a GP run. The symbol = is a "don't care" symbol which stands for a *single* terminal or function. Therefore, a schema H represents programs having the same shape as H and the same labels for the non-= nodes. For example, the schema (AND (= x y) =) represents the programs (AND (OR x y) z), (AND (AND x y) x), etc.

but not (OR (AND x y) z) or (AND (AND x y) (OR x z)).

While this definition is simpler than others, it is still very difficult to model mathematically the effects on schemata of standard crossover. This prompted us to find a more natural form of crossover for GP which was mathematically in tune with our definition of schema. The similarity between our GP schemata and the original GA schemata, suggested to us a new form of crossover for GP, which we called *one-point crossover*, in which the same crossover point is selected in both parents (see Section 2). This is very similar to one-point crossover for bit strings where a common crossover point is selected in both parents and the offspring are produced by swapping the bits on the right or the left of the crossover point. We also chose a simple form of mutation, *point mutation*, in which a function in the tree is substituted with another function with the same arity or a terminal is substituted with another terminal [16].

With these genetic operators we were able to derive very naturally a schema theory [15] which has a considerable explanatory power (see Section 2.1). Indeed, the predictions of the theory have been later corroborated by an experimental study [17] on the creation, propagation and disruption of GP schemata in small populations using the XOR problem. In this paper we experimentally study the performance of our genetic operators and compare them to standard GP on larger parity problems.

The paper is organised as follows. In Section 2 we describe one-point crossover as well as a new variant called *strict one-point crossover* and we discuss their properties, also recalling the main results of our GP schema theory. In Section 3, we present experimental evidence which shows that both forms of one-point crossover compare favourably with standard crossover on the even-3, 4 and 5 parity problems. Finally, we draw some conclusions and we give indications of future work in Section 4.

2 One-Point Crossover

One-point crossover works by selecting a *common* crossover point in (copies of) the parent programs and then swapping the corresponding subtrees like standard crossover. If the parents had always the same size and shape, this operation could be performed in a single stage, by selecting any link as the crossover point. However, in order to account for the possible structural diversity of the two parents, one-point crossover requires two phases:

(a) first the two parent trees are traversed to identify the parts with the same shape, i.e. with the same arity in the nodes encountered traversing the trees from the root node, then

(b) a random crossover point is selected with a uniform probability among the links belonging to the common parts identified in step (a).

Figure 1 illustrates the behaviour of one-point crossover. It is worth noting that the offspring produced inherit the common structure (emphasised with thick lines) of the upper part of the parents. (One-point crossover has some similarity to the strong context preserving crossover operator proposed in [18] but context preserving crossover is less restrictive than one-point crossover as to which links can be selected as crossover points.)

One-point crossover has a very important property: it makes the calculations necessary to model the disruption of GP schemata feasible. This means that it is possible to study in detail its effects on different kinds of schemata and to obtain a schema theorem. This tells us how the GP search proceeds by predicting which areas of the search space have a high probability of being sampled by the programs in a generation, given the programs in the previous one. In the following subsections we summarise our GP schema theorem, we discuss the main theory-related and practical properties of one-point crossover, and we introduce a new kind of one-point crossover.

2.1 GP Schema Theorem

In order to understand the importance of one-point crossover from the theory point of view it is necessary to introduce some additional definitions (see [15] and [17] for a more details on our schema theory). The number of non-= symbols in a schema H is called the *order* $\mathcal{O}(H)$ of the schema, while the total number of nodes in the schema is called the *length* $N(H)$ of the schema. The number of links in the minimum subtree including all the

PARENTS

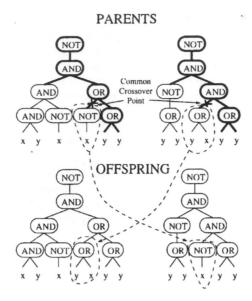

Figure 1: One-point crossover (potential crossover locations are shown in bold).

non-= symbols within a schema H is called the *defining length* $\mathcal{L}(H)$ of the schema. For example, the schema (AND (= y =) x) has order 3 and defining length 3.

Our GP schema theorem provides the following lower bound for the expected number of individuals sampling a schema H at generation $t + 1$ for GP with one-point crossover and point mutation:

$$
\begin{aligned}
E[m(H, t+1)] \quad \geq \quad & m(H, t)\frac{f(H, t)}{\bar{f}(t)}(1 - p_m)^{O(H)} \times \\
& \left\{ 1 - p_c \left[p_{\text{diff}}(t) \left(1 - \frac{m(G(H), t)f(G(H), t)}{M\bar{f}(t)} \right) + \right. \right. \\
& \left. \left. \frac{\mathcal{L}(H)}{(N(H) - 1)} \frac{m(G(H), t)f(G(H), t) - m(H, t)f(H, t)}{M\bar{f}(t)} \right] \right\}
\end{aligned}
$$

where $m(H, t)$ is the number instances of the schema H in the population at generation t, $f(H, t)$ is the mean fitness of the instances of H, $\bar{f}(t)$ is the mean fitness of the programs in the population, p_c is the crossover probability, $E[\cdot]$ is the expected-value operator, p_m is the mutation probability (per node), $G(H)$ is the zero-th order schema with the same structure of H where all the defining nodes in H have been replaced with "don't care" symbols, M is the number of individuals in the population, $p_{\text{diff}}(t)$ is the conditional probability that H is disrupted by crossover when the second parent has a different shape (i.e. does not sample $G(H)$). The zero-order schemata $G(H)$'s represent different groups of programs all with the same shape and size. For this reason we call them *hyperspaces* of programs. We denote non-zero-order schemata with the term *hyperplanes*, as they can be seen as sub-spaces of the spaces of programs identified by different $G(H)$'s.

Our schema theorem is more complicated than the corresponding version for GAs [12, 11, 21]. This is due to the fact that in GP the trees undergoing optimisation have variable size and shape. This is accounted for by the presence of the terms $m(G(H), t)$ and $f(G(H), t)$, which summarise the characteristics of the programs belonging to the same hyperspace in which H is a hyperplane. However, both the theoretical analysis presented in [15] and the experimental work in [17] suggest that after a first phase in which GP really behaves differently from a standard GA, the number of hyperspaces is considerably reduced and GP behaves like a GA, i.e. the GP schema theorem asymptotically tends to the GA schema theorem.

2.2 Properties of One-point Crossover

The most important predicted and observed effect of one-point crossover is that, unlike standard crossover, in the absence of mutation, it makes the population converge quite quickly like a standard GA (in some cases with help from genetic drift). The reason for this is probably that until a large-enough proportion of the population has exactly the same structure in the upper parts of the tree, the probability of selecting a crossover point in the lower parts will be very small. This effectively means that until a common upper structure is found, one-point crossover is actually searching a much smaller space of (approximately) fixed-size structures.[1] Therefore, GP behaves like a GA searching for a partial solution (i.e. a good upper part) in a relatively small search space. This means that the algorithm converges and a common upper part is quickly found, which cannot later be modified unless mutation is present. At that point the search concentrates on slightly lower levels in the tree with a similar behaviour, until level after level the entire population has completely converged. So, one-point crossover transforms a large search in the original space containing programs with different sizes and shapes into a sequence of smaller quick searches in spaces containing structures of fixed size and shape.

An important consequence of the convergence property of GP with one-point crossover is that like in GAs mutation becomes a very important operator to prevent premature convergence and to maintain diversity.

In addition to the theory-related properties mentioned above, one-point crossover offers another very important property from the practice point of view: it does not increase the depth of the offspring beyond that of their parents, and therefore beyond the maximum depth of the initial random population. This can be very useful to avoid the typical undesirable growth of program size (bloating) observed in GP runs (see for example [19]), which slows down the search for solutions and, in some cases, can lead to overfitting. One-point crossover does this without the need of any extra machinery (e.g. parsimony terms in the fitness function). Similarly, one-point crossover will not produce offspring whose depth is smaller than that of the shallowest branch in their parents and therefore than the smallest of the individuals in the initial population. This means that the search performed by GP with one-point crossover and point mutation is limited to a subspace of programs defined by the initial population. Therefore, the initialisation method and parameters chosen for the creation of the initial population can modify significantly the behaviour of the algorithm. For example, if one uses the "full" initialisation method [1] which produces balanced trees with a fixed depth, then the search will be limited to programs with a fixed size. If on the contrary the "ramped half-and-half" initialisation method is used [1], which produces trees of variable shape and size with depths ranging from 0 to the prefixed maximum initial tree depth D, then the entire space of programs with maximum depth D will be searched (at least if the population is big enough).[2]

It should be noted that the initial-population-dependent search performed by GP with one-point crossover might be inefficient or ineffective if the parameters which determine the composition and size of the initial population are not selected properly. For example, with the wrong initial population there might be no 100% correct solutions in the search space.

2.3 Strict One-point Crossover

An interesting variant of one-point crossover, which we call *strict one-point crossover*, behaves exactly like one-point crossover except that the crossover point can be located only in the parts of the two trees which have exactly the same structure (i.e. the same functions in the nodes encountered traversing the trees from the root node). The links eligible as crossover points in strict one-point crossover are a subset of those eligible in standard one-point crossover. Figure 2 illustrates the behaviour of strict one-point crossover. In this case the offspring produced inherit both the structure and the nodes (emphasised with thick lines) of the upper part of the parents.

Strict one-point crossover has the same properties as one-point crossover but it more energetically forces the population to converge. This can be understood considering that until a large-enough proportion of the population has exactly the same nodes in the upper parts of the tree, the search will not be able to proceed to lower levels. Strict one-point crossover transforms the original search into a sequence of quick searches in spaces of structures

---

[1] Obviously the lower parts of the trees moved around by crossover influence the fitness of the fixed-size upper parts, but they are not modified by crossover at this stage.

[2] The "ramped half-and-half" initialisation method used in [1] enforced a minimum depth of 2. In our work we use a minimum depth of 0.

PARENTS

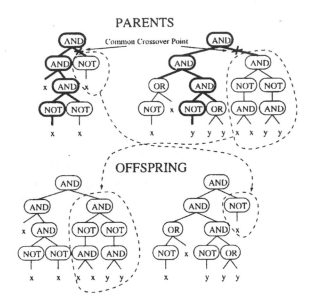

Figure 2: Strict one-point crossover (potential crossover locations are shown in bold).

of fixed size and shape which are even smaller than those used by one-point crossover.

In the case of strict one-point crossover the search seems to proceed very similarly to the search performed by a GA with Dynamic Parameter Encoding (DPE) [20], a technique for overcoming the precision/speed dilemma when encoding real-valued parameters with binary strings. In DPE the resolution of the encoding of one parameter is increased at run time when the most significant bit of such parameter has (nearly) converged in the whole population. A difference is that in GP with strict one-point crossover the search zooms into the subtrees of a converged node automatically, without the need for maintaining global convergence statistics.

The convergence property of the two forms of one-point crossover has been observed in real runs with the XOR problem [17]. Figure 3 shows the diversity in the population (averaged over 10 independent runs) as a function of the generation number for standard crossover and for the two types of one-point crossover in the absence of mutation (the experimental conditions are as in [17]). It is quite clear that standard crossover does not lead to convergence, while one-point crossovers do.

3 Experimental Results

The behaviour of the two forms of one-point crossover introduced in the previous section has been studied and compared to standard crossover in over 3,000 runs on the even-n parity problems with n=3, 4 and 5, which have been extensively studied in the GP literature [1, 2].

An even-n parity problem consists of finding a combination of functions from the set $\mathcal{F}$={OR, AND, NOR, NAND} and terminals from the set $\mathcal{T}$={x1, x2, x3, ..., xn} which returns true if an even number of the n inputs xi is true and false otherwise. The fitness function for this class of problems is simply the number of entries of the truth table of the even-n parity function correctly represented by each program.

Given the importance of the initial population and the expected need for mutation to maintain diversity when using one-point crossover, we decided to test the performance of GP with different initial depths and different point mutation probabilities. In these experiments we used a crossover probability of 0.7, tournament selection with tournament size 7, no depth or size limit (for standard crossover only), and the "ramped-half-and-half" initialisation method. The population size was 1,000, the maximum number of generations was 50. In the tests we

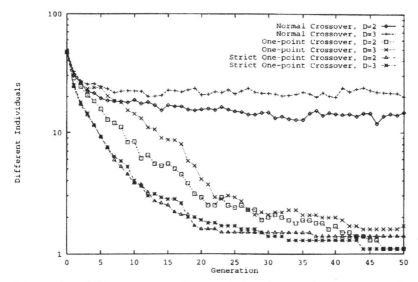

Figure 3: Plot of the number of different programs in a population of 50 individuals vs. generation number for the XOR problem. The data (averaged over 10 runs with different random seeds) for standard crossover, one-point crossover and strict one-point crossover with maximum initial depth $D=2$ and $D=3$ are show.

used the following mutation probabilities per node: $p_m=0$, 1/256, 1/128, 1/64, 1/32, 1/16. For the even-3 parity problem we used initial depths $D=4$ and $D=6$, while for the even-4 parity problem we used $D=6$ and $D=8$. For each combination of parameters we tried standard crossover, one-point crossover, strict one-point crossover and mutation only (when applicable). We repeated 20 runs using different random seeds for each combination of parameters and operators.

To assess the performance of the various operators we used the computational effort E used in the GP literature (E is the minimum number of fitness evaluations necessary to get a correct program, in multiple runs, with probability 99%). We also measured the average size of the solutions found. On the even-n parity problems Koza [1, 2] obtained the results shown in Table 1 (we report them for an easier comparison with our experiments). Table 2 describes the results of our experiments.[3]

The experiments show that the maximum size of the initial population, a parameter largely ignored by the GP literature, has a considerable effect on the computational effort required to solve the problem. This is particularly true for the experiments with the two forms of one-point crossover and with point mutation only, as these operators cannot expand the search space determined by the initial population. For example, the effort for one-point crossover to find solutions to the even-3 parity problem becomes nearly 13 times smaller if the maximum depth is increased from 4 to 6.

As expected from our schema theory and the previous experiments with the XOR problem, the experiments suggest that both forms of one-point crossover suffer from premature convergence in the absence of mutation. For example, no solutions were obtained to the even-4 parity problem in the 80 runs using either form of one-point crossover.

The situation changes considerably if point mutation is present. Indeed, if the maximum depth of the initial population is appropriate and the right amount of point mutation is present, one-point crossover can do up to 10 times better than standard GP without ADFs on the even-3 parity problem and up to 3 times better on the even-4 parity problem. This happens because point mutation can counteract the excessive convergence tendency of one-point crossover. As shown by Table 3, this positive effect cannot be obtained by simply reducing the selective pressure using tournament selection with tournament size 2.

---

[3] Koza enforces a program-depth limit $D=16$ (considering the root node as depth 0), while we do not, uses a crossover probability of 90% rather than 70%, and selects crossover points with non-uniform probability.

Interestingly, point mutation improves performance considerably also when standard crossover is used. For example, the even-4 parity problem becomes 5 times easier with mutation rates as small as 1 node out of 256. Given that standard crossover is very disruptive and does not allow the convergence of the population, it is arguable that point mutation in this case helps settling into the narrow minima of the fitness function which need to be reached with small changes. These are very unlikely produced by standard crossover alone.

Point mutation performs very well on these problems even in the absence of crossover, and in some cases it outperforms GP with crossover (although these results need to be corroborated with larger numbers of runs).

In all cases there seems to be an optimal mutation probability somewhere between 1/128 and 1/32 which is problem and depth dependent. By computing the product of the average size of the solutions obtained with the best mutation rate and the mutation rate for each combination of D and n, it is possible to infer that the ideal mutation probability p_m is very close to 2 divided by the size of the tree. We checked this hypothesis on the even-3, 4 and 5 parity problems using a mutation scheme in which exactly two random nodes are mutated in each individual, i.e. in which the mutation probability is variable. The results of these experiments (averages of 20 runs) are shown in Table 4.

These results seem to suggest that a variable mutation probability is in general very beneficial, in particular for strict one-point crossover. Indeed, with variable mutation probability strict one-point crossover outperforms all other settings tried in our experiments and requires a computational effort up to 10 times smaller than for standard GP without ADFs. For the even-4 parity problem the computational effort is even smaller than for standard GP *with* ADFs.

Given the considerable effect of the maximum initial depth in all the combinations and settings of the operators used in our experiments, we decided to check the effects of the initialisation method, too. Table 5 shows the results obtained using the "full" initialisation method on the even-3 parity problem. Despite the fact that all trees have exactly the same shape (all the functions in the function set have the same arity) and that the search is much more constrained, the effects of starting from a population of larger individuals are striking: in nearly all cases the results are significantly better than those in Table 1. In particular for $D=6$, nearly any choice of operators and mutation rates leads to speed-ups of 10 to 15 times with respect to standard GP, the best results being obtained with variable mutation rates.

4   Conclusions

In this paper we have described two forms of crossover, one-point crossover and the new strict one-point crossover, which, thanks to constraints on the selection of crossover points, transmit to the offspring many of the common features of their parents.

These forms of crossover have several interesting properties. From the theory point of view they allow the derivation of a more explanatory schema theorem in which the effects of crossover on schemata are mathematically modelled. From the practice point of view, one-point crossover eliminates the bloating problem directly and naturally and forces the population to converge like in standard GAs.

In the paper we have presented the first experimental evidence which shows that one-point crossover compares favourably with standard crossover as long as the initial population has the correct depth and premature convergence is prevented by using point mutation. Interestingly, in our study point mutation seemed a very beneficial operator also when used with standard crossover, in particular when the mutation probability was size-dependent. Surprisingly, point mutation did very well on the even-n parity problems even in the absence of crossover. Future research will be necessary to confirm these results for other classes of problems.

Acknowledgements

The authors wish to thank the members of the EEBIC (Evolutionary and Emergent Behaviour Intelligence and Computation) group for useful discussions and comments. This research is partially supported by a grant under the British Council-MURST/CRUI agreement and by a contract with the Defence Research Agency in Malvern.

References

[1] John R. Koza, 1992, *Genetic Programming: On the Programming of Computers by Means of Natural Selection*, MIT Press, Cambridge, Massachusetts.

[2] John R. Koza, 1994, *Genetic Programming II: Automatic Discovery of Reusable Programs*, MIT Press, Cambridge, Massachusetts.

[3] K. E. Kinnear, Jr., Ed., 1994, *Advances in Genetic Programming*, MIT Press, Cambridge, Massachusetts.

[4] John R. Koza, David E. Goldberg, David B. Fogel, and Rick L. Riolo, Eds., 1996, *Genetic Programming 1996: Proceedings of the First Annual Conference*, MIT Press, Cambridge, Massachusetts.

[5] Peter J. Angeline and K. E. Kinnear, Jr., Eds., 1996, *Advances in Genetic Programming 2*, MIT Press, Cambridge, MA, USA.

[6] Riccardo Poli, 1996, Genetic programming for image analysis, in *Genetic Programming 1996: Proceedings of the First Annual Conference*, Stanford University, CA, USA, July, pp. 363–368.

[7] Riccardo Poli and Stefano Cagnoni, 1997, Genetic Programming with User-Driven Selection: Experiments on the Evolution of Algorithms for Image Enhancement, in *Genetic Programming 1997: Proceedings of the Second Annual Conference*, Stanford University, CA, USA, July, pp. 269–276.

[8] Riccardo Poli, 1997, Discovery of symbolic, neuro-symbolic and neural networks with parallel distributed genetic programming, in *3rd International Conference on Artificial Neural Networks and Genetic Algorithms, ICANNGA'97*, Norwich, UK.

[9] Riccardo Poli, 1997, Evolution of recursive transistion networks for natural language recognition with parallel distributed genetic programming, in *Proceedings of AISB-97 Workshop on Evolutionary Computation*, Manchester, UK.

[10] William B. Langdon, 1996, A bibliography for genetic programming, in *Advances in Genetic Programming 2*, Peter J. Angeline and K. E. Kinnear, Jr., Eds., appendix B, pp. 507–532, MIT Press, Cambridge, MA, USA.

[11] John Holland, 1992, *Adaptation in Natural and Artificial Systems*, MIT Press, Cambridge, Massachusetts, second edition.

[12] David E. Goldberg, 1989, *Genetic Algorithms in Search, Optimization, and Machine Learning*, Addison-Wesley, Reading, Massachusetts.

[13] Una-May O'Reilly and Franz Oppacher, 1994, The troubling aspects of a building block hypothesis for genetic programming, in *Foundations of Genetic Algorithms 3*, Estes Park, Colorado, USA, July, pp. 73–88.

[14] P. A. Whigham, 1995, A schema theorem for context-free grammars, in *1995 IEEE Conference on Evolutionary Computation*, Perth, Australia, Nov, pp. 178–181.

[15] Riccardo Poli and W. B. Langdon, 1997, A new schema theory for genetic programming with one-point crossover and point mutation, in *Genetic Programming 1997: Proceedings of the Second Annual Conference*, Stanford University, CA, USA, July, pp. 278–285.

[16] Ben McKay, Mark J. Willis, and Geoffrey W. Barton, 1995, Using a tree structured genetic algorithm to perform symbolic regression, in *First International Conference on Genetic Algorithms in Engineering Systems: Innovations and Applications, GALESIA*, Sheffield, UK, Sept, pp. 487–492.

[17] Riccardo Poli and W. B. Langdon, 1997, An experimental analysis of schema creation, propagation and disruption in genetic programming, Technical Report CSRP-97-8, University of Birmingham, School of Computer Science, Feb., To be presented at ICGA-97.

[18] Patrik D'haeseleer, 1994, Context preserving crossover in genetic programming, in *Proceedings of the 1994 IEEE World Congress on Computational Intelligence*, Orlando, Florida, USA, June, pp. 256–261.

[19] W. B. Langdon and R. Poli, 1997, Fitness causes bloat, in *Proceedings of the Second On-line World Conference on Soft Computing (WSC2)*, Internet, June, 1997.

[20] N. N. Schraudolph and R. K. Belew, 1992, Dynamic parameter encoding for genetic algorithms, *Machine Learning*, **9**, pp. 9–21.

[21] Darrel Whitley, 1993, *A genetic algorithm tutorial*, Tech. Rep. CS-93-103, Department of Computer Science, Colorado State University.

Table 1: Computational effort E and average solution size (in parenthesis) for standard GP with and without ADFs reported by Koza in [1, 2].

Even-3, 4 and 5 Parity Problem (Koza's Results)

Population Size	Even-3	Even-4	Even-5
4,000, no ADFs	80,000 (N/A)	1,276,000 (N/A)	7,840,000 (N/A)
4,000, with ADFs	N/A	80,000 (N/A)	152,000 (N/A)
16,000, no ADFs	96,000 (45)	384,000 (113)	6,528,000 (300)
16,000, with ADFs	64,000 (48)	176,000 (60)	464,000 (157)

Table 2: Computational effort E and average solution size (in parenthesis) as a function of the genetic operators used, the mutation probability and the maximum depth of the initial programs.

Even-3 Parity Problem

Depth	p_m	Normal Crossover	1-pt Crossover	Strict 1-pt Crossover	Mutation Only
4	0	52,000 (45)	308,000 (24)	No Solution	N/A
4	1/256	39,000 (63)	264,000 (27)	810,000 (31)	810,000 (31)
4	1/128	48,000 (38)	110,000 (27)	1,320,000 (31)	396,000 (31)
4	1/64	32,000 (46)	170,000 (26)	315,000 (29)	399,000 (29)
4	1/32	31,000 (42)	128,000 (27)	105,000 (30)	147,000 (31)
4	1/16	54,000 (54)	96,000 (26)	133,000 (28)	86,000 (30)
6	0	24,000 (86)	24,000 (64)	270,000 (88)	N/A
6	1/256	16,000 (88)	27,000 (77)	42,000 (77)	54,000 (92)
6	1/128	15,000 (101)	18,000 (72)	28,000 (75)	44,000 (86)
6	1/64	8,000 (98)	10,000 (81)	8,000 (87)	25,000 (79)
6	1/32	16,000 (100)	14,000 (77)	9,000 (87)	21,000 (92)
6	1/16	19,000 (82)	42,000 (67)	26,000 (67)	28,000 (80)

Even-4 Parity Problem

Depth	p_m	Normal Crossover	1-pt Crossover	Strict 1-pt Crossover	Mutation Only
6	0	1,276,000 (148)	No Solution	No Solution	N/A
6	1/256	238,000 (215)	638,000 (109)	725,000 (119)	880,000 (111)
6	1/128	216,000 (168)	507,000 (98)	357,000 (112)	220,000 (122)
6	1/64	195,000 (154)	224,000 (114)	136,000 (105)	198,000 (116)
6	1/32	611,000 (193)	598,000 (91)	360,000 (111)	510,000 (83)
6	1/16	No Solution	No Solution	No Solution	No Solution
8	0	812,000 (271)	No Solution	No Solution	N/A
8	1/256	126,000 (396)	189,000 (296)	196,000 (313)	319,000 (370)
8	1/128	120,000 (382)	140,000 (296)	129,000 (302)	144,000 (359)
8	1/64	170,000 (377)	144,000 (287)	154,000 (275)	105,000 (210)
8	1/32	1,131,000 (188)	329,000 (112)	500,000 (127)	385,000 (151)
8	1/16	No Solution	No Solution	No Solution	No Solution

Table 3: Computational effort E and average solution size as a function of the genetic operators used, with and without point mutation, for two different maximum depths of the initial programs when the tournament size is reduced to 2.

Even-4 Parity Problem (Tournament Size = 2)

Depth	p_m	Normal Crossover	1-pt Crossover	Strict 1-pt Crossover	Mutation Only
6	0	697,000 (181)	No Solution	No Solution	N/A
6	1/128	No Solution	4,320,000 (105)	2,024,000 (87)	No Solution
8	0	1,392,000 (439)	3,330,000 (199)	No Solution	N/A
8	1/128	1,363,000 (361)	3,780,000 (115)	1,421,000 (265)	No Solution

Table 4: Computational effort E and average solution size (in parenthesis) as a function of the genetic operators used and the maximum depth of the initial programs in the presence of variable point-mutation probability.

Even-3 Parity Problem (Variable Mutation Probability)

Depth	p_m	Normal Crossover	1-pt Crossover	Strict 1-pt Crossover	Mutation Only
4	2/Size	48,000 (61)	44,000 (28)	51,000 (30)	64,000 (31)
6	2/Size	12,000 (105)	11,000 (92)	8,000 (101)	8,000 (106)

Even-4 Parity Problem (Variable Mutation Probability)

Depth	p_m	Normal Crossover	1-pt Crossover	Strict 1-pt Crossover	Mutation Only
6	2/Size	156,000 (283)	210,000 (121)	99,000 (119)	144,000 (127)
8	2/Size	108,000 (518)	168,000 (334)	78,000 (353)	196,000 (328)

Even-5 Parity Problem (Variable Mutation Probability)

Depth	p_m	Normal Crossover	1-pt Crossover	Strict 1-pt Crossover	Mutation Only
8	2/Size	Not tested	Not tested	1,232,000 (489)	Not tested
9	2/Size	Not tested	Not tested	730,000 (660)	Not tested

Table 5: Computational effort E and average solution size as a function of the genetic operators used and the maximum depth of the initial programs when the "full" initialisation method is used.

Even-3 Parity Problem ("Full" Initialisation)

Depth	p_m	Normal Crossover	1-pt Crossover	Strict 1-pt Crossover	Mutation Only
4	0	28,000 (45)	80,000 (31)	220,000 (31)	N/A
4	1/256	19,000 (42)	70,000 (31)	78,000 (31)	572,000 (31)
4	1/128	16,000 (41)	48,000 (31)	121,000 (31)	147,000 (31)
4	1/64	14,000 (43)	63,000 (31)	36,000 (31)	84,000 (31)
4	1/32	18,000 (40)	48,000 (31)	22,000 (31)	31,000 (31)
4	1/16	33,000 (52)	39,000 (31)	30,000 (31)	26,000 (31)
4	2/Size	34,000 (66)	40,000 (31)	54,000 (31)	42,000 (31)
6	0	7,000 (129)	18,000 (127)	30,000 (127)	N/A
6	1/256	10,000 (131)	6,000 (127)	7,000 (127)	12,000 (127)
6	1/128	8,000 (134)	6,000 (127)	6,000 (127)	8,000 (127)
6	1/64	6,000 (133)	6,000 (127)	5,000 (127)	7,000 (127)
6	1/32	8,000 (127)	6,000 (127)	6,000 (127)	5,000 (127)
6	1/16	10,000 (128)	8,000 (127)	9,000 (127)	7,000 (127)
6	2/Size	5,000 (140)	7,000 (127)	5,000 (127)	5,000 (127)

Evolutionary Tabu Search For Geometric Primitive Extraction

Jinxiang Chai[1], Tianzi Jiang[2], Song De Ma[3]

National Laboratory of Pattern Recognition, Institute of Automation,
Chinese Academy of Science, Beijing 100080, P.R. China

[1] *E-mail: chaij@prlsun3.ia.ac.cn*
[2] *E-mail: jiangtz@prlsun6.ia.ac.cn*
[3] *E-mail: masd@prlsun2.ia.ac.cn*

Keywords: Geometric primitive, Evolutionary Tabu Search, Tabu Search, evolution algorithm.

Abstract

Many problems in computer vision can be formulated as an optimization problem. Developping the efficient global optimizational technique adaptive to the vision proplem becomes more and more important. In this paper, we present a geometric primitive extraction method, which plays a crucial role in content-based image retrieval and other vision problems. We formulate the problem as a cost function minimization problem and we present a new optimization technique called Evolutionary Tabu Search (ETS). Genetic algorithm and Tabu Search Algorithm are combined in our method. Specifically, we incorporates "the survival of strongest" idea of evolution algorithm into tabu search. In experiments, we use our method for shape extraction in images and compare our method with other three global optimization methods including genetic algorithm, simulated Annealing and tabu search. The results show that the new algorithm is a practical and effective global optimization method, which can yield good near-optimal soultions and has better convergence speed.

1 Introduction

Extracting predefined shape from geometric data is an important problem in the field of model-based vision because it is a prerequisite to solving other problems in model-based vision, such as pose determination, model building, object recognition, and so on. The most commonly used method of geometric primitive extraction is the Hough transforn (HT). The HT divides parameter space of the geometric primitive into cells (usually rectangular) by quantizing each dimension into a fixed number of intervals. Each datum point adds a vote to every cell whose parameters are such that the primitive associated with that cell passed through the point. After all the points have voted, the cell which have a number of votes greater than a threshold are marked. For each such cell the associated geometric primitive is taken as a description of the points that voted for the cell, and this primitive is said to be extracted from the data[1]. The HT has been show to be equivalent to template matching, where the templates are defined by each of the cells in parameter space[2]. Performing template matching in this way is time effcient, but space inefficient. The space requirements are proportional to the number of the cells, and this number is an exponential function of the dimension of the parameter space. This means that unless the quantization of the parameter space is coarse, the HT can be pratically used for primitives with at most two degrees of the fredom. In fact, the majority of applications of the HT are for line extraction[3].

Recently, Roth *et al*[4] proved that extracting the best geometric primitive from a given set of geometric data is equivalent to finding the optimum value of a cost function. Once it is understood that primitive extraction is such an optimization problem, the use of any technique for tackling optimization problem suggests itself. The objective function of the global optimization problem for geometric primitive extraction has potentially many local minima. Conventional local search minimization techniques are time consuming and tend to converge to whichever local minimum they first encounter. These methods are unable to continue the search after a local minimum is reached. The key requirement before any global optimization method is that it must be able to avoid entrapment in local minimal and continues the search to give a near-optimal final solution whatever the initial conditions. It is well known that *Simulated Annealing*[5,6](SA) and *Genetic Algorithm*[7,8] (GA) meet this requirement. Some researchers suggested to solve this problem using the Genetic Algorithms[9,10].

In this paper we develop an efficient algorithm based on the combination of tabu search and evolution theory. This new global optimization method has the ability to find the global optimum, which not only keeps the advantages of Tabu Search and Genetic Algorithms, but also overcomes some of their shortages. Specifically, by comparing our algorithm with the existing other global optimization methods (such as Genetic Algorithm, Simulated Annealing and Tabu Search), we find that the ETS is more practical and effective, which also yields good near-optimal solutions and has better convergence speed. The rest of this paper is organized as follows. Section 2 devotes to the background of our problem. Section 3 gives a general description of our new algorithm for geometric primitive extraction. The statement of ETS is presented in Section 4. Section 5 devotes to the experimental results. We give our conclusion in Section 6.

2   Primitive Extraction and Minimal Subsets

In this section, we briefly review facts of geometric primitive extraction and the definition of minimal subsets. We refer the readers to Roth *et al*[4] for the details.

A geometric primitive is a curve or surface which can be described by an equation with a number of free parameters. The input to a primitive extraction algorithm consists of N geometric data points in two or three dimensional Cartesian space, which are labeled $p_1, p_2, \cdots, p_N$, along with the equation defining the type of geometric primitive to be extracted. We assume that this defining equation is an implicit form $f(\mathbf{p}, \mathbf{a}) = 0$. Here $\mathbf{p}$ is the datum points, and $\mathbf{a}$ defines the parameter vector for this particular primitive. This assumption is not restrictive because it has been shown that the parametric curves and surfaces used to define the parts in CAD databases can be converted to implicit for[11]. The output consists of the parameter vector $\mathbf{a}$ of the best primitive, along with the subset of the geometric data that belongs to this primitive. We define the residual r_i as the closest distance of ith point of the geometric data to the curve or surface. Given that residuals $r_1, r_2, \cdots, r_N$ have been calculated, then extracting a single instance of a geometric primitive is equivalent to finding the parameter vector $\mathbf{a}$ which minimizes the value of a cost function $h(r_1, r_2, \cdots, r_N)$. Denote $f(\mathbf{a}) = h(r_1, r_2, \cdots, r_N)$. Therefore, extracting the best geometric primitive from a given set of geometric data is equivalent to solving the following global optimization problem:

$$\min_{\mathbf{a} \in A} f(\mathbf{a}). \tag{1}$$

where A is a set of feasible solutions. A *minimal subset* is the smallest number of points necessary to define a unique instance of a geometric primitive. It is possible to convert from a minimal subset of points to the parameter vector $\mathbf{a}$ for a wide variety of geometric primitives[4].Since only values of the parameter vector defined by these minimal subsets are potential solutions to the extraction problem, we could search the best solution in a smaller space.

3   Evolutionary Tabu Search

In this section, we will state the thought of the Evolutionary Tabu Search. For the sake of completeness, we briefly review other two global optimization algorithms, including Genetic Algorithm and Simulated Annealing. All of them can be presented as methods for finding a global optimization in the presence of local optimum.

3.1   Simulated Annealing

Simulated Annealing is a stochastic optimization algorithm based on the physical analogy of annealing a system of molecules to its ground state. Originally developed by Kirkpatrick *et al*[5,22], the method has been successfully applied to a variety of hard optimization problems in different fields. Starting from initial configuration X_0 which can be generated randomly, the simulated annealing procedure generates a stationary Markov chain of configuration X_k, A objective function $f(X_k)$, which is to be minimized, is used to evaluate the configuration X_k. During the simulated annealing process, for a given present configuration X_k, the algorithm generates a candidate configuration Y_k. The decision of selecting either configuration X_k or configuration Y_k as the next state X_{k+1} for the following iteration is not made deterministically via a straight-forward comparison of the objective function values of the configuration X_k and Y_k, but stochastically, based on a sampling of the Bolzamann distribution via Mentropolis function. If configuration Y_k has a objective function value smaller than that of configuration X_K, configuration Y_k is selected as configuration X_{k+1}. If configuration Y_k has a objective function value larger than that of configuration X_k, configuration Y_k is selected as X_{k+1} with probability p determined by the Metropolis function $p = exp(-max(\Delta F(Y_k, X_k), 0)/T_k)$, where $\Delta F(Y_k, X_k) = F(Y_k) - F(X_k)$. The set T_k is called an *annealing schedule*, and is a sequence of strictly monotonically decreasing nonnegative numners such that $T_1 > T_2 > \cdots > T_n$ and $\lim_{k\to\infty} T_k = 0$. The asymptotic convergence(i.e. in the limit $k \to \infty$) is guaranteed for a logarithmic annealing schedule of the form $T_k = T_1/(1 + lnk)$ where $k \geq 0$.

3.2 Genetic Algorithm

Genetic Algorithms(GA), first developed by Holland[12], are stochastic optimization techniques that mimic the principles of natural evolution. According to the genetic evolution theory, stronger and fitter individuals have better chances of survival than weaker ones. Offsprings are produced by parents using interesting genetic information from both parents, or strings of chromosomes are inherited from both parents. Possible solutions of the optimization problems resemble chromosomes in the natural process, and producing new solution resembles producing offsprings with certain chromosomes. The objective function in the optimization problem is evaluated in accordance with "survival of the strongest" principle in the natural evolution process. The chromosomes are manipulated by genetic operators to try to simulated the effects of natural evolution. The basic four operators are Reproduction, Crossover, Mutation, Inversion, the first three resemble the natural operators, while the last is done for merely improving the simulation results. These genetic operators are discussed in detail in standard references on genetic algorithms[7,8].

3.3 Tabu Search

Here we briefly review some notations of Tabu Search[13,14,15] and outline the basic steps of the Tabu Search procedure for solving optimization problems. Tabu Search is a metaheuristic that guides a local heuristic search procedure to explore the solution space beyond local optimum. It is different from the well-known hill-climbing local search techniques because Tabu Search allows moves out of a current solution that makes the objective function worse in the hope that it eventually will achieve a better solution. It is also different from the Simulated Annealing and Genetic Algorithm because the Tabu Search includes a memory mechanism. According to Glover's idea, in order to solve a optimization problem using Tabu Search, the following components must be defined.

Configuration: *Configuration* is a solution or an assignment of values of variables.

Move: A *move* characterizes the process of generating a feasible solution to the problem that is related to the current solution (i.e. a move is a procedure by which a new solution is generated from the current one).

Neighborhood: A *neighborhood* of the solution is the collection set of all possible moves out of a current configuration. Note that the actual definitions of the neighborhood depend on the particular implementation and the nature of problem.

Tabu Conditions: In order to avoid a blind search, tabu search technique uses a prescribed problem-specific set of constraints, known as *tabu conditions*. They are certain conditions imposed on moves which make some of them forbidden. These forbidden moves are known as *tabu* . It is done by forming a list of certain size that records these forbidden moves. This is known as *tabu list*.

Aspiration Condition: These are rules that override tabu restrictions, that is, if a certain move is forbidden by tabu restriction, then the aspiration criterion, when satisfied, can make this move allowable.

With the above basic components, the Tabu Search algorithm can be described as follows.

(i) Start with a certain (current) configuration and evaluate the criterion function for that configuration.

(ii) Follow a neighbor of the current configuration, that is, a set of candidate moves. If the best of these moves is not tabu or if the best is tabu, but satisfies the aspiration criterion, then pick that move and consider it to be the new current configuration; otherwise pick the best move that is not tabu and consider it to be the new current configuration.

(iii) Repeat (i) and (ii) until some termination criteria are satisfied.

The best solution in the final loop is the solution obtained by the algorithm. Note that the move picked at a certain iteration is put in the tabu list so that it is not allowed to be reversed in the next iterations. The tabu list has a certain size, and when the length of the tabu reaches that size and a new move enters that list, then the first move on the tabu list is freed from being tabu and the process continues (i.e. the tabu list is circular);

3.4 Evolutionary Tabu Search

The Evolutionary Tabu Search takes advantages of the GA and Tabu Search, and it has two level selections, which are respectively called the first-level selection and the second-level selection. We propose the use of Tabu search in the first level, and evolution ideas[8,16] in the second level.

The ETS technique starts with a guess of N likely candidates, actually chosen at random in the search space. These candidates are the so-called *parents*. Initially each *parent* can generate a number of children, say NTS, which consists a family. The children in the same family(i.e., generated from the same *parent*) constitute the first level selection, then we use Tabu Search to select the child as *parent* for the next generation. This selection creates the *parents* for the next generation. The second-level selection is the competition between the families. The number of children that should be generated in the next generation depends on the results of the second-level selection. This second-level selection actually provides a measure of the fitness of each family. Instead of using the objective values of a single point that might be considerable biased, we define a fitness value based on the objective values of all the children in the same family for that measure. The number of children allocated to each family for the next generation is proportional to their fitness values, but the total number of the next generation's children is still constant. In such way, fitter individuals have better survival chance. It has been show that eventually only one family survives, which is usually the best one. The procedures of the first-level and second-level selection continue until a certain number of iterations have been reached or an acceptable solution has been found. It is the effect of the second-level competition that gives measure of the regional information. In fact, The fitness value provides the information of how good the region is. If the region is found to contain a higher fitness value, we allocated more attention to search in that region.

4   New Algorithm for Geometric Primitive Extraction

In this section, we apply ETS to extract geometric primitives. The input is N geometric data points. Each has an associated index, a number from 1 to N. Then a minimal subset is identified by the indices of its member points. Let m be the size of minimal subset of the particular primitive we want to extract. Let $I = (I_1, \cdots, I_m)$ denote the minimal subset of the size m, where I_i corresponds to the indices of its member points. For the sake of convenience, we assume that $I_i < I_j$ for $i < j, i, j = 1, \cdots, m$. For any minimal subset I, the objective function $f(I)$ is defined as follows:

$$f(I) = \sum_{i=1}^{N} s(r_i^2) \tag{2}$$

where r_j denotes the minimum distance between $i - th$ data point and geometric primitive, and s is step function; $s = 0$ if r_i is greater than or equal to the template width, and $s = 1$ otherwise. This objective function counts the number of points within a fixed distance of the geometric primitive. Moreover, it effectively matches a small template around this primitive to the geometric data. Therefore, our task is to find the maximum of objective function. Therefore, our task becomes to find the maximum of the objective function.

Let I_c, I_t and I_b denote the current, trial and best configurations (minimal subsets) and f_c, f_t and f_b denote the corresponding current, trial and best objective function values, respectively. As described in the previous section, we operate with a configuration which is known as the current solution I_c and then through moves which were explained in what follows, we generate trial solutions I_t. As the algorithm proceeds, we also save the best solution found so far which is denoted by I_b. Corresponding to these configurations, we also operate with the objective function values f_c, f_t and f_b, respectively.

Before stating the two level selection algorithm, we define the basic components in the geometric primitive extraction.

Configuration : Configuration is denoted by $I = (I_1, \cdots, I_m)$, where I_i corresponds to the indices of its member points.

Neighborhood : We first give a definition about neighbor-point. Two points is called neighbor-point or is within the Move Distance(MD) if their Euclidean distances is less than the given value of MD. For Configurations defined by $I = (I_1, \cdots, I_m)$, One configuration moved to another neighbor configuration only if each point in it moves to its neighbor-point.

Tabu element: We assume that the number of the parameter of a specific geometric primitive is m. For example, m is equal to three if geometric primitive is circle, which includes radius(R) and the coordinate of the center(X_c, Y_c). Therefore, we can use a vector A($a_1, \cdots, a_m$) to denote specific geometric primitive, where $a_i(i = 1, \cdots, m)$ is parametric variable. Now, our tabu element can be defined as a vector B($b_1, \cdots, b_m$), $b_i(i = 1, \cdots, m)$ is $i - th$ parametric variable of geometric primitive.

Tabu condition: Assume that the parametric vector of current configuration is $A_t(a_1, \cdots, a_m)$ and tabu element B($b_1, \cdots, b_m$),we can define tabu condition as follows: For $i = 1, \cdots, m$,if $\|a_i - b_i\| \leq \epsilon$, where ϵ is given threshold, then current configuration is tabu.

Aspiration Condition: If current configuration satisfies the aspiration condition, then it is no longer tabu even if it satisfies the tabu condition. we define aspiration condition as follows: $f_t > f_b$, where f_t denotes the corresponding trial objective function value and f_b denotes the best objective function values according to equation 2.

The main steps of our ETS algorithms are as follows:

Step 1 Randomly select N_0 parents(i.e.initial families). For each family, select $NTS = NTS_0$(initial number of children), $\varepsilon = \varepsilon_0$(initial error value) and $MD = MD_0$(initial move distance). Set the following parameters: $MTLS$ (tabu list size), P (probability threshold). Set the counter TLL(tabu list length) = 0.

Step 2 For each family, use Tabu Search(details are given in section 4.1) to creat the child, then to find the parent of the next generation.

Step 3 According to the definition of the family's fitness function, compute the fitness of each family.

Step 4 For each family, compute the number of the children in the next generation.

Step 5 According to some specific strategy, change the Move Distance(MD)and error value(ε). Goto step 2.

Step 6 Until an acceptable solution has been found or until a certain number of iterations(IMAX) has been reached.

4.1 Tabu search implementation

Basic procedure can be described as follows:

Step 1 Let I_c be parent of one family, and f_c be the corresponding objective function value computed using equation (2). For the first generation, Let $I_b = I_c$ and $f_b = f_c$.

Step 2 Generate NTS children $I_t^1, I_t^2, \cdots, I_t^{NTS}$ (see Remark 1) and evaluate their corresponding objective function values $f_t^1, f_t^2, \cdots, f_t^{NTS}$ and go to Step 3.

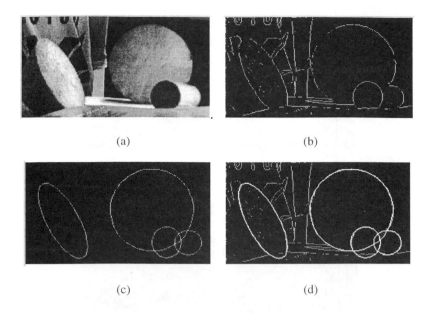

(a)   (b)

(c)   (d)

Figure 1: Extracting multiple ellipses: (a)real image. (b)edge pixel.
(c)extracted ellipses. (d)extracted ellipses superimposed on edge pixel.

Step 3 Arrange $f_t^1, f_t^2, \cdots, f_t^{NTS}$ in a descending order, and denote them by $f_t^{[1]}, f_t^{[2]}, \cdots, f_t^{[NTS]}$. If $f_t^{[1]}$ is not tabu, or if it is tabu but $f_t^{[1]} > f_b$, then let $I_c = I_t^{[1]}$ and $f_c = f_t^{[1]}$, and go to Step 4; otherwise, let $I_c = I_t^{[L]}$ and $f_c = f_t^{[L]}$, where $f_t^{[L]}$ is the best objective function of $f_t^{[2]}, \cdots, f_t^{[NTS]}$ that is not tabu and go to Step 4. If all $f_t^{[1]}, f_t^{[2]}, \cdots, f_t^{[NTS]}$ are tabu, we change MD according to the specific strategy. Go to step 2.

Step 4 Insert new tabu element at the bottom of the tabu list and let $TLL = TLL + 1$ (if $TLL = MTLS + 1$, delete the first element in the tabu list and let $TLL = TLL - 1$). If $f_b < f_c$, let $I_b = I_c$ and $f_b = f_c$.

Remark 1 *Given a current solution I_c, one can generate a trial solution using several strategies. We use the following strategy: Given I_c and a probability threshold P, for $i = 1, 2, \cdots, m$, draw a random number $R_i \sim u(0, 1)$, where $u(0, 1)$ is the uniform distribution on interval $[0, 1]$. If $R_i < P$, then $I_t(i) = I_c(i)$; otherwise draw randomly point $\hat{I}$ from the set including all the $I_c(i)$ neighbor-points and let $I_t(i) = \hat{I}$.*

5   Experimental Results

In this section, we present our experiment results on extracting various kinds of geometric primitives. We will also make some comparison of our method with other three global optimization methods: Genetic Algorithm(GA), Simulated Annealing(SA) and Tabu Search(TS).

As mentioned in the previous section, ETS has four important parameters $MTLS$, N_0, NTS_0 and P. The tabu list enables the algorithm to have short term memory. A large tabu list size allows more diversification while a small list size makes the algorithm more forgetful, i.e., allows intensification to happen. Determining the list size is a non-trivial problem. Glover[17] suggests using a tabu list size in the range $[\frac{1}{3}n, 3n]$, where n is the size of problem. Recently, some researchers[18, 19] propose variable tabu list size(tabu tenure). The next important parameter is the size of the initial family(N_0). If it is too small, it will resemble Tabu Search. Especially, when N_0 equals to 1, then ETS becomes Tabu Search. It is too large, then convergence speed will become worse. Thus selecting this value is a tradeoff between the final extraction result and convergence speed. The third parameter is the initial number of children(NTS_0), which also affect the convergence speed. In our experiments, $NTS_0 \in [2, 8]$. The

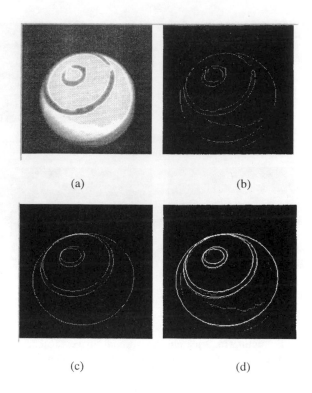

Figure 2: Extracting multiple ellipses: (a)real image. (b)edge pixel. (c)extracted ellipses. (d)extracted ellipses superimposed on edge pixel.

last parameter is P (Pribability threshold) . This parameter controls the shake-up that is performed on a certain solution to produce a neighbor. the higher the value of P, The less shake-up is allowed and consequently the closer the neighbor (the solution obtained after move) to the current solution, and *vice verse*. In our case, we have found that $p \in [0.3, 0.5]$ to be well suited our purpose.

To extract primitives from real images, the first step is to obtain the geometric data, which could be obtained by active sensors or edge detection. In our experiment, geometric data are produced by zero-crossing edge detector[20]. To extract multiple primitives, we just repeat the algorithm presented in the above section. Following the first application of the algorithm, the data points belonging to the first geometric primitive are removed, and the next application of the algorithm has as input the remaining geometric data points. A more complex situation is the extraction of not only multiple primitives, but also different types of primitives. In this case, we apply the algorithm for circles and ellipses simultaneously on the same geometric data. The best primitive is extracted and the algorithm repeated.

Extracting circles and ellipses using HT is still an active area of research. When the number of parameters of these primitives is higher than two (3 for circle and 5 for ellipse), it is space inefficient and therefore difficult for HT to perform the extraction of circle, especially the ellipse. Our approach has no such difficulty, as can be seen from Fig.1 and Fig.2.

Fig.1 shows the extraction of four ellipses (here circle is considered as a special kind of ellipse). This example is significant in that: (1) the background is complex and the ellipses are occluded; (2) the small ellipse in the right-down corner is successfully extracted, which is very difficult for HT to extract. The last step of HT is peak detection and it is difficult in detecting a very small real peak. Case becomes more serious when the original images are contaminated by noise.

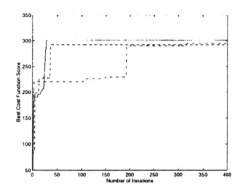

Figure 3: Cost function value of best primitive in Figure
3 versus number of iteration.

Fig.2 shows the extraction of the five adjacent ellipses. The fact that the ellipses are very close together makes this a particularly difficult example for HT method[21] (because of the coarse quantization necessary for HT to deal with higher dimensional spaces), while our global optimization algorithms can fulfill the extracting task quickly and correctly, as can be seen from this figure.

Now, we compare ETS with other three global optimization algorithms, including GA, SA and TS. Fig.3 gives the comparison results of four algorithms in the extraction of the optimal ellipse in fig.2. For comparison, each iteration consists of 40 times of objective function evaluation in all of the four algorithms in our experiment. From the experimental results, we can see that ETS has better convergence speed than the other three algorithms. At the same time, the final best cost function value indicates that ETS and GA yields better result than SA and TS. In addition, it is obvious that ETS algorithm can be implemented in parallel.

Note: In Figure 3, sold line is ETS, dashed line is Tabu Search, dotted line is Genetic algorithm, dashdot line is Simulated Annealing.

6   Conclusion

Extracting predefined geometric primitives from geometric data is an important problem in the field of model-based vision because it is a prerequisite to solving other problems in model-based vision, such as pose determination, model building, and object recognition. In this paper, we propose a novel method for extracting the geometric primitives from geometric data. Specifically, by incorporating the idea of evolution ideas into Tabu Search, we propose a very powerful optimization technique called Evolutionary Tabu Search(ETS). For the geometric primitive extraction, the ETS can discover a very good near-optimum solution after examining an extremely small fraction of possible solutions. In our experiment, we find that the ETS has satisfactory convergence speed and yields good near-optimum solution. As a matter of fact, our new algorithm can obtain better results comparing with the Genetic Algorithm and Simulated Annealing for many test pictures. Furthermore, ETS, which absorbs advantages of evolution algorithm, makes Tabu Search become one special form of it. Now, we are considering the application of Tabu Search to other problems in computer vision.

References

[1] M. D. Levine, Vision in Man and Machine, 1985, McGraw-Hill, New York.

[2] G. Stockman and A.Agrawala, 1979, Equuivalence of Hough transform to template matching, Comm. ACM **20**, 820-822.

[3] T. Risse, 1989, Hough transform for line recognition: complexity of evidence accumulation and cluster detection, Comput. Vision Graphics Image Process. **46**, No.3,327-345.

[4] G. Roth and M.D. Levine,1993, Extracting Geometric Primitives, CVGIP: Image Understanding, **58**, 1-22.

[5] S. Kirkpatrick, C.D. Gelatt Jr., 1983, and M.P. Vecchi, Optimization by Simulated Annealing, Science, **220**:671, 621-680.

[6] E. Aarts and J. Korst, 1989, Simulated Annealing and Boltzmann Machine. New York: Wiley.

[7] D.E. Goldberg, 1989, Genetic Algorithms in Search, Optimization and Machine Learning, Addison-Wesley.

[8] Z.Michalewicz, Genetic Algorithms+Data Structures=Evolution Programs, Springer-Verlag, 1992.

[9] G. Roth and M.D. Levine, 1994,Geometric Primitive Extraction Using a Genetic Algorithm, IEEE Trans. on Pattern Analysis and Machine Intelligence, **16**:9, 901-905.

[10] E.Lutton and P. Martinez, 1994, A Genetic Algorithm for the Detection of 2D Geometric Primitive in Images, Proc.ICPR'94, Vol.1, 526-528.

[11] T.W. Sederberg and D.C. Anderson, 1984, Implicit Representation of parametric Curves and Surfaces, Computer Vision, Graphics, and Image Processing, **28**: 72-84.

[12] J. Holland, 1975, Adaptation in Natural and Artificial Systems, University of Michigan Press, Ann Arbor.

[13] F. Glover, 1993, Tabu Search, *in* Modern Heuristic Techniques for Combinatorial Problems, C.R. Reeves ed., John Wiley & Sons, Inc.

[14] K.S.Al-Sultan, 1995, A Tabu Search Approach to the Clustering Problem, Pattern Recognition, **28**:9,1443-1451.

[15] D.Cvijovic and J.Klinowski, 1995, Taboo Search:An Approach to tha Multiple Minima problem, Science, Vol.267, 664-666.

[16] P.Yip and Y.H.Pao, (1995), Combinational Optimization with Use of Evolutionary Simulated Anneaing, IEEE Trans. on Neural Network,**6**:2, 290-295.

[17] F.Glover, 1990, Artificial intelligence, heuristic frameworks and tabu search, Manag. Decis. Econom., **11**:365-375.

[18] R. Battiti and G. Tecchiolli, 1994, The reactive tabu search, ORSA Journal on computing, **6**:126-140.

[19] J. Xu, Steve Chiu and F.Glover, 1996, Fine-tuning a tabu search algorithm with statistical tests, Technical Report, University of Colorado at Boulder.

[20] S.D. MA and B.Li, 1996, Multiscale derivative computation, to appear in *Image Vision Comput.*.

[21] H. Yuen, J.Illingworth, and J. Kittler, 1989, Detecting partically occluded ellipses using the Hough Transform, *Image Vision Comput.*7.

Parallel Genetic Algorithms in the Optimization of Composite Structures

Erik D. Goodman[1], Ronald C. Averill[2], William F. Punch, III[3], and David J. Eby[2]

Genetic Algorithms Research and Applications Group, Michigan State University

[1]*Case Center for Computer-Aided Engineering and Manufacturing, goodman@egr.msu.edu*

[2]*Department of Materials Science and Mechanics, averill@egr.msu.edu*

[3]*Department of Computer Science, punch@cps.msu.edu*

Keywords: Automated Design, Genetic Algorithms, Composite Material, Laminated Structure, Optimization

Abstract

Genetic Algorithms (GAs) are a powerful technique for search and optimization problems, and are particularly useful in the optimization of composite structures. The search space for an optimal composite structure is generally discontinuous and strongly multimodal, with the possibility for many local sub-optimal solutions or even singular extrema. These facts severely limit gradient-type approaches to optimization, bringing this broad class of problems under scrutiny for application of GAs. Examples described here of the successful use of parallel GAs to design composite structures by the authors include energy-absorbing laminated beams [1], airfoils with tailored bending-twisting coupling [2], and flywheel structures [3]. Optimal design of laminated composite beams was performed using a GA with a specialized finite element model to design material stacking sequences to maximize the mechanical energy absorbed before fracture. An initial GA approach to the optimal design of a specialized, idealized composite airfoil is now being refined for a practical application. The optimum stacking sequence to produce a desired twisting response while minimizing weight, maximizing in-plane stiffness and maintaining acceptable stress levels is determined. The GA has also been used to maximize the Specific Energy Density (SED) of composite flywheels. SED is defined as the amount of rotational energy stored per unit mass. Optimization of SED was achieved by allowing the GA to search for various flywheel shapes and allowing the GA to pick material sequences along the radius of the flywheel.

1. Introduction

Laminated composite construction of panels and other structural elements are used for many applications in the aerospace, automotive, civil, and defense industries. Multi-layer and sandwich construction offer many opportunities for analysts and designers to optimize structures for a particular or even multiple tasks, but this flexibility results in a very large number of often discrete design variables, each with a complex dependence on the others. In addition, the design space may contain many local extrema, and there may be many designs that meet or very nearly meet the design criteria. Thus, there is a need for design techniques that can cost-effectively identify near optimal designs.

Gradient-based optimization techniques have been applied successfully to many shape optimization problems (e.g. [4-6]). However, these methods have several drawbacks. First, they tend to find quickly and get stuck on local extrema [6]. In addition, gradient methods are not suitable for finding singular extrema, or for optimizing truly discrete problems such as the location and geometry of ply dropoffs (terminations) or the material type and orientation of plies in composite structures.

Optimization methods based on genetic algorithms (GAs) have recently been applied to various structural problems [1-3,7-12], and have demonstrated the potential to overcome many of the problems associated with gradient-based methods. However, the need of GAs to evaluate many alternative designs often limits their application to problems in which the design space can be made sufficiently small, even though GAs are most effective when the design space is large. Alternatively, the computational cost per evaluation can be reduced, but that often limits allowable design complexity. A solution to the above dilemma is to perform the GA analysis in a parallel computing environment, where full advantage can be taken of the low communication requirements of GAs, and to use specialized models allowing sufficiently detailed representation without excessive computational requirements.

2. Parallel Genetic Algorithms

Two problems associated with GAs are their computational intensity and their propensity to converge prematurely. An approach that ameliorates both of these problems is parallelization of GAs (PGAs), which also produces a more realistic model of nature than a single large population. PGAs both decrease processing time and better explore the search space. Unlike some specialized sequential GAs which pay a high computational cost for maintaining subpopulations based on similarity comparisons (niching techniques, etc.), PGAs maintain multiple, separate subpopulations which may be allowed to evolve nearly independently. This allows each subpopulation to explore different parts of the search space, each maintaining its own high-fitness individuals and each controlling how mixing occurs with other subpopulations, if at all.

2.1 Coarse-Grain GAs

Coarse-grain GAs (cgGAs) [13] maintain independent subpopulations (often referred to as "islands") which occasionally exchange solutions. The frequency of migration among subpopulations is typically small, and is selected so as to achieve a problem-specific balance between combining good schemata (building blocks) discovered in different subpopulations and allowing the subpopulations to search relatively independently (*i.e.*, promoting diversity). Island parallel GAs have been shown to outperform "serial" GAs dramatically in many contexts [13]. We have categorized cgGAs according to three characteristics: migration method (isolated island, synchronous island, or asynchronous island), connection scheme (static or dynamic), and node homogeneity (homogenous or heterogeneous).

2.2 Injection Island GAs (iiGAs)

We have developed a new PGA architecture called injection island GAs (iiGAs) [13]. iiGAs are a class of asynchronous, static- *or* dynamic-topology, heterogeneous GAs.

The concept of discretizing the topology of a structure for analysis by numerical solution techniques such as the finite element method is by now a familiar one. Often, the initial attempts to analyze a structure involve the use of a coarse discretization (mesh) in which only the critical features of the structure are identified and modeled. In subsequent analyses, progressively finer discretizations may be employed to obtain more accurate local variations of deformation or stress and to determine the effect of these local variations on the overall structural response. In this stepwise approach, additional information about the solution is obtained at each stage of model refinement, and this information is used to guide the development of more refined models, if they are necessary. Often, moderately refined models will provide the majority of needed information, including accurate predictions of the overall structural response and first approximations of local variations.

Where possible, it is both logical and beneficial to take a similar approach in optimal design of structures as well as in other optimization problems. More specifically, the spatial distribution of certain design variables and even the values of design variables themselves can be represented or discretized at various levels of refinement. Optimization using the coarse representations may then provide a gross estimation of a good design that can be improved by performing further optimizations with more refined representations of the structure or its design variables. This is the basis of iiGAs, and their implementation is described below.

iiGA Heterogeneity

GA problems are typically encoded as an n-bit string which represents a complete solution to the problem. However, for many problems, the resolution of that bit string can be allowed to vary. That is, we can represent those n bits in n' bits, $n' < n$, by allowing one bit in the n'-long representation to represent r bits, $r > 1$, of the n-long bit representation. In such a translation, all r bits take the same value as the one bit from the n'-long representation and vice-versa. Thus the n'-long representation is an abstraction of the n-long representation. More formally, let

$$n = p \times q$$

where p and q are integers, $p, q \geq 1$.

Once p and q are determined, we can re-encode a *block* of bits $p' \times q'$ as 1 bit if and only if

$$p = l \times p', \quad q = m \times q'$$

where l and m are integers, $l, m \geq 1$. Such an encoding has the following basic properties,

    (i) The smallest block size is 1x1, and the corresponding search space is 2^n.

    (ii) The largest block size is pxq, and the corresponding search space is $2^1 = 2$.

    (iii) The search space with a block size $p'xq'$ is $2^{p/p'} \times 2^{q/q'}$.

An iiGA has multiple subpopulations that encode the same problem using different block sizes. Each generates its own "best" individual separately.

iiGA Migration Rules

An iiGA may have a number of different block sizes being used in its subpopulations. To allow interchange of individuals, we only allow a one-way exchange of information, where the direction is from a low resolution to a high resolution node. Solution exchange from one node type to another requires translation to the appropriate block size, which is done without loss of information from low to high resolution. One bit in an n'-long representation is translated into r bits with the same value in an n-long representation. Thus all nodes inject their best individual into a higher resolution node for "fine-grained" modification. This allows search to occur in multiple encodings, each focusing on different areas of the search space.

More formally, we note that node x with block size $p_1 x q_1$ can pass individuals to node y with block size $p_2 x q_2$ iff:

$$p_1 = j \times p_2 \ , \quad q_1 = k \times q_2$$

where j, k are integers, $j, k \geq 1$. This establishes a hierarchy of exchange, where node x (lower resolution) is the parent of node y (higher resolution) and node y is the child of node x. The direct child is the child with the largest block size; others are "heirs" of the lower-resolution "ancestors." Based on this general migration rule, we have designed the four following static topologies:

•Simple iiGAs

Each node passes individuals to only one node, the node with the highest resolution (a block size of 1x1).

•Complete iiGAs

Each node passes individuals to all of its heirs (of higher resolution) in the hierarchy.

•Strict iiGAs

Each node passes individuals only to its direct children.

•Loose iiGAs

Each node passes individuals to both its direct children and the node with the highest resolution (block size 1x1).

iiGA Advantages

iiGAs have the following advantages over other PGAs.

(i) Building blocks of lower resolution can be directly found by search at that resolution. After receiving lower resolution solutions from its parent node(s), a node of higher resolution can "fine-tune" these solutions.

(ii) The search space in nodes with lower resolution is proportionally smaller. This typically results in finding "fit" solutions more quickly, which are injected into higher resolution nodes for refinement.

(iii) Nodes connected in the hierarchy (nodes with a parent-child relationship) share portions of the same search space, since the search space of parent is contained in the search space of child. Fast search at low resolution by the parent can potentially help the child find fitter individuals.

(iv) iiGAs embody a divide-and-conquer and partitioning strategy which has been successfully applied to many problems. Homogeneous PGAs cannot guarantee such a division since crossover and mutation may produce individuals that belong to many subspaces, i.e., the divisions cannot be maintained. In iiGAs, the search space is fundamentally divided into hierarchical levels with well defined overlap (the search space of the parent is contained in the search space of the child). A node with block size $r = p' \times q'$ only searches for individuals separated by Hamming distance r.

(v) In iiGAs, nodes with smaller block size can find the solutions with higher resolution. Although Dynamic Parameter Encoding (DPE) [14] and ARGOT [15] also deal with the resolution problem, using a zoom or inverse zoom operator, they are different from iiGAs. First, they are working at the phenotype level and only for real-valued parameters. iiGAs divide the string into small blocks regardless of the meaning of each bit. Second, it is difficult to establish a well-founded, general trigger criterion for zoom or inverse zoom operators in PDE and ARGOT. Furthermore, the sampling error can fool them into prematurely converging on suboptimal regions. Unlike PDE and ARGOT, iiGAs search different resolution levels in parallel and eliminate the risk of zooming into the wrong target interval, although there remains some risk that search will prematurely converge on a suboptimal region.

2.3 Parallel Computational Environment

The parallel environment in which we first implemented these ideas was based on a modification to GAucsd [16] established by using a modified P4 [17] on a network of Sparc 10 workstations, and has now been more flexibly realized in the GALOPPS [18] toolkit developed by the authors.

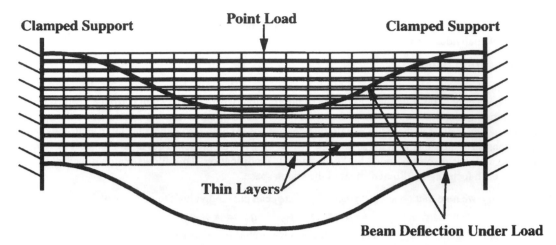

Figure 1. Clamped-clamped laminated beam with center point load.

3. Composite Applications

3.1 Laminated Beam

The goal of this effort is to develop tools for designing 3D large scale laminated composite panels such as might be found in aerospace, automotive, civil, and marine structures. In order to demonstrate and exercise the capabilities of our models, however, the first application presented here will focus on the design of laminated composite beams, a 2-dimensional problem. In particular, the current application is the design of laminated beams in order to maximize their energy-absorption characteristics, such as might be required in armor plating for tanks or bumpers for automobiles. A simplification is made in that only quasi-static loading is considered.

It is hypothesized that energy absorption can be increased by placing thin compliant layers between the fiber-reinforced composite layers. The purpose of the thin compliant layers is to modify the load path characteristics of the structure for a given applied load to increase the amount of energy the beam can absorb before failure. These compliant layers are a form of "damage" between two layers, and they allow the adjacent layers to "slide" relative to one another under a given load. This sliding mechanism may absorb energy but might also increase the local stresses and reduce the strength of the beam. Thus, the role of the GA is to identify the stacking sequence of the composite layers along with the locations and sizes of the thin compliant layers so as to maximize the amount of energy that can be absorbed before failure. The structural configurations under consideration thus contain complex local material arrangements that significantly affect the local stress/strain state as well as the global deflection response of the structures, both of which are used to evaluate and rank each possible design.

The beam's energy absorption capacity is determined by finding the maximum load the beam can carry and the associated center deflection (see Figure 1) before failure occurs, as predicted by a stress-based failure criterion. More specifically, the center deflection and inplane normal stresses in each design element are computed for an applied unit load. These quantities are then scaled linearly to a level at which the inplane normal stress in one or more design elements is equal to the allowable normal stress of the material in that element. The product of the applied load and center deflection at failure, or the work done by the applied load, is then taken as a direct measure of the amount of energy absorbed by the beam before failure.

Results were obtained using both ring topology and island injection topology PGAs. In both cases, ten computational nodes were used, each containing a population of one hundred designs. Thus, the total number of design evaluations per generation in each experiment was the same in each case, allowing a direct evaluation to be made concerning the effect of PGA topology on convergence rate. The node connectivity graphs for the two experiments are shown in Figure 2. In Figure 2b, the topology labeled "island injection" is actually a hybrid topology, with low resolution nodes injecting solutions into a set of high resolution nodes that themselves comprise a ring topology. The differences in convergence characteristics between the two PGA topologies are striking. The results from the ring topology exhibit a classical asymptotic approach to convergence, with indistinguishable differences among the subpopulations.

When using the island injection topology (see Figure 3), converged designs have approximately the same fitness as those obtained using the ring topology. However, the nature of convergence in these two cases is quite different. Referring to Figure 3, it can be seen that, at each stage of the analysis, there is generally a large variation in

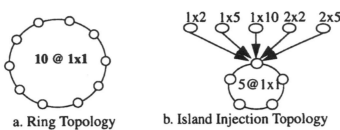

a. Ring Topology   b. Island Injection Topology

Figure 2. (a) Ring and (b) island injection PGA topologies.

the fitness of the best individual designs in each node. This is due to the independent search of several nodes at various levels of resolution of the design. More importantly, however, is the rate of convergence, especially during the early stages of the analysis. Consider the number of evaluations required to attain a design with a predicted energy absorption of 5000 Nm (see comments below regarding the predicted energy absorptions). When the ring topology is used, approximately 17,000 evaluations are required to attain this level of fitness, while only approximately 9,000 evaluations are required when the island injection topology is used. The increased rate of convergence at earlier stages of the analysis is even more remarkable, and is easily explained as follows. When a lower resolution of the design is used, the size of the design space is reduced significantly, allowing a faster search for good designs at that resolution. When a node reaches a converged state, it is reinitialized and the search at that resolution is begun anew, as manifested by the "cliffs" on the graph. If the lower resolution discretizations of the design are appropriate --that is, if they are capable of representing good building blocks for the final design -- then nodes searching at these resolutions will produce good designs very rapidly and inject their best designs into the nodes in which the design is represented at a higher resolution, essentially for refinement. In the present case, it is seen in Figure 3 that the node searching at a 2x2 resolution leads the search until approximately 17,000 total evaluations are completed, at which point this node has converged, and the nodes searching at 1x1 resolution are able to find even better designs.

Sample designs are shown in Figure 4. The designs, while slightly different, have very similar levels of predicted energy absorption. This illustrates one of the advantages of genetic algorithms over other design techniques -- several, and possibly many, good designs are generated from which the designer can choose the best one based on other criteria such as manufacturability, cost, etc. Also note the similarity of the designs in Figure 4b-4d to the design in Figure 4a, which was obtained through search at a 2x2 resolution. For reasons of simplicity of design, the beam in Figure 4a may even be the design of choice for a given application. Detailed finite element analyses of selected designs using a commercial finite element code [19] and the current model with a more refined mesh have confirmed the validity of the designs and the model used to obtain them.

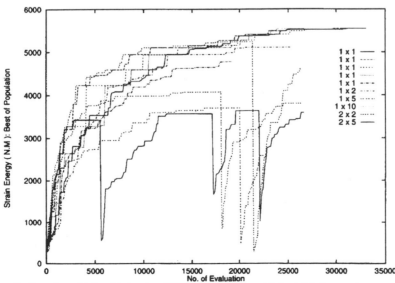

Figure 3. Results of island injection PGA design runs, which yielded superior results to the PGA ring topology. (Sharp cliffs result when converged subpopulations are "restarted".)

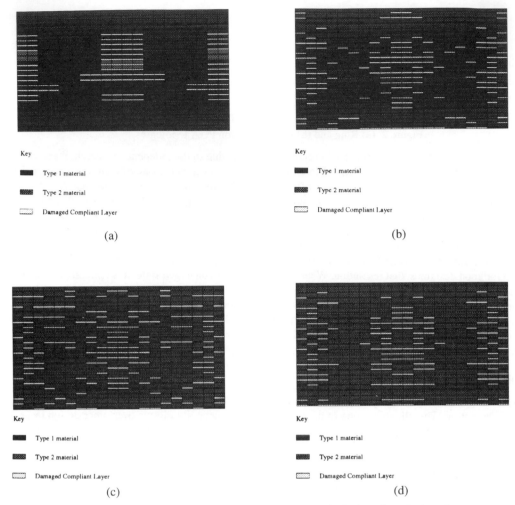

Figure 4. (a) shows best design from the 2x2 subpopulation; (b-d) show best designs from the 1x1 (highest resolution) ring of subpopulations, into which the 2x2 and other designs were injected. 4(b) was the best design produced.

The effectiveness of the island injection PGA architecture for search in moderate-dimensionality (960-bit) design spaces has been demonstrated. Its use is enabled by the availability of a sufficiently accurate and computationally efficient finite element model of a laminated composite beam, since the amount of computation required using a traditional FEA model would still be prohibitively high. The efficiency of the technique shows promise for extension to more realistic design problems involving more complex structures. Such problems can be addressed on a distributed workstation network using the techniques demonstrated here. Our earlier work [13] has showed that, for the problem at hand, using iiGA on a distributed workstation network resulted in approximately linear speedup of the search process, and this fact means that a contemporary workstation network can solve problems one order of magnitude more complex than this one in a timeframe of 1-2 days.

3.2 Composite Airfoil

During flight, aircraft wings are subjected to aerodynamic loading which causes bending and twisting to occur. The twisting load is due to a pressure differential across the airfoil in which pressure is greater at the leading edge than the trailing edge. Many wings are designed with a smaller (typically up to 5 degrees) angle of attack at the tip than at the root ("washout"). Wing twist is built into an aircraft to reshape the spanwise lift distribution to approximate an ellipse and to prevent tip stall [20]. At a given lift coefficient the lift distribution can be optimized by correct choice of initial washout. But when aerodynamic loads cause more pressure to be applied to the leading edge, an upward deflection relative to the trailing edge effectively increases the angle of attack. This work seeks to counteract that effect by engineering a twisting behavior opposite to that caused by the aerodynamic

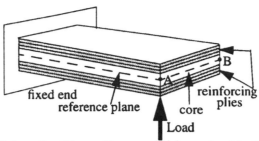

Figure 5. Schematic of the cantilever sandwich panel and loading conditions

loads, so the trailing edge deflects more strongly than the leading edge. This can be achieved using composite sandwich structures in which inherent bending-twisting coupling allows tailoring of the twisting response. For a given loading the objective was to determine the optimum layup (orientation and number of plies in the top and bottom face sheets) which maximizes opposite wing twist while minimizing weight, subject to stiffness and ply clustering constraints. Ultimately, a revised goal will be to achieve a specified amount of opposite wing twist for a given loading condition, rather than to maximize it. The model used is a rectangular cantilever sandwich panel, intended to represent an idealized aircraft wing (see Figure 5). For this demonstration problem, no attempt was made to model exactly actual aircraft wing geometries or loading.

The experiments done compared three different GA topologies -- a single node with a population of 1400, a ring topology with 7 subpopulations having population size of 200 each, and an island injection topology containing 7 subpopulations, each with a population size of 200. In all three cases, a total population size of 1400 is used. Only single population and iiGA results are presented here, due to space limitations. Figure 6 shows single-node results, in which fitness begins to asymptote after about 50 generations. Results from the ring architecture showed each node achieving approximately the same performance after relatively few generations and convergence at around 50 generations. Figure 8b shows the island injection architecture used in this study, and typical results are shown in Figure 7. GALOPPS was used with a two point crossover rate of 0.5, mutation rate of 0.001 per bit, a crowding factor of 3, incest reduction (a form of mate selection) of 3, and tournament selection. The crowding factor and incest reduction were used to help maintain population diversity. Core thickness and angular resolution for the plies in the four top level (lowest resolution) subpopulations is 4 bits, 5 bits for the two middle subpopulations, and 6 bits for the final subpopulation (only one subpopulation at this level was used).

While Figure 2 shows a typical iiGA configuration, a simpler 7-subpopulation tree was used for the preliminary testing (see Figure 8) because of computational resource limitations. In this configuration, the bottom level (highest resolution) contains only one subpopulation into which both moderate level subpopulations inject migrants. Due to the dimensionality of the problem being relatively small, both the iiGA and the ring architecture approaches were able to find a near-optimal solution within approximately 100 generations. Thus additional complexity must be added to the problem to allow conclusive comparison of the different GA architectures. However, certain observations can still be made. Figure 7 shows a typical result of the iiGA, in which migration occurred after each set of 3 generations. The seven curves represent the seven subpopulations. The behavior expected is that the lower-refinement subpopulations, which are working in a smaller search space, will initially make more rapid progress, providing good "building blocks" to the moderate-refinement subpopulations, which should exceed their performance due to a more refined search space. Finally, the most refined subpopulation(s) should receive good building blocks from the moderate-resolution level, and should achieve the best solutions. However, in the small search space represented here, this behavior is compressed into about the first 10 generations. By generation 3, after one generation of "processing" the first migrants from the lowest level, the middle level subpopulations (triangles and asterisks on the graph) have exceeded their performance. Then at about 6 generations, the highest refinement subpopulation (represented by diamonds and dotted line) dominates and continues to outperform the others for the remainder of the run. This run, while outperforming a set of 7 subpopulations at the highest level of refinement using identical parameters and population sizes in a ring topology, reached the same optimum, if allowed to continue to around generation 100. But this example, with only 200 individuals working at the full level of refinement, is eventually beaten by even a single-population GA. However, architectures such as shown in Figure 2(b) do not suffer that problem. The key characteristic of the injection architecture for compute-bound problems is an increased rate of improvement early in the run. For example, note in Figures 6 and 7 that the iiGA produces a design with fitness of 7000 within about 15 generations, while the single node approach requires at least 20 generations to attain a design with the same level of fitness. Another advantage of the injection architecture is that once the low-resolution subpopulations have converged, they can be either "frozen", avoiding additional function evaluations, or perhaps partially reinitialized, broadening the search.

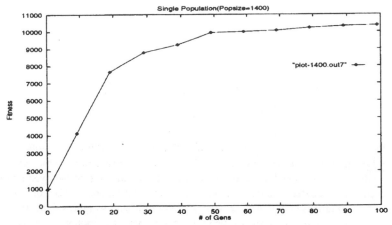

Figure 6. Results from single-node architecture, 1400 individuals.

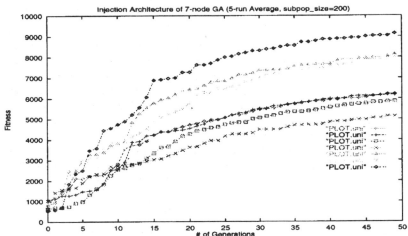

Figure 7. Results from seven-node island injection architecture, each 200 individuals

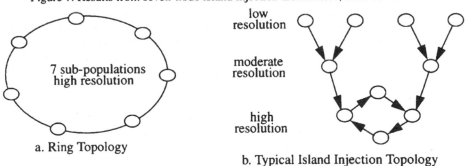

a. Ring Topology

b. Typical Island Injection Topology

Figure 8. Comparison of levels of refinment and paths for migration among subpopulations.

3.3 Composite Flywheel

There are numerous applications of flywheels in the area of energy conservation systems. Vehicles can use fly-wheels during braking for capturing energy lost during deceleration. Another practical application is energy storage in low earth orbit satellites, in which photoelectric cells are exposed to 60 minutes of light, followed by 30 minutes of darkness during which stored energy must be used. The flywheel is well-suited to such applications due to high cyclic lifetimes, longtime reliability and high specific energies. Large flywheels could be used in energy plants.

The fitness function is the Specific Energy Density (SED) of the flywheel, which is the amount of rotational energy stored per unit weight. The "fitness" of each flywheel was evaluated with an axisymmetric finite element code that calculates the 3-dimensional static stresses and strains produced in a flywheel at a constant angular velocity. Even for isotropic materials, this model does not make the assumption that the optimal design is one with equal stresses throughout, a common assumption that facilitates analytical solutions. The maximum allowable angular velocity before failure was determined using the maximum stress criterion for isotropic flywheels, and the maximum strain criterion for composite flywheels. The thickness of each ring varies linearly in the radial direction and a diverse set of material choices exists for each ring. A typical annular flywheel is shown in Figure 9. Optimizing isotropic flywheels using the island injection GA seems quite promising. To test the GA, the constant stress shape was first sought. Figure 10 shows that the GA rapidly captures the shape of the isotropic flywheel within 47 generations. Figure 11 compares the GA solution to the analytical constant stress solution. The power of the GA was shown in Figure 12 in the design of a solid isotropic flywheel with stress free edges with an SED that is 55% higher than that of a constant stress flywheel. The commonly used rotor shape of the constant stress flywheel is not optimal when a flywheel has stress free edges. These important findings of the isotropic flywheel can be directly applied to the GA design of composite flywheels. Full results for composite flywheel optimization using island injection GAs will be completed in the near future.

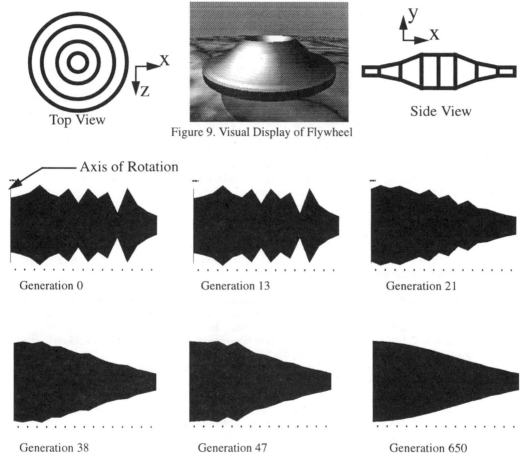

Figure 9. Visual Display of Flywheel

Figure 10. The Evolution of the Constant Stress Flywheel

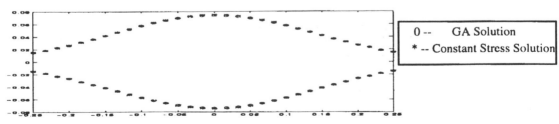

Figure 11. GA Vs. Constant Stress Solution

208

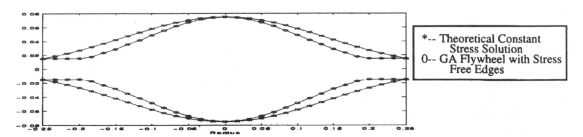

Figure 12. Constant Stress Vs. GA Designed Flywheel with Stress Free Edges

References

[1]   R.C. Averill, W. F. Punch, E. D. Goodman, S.-C. Lin, Y. C. Yip, Y. Ding, 1995, "Genetic Algorithm-Based Design of Energy Absorbing Laminated Composite Beams", ASME Design Engin. Tech. Conf., Boston.

[2]   B. Malott, R. C. Averill, E. D. Goodman, Y. Ding, W. F. Punch, 1996, "Use of Genetic Algorithms for Optimal design of Laminated Compostite Sandwich Panels with Bending-Twisting Coupling", AIAA/ASME/ASCE/AHS/ASC 37th Structures, Structural Dynamics and MAterials Conference, Salt Lake City, Utah.

[3]   D. Eby, R. C. Averill, W. Punch, O. Mathews, E. Goodman, 1997, "An Island Injection GA for Flywheel Design Optimization", Proc. EUFIT '97, Aachen, Germany (forthcoming).

[4]   C.A. Soto and A.R. Diaz, 1993, "Optimum layout and shape of plate structures using homogenization," in *Topology Design of Structures*, M.P. Bendsoe and C.A. Mota Soares, eds., pp. 407-420.

[5]   K. Suzuki and N. Kikuchi, 1990, "Shape and topology optimization by a homogenization method," in *Sensitivity Analysis and Optimization with Numerical Methods*, AMD-Vol. 115, ASME, pp. 15-30.

[6]   K. Suzuki and N. Kikuchi, 1991, "A homogenization method for shape and topology optimization," *Comp. Meth. Appl. Mech. Eng.*, **93**, 291-318.

[7]   E. Sandgren, E. Jensen, and J.W. Welton, 1990, "Topological design of structural components using genetic optimization methods," in *Sensitivity Analysis and Optimization with Numerical Methods*, S. Saigal and S. Mukherjee, eds., AMD-Vol. 115, ASME, pp. 31-43.

[8]   C.D. Chapman, K. Saitou, and M.J. Jakiela, 1993, "Genetic algorithms as an approach to configuration and topology design," in Proc. *1993 ASME Design Automation Conference*, Albuquerque, New Mex., Sept.

[9]   S. Nagendra, R.T. Haftka, and Z. Gurdal, 1992, "Stacking sequence optimization of simply supported laminates with stability and strain constraints," *AIAA Journal*, **30**, 2132-2137.

[10]   R. LeRiche and R.T. Haftka, 1993, "Optimization of laminate stacking sequence for buckling load maximization by genetic algorithm," *AIAA Journal*, **31**, 951-956.

[11]   S. Nagendra, R.T. Haftka, and Z. Gurdal, 1993, "Design of blade stiffened composite panels by a genetic algorithm approach," in *Proceedings of the 34th AIAA/ASME/AHS SDM Conference*, La Jolla, CA, April 19-22, pp. 2418-2436.

[12]   M. Leung and G.E. Nevill, Jr., 1994, "Genetic algorithms for preliminary 2-D structural design," in *Proceedings of the 35th AIAA/ASME/AHS SDM Conference*, Hilton Head, SC, April 18-20.

[13]   S.-C. Lin, W.F. Punch, and E.D. Goodman, 1994, "Coarse-grain parallel genetic algorithms: categorization and analysis," IEEE Symposium on Parallel and Distributed Processing, pp.27-36.

[14]   N. Schraudolph and R. Belew, 1992, "Dynamic Parameter Encoding for Genetic Algorithms," *Machine Learning*, June, pp. 9-21.

[15]   C. G. Shaefer, 1987,"The ARGOT Strategy: Adaptive Representation Genetic Optimizer Technique," *Proceedings of the Second International Conference on Genetic Algorithms*, July, pp. 50-55.

[16]   N. Schraudolph and J. Grefenstette, 1992, *A User's Guide to GAucsd 1.4*, July.

[17]   R. Butler and E. Lusk, 1992, *User's Guide to the P4 Programming System*.

[18]   E. Goodman, 1994, GALOPPS, *The Genetic ALgorithm Optimized for Portability and Parallelism System*, Tech. Rept. #94-5, MSU GARAGe, Michigan State University, 100pp.

[19]   MARC User's Guide: User Information, 1994, MARC Analysis Research Corporation, Palo Alto, CA.

[20]   P.C. Yang, C.H. Norris, , and Y. Stavsky, 1966, Elastic Wave Propagation in Heterogeneous Plates, *International Journal of Solids*, Vol. 2, pp. 665-684.

Part 5: Decision Support, Constraints and Optimisation

Papers:

$\mathcal{M}_{ijn}$ Mutation Operator for Aerofoil Design Optimisation

I. De Falco, A. Della Cioppa, A. Iazzetta and E. Tarantino

Institute for Research on Parallel Information Systems (IRSIP)
National Research Council of Italy (CNR)
Via P. Castellino, 111
80131 Naples - Italy
email: ivan@irsip.na.cnr.it

Keywords: Evolutionary Algorithms, Co–mutation, $\mathcal{M}_{ijn}$, Aerofoil Optimisation, Inverse Design.

Abstract

A new mutation operator, called $\mathcal{M}_{ijn}$, capable of operating on a set of adjacent bits in one single step, is introduced. Its features are examined and compared against those of the classical bit–flip mutation. A simple Evolutionary Algorithm, $\mathcal{M}$–EA, is described which is based only on selection and $\mathcal{M}_{ijn}$. This algorithm is used for the solution of an industrial problem, the Inverse Aerofoil Design optimisation, characterised by high search time to achieve satisfying solutions, and its performance is compared against that offered by a classical binary Genetic Algorithm. The experiments show for our algorithm a noticeable reduction in the time needed to reach a solution of acceptable quality, thus they prove the effectiveness of the proposed operator and its superiority to GAs for the problem at hand.

1. Introduction

SINCE early 60's both in the USA and in Europe researchers tried to understand the relative importance of crossover and mutation operators in the execution of an Evolutionary Algorithm (EA). One of the first attempts to use mutation alone for evolution of good solutions to a given problem was made by Friedberg [1] who tried to evolve finite state machines. With time two different opinions became widespread in the scientific community and divided it into two schools. On the one hand, people like Fogel et al. in the USA, Rechenberg and Schwefel in Germany supposed mutation to be the key operator for evolution, or at least of high importance, so they used just selection and mutation to evolve a population. This opinion has historically led to the designs of the Evolutionary Programming (EP) [2] and of the Evolution Strategies (ES) [3, 4], and to their continuous improvement in recent years. On the other hand, the other school believes that crossover does play the vital role and mutation is seen as being only a background operator. This can be found in Genetic Algorithms (GAs). Among these researchers we can cite Holland [5], De Jong [6], Goldberg [7], and, more recently, Culberson [8] and his GIGA based only on crossover. Presently, however, it is growing more and more within the EA community the awareness of the importance of mutation, and interest is increasing in a closer consideration of its features. Jones [9] implemented an operator he called macromutation as follows: the genome is viewed as a ring, two distinct points in the ring are randomly selected with uniform probability and the loci in the smaller section are set to random alleles. When applied to a problem set, this macromutational hillclimber in all but one case turned out to be better than standard GAs and a bit–flipping hillclimber. He argued that the conclusion that crossover is useful since it facilitates the building blocks exchange between individuals is not justified. So, researchers are striving to precisely measure and compare the effectiveness of these two operators, and to find their strengths and their limits as well [10, 11]. As W. Spears states [12] *"mutation and crossover are simply two forms of a more general exploration operator ... the current distinction between crossover and mutation is not necessary, or even desirable, although it may be convenient, ... it may be possible to implement one general operator that can specialise to mutation, crossover, or any variation in between"*.

Our recent research fits well in this frame; in fact, we have decided to focus our attention on the mutation behaviour, and to attempt to formalise an evolution model based on selection and mutation only. Furthermore, keeping in mind the work carried out by many other researchers, we have aimed to design a new mutation operator which could exhibit, apart from the features usual of such an operator, also other ones whose importance has been stressed in the recent literature, like those typical of crossover, so as to save good building blocks. The basic idea underlying our work is the design of a mutation operator allowing us to take care of the information shared between adjacent bits and among neighboring ones, since in many cases they are not completely independent one from another, rather they as a whole encode in the most general case real variables. What is really important for us is that such an operator should be capable of mutating a number of adjacent bits at the same time. With this we do not mean we are looking for something which simply chooses randomly a substring and applies flip–mutation to all of the therein contained bits. What we are looking for is something completely different, which applied to a substring mutates it trying to save the implicit information contained in the fact that such bits are adjacent. We may call such a mutation *co-mutation*. The co–mutation operator we have developed is the $\mathcal{M}_{ijn}$, and, starting from it, we have designed a new Evolutionary Algorithm, the $\mathcal{M}$–EA. This idea of co-mutation takes origin from work of one of the authors in recursive multicomputer topologies devised as an alternative to hypercubes. It somehow resembles that of *macromutation*, introduced in Evolution Theory by Professor R. Goldschmidt of the University of California at Berkeley in the 40's. According to this theory [13], there exist in nature several complex structures from mammalian hair to hemoglobin which could not have been produced by the accumulation and selection of lots of small successive mutations, and darwinian evolution could account for no more than variations within the species boundary, so evolution needs every now and then single jumps as macromutations. More specifically, this means the appearance of previously unknown *patterns* of genes. This idea met strong opposition from Darwinians.

With this paper not only do we wish to introduce this new operator, we also intend to test its effectiveness on a typical industrial problem. The application chosen is the Inverse Aerofoil Design Optimisation as met in Aerodynamics. It is becoming quite common to read of researchers using binary Genetic Algorithms for its solution. The quality of the resulting aerofoils is satisfying, unfortunately the time needed to find them is quite high. We hope to reduce this search time by using $\mathcal{M}$–EA.

Our paper is outlined as follows: Section 2 contains our proposed model for mutation, and a formalisation of this model, together with a description of some of its features compared against those offered by the usual bit–flip mutation. The $\mathcal{M}$–EA, based on this operator, is introduced. In Section 3 we briefly mention the state of the art in Aerodynamic Aerofoil Design Optimisation, and we focus attention on the continuously increasing interest in Evolutionary Algorithms within this field. In Section 4 we provide the reader with results achieved by both our algorithm and classical binary Genetic Algorithms on the Inverse Design Problem, and we compare these results in terms of both quality and search time. Section 5 contains our conclusions, and a description of our foreseen future work which will have to be done in the near future to further increase the impact of $\mathcal{M}_{ijn}$ model in Aerofoil Design.

2. The $\mathcal{M}$–EA algorithm

2.1. $\mathcal{M}_{ijn}$ Mutation

Let us consider a generic alphabet $\mathcal{A}$ composed by $s \geq 2$ different symbols, $\mathcal{A} = \{a_1, \ldots, a_s\}$ and strings σ over the universe $\Sigma = \mathcal{A}^\ell$ (the set of all the strings σ of a given finite length ℓ).

In the following we shall simply denote with σ a generic string $\sigma_{\ell-1} \ldots \sigma_0$ of any length ℓ, where $\sigma_q \in \mathcal{A} \ \forall q \in \{0, \ldots, \ell-1\}$. If we want to specify the i-th symbol of the alphabet occurring in the generic position q of the string σ we shall use $\sigma_{q,i}$.

Let's start by describing in words the $\mathcal{M}_{ijn}$ behaviour. Our operator can be represented as a two phase–algorithm, the first phase consisting in search and the second in replacement. In the search phase, we randomly choose a point p in the string σ, and we consider $sigma_p$, let it be equal to a_i ($\sigma_p = a_i$). Then, starting from p, we go leftbound and we examine the contents of the adjacent position, σ_{p+1}; if it is equal to a_i we move leftbound until we find a position k whose contents is different from the previously found values, let's say it is a value a_j ($\sigma_k = a_j$). Now the search phase has ended and we start the replacement phase: σ_k becomes a_i and $\sigma_{k-1}, \ldots, \sigma_p$ all b ecome a_j.

Then, formally, the $\mathcal{M}_{ijn}$ operator is the following:

$$\mathcal{M}_{ijn} \colon \sigma \in \Sigma, p \in \{0, \ldots, \ell-1\} \longrightarrow \sigma' \in \Sigma' \subset \Sigma \tag{1}$$

where p is randomly chosen and $p = 0$ means the rightmost position in σ, while $p = \ell - 1$ means the leftmost one.

Denoting with $\sigma_{q,i}^z$ the generic sequence of z equal symbols a_i starting from the position q and going left, we have that the application of $\mathcal{M}_{ijn}$ to p when $n - 1$ equal symbols are met according to the above procedure yields for $n < \ell - p - 1$:

$$\sigma = \sigma_{\ell-1} \ldots \sigma_{p+n} \boxed{\sigma_{p+n-1,i} \sigma_{p,j}^{n-1}} \sigma_{p-1} \ldots \sigma_0 \xrightarrow{\mathcal{M}_{ijn}} \sigma' = \sigma_{\ell-1} \ldots \sigma_{p+n} \boxed{\sigma_{p+n-1,j} \sigma_{p,i}^{n-1}} \sigma_{p-1} \ldots \sigma_0 \tag{2}$$

while for $n = \ell - p - 1$:

$$\sigma = \boxed{\sigma_{p,j}^{\ell-p}} \, \sigma_{p-1} \ldots \sigma_0 \quad \xrightarrow{\mathcal{M}_{ijn}} \quad \sigma' = \boxed{\sigma_{p,k}^{\ell-p}} \, \sigma_{p-1} \ldots \sigma_0 \tag{3}$$

with $a_k \neq a_j$ randomly chosen in $\mathcal{A}$.

Let us give some explanatory examples. Let us consider the binary case, and the following 8-bit string $\sigma \equiv 01011010$. The length of 8 means that p can vary between 0 and 7. Let us suppose that the application point p is randomly chosen as 3, i.e. $p = 3$. In this case $\sigma_3 = 1$, so we have to go leftbound and look for the first occurrence of a 0 in σ. For $p = 4$ we find $\sigma_4 = 1$, so we further move leftbound; for $p = 5$ we have that $\sigma_5 = 0$ so we stop. Then we replace the string components from 3 to 5: precisely, in the 3rd and in the 4th position, where a 1 was contained, we insert a 0, and in the 5th position, where a 0 was held, we put a 1. In conclusion, in such a case the new string obtained after the application of $\mathcal{M}_{ijn}$ is 01$\boxed{\textbf{100}}$010, where the substring in bold is the part of σ modified according to $\mathcal{M}_{ijn}$. As a further example, $\mathcal{M}_{ijn}$ applied to the string $\sigma \equiv 01000011$ with $p = 3$, instead, yields 0$\boxed{\textbf{0111}}$011 (again, we have marked in bold the modified part of σ).

All of this can be written in form of a procedure as follows:

Procedure $\mathcal{M}_{ijn}$ Mutation Operator
begin
    choose randomly a value p ($0 \leq p \leq \ell - 1$); let us suppose that $\sigma_p = a_i$;
    $index = p$;
    /* start search phase */
    repeat
       $index = index + 1$;
    until $\sigma_{index} <> a_i$;
    /* end search phase and start replacement phase */
    $save = \sigma_{index}$;
    $\sigma_{index} = \sigma_p$;
    for $count = (index - 1)$ to p step 1 do
       $\sigma_{count} = save$;
    od
    /* end replacement phase */
end.

The (3) takes place when, going leftbound, we reach the leftmost bit of the string σ. This means that we have met a long sequence of all equal elements, so that we have a_i but we have not been able to find a_j. This introduces some randomness in the general s-ary case, but this is no longer true in a binary alphabet, where the only allowable symbol is the one not present in the substring under examination.

As an example of this situation, let us consider the 8-bit-long binary string $\sigma \equiv 00011010$ and the randomly chosen application point $p = 5$; we have in this case that $\sigma_5 = 0$, so we go leftbound looking for the first occurrence of a 1. For $p = 6$ we have that $\sigma_6 = 0$, so we further proceed leftbound, and we find $\sigma_7 = 0$. We have reached the end of the string, and no occurrence of a 1 has been met. In such a case the string becomes $\boxed{\textbf{111}}$11010 (as usual, we have represented in bold the part of σ modified by the application of $\mathcal{M}_{ijn}$). As a further example of this special situation, if, instead, we apply $\mathcal{M}_{ijn}$ to $\sigma \equiv 11010011$ and $p = 7$ we have a new string like this: $\boxed{\textbf{0}}$1010011 (the only modified bit of σ is the leftmost one, represented in bold).

As it can immediately be seen the $\mathcal{M}_{ijn}$ mutation changes at least two symbols at the same time apart from the case $p = \ell - 1$ in which only the leftmost bit is changed, so it is definitely something different from the classical Bit–Flip Mutation (BFM) used in GAs.

2.2. Features of $\mathcal{M}_{ijn}$ operator

This subsection gives a description of some features of $\mathcal{M}_{ijn}$. Some of them are quite intuitive, while others need calculations. We have decided, for the sake of conciseness, to omit the numerical demostrations.

From now on we shall discuss by taking into account the binary alphabet, i.e. $\mathcal{A} = \{0, 1\}$. For the sake of simplicity let us suppose to work with a string σ of ℓ bits representing just one and only one integer or real variable whose variation range is divided into 2^ℓ steps. In the following we shall call *value* of the string σ the integer number between 0 and $2^\ell - 1$ corresponding to the string encode and *jump length* the difference between the values represented by σ' and σ, respectively.

For the BFM actual implementation we may either vary one and only one bit or anyway consider the mutation rate equal to $1/\ell$ so that mutation will yield on average the variation of one bit.

substring	BFM	sign	$\mathcal{M}_{ijn}$	sign	substring	BFM	sign	$\mathcal{M}_{ijn}$	sign
10	11	+	01	−	011	010	−	100	+
01	00	−	10	+	1000	1001	+	0111	−
100	101	+	011	−	0111	0110	−	1000	+

Table 1. The jump directions for BFM and $\mathcal{M}_{ijn}$.

Feature 1. Starting from any given string σ of length ℓ, $\mathcal{M}_{ijn}$ allows to reach in one step exactly ℓ different strings.

Feature 2. Starting with any given string σ it is always possible to reach any other string σ' by successive applications of $\mathcal{M}_{ijn}$.

Let us suppose for the moment that during its application $\mathcal{M}_{ijn}$ will not run until the leftmost bit of the string (otherwise it takes place what is described below in Feature 4).

Feature 3. Given a string σ and a randomly chosen position p, the application of $\mathcal{M}_{ijn}$ in p yields a jump of length 2^p, equal in module to that offered by BFM, but with opposite sign.

This can be seen with examples as shown in Table 1. In it, we report only the substring of σ from $\mathcal{M}_{ijn}$ application point p going left, until the first occurrence of a symbol other than that present in σ_p.

As a conclusion from these latter properties we have that the jump length for both BFM and $\mathcal{M}_{ijn}$ is one among $1, 2, 4, 8, 16, \ldots, 2^{(\ell-1)}$.

Let us now remove the hypothesis we made above, i.e. that during its application $\mathcal{M}_{ijn}$ will not run until the leftmost bit of the string. We wish now to consider this case, mathematically represented by the (3). In such a situation the result of the application of the $\mathcal{M}_{ijn}$ operator leads to jumps longer than those normally allowed, i.e. greater than 2^p if $\mathcal{M}_{ijn}$ is applied to the p-th bit. Let us call such jumps *long jumps*. An example of such a long jump is a string σ (of length 4) consisting of 0001 (i.e. a value of 1) and $\mathcal{M}_{ijn}$ applied to the $p = 2$ leading to 1101 corresponding to a value 13. In this case a jump of 12 has been performed. It is clear that in all the situations leading to long jumps there is a sequence of equal bits in the leftmost part of the string, the MSB included. Let us call *train* such a sequence and let us call τ its length, that is the number of equal bits.

Feature 4. The length of a long jump caused by $\mathcal{M}_{ijn}$ application to k bits is equal to

$$\sum_{p=\ell-k}^{\ell-1} 2^p$$

We wish to determine how many such jumps there exist as a function of ℓ.

Feature 5. $\mathcal{M}_{ijn}$ allows to perform a number of long jumps γ equal to:

$$\gamma = 2 \sum_{p=1}^{\ell-1} 2^{(\ell-p-1)} = 2^\ell - 2$$

Feature 6. The relative number of long jumps ρ is equal to:

$$\rho = \frac{number\ of\ long\ jumps}{total\ number\ of\ possible\ jumps} = \frac{\gamma}{\ell\,2^\ell} = \frac{1}{\ell}\left(1 - \frac{1}{2^{(\ell-1)}}\right)$$

From the above formula we can evaluate ρ as a function of the number of bits n_b representing the variable. This leads to the following property.

Feature 7. The probability of performing long jumps decreases as n_b increases and it tends to $1/n_b$.

n_b	2	3	4	5	6	7	8	9	10	16	17	20
ρ	25.00%	25.00%	21.87%	18.75%	16.14%	14.06%	12.40%	11.06%	9.98%	6. 24%	5.88%	4.99%

Table 2. The behaviour of ρ (in percentage) as a function of n_b.

In Table 2.2. we report ρ (in percentage) as a function of n_b. In our opinion one of the most interesting points in our operator is the fact that it allows long jumps so as to let the search reach very far points from where the search currently is. Stated another way, $\mathcal{M}_{ijn}$ allows to perform longer jumps which are not possible in BFM. In fact while BFM can jump only in a limited number of ways, given by the 2^p formula, $\mathcal{M}_{ijn}$ has at its disposal, apart from those usual for BFM, also some longer than the maximum allowed by BFM. This means that $\mathcal{M}_{ijn}$ has about the same capabilities of local search as ordinary mutation has (in Table 2.2. from 75% to 95% of the mutations will yield the same jumps), and it has also the possibility of performing jumps to quite far regions in the search space which cannot be reached by BFM. In this way there is a balance between local investigation and far travels.

Feature 8. $\mathcal{M}_{ijn}$ allows to modify more variables at the same time and in a related way.

In fact, if σ is 8–bit long and it encodes for two variables x_1 and x_2 (4 bits each), in the situation $\sigma \equiv 00111010$ ($x_1 = 3$ and $x_2 = 6$) and $p = 3$ we have σ' *equiv*01000010 ($x_1 = 4$ and $x_2 = 2$).

2.3. The $\mathcal{M}$–EA algorithm

Our Evolutionary Algorithm $\mathcal{M}$–EA [14] works as follows: we start with a randomly generated initial population of, say, N elements each with ℓ positions. Each position contains a symbol in an s–ary alphabet. We let this population undergo selection according to the fitness and we apply $\mathcal{M}_{ijn}$ as the only operator affecting the strings in the population. To each string $\mathcal{M}_{ijn}$ is applied once and only once. There is no *a priori* chosen selection scheme, rather we may choose one among the most commonly used (proportional, truncation, tournament, exponential, and so on). Elitism may be taken into account. In the remainder of this paper we make reference to a truncation selection scheme. The basic scheme for $\mathcal{M}$–EA is described in the following.

```
Procedure M-Evolutionary Algorithm
begin
    t=0;
    initialise randomly P(t) with P elements;
    evaluate P(t) by using fitness function;
    while not terminated do
        for j=1 to P do
            select randomly one element among the best T% in P(t);
            mutate it;
            evaluate the obtained offspring;
            insert it into P'(t);
        od
        P(t + 1) = P'(t);
        t=t+1;
    od
end.
```

3. Overview of Evolutionary Aerofoil Design

It is well known that in Aerofoil Design skilful designers can efficiently solve the problem of finding a good shape satisfying all of the requirements for one operating condition; as soon as we have to deal with more than one such condition the design becomes less and less efficiently solvable by using only human mind capabilities, since too many conditions and constraints must be handled at the same time. It must be remarked here that the usage of the classical gradient-based optimisation methods leads to a boring trial-and-error work, since the procedures based on this will get stuck in any local optimum they meet; every time this takes place, the program must be restarted from another randomly chosen initial point in the search space. Because of this, and due to the limitations of mathematical methods, new approaches to Design Optimisation are welcome.

Usually, this shape design has been faced by using stream function based methods [16], numerical optimisation [15] or control theory [17]. Further methods based on expert systems have also been introduced, like in [18]. Unfortunately these methods show in this problem their critical aspects such as the use of the experience acquired in previous designs and the difficulty in the synthesis of innovative solutions, aspects that must be carefully considered in any aerodynamic optimisation problem.

Quite recently, Genetic Algorithms (GAs) are becoming more and more widely used in aeronautical problems, including parametric and conceptual design of aerocraft, preliminary design of turbines, topological design of nonplanar wings, and aerodynamic optimisation using CFD. Like in the above mentioned fields, there is an increasing interest for EAs also in Aerofoil Design Optimisation, both direct and inverse. As an example, Obayashi and Takanashi [19] have implemented a GA–based tool to optimise target pressure distributions for inverse design methods; once target pressure distributions are obtained, corresponding aerofoil/wing geometries can be computed by an inverse design code coupled with a Navier–Stokes solver. Design examples indicate that their optimisation algorithm is efficient and that supercritical wing shapes are reproduced by the simulated evolution. Quagliarella and Della Cioppa [20] have developed a transonic aerofoil aerodynamic design method based on a GA coupled to a full potential flow field solver. They have investigated the effectiveness of such a method for the optimisation of both a one operating condition design problem and a two operating conditions one. Tong [21] has proposed the usage of GAs in the aerospace field to solve complex design problems. He has used these algorithms together with expert system technology and numerical programs to develop the preliminary design of aerocraft engine turbines. Mosetti and Poloni [22] have implemented a design tool that couples an *ad hoc* designed GA with a Navier–Stokes flow field solver. Gage and Kroo [23] have investigated GAs as an alternative to conventional optimisation for the preliminary design of wings; their conclusion is that GAs promise to be a useful tool for the exploration of difficult design domains like the one examined.

As it results from the above examples, the classical binary GAs are being more and more often used as an optimisation means in aerofoil design. We wish to evaluate the effectiveness of GAs usage in Aerofoil Design by comparing their results against those achieved by $\mathcal{M}$–EA.

4. Results in Aerofoil Design Optimisation

In order to compare the behaviour of the $\mathcal{M}$–EA against that offered by binary–valued GAs on the Aerofoil Design Optimisation, we have developed two algorithms, one based on GAs and another on $\mathcal{M}$–EA, with the aim to apply them to and the Inverse Design. Just to recall the meaning of the problem, in Inverse Design we are assigned a pressure distribution along an aerofoil shape, and we have to, starting from an initial aerofoil shape, reconstruct the target one showing the given pressure distribution. Of course, the closer we get to the goal configuration, the better the search technique. We have used as testbed the reconstruction of the Korn aerofoil for subsonic inviscid flow starting from the NACA64A410 aerofoil. The operating conditions considered for the design task are the following: Mach number M=0.75 and angle of attack α equal to zero. The objective function used is the classical referred to in literature for this kind of problems, that is:

$$f(\mathbf{x}) = \oint_{\Gamma} \left(C_p(\mathbf{x}, s) - C_p^t(s) \right)^2 ds \tag{4}$$

where C_p and C_p^t are respectively the current and the target pressure distribution at the operating condition and s is the arc length along the C_p shape Γ. The fitness function of a generated aerofoil is proportional to the inverse of the objective function, i.e. $\Phi(\mathbf{x}) = \frac{k}{f(\mathbf{x})}$, so the problem becomes a typical maximisation problem and a fitness of infinite value would mean that the two C_p shapes are absolutely identical. A very important issue is the representation of the geometry of the aerofoil. We have used for this problem a B–spline representation [25], by means of which we define a closed curve (the aerofoil) with a set of m points. Such a curve varies as a function of a parameter called degree of the curve (the higher the degree the softer the curve, without spikes, and the farther the shape from the control points). If we let these control points slightly move within given ranges, we will obtain new curves, so new aerofoils. The parameters to be optimised are, thus, the positions of these control points. There are two reasons to choose the B–spline representation. Firstly, it allows to generate curves in a general way and, if it is needed, permits to set some control points and to change the remaining ones. Secondly, this representation is particularly interesting for our task because it allows a local optimisation. From a

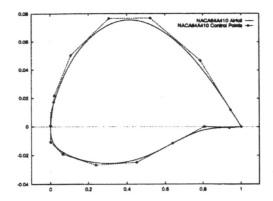

Fig. 1. The NACA64A410 aerofoil and its B–spline representation.

mathematical point of view, a curve generated by using the vertices of a defining polygon is dependent on some interpolation or approximation scheme establishing the relationship between the curve and the polygon. We have defined a set of control points for both the upper and the lower aerofoil shape:

$$\mathbf{B_0} = \{B_{1x_0}, B_{1y_0}, \ldots, B_{mx_0}, B_{my_0}\} \tag{5}$$

and for each point the following variation has been allowed:

$$\mathbf{W} = \{W_{1x}, W_{1y}, \ldots, W_{mx}, W_{my}\} \tag{6}$$

Hence, the variation law that has been defined for each control point is the following:

$$B_{ix} = B_{ix_0} + W_{ix} \text{ and } B_{iy} = B_{iy_0} + W_{iy}$$

The aerofoil shape has been defined using two open B–spline curves of degree 4 with 9 control points for both the upper and the lower surface. The initial and final points, i.e. $B_{1_0} = (0.0, 0.0)$ and $B_{9_0} = (1.0, 0.0)$, have been set on both the curves to close the aerofoil and the remaining ones are let free to move. So, we have 14 control points corresponding to 28 coordinates to be optimised. The variation W_i of each coordinate is in the range $[-0.04, 0.04]$. This is a very general approach because no specific knowledge of the desired final shape has been used, differently from other approaches which employ a different range for each point. Thus, we have computed the vertices of the polygon which defines the NACA64A410 geometry using it as the initial polygon for the genetic optimisation. In Fig. 1 the usage of a B–spline to represent the NACA64A410 shape is shown.

The behaviour of GAs and of $\mathcal{M}$–EA on this problem is described in next subsections.

4.1.   GA usage

The GA uses binary encoding for the continuous parameters to be optimised. The population size chosen is 120. A one–point crossover ($p_c = 0.8$) and a mutation operator ($p_m = 0.005$) are used; moreover a 2–elitist strategy has been taken into account. A preliminary set of experiments has been performed with the aim to find the most suitable number of bits for the GA gene representation. The experiments performed by using 8, 16 and 32 bits have indicated that an 8-bit encoding is the most appropriate for the problem under examination and that the evolution gets slower as the number of bits grows. The decision has been taken to stop the evolution when a fitness value equal to or higher than 70,000 is reached. In Fig. 2 the best among 10 runs is reported. Namely, in Fig. 2(a) the evolution process in terms of maximum and average fitness within the population is depicted; Fig. 2(b) and Fig. 2(c) report respectively the target and the computed aerofoils and the relative distributions of the pressure coefficient C_p. The number of generations averaged over 10 runs needed to reach the final fitness value is 190, though in the best run it is about 150.

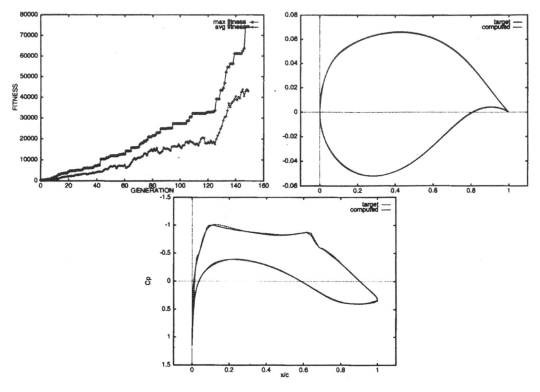

Fig. 2. The results obtained in Inverse Design by means of a sequential GA.

4.2. M–EA usage

In this case we have used a population size of 120 individuals, truncation selection, an 8–bit encoding and 1–elitism. Preliminary sets of runs have allowed to determine the best range for the threshold T%. This has turned out to be [20%, 50%], with the best performance shown for T%=40%.

Ten runs have been made with a value of T% equal to 40%. The final value chosen is the same for GAs, i.e. 70,000. In Fig. 3(a) we report the best fitness and the average fitness for the best run, and in Fig. 3(b) the target and the computed shapes are given. Fig. 3(c), finally, shows the pressure distributions along the target and the computed aerofoils.

4.3. Comparison

Firstly it should be noted that, since each individual represents variables with 8 bits, the probability of performing a long jump in M–EA is, according to Table 2, 12.40%. At a first glance this quantity might seem not high enough. In spite of this, however the M–EA performance is by large better than that offered by GA, and this holds for both the best and the average results on 10 runs. In fact, comparison between Figg. 2(a) and 3(a) shows that M–EA is able to obtain a final solution equivalent in quality to that achieved by GAs in a lower number of generations (88 generations are needed in this best case, and 115 on average, against 150 and 190 respectively required by GAs). We have wondered whether this result could depend on the truncation selection used, so we have performed some runs in which M–EA used roulette wheel. The results have been similar to those obtained with thresholding. This has allowed us to state that the reduction in generations needed is due to the operator.

This reduction in the number of generations is highly important especially in the Aerofoil Design Problem. In fact, the evaluation time for each individual is the time needed to run aerodynamic code, in our case a full–potential non–conservative code; even for this relatively simple code the execution time is about 15 seconds on an IBM Risc 6000 or on a single node of a Meiko Computing Surface CS2. This means that in order to evaluate the total execution time we can neglect the time due to the genetic phases with respect to that required by the fitness evaluation, which turns out to be by large the most–consuming part. Therefore, a reduction in the number of evaluations needed to reach a solution with acceptable

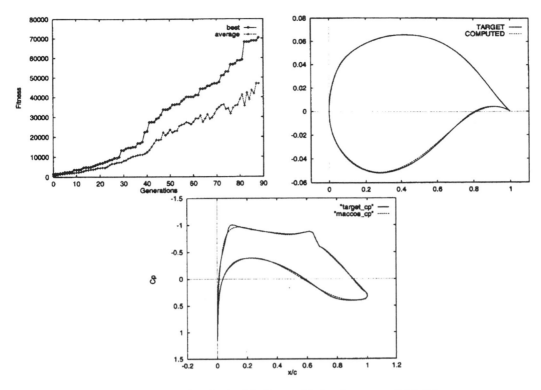

Fig. 3. The results obtained in Inverse Design by means of $\mathcal{M}$–EA.

quality is highly desirable in this field. In our case, if we take into account the number of evaluations of the individuals *noe* (with reference to the average values) we find that for the GA we have:

$$noe_{GA} = 190 \cdot 120 = 22,800$$

and an execution time of about:

$$t_{GA} = 22,800 \cdot 15 = 342,000s = 95h$$

while for the $\mathcal{M}$–EA we have:

$$noe_{\mathcal{M}\text{-EA}} = 115 \cdot 120 = 13,800$$

and an execution time of about

$$t_{\mathcal{M}\text{-EA}} = 13,800 \cdot 15 = 207,000s = 57.5h$$

thus a speed–up s of

$$s = \frac{95}{57.5} = 1.65$$

has been obtained, resulting in a solution of good quality obtained in less than two and a half days with respect to almost four days needed by GA. In conclusion, our $\mathcal{M}$–EA allows to obtain a substantial reduction in search time, and can therefore be exploited in the industrial environment under account.

5.   Conclusions

Within this paper we have introduced a brand–new mutation operator, the $\mathcal{M}_{ijn}$, and the $\mathcal{M}$–EA Evolutionary Algorithm based on it. We have described some of the properties of this operator. We have then taken into account an industrial problem, the Aerofoil Design Problem, and we have evidenced by looking at the available literature how the application of Evolutionary Algorithms may lead to a substantial improvement in the quality of the obtained design while, at the same time, reducing the time needed to perform the search. Nonetheless, the execution time is quite high, so that a reduction in this serch time is welcome. We have then compared the results achieved by $\mathcal{M}$-EA against those offered by binary Genetic Algorithms on the aforementioned problem. A remarkable conclusion coming out from our experimental results is that $\mathcal{M}$-EA can profitably be exploited in Inverse Design, as they have proved its superiority in this problem with respect to classical discrete Genetic Algorithms, widely used in this area. The superiority consists in a noticeable reduction in search time to achieve good–quality solutions.

As regards future work, we know that EAs can be easily parallelised. We have already implemented a coarse–grained parallel GA on a cluster of 4 RISCs, and we have used it in the Direct Aerofoil Design Problem. By doing so we have obtained a speed–up of 3.7 with respect to the sequential case [24]. We wish now to use the same coarse– grained parallel framework for $\mathcal{M}$-EA to implement a parallel version for it and run both the parallel versions on the Inverse Design Problem. We hope that in this way we can further reduce the search time. Furthermore, we plan to investigate whether the utilisation of k-ary alphabets may provide us with speed-up with respect to the binary case.

References

[1]   Friedberg, R.M., 1958, A Learning Machine, Part I, *IBM Journal of Research and Development,***2**.

[2]   Fogel, L.J., Owens, A.J., and Walsh, M.J., 1966, *Artificial Intelligence through Simulated Evolution*, Wiley, New York.

[3]   Rechenberg, I., 1973, *Evolutionsstrategie: Optimierung Technischer Systeme nach Prinzipien der Biologischen Evolution*, Frommann-Holzboog, Stuttgart.

[4]   Schwefel, H.P., 1977, *Numerical Optimization of Computer Models*, John Wiley and Sons, New York.

[5]   Holland, J.H., 1975, *Adaptation in Natural and Artificial Systems*, University of Michigan Press, Ann Arbor.

[6]   De Jong, K.A., 1975, *An analysis of the behavior of a class of genetic adaptive systems*, Doctoral Thesis, Department of Computer and Communication Sciences, University of Michigan, Ann Arbor.

[7]   Goldberg, D.E., 1989, *Genetic Algorithms in Search, Optimization and Machine Learning*, Addison-Wesley, Reading, Massachussetts.

[8]   Culberson, J., 1993, Crossover versus Mutation: fueling the debate: TGA versus GIGA, *Proceedings of the Fifth International Conference on Genetic Algorithms* Morgan Kauffmann, pp. 632–639.

[9]   Jones, T., 1995. Crossover, Macromutation, and Population-based Search. *Proceedings of the Sixth International Conference on Genetic Algorithms.* Morgan Kauffmann.

[10]   Mitchell, M., Holland, J.J, and Forrest, S., 1992, When will a Genetic Algorithm outperform Hill Climbing?, *Advances in Neural Information Processing Systems 6* Morgan Kaufmann.

[11]   Horn, J., Goldberg, D.E., and Kalyanmoy, D., 1994, Long Path Problems, *Parallel Problem Solving from Nature - PPSN III*, Springer-Verlag, pp. 149–158.

[12]   Spears, W.M., 1993, Crossover or Mutation?, *Proceedings of the Foundations of Genetic Algorithms*, vol.2, San Mateo, CA, Morgan Kauffmann, pp. 221–237.

[13]   Goldschmidt, R.B., 1956, Portraits from memory.

[14]   De Falco, I., 1997, An introduction to Evolutionary Algorithms and their application to the Aerofoil Design Problem – Part I: the Algorithms, *von Karman Lecture Series on Fluid Dynamics*, Bruxelles, Belgium, April 1997.

[15]   Vanderplaats, G.N., 1984, *Numerical Optimization Techniques for Engineering Design: with applications*, Mc Graw Hill, New York.

[16]   Dulikravich, G.S., 1991, *Aerodynamic shape design and optimization*, Tech. Rep. 91–0476, AIAA paper.

[17]   Jameson, A., 1988, *Aerodynamic design via Control Theory*, Tech. rep. 88–64, ICASE.

[18]   Tong, S.S., 1985, *Design of aerodynamic bodies using artificial intelligence/expert system technique*, Tech. Rep. 85–0112, AIAA Paper.

[19]   Obayashi, S, and Takanashi, S., 1995, Genetic Optimization of Target Pressure Distributions for Inverse Design Methods, *Proceedings of the 12th AIAA Computational Fluid Dynamics Conference*, San Diego, CA, Jun 19-22 1995.

[20]   Quagliarella, D., and Della Cioppa, A., 1994, Genetic Algorithms Applied to the Aerodynamic Design of Transonic Airfoils, *Proceedings of the 12th AIAA Applied Aerodynamics Conference*, Colorado Springs, CO, USA, Jun, 1994, AIAA-94-1896-CP, pp. 686-693.

[21]   Tong, S.S., Powell, D., and Skolnick, M., 1989, Engeneous: domain independent, machine learning for design optimization, *Proceedings of the third International Conference on Genetic Algorithms*, (J. D. Schaffer editor), M. Kauffmann Publishers, pp. 151–159.

[22]   Mosetti, G., and Poloni, C., 1993, Aerodynamic shape optimization by means of a genetic algorithm, *Proceedings of the fifth International Symposium on Computational Fluid Dynamics*, Sendai, Japan.

[23]   Gage, P., and Kroo, I., 1993, *A Role for Genetic Algorithms in a Preliminary Design Environment*, Stanford University Technical Report.

[24]   De Falco, I., Del Balio, R., Della Cioppa, A. and Tarantino, E., 1995, A Parallel Genetic Algorithm for Transonic Airfoil Optimisation, *Proceedings of the Second IEEE International Conference on Evolutionary Computing*, Perth, University of Western Australia, Australia, Nov. 1995.

[25]   Rogers, D.E., 1989, *Mathematical Elements for Computer Graphics*, Addison-Wesley, Reading, Massachussetts.

Design Problems with Soft Linear Constraints

Marián Mach

Dept. of Cybernetics and Artificial Intelligence, Technical University of Košice,
Letná 9, 041 20 Košice, Slovak Republic, machm@ccsun.tuke.sk

Keywords: Constraint satisfaction, soft constraints, genetic algorithms, penalty functions

Abstract

A broad class of design problems can be represented as a class of problems with linear constraints. Several approaches can be used to solve such problems. Unfortunately, most of them cannot process soft constraints which do not limit the solution space but introduce only a measure of acceptance for the space of possible solutions.

A simple framework based on genetic algorithms with penalty functions is presented. If a solution violates some constraint, it is penalized according to the importance of this constraint and the degree of its violation. The penalty functions are combined with an optimization function into the information used to guide the search.

A set of parameters enables to model different types of human processing of soft constraints. Since the number of these parameters can cause difficulties, a simple heuristic how to set them is presented.

1. Introduction

A broad class of design problems can be represented as a class of problems with linear constraints. One example of this class is the problem how to combine available ingredients containing different amounts of required elements in order to obtain a final mixture (which can be optimal in some sense) while ensuring the proper balance of particular elements in this mixture. More formally, it can be defined as follows:

- a set of variables $X = \{x_1, x_2, \ldots, x_n\}$,

- a set of linear constraints $C = \{c_1, c_2, \ldots, c_m\}$. Each constraint has the form

$$MIN_i \leq a_{i1}x_1 + a_{i2}x_2 + \ldots + a_{in}x_n \leq MAX_i$$

- a function $f(x_1, x_2, \ldots, x_n)$ which is a subject of optimization.

The definition covers both optimization and non-optimization problems (in this case no potential solution is preferred to others – it can be reflected by $f(x_1, \ldots, x_n) = const$). Constraints define an area of interest which is searched to find a solution simultaneously satisfying these constraints and optimizing the function f. Thus, the constraints guarantee the functionality of the final solution and the optimization function represents some desirable properties of this solution.

Such problems can be easily solved by various methods from the field of operation research (e.g. linear programming) or artificial intelligence (e.g. constraint logic programming). Unfortunately, these methods can handle only *hard* constraints – a solution is acceptable only if it satisfies all given constraints. These methods are able to produce a solution only if all constraints can be satisfied. They fail if any constraint cannot be satisfied (e.g. the problem is overconstrained). In this case they do not provide either any solution or information how "close" to the success they were.

On the other hand, the case $m \geq n$ is quite common in practice. This fact (together with the independence of constraints from each other in general) can result in overconstrained problems with no solution.

In many cases an acceptable solution can be found despite the violation of some constraints. The violation of such constraints (called *soft* constraints) only signalizes the decreasing quality of this solution – the solution itself remains acceptable. The more violated a soft constraint, the poorer quality of the solution (but the solution is still acceptable). These soft constraints can be often ordered according to their importance.

2. Framework for handling soft constraints

In various papers on genetic algorithms several approaches to handling constraints can be found [1]. We have used the one based on penalty functions. This approach transforms problems with constraints into search problems within the space of all possible solutions (both acceptable and non-acceptable). Only limited information is used – the information about the overall fitness of particular solutions. This overall fitness value expresses the balance between desirable properties of the solution and its functionality (represented by the degree of satisfaction or violation of a given set of constraints) [2].

The population of potential solutions is generated in a random way without considering the constraints. If a solution violates some constraints, its fitness is penalized according to the importance of these constraints. The final fitness function combines the optimization function f (in general, it can be modified in order to reflect a particular optimization aim, eg. by scaling, reversing, etc.) with the overall penalty function P. It has the following form:

$$\alpha F(f(x_1, x_2, \ldots, x_n)) + \beta P(x_1, x_2, \ldots, x_n) \tag{1}$$

The goal is to find a solution which minimizes this function (the aim is to minimize the penalty). Coefficients α and β enable to express the balance between the importance of the global optimization and the satisfaction of all constraints. The function F represents all modifications of the optimization function f. The optimization part corresponds to the fitness of the solution without considering any constraint. The penalty part represents the degree of satisfaction or violation of a given set of constraints.

The overall penalty function considers the importance of particular constraints (some of them can be *more soft* than others) in the form:

$$P(x_1, x_2, \ldots, x_n) = \sum_{i-1}^{m} \gamma_i p_i(x_1, x_2, \ldots, x_n) \tag{2}$$

where $p_i(x_1, x_2, \ldots, x_n)$ represents the i-th individual penalty function (corresponding to the i-th constraint).

The more important (i.e. *less soft*) the i-th constraint, the greater the coefficient γ_i is and the heavier penalty p_i is added to the fitness of the solution.

The value of an individual penalty function depends on the satisfaction or violation of the corresponding constraint. This function can be of different forms and it should enable to handle two different types of constraint violation (the value below the lower threshold MIN_i or the value above the upper threshold MAX_i) separately. Different forms of individual penalty functions together with the constants in the formulas above represent several degrees of freedom which enable to model different types of human processing of soft constraints.

Relations among different properties of the solution can be modified in several ways. Coefficients α, β, γ_i, and the forms of individual penalty functions correspond to one another. For instance, changing the weight β of the overall penalty function has exactly the same result as the same change of the constraint importances (represented by γ_i coefficients) or a proper modification of individual penalty function shapes (slopes of the parts corresponding to the violation of particular constraints).

2.1. Individual penalty functions

If a penalty-based method is used to solve constrained problems, the construction of penalty functions is a key issue in the application of genetic algorithms. Some guidelines can be found in literature (e.g. [3]). We have adopted the following policy:

The penalty for a satisfied constraint should be zero. If a constraint is violated, the penalty for this violation should be non zero – its value depends on the degree of violation.

In general, an individual penalty function which is associated with a constraint has the form depicted in Figure 1.

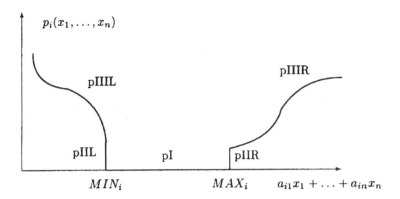

Figure 1. General form of an individual penalty function

It consists of five parts. Various lengths of these parts and various forms of the border parts $pII*$ enable to model different approaches to constraint violations. The x-axis is divided into three intervals. The central interval $< MIN_i, MAX_i >$ represents satisfaction of the corresponding constraint and the other intervals $(-\infty, MIN_i)$ and (MAX_i, ∞) stand for two different ways of violation of this constraint.

All solutions belonging to the central part pI of an individual penalty function have the same degree of acceptance – they satisfy the constraint associated with this penalty function and thus there is no need to penalize them. If for some constraint $MIN_i = MAX_i$ is true, the length of pI is zero.

Two parts $pII*$ represent willingness to accept a solution which is outside the interval defined by a given constraint. Their length is non zero in two cases:

- if there is really some stepwise change of the degree of acceptance resulting from a physical interpretation of the given constraint, or

- if the transition of the solution from inside $< MIN_i, MAX_i >$ to outside this interval should be related with a step-like decrease of the chance to accept such solution for psychological reasons.

Otherwise, there is no reason for the presence of the parts $pII*$ and thus their lengths equal zero. On the other hand, if a solution violating a particular constraint is not acceptable (the constraint is *hard*) the length of both parts $pII*$ approaches infinity. If the importances of two possible ways of constraint violation are not the same, the lengths of $pII*$ should be different.

All solutions located outside the interval $< MIN_i, MAX_i >$ are penalized in some degree. The more severe violation, the lower acceptability the solution has. Parts $pIII*$ represent this fact. The increase of the distance between a solution and the interval $< MIN_i, MAX_i >$ corresponding to the increase of penalty defined by $pIII*$. If the importances of two ways of constraint violation are not the same, the slopes of $pIII*$ differ from each other.

A solution satisfying a given constraint (i.e. located within $< MIN_i, MAX_i >$) is identical with the solution which is looked for if there is no solution which:

- satisfies this constraint and has better value of the optimization function, or

- violates the given constraint but its value of the optimization function is significantly better.

Some solution S violating a given constraint is more acceptable than the best solution among those satisfying this constraint if its value of the optimization function is significantly better. In this case the proportion of the difference between the values of the optimization function corresponding to these two solutions to the degree of violation of this constraint by the solution S is greater than or equal to a threshold.

If the border parts $pIII*$ are represented by linear functions, this threshold is independent on the distance to $< MIN_i, MAX_i >$. Usually, it is beneficial to use some convex functions on behalf of $pIII*$ (the greater distance, the greater threshold). We have tested several types of functions, e.g. quadratic, cubic, power, and exponential. All these functions have one disadvantage – their minimal slope is 0. That is why sometimes it may be useful to combine them with a linear function – it prevails in the area close to $< MIN_i, MAX_i >$ and guarantees some required minimal threshold. Nonlinear functions dominate in the area with greater degree of constraint violation and ensure the threshold to be dependent on the distance to $< MIN_i, MAX_i >$ in an appropriate way.

It is possible to obtain various forms of individual penalty functions (various forms can be used simultaneously for different constraints in the same problem). Three basic forms are depicted in Figure 2. The third form represents an equality constraint. In all three cases the value MIN_i is more important than MAX_i.

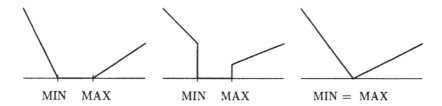

Figure 2. Examples of individual penalty functions

2.2. Overall penalty function

Each constraint divides the space of possible solutions into two subspaces: one with no penalty and the other with variable penalty (the greater distance to the subspace without penalty, the greater penalty). In this way constraints define more or less acceptable parts of the space of all possible solutions. The intersection of all subspaces of the first type (if it exists) represents the most acceptable part of the solution space. A subspace covered by smaller number of subspaces of the first type represents less acceptable subspace. A typical distribution of constraints is depicted in Figure 3 (for the sake of simplicity only two variables are considered).

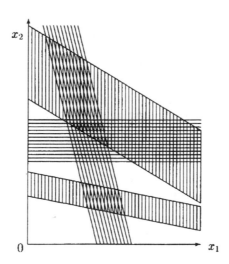

Figure 3. Example of a constraint distribution

The overall penalty function is given by (2). Its final shape depends on the shapes of used individual penalty functions and on the location of particular constraints. Typically, it has one global minimum and zero or a few local minima (every minimum can consist of one or more "neighbouring" solutions). The overall penalty function can be of various shapes but from the point of genetic algorithms it is considered to be relatively simple. An example with one local minimum is depicted in Figure 4 where individual penalty functions have the form according to the second case in Figure 2 and the location of constraints is the same as depicted in Figure 3.

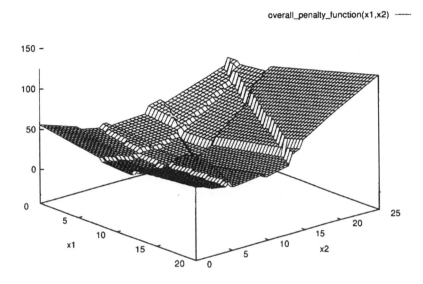

overall_penalty_function(x1,x2) ——

Figure 4. Overall penalty function landscape

Because of the way individual penalty functions are put together the global minimum of the overall penalty function can be located not only in the area covered by as many constraints as possible (in an ideal case all constraints are involved) but in the area with the best location regarding to the distance between this area and all constraints (this area can be covered by no constraint at all) too. In this way the global minimum represents a compromise among all constraints.

The importances of particular constraints are represented by a set of weight coefficients $\{\gamma_i, i = 1, \ldots, m\}$ which enable to differentiate between less and more important constraints. Unfortunately, an exact quantitative meaning of these coefficients cannot be given – the shape of the overall penalty function heavily depends on the location of these constraints. In general, the greater value of a coefficient, the more important the constraint corresponding to this coefficient and the higher penalty. Therefore, the greater weight of a constraint, the closer the global minimum of the overall penalty function to this constraint.

2.3. Final fitness function

The final fitness function is designed as a combination of the optimization function (in general, it can be modified in various ways) and the overall penalty function. Weight coefficients enable to differentiate between importances of these two parts. Their setting heavily influences the location of the global minimum of the fitness function. Different values of the coefficients α a β can result in the following cases:

- two solutions can have the same fitness using one set of weight values while using another set of weights forces the fitness values of these solutions to be different, or

- one solution which is better than other solutions using one set of weight values can become worse when using another set of weights.

During the search for the final solution the proportion of α and β can be constant (the shape of fitness landscape is static) or it can vary (fitness landscape can change its shape – it can result in a modification of the minima locations). The latter enables to reflect different search strategies within different phases of the search for the final solution.

If the global minima of the optimization function and the overall penalty function are different, then the global minimum of the fitness function is located somewhere between them. Its location depends on the values α, β, and the shapes of these functions. In general, the greater the weight corresponding to the overall penalty function, the less acceptable solutions which do not satisfy all constraints (the global minimum of the fitness function becomes closer to the global minimum of the overall penalty function).

3. Heuristic parameter setting

A set of parameters (defining the form of penalty functions and the way these functions and the optimization function are combined together) enables to model different types of human processing of soft constraints. On the other hand, the number of these parameters can cause difficulties since their setting requires a lot of domain specific information.

In order to reduce the amount of information required from domain experts a simple heuristic can be used to set (some of) these parameters. This setting can be accepted as a definite one or it can represent an initial setting which should be further modified using additional information from domain experts or empirical tests.

Design of an individual penalty function corresponding to some constraint consists of two steps: design of its shape and an appropriate parameter setting. The first one can be replaced by the selection of one shape from those presented in Figure 2. This selection can be done using the information about the values MIN_i and MAX_i and the presence or absence of a stepwise change of the degree of acceptance. To perform the second step requires some additional information from domain experts. In order to simplify the design of individual penalty functions the same form of the penalty function can be used for all of them.

An example of a simple heuristic setting of parameters α and β enabling to combine the overall penalty function and the optimization function can be found in [4].

The weight coefficients γ_i representing the importances of individual constraints can be set in different ways using more or less sophisticated methods. Two simple heuristics to perform this setting are presented. They are based on the distribution of all given constraints into several groups. Each group represents a set of constraints which are equivalent – if two constraints are members of the same group, their importances are the same. Thus, each group defines an equivalence class instances of which are of the same importance. In contrast, if two constraints do not belong to the same group, one of them is more important than the other one.

Both of the presented heuristics address the complexity of the parameter setting by reducing the number of variables to one. The first heuristic ignores the real distribution of constraints and replaces it by only two possible locations of constraints. Let us consider only one variable, the same symmetrical shape of all individual penalty functions resembling the third case in Fig. 2, and quadratic functions with the same slopes replacing the parts $pIII*$ of individual penalty functions. In this case the overall penalty function has only one minimum. The location of this minimum is given by a weighted average of constraint locations l_i:

$$l_{min} = \frac{\sum\limits_{i=1}^{m} \gamma_i l_i}{\sum\limits_{i=1}^{m} \gamma_i} \tag{3}$$

If the importances of two constraints are the same, the minimum of the overall penalty function (consisting of only two individual penalty functions corresponding to these two constraints) should be located exactly in the middle of the area between these two constraints or, more exactly, between their locations.

If some constraint is more important than the other, the minimum should be closer to the location of this constraint. The proportion of importances of these constraints can be expressed as a reversed proportion of distances between the minimum and the locations of both constraints. The similar approach can be used when considering the balance between the importances of a particular constraint and a set of constraints, or between the importances of two sets of constraints.

Thus, if there are two disjunctive sets of constraints – a set of m_1 constraints $C_1 = \{c_i, i = k, \ldots, k + m_1 - 1\}$ with weight values $\{\gamma_i, i = k, \ldots, k + m_1 - 1\}$ (all constraints are located in l_1) and a set of m_2 constraints $C_2 = \{c_i, i = l, \ldots, l + m_2 - 1\}$ with weights $\{\gamma_i, i = l, \ldots, l + m_2 - 1\}$ (all constraints within this set share the same location l_2), the proportion of importances of C_1 and C_2 is:

$$\frac{\sum\limits_{i=k}^{k+m_1-1} \gamma_i}{\sum\limits_{i=l}^{l+m_2-1} \gamma_i} \tag{4}$$

If this proportion should be equal to q, it can be necessary to update weights of constraints (to put the minimum of the overall penalty function into the required position). To do this (and to preserve the ratio of importances for each pair of constraints from C_1 or C_2), it is enough to increase the weight of each constraint c_i from C_1 by:

$$\gamma_i \left(q \frac{\sum\limits_{i=l}^{l+m_2-1} \gamma_i}{\sum\limits_{i=k}^{k+m_1-1} \gamma_i} - 1 \right) \tag{5}$$

or from C_2 by:

$$\gamma_i \left(\frac{1}{q} \frac{\sum\limits_{i=k}^{k+m_1-1} \gamma_i}{\sum\limits_{i=l}^{l+m_2-1} \gamma_i} - 1 \right) \tag{6}$$

These formulas can be used to balance importances of two constraints ($m_1 = m_2 = 1$), a constraint and a set of constraints ($m_1 = 1$ and $m_2 > 1$), or two sets of constraints ($m_1 > 1$ and $m_2 > 1$).

The other method for the setting of coefficients γ_i is slightly more complicated than the previous one since it considers exact locations of particular constraints. Let us consider only the j-th variable x_j. The other conditions remain the same as given above. Therefore, the i-th constraint is reduced into the form:

$$M_i = MIN_i \leq a_{ij}x_j \leq MAX_i = M_i \tag{7}$$

and it is located at M_i/a_{ij}. The exact location of the minimum of the overall penalty function is given as follows:

$$l_{min} = \frac{\sum\limits_{i=1}^{m} \gamma_i a_{ij} M_i}{\sum\limits_{i=1}^{m} \gamma_i a_{ij}^2} \tag{8}$$

As opposed to the previous method for parameter setting, if the importances of two constraints are the same, the minimum of the overall penalty function (corresponding to these two constraints) should be located in such a way that the degrees of violation of both constraints are the same. The more important a constraint, the smaller its violation. Therefore, if there is a constraint c_k with weight value γ_k and a set of m constraints $C_2 = \{c_i, i = k + 1, \ldots, k + m\}$ with weights $\{\gamma_i, i = k + 1, \ldots, k + m\}$, the proportion of importances of c_k and C_2 is:

$$\frac{\sum_{i=k+1}^{k+m} (-1)^{s_i} (a_{ij} l_{min} - M_i)}{(-1)^{s_k} (a_{kj} l_{min} - M_k)} \tag{9}$$

where s_i is defined according to:

$$s_i = \begin{cases} 0 & a_{ij} l_{min} \geq M_i \\ 1 & otherwise \end{cases} \tag{10}$$

If this proportion should be equal to q, it can be necessary to modify the minimum of the overall penalty function l_{min}. It can be done by increasing the weight γ_k by:

$$\frac{1-p}{\frac{p}{\sum_{i=k}^{k+m} \gamma_i \frac{a_{ij}^2}{a_{kj}^2}} - \frac{1}{\sum_{i=k}^{k+m} \gamma_i \frac{a_{ij}}{a_{kj}} \frac{M_i}{M_k}}} \tag{11}$$

where p is defined by:

$$p = \frac{\sum_{i=k}^{k+m} \gamma_i a_{ij}^2}{\sum_{i=k}^{k+m} \gamma_i a_{ij} M_i} \frac{q(-1)^{s_k} M_k - \sum_{i=k+1}^{k+m} (-1)^{s_i} M_i}{q(-1)^{s_k} a_k - \sum_{i=k+1}^{k+m} (-1)^{s_i} a_{ij}} \tag{12}$$

The final weight γ_k is valid (and the presented heuristic can be used to set it) only if it remains greater than zero.

These formulas can be used to balance properties of two constraints ($m = 1$) or a constraint and a set of constraints ($m > 1$).

Since now it is possible to obtain n different weights for each individual penalty function (each corresponds to considering only one variable from the set of n variables), the final weights can be calculated as average values.

4. An example

The proposed approach to handling soft constraints has been tested on the design of optimal feed ration from a given set of feeds. The goal is to combine the available feeds in order to obtain the least expensive feed ration while ensuring the proper balance of particular elements in this ration. The module dedicated to the design of feed ration has been implemented in the form of a genetic algorithm (in a close collaboration with a domain expert in veterinary medicine). This module completes KRAVEX – the expert system enabling to solve nutrition problems in connection with the health and production of dairy cows [5].

Variables represent amounts of particular feeds which can be used to compose a proper feed ration. The price of the final feed ration plays the role of the optimization function. This function has the following linear form:

$$price_1 x_1 + price_2 x_2 + \ldots + price_n x_n \tag{13}$$

The aim of the optimization is to find a minimum of the overall price.

Constraints represent a required structure of the designed feed ration. The problem includes 23 soft constraints. They correspond to the requirements for quantities of relevant elements (minerals, vitamins, and some additional parameters like structural fibre contents or dry matter) in the designed feed ration. Originally, they have the form of equalities (demand standards). Designers commonly replace each equality with a pair of inequalities (covering the original equality) representing less strict requirement than the original constraint (one possible value is replaced with an interval). The boundaries of these new intervals used to be set for 100 and 120 percent of the norm values.

The variability of possible forms which can be used to design individual penalty functions causes difficulties to our domain expert. Since he is not able to employ this variability, all individual penalty functions have the same uniform shape. Their setting is based on the following requirements:

- different constraint violations (a value below the lower threshold or above the upper threshold) should be penalized in a different way – the lower threshold MIN_i is more important than the upper threshold MAX_i,

- there should be some minimal penalty if a constraint is (although slightly) violated – but the expert was not able to quantify this minimal penalty, and

- the increase of penalty should be faster than linear regarding to the increase of the degree of constraint violation.

Therefore, all individual penalty functions have the same form similar to the second case depicted in Figure 2 with the exception that quadratic functions are used on behalf of the border parts $pIII*$.

All constraints can be divided into several groups in accordance with their importances (all members of one group share the same importance). The importances of members of different groups are in a given proportion. This partition enables to set weights γ_i.

The other parameters (e.g. coefficients α and β) have been adjusted in an experimental way during a number of experiments with real data.

The result of the optimization process depends on the collection of available feeds. Improper set of feeds can cause that an attempt to balance all elements can produce a compromise of poor quality - no one element is balanced. Since this situation is quite frequent, the user has the possibility of limiting the number of optimized elements.

Such a limited optimization is presented in Figure 5 - four feeds are combined to create feed ration while only six elements are taking into account. The figure presents amounts of these elements (in percentages of the required norm values) corresponding to the best individual.

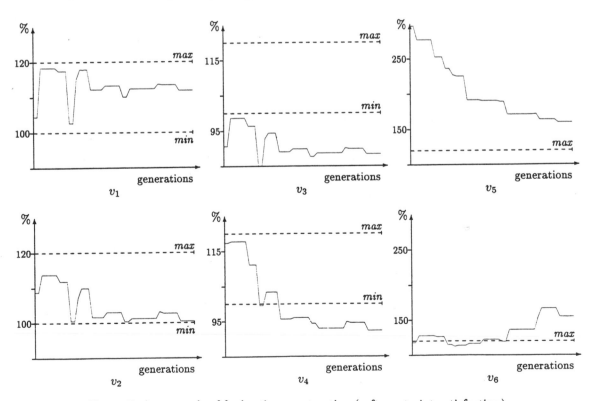

Figure 5. An example of feed ration construction (soft constraint satisfaction)

These six elements (represented by six soft constraints) are divided into three pairs: synthetic parameters v_1 and v_2 (NEL and PDIN), basic elements v_3 and v_4 (dry matter and fibre contents), and minerals v_5 and v_6 (calcium and phosphorus). The importances of these pairs of elements (represented by weights γ_i) are 100, 70, and 30, respectively.

In general, the higher importance of a parameter, the closer the value of this parameter to the required values. The most important parameters v_1 and v_2 are kept within no-penalty range during the whole search. Less important v_3 and v_4 are allowed to get small penalties (they have converged close to values with no penalty). On the other hand, the final penalties corresponding to the least important elements v_5 and v_6 are quite high.

5. Conclusions

The paper addresses a problem of soft constraints which represent desirable features of the final solution. A critical issue of the presented framework for the processing of these constraints is the high number of parameters requiring a great amount of domain specific information in order to set them properly.

To make the presented approach more applicable it is necessary to develop various parameter setting heuristics based on specific features of particular problem types which require only limited information from domain experts. As an example, two simple heuristics for the determination of importances of individual constraints are presented. The only information they require in order to determine the importances of all constraints is the description of equivalence classes defined over these constraints and relations among these classes.

Many alternatives to further reducing the amount of the information required from domain experts remain to be explored, including more sophisticated methods for design of the overall penalty function considering the actual (non-reduced) distribution of constraints and different individual penalty functions corresponding to different constraints.

Acknowledgements

This research is supported by the Grant Agency of the Slovak Republic under the grant No. 1/1686/94. The author thanks Dr. T. Sabol for his enlightening comments on this work and assistance in the preparation of the manuscript.

References

[1] Michalewicz, Z., and Janikow, C.Z., 1991, Handling constraints in genetic algorithms, *Proc. of the 4th Int. Conf. on Genetic Algorithms, San Diego, USA, July*, pp. 151-157.

[2] Mach, M., 1996, Using genetic algorithms for problems with soft constraints, *Proc. of the 3rd Int. Workshop on Artificial Intelligence, Brno, Czech Republic, September*, pp. 173-174.

[3] Richardson, J.T., Palmer, M.R., Liepins, G., and Hilliard, M., 1989, Some guidelines for genetic algorithms with penalty functions, *Proc. of the 3rd Int. Conf. on Genetic Algorithms, George Mason University, USA, June*, pp. 191-197.

[4] Straka, P., 1996, *Feed ration assembling using genetic algorithms*, MSc Thesis, Technical University, Košice, Slovakia.

[5] Naď, P., Rosival, I., and Drozdová. J., 1995, An expert system in the prevention of diseases caused by improper nutrition in ruminants, *Book of Abstracts of the 46th Annual Meeting of the European Association for Animal Production, Prague, Czech Republic, September*, pp. 91-91.

Finding Acceptable Solutions in the Pareto-Optimal Range using Multiobjective Genetic Algorithms

P. J. Bentley[1] and J. P. Wakefield[2]

*[1]Department of Computer Science, University College London,
Gower Street, London WC1E 6BT, UK.
Tel. 0171 391 1329 P.Bentley@cs.ucl.ac.uk (corresponding author)
[2]Division of Computing and Control Systems, School of Engineering,
University of Huddersfield, Huddersfield HD1 3DH, UK.
Tel. 01484 472107 J.P.Wakefield@hud.ac.uk*

Keywords: multiobjective optimization, Pareto-optimal distributions, acceptable solutions, genetic algorithm

Abstract

This paper investigates the problem of using a genetic algorithm to converge on a small, user-defined subset of *acceptable* solutions to multiobjective problems, in the Pareto-optimal (P-O) range. The paper initially explores exactly why separate objectives can cause problems in a genetic algorithm (GA). A technique to guide the GA to converge on the subset of acceptable solutions is then introduced.

The paper then describes the application of six multiobjective techniques (three established methods and three new, or less commonly used methods) to four test functions. The previously unpublished distribution of solutions produced in the P-O range(s) by each method is described. The distribution of solutions and the ability of each method to guide the GA to converge on a small, user-defined subset of P-O solutions is then assessed, with the conclusion that two of the new multiobjective ranking methods are most useful.

1. Introduction

The genetic algorithm (GA) has been growing in popularity over the last few years as more and more researchers discover the benefits of its adaptive search. Many papers now exist, describing a multitude of different types of genetic algorithm, theoretical and practical analyses of GAs and huge numbers of applications for GAs [7,8]. A substantial proportion of these applications involve the evolution of solutions to problems with more than one criterion. More specifically, such problems consist of several separate objectives, with the required solution being one where some or all of these objectives are satisfied to a greater or lesser degree. Perhaps surprisingly then, despite the large numbers of these multiobjective optimization applications being tackled using GAs, only a small proportion of the literature explores exactly how they should be treated with GAs.

With single objective problems, the genetic algorithm stores a single fitness value for every solution in the current population of solutions. This value denotes how well its corresponding solution satisfies the objective of the problem. By allocating the fitter members of the population a higher chance of producing more offspring than the less fit members, the GA can create the next generation of (hopefully better) solutions. However, with multiobjective problems, every solution has a number of fitness values, one for each objective. This presents a problem in judging the overall fitness of the solutions. For example, one solution could have excellent fitness values for some objectives and poor values for other objectives, whilst another solution could have average fitness values for all of the objectives. The question arises: which of the two solutions is the fittest? This is a major problem, for if there is no clear way to compare the quality of different solutions, then there can be no clear way for the GA to allocate more offspring to the fitter solutions.

The approach most users of GAs favour to the problem of ranking such populations, is to weight and sum the separate fitness values in order to produce just a single fitness value for every solution, thus allowing the GA to determine which solutions are fittest as usual. However, as noted by Goldberg: "...there are times when several criteria are present simultaneously and it is not possible (or wise) to combine these into a single number." [7]. For example, the separate objectives may be difficult or impossible to manually weight because of unknowns in the problem. Additionally, weighting and summing could have a detrimental effect upon the evolution of acceptable solutions by the GA (just a single incorrect weight can cause convergence to an unacceptable solution). Moreover, some argue that to combine separate fitnesses in this way is akin to comparing completely different criteria; the question of whether a good apple is better than a good orange is meaningless.

The concept of Pareto-optimality helps to overcome this problem of comparing solutions with multiple fitness values. A solution is Pareto-optimal (i.e., Pareto-minimal, in the Pareto-optimal range, or on the Pareto front) if it is *not dominated* by any other solutions. As stated by Goldberg [7]:

Definition 1. A vector x is partially less than y, or $x <p y$ when:
$$(x <p y) \Leftrightarrow (\forall_i)(x_i <= y_i) \wedge (\exists_i)(x_i < y_i)$$

x *dominates* y iff $x <p y$.

However, it is quite common for a large number of solutions to a problem to be Pareto-optimal (and thus be given equal fitness scores). This may be beneficial should multiple solutions be required, but it can cause problems if a smaller number of solutions (or even just one) is desired. Indeed, for many problems, the set of solutions deemed acceptable by a user will be a small sub-set of the set of Pareto-optimal solutions to the problems [4]. Manually choosing an acceptable solution can be a laborious task, which would be avoided if the GA could be directed by a ranking method to converge only on acceptable solutions. For this work, an *acceptable solution* (or champion solution) is defined:

Definition 2. A solution is an *acceptable solution* if it is Pareto-optimal and it is considered to be acceptable *by a human*.

Consequently, this paper will investigate the problem of using a genetic algorithm to converge on a small, user-defined subset of acceptable solutions to multiobjective problems, in the Pareto-optimal (P-O) range.

The paper will initially focus on the difficulties posed by multiobjective problems to genetic algorithms. A technique to guide the GA to converge on the smaller subset of acceptable solutions will then be introduced. In the light of this, six different ranking methods will be described: three commonly used methods ('sum of weighted objectives', 'non-dominated sorting', and 'weighted maximum ranking' - based on Schaffer's VEGA [11]), and three new, or less commonly used methods ('weighted average ranking', 'sum of weighted ratios', and 'sum of weighted global ratios').

This paper will then describe the application of these six multiobjective techniques to four established test functions, and will examine the previously unexplored distribution of solutions produced in the P-O range(s) by each method. The distribution of P-O solutions and the ability of each method to guide the GA to converge on a small, user-defined subset P-O solutions will then be assessed.

2. Background

Existing literature seems to approach this ranking problem using methods that can be classified in one of three ways: the aggregating approaches, the non-Pareto approaches and the Pareto approaches.

Many examples of aggregation approaches exist, from simple 'weighting and summing' [7,15] to the 'multiple attribute utility analysis' (MAUA) of Horn and Nafpliotis [9]. Of the non-Pareto approaches, perhaps the most well-known is Schaffer's VEGA [11,12], who (as identified by Fonseca [3]) does not *directly* make use of the actual definition of Pareto-optimality. Many other non-Pareto methods have been proposed (e.g. by Linkens [5], Ryan [10] and Sun [14]). Finally the Pareto-based methods, proposed first by Goldberg [7] have been explored by researchers such as Horn [9] and Srinivas [13].

In addition, many researchers are now introducing 'species formation' and 'niche induction' in an attempt to allow the uniform sampling of the Pareto set (e.g. Goldberg [7] and Horn [9]). For a comprehensive review, see the paper by Fonseca and Fleming [3].

3. Range-Independence

Upon consideration, it seems that the problems caused by multiple objectives within the evolutionary search process of the GA have more to do with mathematics than evolution. Throughout the evolution by the GA, every separate objective (fitness) function in a multiobjective problem will return values within a particular range. Although this range may be infinite in theory, in practice the range of values will be finite. This 'effective range' of every objective function is determined not only by the function itself, but also by the domain of input values that are produced by the GA during evolution. These values are the parameters to be evolved by the GA and their exact values are normally determined initially by random, and subsequently by evolution. The values are usually limited still further by the coding used, for example 16 bit sign-magnitude binary notation per gene only permits values from -32768 to 32768. Hence, the *effective range* of a function can be defined:

Definition 3.	The *effective range* of $f(x)$ is the range from $\min(f(x))$ to $\max(f(x))$ for all values of x that are actually generated by the GA, and for no other values of x.

Although occasionally the effective range of all of the objective functions will be the same, in most more complex multiobjective tasks, every separate objective function will have a different effective range (i.e., the function ranges are noncommensurable [12]). This means that a bad value for one could be a reasonable or even good value for another, see fig. 1. If the results from these two objective functions were simply added to produce a single fitness value for the GA, the function with the largest range would dominate evolution (a poor input value for the objective with the larger range makes the overall value much worse than a poor value for the objective with the smaller range).

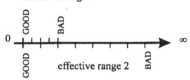

Figure 1. Different effective ranges for different objective functions (to be minimized)

Thus, the only way to ensure that all objectives in a multiobjective problem are treated equally by the GA is to ensure that all the effective ranges of the objective functions are the same (i.e., to make all the objective functions commensurable), or alternatively, to ensure that no objective is directly compared to another. In other words, either the effective ranges must be converted to make them equal, and a range-dependent ranking method used, or a range-independent ranking method must be used. Typically, range-dependent methods (e.g., 'sum of weighted objectives', 'distance functions', and 'min-max formulation') require knowledge of the problem being searched to allow the searching algorithm to find useful solutions [13]. Range-independent methods require no such knowledge, for being independent of the effective range of each objective function makes them independent of the nature of the objectives and overall problem itself. Hence, a ranking method should not just be independent of individual applications (i.e., problem independent), as stated by Srinivas [13], it should be independent of the effective ranges of the objectives in individual applications (i.e., range-independent). Multiobjective ranking methods that are *range-dependent* or *range-independent* can be defined:

Definition 4.	Given the objective functions of a problem:	$f_{1..n}(x)$
	and a set of solution vectors to the problem:	$\{s_1, s_2, \dots, s_m\}$

A multiobjective ranking method is *range-dependent* if the fitness ranking of $\{s_1, s_2, \dots, s_m\}$ defined by the method changes when the effective ranges of $f_{1..n}(x)$ change.

A multiobjective ranking method is *range-independent* if the fitness ranking of $\{s_1, s_2, \dots, s_m\}$ defined by the method *does not* change when the effective ranges of $f_{1..n}(x)$ change.

Because range-independent ranking methods are independent of the problem, they require no weights to fine-tune in order to allow them to rank solutions appropriately into order of overall fitness for a GA. This is a significant advantage over range-dependent methods [1], allowing the same multiobjective GA to be used, unchanged, for a number of different multiobjective problems. Consequently, it would seem that range-independent ranking methods are the most appropriate type of ranking method to use in a general-purpose multiobjective GA.

4. Importance

In addition to being range-independent, there is another significant, and usually overlooked property that a good ranking method should have: the ability to increase the 'importance' of some objectives with respect to others in the ranking of solutions, to allow search to be directed to converge on *acceptable* solutions. *Importance* can be defined:

> **Definition 5.** *Importance* is a simple way to give a ranking method additional problem-specific information, in order to direct a GA to converge on acceptable solutions within a smaller subset of the Pareto-optimal range, by favouring those solutions closer to the optima of functions with increased importance, in proportion to this increased importance.

It has been known for some time that the quality of solutions to complex search problems can be improved by increasing the importance of a particular part or objective of the problem [2,6]. This is often achieved either by introducing objectives to the search algorithm one at a time (or in distinct 'stages') with the most important first, or by simply weighting the most important objectives more heavily. Indeed, experience shows that many users of GAs and the 'sum of weighted objectives' ranking method are inadvertently increasing the importance of certain objectives without being aware of it, as they fine-tune their weights to improve evolution. In other words, the dual nature of these weights (i.e., the fact that each weight can not only equalise the effective ranges of objectives, but also define increased importance for objectives), is often overlooked.

Intentionally determining which objectives are more important in a problem can be a matter of debate, but to improve evolution time, it seems that often the best results are gained by making the most difficult to satisfy objectives the most important. However, some problems demand that certain objectives have differing levels of importance just to allow evolution of an acceptable solution. (For example, the optimization of an electronic device has the design criteria: cost, speed, size and power consumption. For some devices, a low cost is overwhelmingly important, for others, a high speed is of greatest importance.)

Consequently, it is clear that importance is an essential tool to help the evolution of acceptable solutions. What is perhaps less clear, is how the concept of importance should be implemented within multiobjective ranking methods.

One way to allow the definition of importance within aggregation-based ranking methods is to take advantage of the fact that these methods usually guide the GA to converge upon a single 'best compromise' solution. For the purposes of this paper, the *best compromise* solution is defined:

> **Definition 6.** A *best compromise* solution is the solution with the sum of (weighted) objective fitnesses minimized.

By weighting appropriate objectives with importance values, this best compromise solution can be made the same as (or at least moved into the vicinity of) the required solution, allowing the GA to converge directly to an acceptable solution. Thus, producing a single best compromise solution is not always a disadvantage.

Nevertheless, the more favoured ranking methods do not employ aggregation (and typically are range-independent). They are usually used with some form of niching and speciation method to allow the GA to generate not one, but a range of non-dominated P-O solutions. (Niching can also help the quality of solutions by preventing excessive competition between distant solutions [7].) The user is then required to select the preferred solution from this range of different solutions.

However, particularly for problems with many objectives, only a small proportion of P-O solutions may be acceptable solutions. This means that even when hundreds of different solutions are generated by the GA, there can be no guarantee that an acceptable solution will be among them. Moreover, for such large problems, it is not always feasible to allow the user to pick the preferred solution from a truly representative range of P-O solutions: the number to be considered may be too large. Thus, the ranking method needs further information, to guide the algorithm to converge more closely to acceptable solutions *within* the range of P-O solutions. This information is 'importance' - by specifying which objectives must be satisfied more than others, the GA can converge more closely to acceptable solutions, not just P-O solutions.

Unfortunately, there is no easy way to increase the importance of one objective in relation to another, without the two objectives being directly compared to each other. In other words, whilst it is simple to specify increased importance with a range-dependent aggregation method such as 'sum of weighted objectives' (just increase the weights), with a range-independent method such as 'non-dominated sorting', specifying importance is more complex. (Fonseca forces a kind of importance with his 'preference articulation' method [4], but this requires detailed knowledge of the ranges of the functions themselves, and is not a continuous guide to evolution.) Thus, alternative methods of ranking multiobjective solutions are required, that are ideally range-independent and allow the easy specification of importance, to enable the GA to converge on the subset of acceptable solutions.

5. Multiobjective Ranking Methods

There follows descriptions of six different ranking methods. The first three are the most commonly used methods: the range-dependent 'weighted sum' (aggregation) method, the range-independent Pareto non-dominated sorting, and a range-independent method based on Schaffer's VEGA [11,12]. The last three are new range-independent methods, developed in an attempt to allow importance to be specified with such methods. The techniques used within these methods are not new, but they have as yet been rarely used to rank multiobjective populations within a genetic algorithm.

Method 1:   Sum of Weighted Objectives (SWO)
This is perhaps the most commonly used method because of its simplicity. All separate objectives are weighted to make the effective ranges equivalent (and to specify importance) and then summed to form a single overall fitness value for every solution. These values are then used by the GA to allocate the fittest solutions a greater chance of having more offspring. (Because of the similarity in nature and performance between this method and many of the other 'classical' methods [13], only this classical method will be explored.)

Method 2:   Non-Dominated Sorting (NDS)
Described by Goldberg [7], this range-independent method and variants of it are commonly used. The fitnesses of the separate objectives are treated independently and never combined, with only the value for the same objective in different solutions being directly compared. Solutions are ranked into 'non-dominated' order, with the fittest being the solutions dominated the least by others (i.e., having the fewest solutions partially less than themselves). These fittest can then be allocated a greater probability of having more offspring by the GA.

Method 3:   Weighted Maximum Ranking (WMR)
This ranking method is based on Schaffer's VEGA [11,12]. WMR forms lists of fitness values of each solution for each objective. The fittest n solutions from each list are then extracted, and random pairs are selected for reproduction. Importance levels can be set by weighting appropriate fitness values for solutions. Note that the additional heuristic used by Schaffer to encourage 'middling' values [11] was not implemented in WMR.

Method 4:   Weighted Average Ranking (WAR)
This is the first of the alternative ranking methods proposed. The separate fitnesses of every solution are extracted into a list of fitness values for each objective. These lists are then individually sorted into order of fitness, resulting in a set of different ranking positions for every solution for each objective. The average rank of each solution is then identified, with this value allowing the solutions to be sorted into order of best average rank. Thus, the higher an average rank a solution has, the greater its chance of producing more offspring. Since all objective fitnesses are treated separately, this method is range-independent. This technique allows the specification of importance by the weighting of average ranking values for each solution.

Method 5:   Sum of Weighted Ratios (SWR)
This is the second of the ranking methods proposed for GAs and is basically an extension to SWO (method 1). The fitness values for every objective are converted into ratios, using the best and worst solution in the current population for that objective every generation. More specifically:

$$fitness\_ratio_i = \frac{(fitness\_value_i - \min(fitness\_value))}{(\max(fitness\_value) - \min(fitness\_value))}$$

This removes the range-dependence of the solutions, and they can be weighted (for the setting of importance) and summed to provide a single fitness value for each solution as with the first method.

Method 6:   Sum of Weighted Global Ratios (SWGR)
This method is the third of the proposed ranking methods for GAs, and is a variation of SWR (method 5). Instead of the separate fitnesses for each objective in every solution being converted to a ratio using the *current* population best and worst values, the *globally* best and worst values are used. Again the importance of individual objectives can be set by weighting the appropriate values.

6. Application of the Ranking Methods

6.1. Test Functions

To explore and compare the distributions of solutions generated by the six ranking methods, they were applied in turn to four different test functions: F_1 to F_4. The first three are identical to those used by Schaffer [11,12], whilst F_4 is identical to Fonseca's f_1 [4]. Each function was chosen to represent a different class of function (i.e., each has different numbers of P-O ranges and/or best compromise solutions). All functions are to be minimized, see Fig. 2.

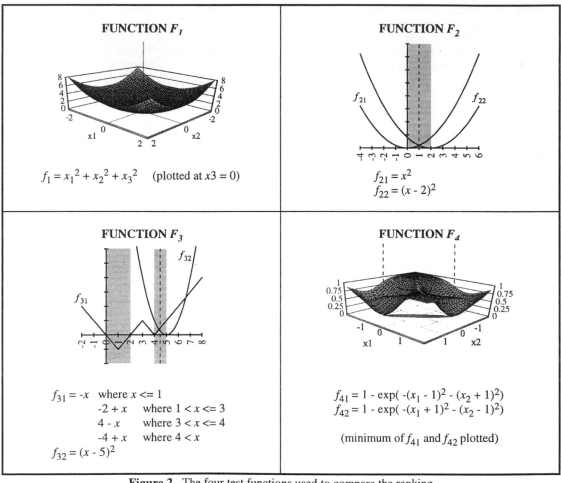

FUNCTION F_1

$f_1 = x_1^2 + x_2^2 + x_3^2$ (plotted at $x3 = 0$)

FUNCTION F_2

$f_{21} = x^2$
$f_{22} = (x - 2)^2$

FUNCTION F_3

$f_{31} = -x$ where $x <= 1$
$\quad\quad -2 + x$ where $1 < x <= 3$
$\quad\quad 4 - x$ where $3 < x <= 4$
$\quad\quad -4 + x$ where $4 < x$
$f_{32} = (x - 5)^2$

FUNCTION F_4

$f_{41} = 1 - \exp(-(x_1 - 1)^2 - (x_2 + 1)^2)$
$f_{42} = 1 - \exp(-(x_1 + 1)^2 - (x_2 - 1)^2)$

(minimum of f_{41} and f_{42} plotted)

Figure 2. The four test functions used to compare the ranking
(P-O ranges shown by grey shaded regions and best compromise solutions marked with dotted lines).

All six methods were used with a basic genetic algorithm using binary coding, a population size of 50, and running for 100 generations. Probability of crossover was 1.0, probability of mutation was 0.01. Although this GA used elitist selection techniques, with all of the ranking methods described in this paper it is possible to use alternatives.

The distributions produced by methods 1-6 for each function were calculated by running the GA between 1,000 and 10,000 times (1000 runs for F_1, 2000 runs for F_2 and F_3, and 10000 runs for F_4). It was assumed that the distribution of solutions produced by a series of runs of this algorithm would not differ significantly from the distribution of solutions obtained by an algorithm with niching or other speciation techniques.

6.2. Evolved Results: F_1

The first experiment performed with each method was simply to allow the GA to minimize F_1. This function was used to validate that each method would rank solutions to single-objective problems correctly (as was done for

VEGA by Schaffer [12]. As expected, every method allowed the GA to converge on, or very near to, the optimal solution of (0,0,0), every time. (The distributions of solutions for this function are all at a single point and hence are not shown.)

6.3. Evolved Results: F_2

The next experiment involved minimising F_2. To give some idea of the quality and distribution of solutions, 2,000 test runs were performed for each method. All methods allowed the GA to produce P-O solutions every time, however, as fig. 3 shows, the distribution of these solutions on the Pareto front for this function are very different for each method. SWO and SWGR both produced solutions very close to or exactly the best compromise value of 1.0. SWR also favoured this value, but with a larger 'spread', with the numbers of solutions produced falling almost logarithmically the further from the best compromise value they were. NDS showed a fairly even distribution throughout the P-O range, and WMR favoured solutions at either function optima, with nothing in between. WAR gave the most unexpected and fascinating distribution, with solutions close to each optima and close to the best compromise value being favoured, all other P-O values being less commonly produced, see fig. 3.

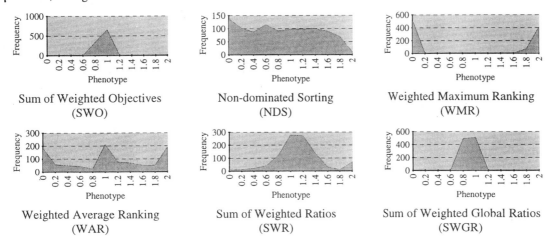

Sum of Weighted Objectives (SWO)   Non-dominated Sorting (NDS)   Weighted Maximum Ranking (WMR)

Weighted Average Ranking (WAR)   Sum of Weighted Ratios (SWR)   Sum of Weighted Global Ratios (SWGR)

Figure 3. Distributions of solutions within the Pareto-optimal range for function F_2.

Additionally for F_2, the average solution of each method was calculated to give an indication of how balanced these distributions were. In other words, no matter what value(s) of P-O solution were favoured, the mean value for F_2 should be the centre value of 1.0. Table 1 (F_2 test 1) shows that all methods produced mean solutions close to 1.0.

	Best Compromise	SWO	NDS	WMR	WAR	SWR	SWGR
F_2 test 1	1.0	1.00922	0.93999	0.97595	1.10226	1.21556	0.98763
F_2 test 2	1.0	2.01459	0.85992	0.99532	1.17007	1.22672	0.98825
F_2 test 3	1.333	1.37837	N/A	1.45757	2.01466	1.66141	1.310

Table 1. Average solutions for each ranking method in F_2 tests 1-3.

Two further tests were performed using F_2. For the second test, f_{21} was temporarily changed to:
$$f_{21} = x^2 / 1000$$
to investigate the range-independence (or lack of it) for each method. As Table 1 (F_2 test 2) shows, after 2000 test runs for each method, SWO (method 1) clearly demonstrates its range-dependence by converging, on average, to the optimal of f_{22} instead of near to 1.0. All other methods show their range-independence by continuing to give mean solution values close to 1.0.

Finally, for the third test with F_2, the importance of f_{22} was doubled for every method capable of supporting importance (the two objectives being otherwise unchanged from the first test). By increasing the importance, the best compromise solution (i.e., the minimum of weighted and summed objectives) is changed from 1.0 to 1.333. Only three methods: SWO, SWR and SWGR, all successfully produced values close to this new desired value (see Table 1, F_2 *test 3*). NDS does not support importance, and WMR just doubled the frequency of optimal solutions to f_{22} (giving a deceptive mean solution), without actually producing any values between the two function optima. Finally, and quite unexpectedly, WAR simply converged every time to the optimal of f_{22}.

Upon investigation, it emerged that WAR does not permit the specification of gradual importance values. It was expected that increasing the weighting of the ranking value for more important objectives would introduce some level of additional importance for these objectives. Interestingly though, in practice it does not appear to be possible to gradually increase 'importance' values: either all objectives are treated equally, or the objective with the increased weight dominates all other objectives completely. Somewhat counter-intuitively, it seems that no matter how large or small an increase is made to a weight, that objective will dominate all others.

6.4. Evolved Results: F_3

Experiments were then performed using F_3 with each method in turn. The function F_3 is significant since it has two disjoint P-O ranges. Nevertheless, the distributions of solutions for this function were surprisingly consistent with those for F_2, see fig. 4. As before, SWO and SWGR almost always converged to solutions near to the best compromise value of 4.5 (for F_3). Again, SWR favoured the best compromise solution with a slightly larger 'spread', but this time some solutions close to the optimal of f_{31} were also produced. NDS gave a fairly even distribution of solutions within the two P-O ranges, and WMR again only generated solutions at the optima of the two objectives, with none in between. Finally, WAR showed its highly unusual distribution once more, by favouring solutions close to the optima of both objectives (including both minima of the multimodal objective f_{31}), and the best compromise solution to a lesser degree.

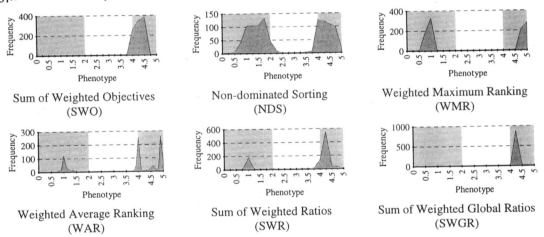

Figure 4. Distributions of solutions within the P-O ranges (shown by grey shaded regions) for function F_3.

6.5. Evolved Results: F_4

Finally, experiments were performed using F_4 with each method in turn. Again, consistent distributions of solutions were obtained, see fig. 5. It should be noted that F_4 is a significant type of function because solutions between the optima of the two objectives are worse than at one optima or the other. This results in two equal best compromise solutions, one at each optima. Hence, although SWO and SWGR this time showed two peaks of distribution, these lie on the best compromise solutions, just as before. Once again, SWR favoured the best compromise solutions with a slightly larger 'spread'. As before, WMR favoured the two optima of the functions with nothing in between. NDS again produced a distribution of solutions covering the entire P-O range, but for this function an unexpected and unwelcome bias towards the middle of the range was evident (where most solutions are very poor). Finally, WAR showed its typically unusual distribution, again favouring values close to the optima of the objectives (and the best compromise solutions, as they are the same for F_4), with other Pareto-optimal values being favoured less.

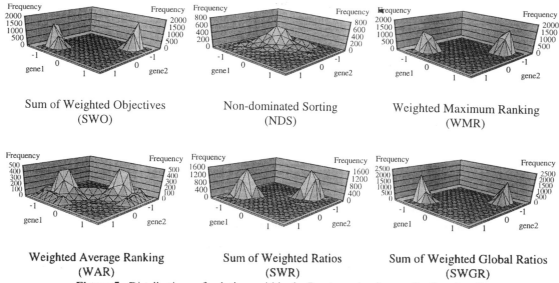

Figure 5. Distributions of solutions within the Pareto-optimal range for function F_4.

6.6. Assessing the Distributions

It should be stressed that all six of the ranking methods allow a GA to produce almost nothing but Pareto-optimal solutions. It is clear, however, that the distribution of these solutions within the Pareto-optimal range is a highly significant factor in determining whether an acceptable solution will be produced.

As the results of the tests show, each ranking method consistently seems to favour certain types of P-O solution, based upon three factors: the Pareto range(s), the separate optimum or optima for each objective and the best compromise solution(s) of the function. These patterns of distributions remain consistent even with more unusual functions with multiple Pareto-ranges (F_3) and multiple best compromise solutions (F_4).

Upon consideration, these distributions are explicable. The three aggregation-based ranking methods: SWO, SWR and SWGR must inevitably favour the best compromise solution(s) to a problem, by definition. (The best compromise solution is the solution with *sum* of weighted objectives minimized, so any ranking method that sums objectives in any way, should have a convergence related to the best compromise value.) NDS gives all non-dominated solutions equal rank, so a fairly even distribution throughout the P-O range is to be expected. WMR bases the fitness of a solution on the maximum rank the solution has for any single objective, so this predictably will result in the generation of solutions only at the optimal of one objective, with nothing in between (a high rank equates to a good value for that objective). Finally, even the unexpected distributions of WAR are explicable. WAR bases the fitness of a solution on the average rank for every objective. This means that a solution with a very high rank for one objective and a low rank for another will be judged equally fit compared to a solution with 'middling' ranks for two objectives. In other words, solutions close to optima of objective functions will be favoured, as will solutions close to the best compromise solution(s).

The results show that NDS, WMR and the new WAR method all give potentially useful distributions of solutions for applications where multiple solutions are required, with predictable, but immovable biases. In contrast, SWO forces the GA to converge on a single solution as close to the best compromise solution as possible, and does allow this bias of P-O solutions to be altered by the user. Unfortunately, because it is range-dependent, its weights must be laboriously set by trial and error in order to define the location of the best-compromise solution(s) in the Pareto range. However, the methods SWR and SWGR both generate solutions in the vicinity of the best-compromise solution(s), and being range-independent, they allow the location of this bias to be easily defined by specifying relative importance values for the objectives. In other words, these two methods allow the location of a subset of acceptable solutions in the P-O range to be defined by specifying which objectives of the problem are more important. The size of the subset depends on which method is used. Hence, for problems where a range of acceptable solutions are desired, biased in favour of those objectives with increased importance, SWR is a suitable choice. For problems where a smaller range, or a even single acceptable solution is desired, SWGR is a suitable choice.

7. Conclusions

This paper investigated the problem of using a genetic algorithm to converge on a small, user-defined subset of *acceptable* solutions to multiobjective problems, in the Pareto-optimal range.

Multiobjective fitness functions cause problems within GAs because the separate objectives have unequal *effective ranges*. If the multiobjective ranking method is not *range-independent*, then one or more objectives in the problem can dominate the others, resulting in evolution to poor solutions.

The concept of *importance* introduced in this paper, allows the GA to converge on a smaller subset of acceptable P-O solutions. Giving certain objectives in a problem greater importance allows ranking methods to generate not just non-dominated solutions, but smaller subsets of acceptable non-dominated solutions at user-defined locations in the Pareto front.

The significance of range-independence and importance in multiobjective ranking methods was shown by the distributions of solutions generated by six methods, applied to four established test functions. The only range-dependent method, SWO, was found to be incapable of coping with objectives with incompatible effective ranges. The three range-independent methods: NDS, WMR and WAR all produced consistent, and sometimes unusual distributions of P-O solutions, making them potentially useful for some problems. However, only the two new range-independent methods that supported importance: SWO and SWGR, had useful distributions and allowed the bias of their distributions to be easily alterable. Indeed, because of these results, SWGR was chosen to be used within a generic evolutionary design system, which has since been used to tackle a wide range of different solid object design problems (involving the minimization of numerous different multiobjective functions) with great success [1].

References

[1] Bentley, P. J., 1996, *Generic Evolutionary Design of Solid Objects using a Genetic Algorithm*. Ph.D. Thesis, University of Huddersfield, Huddersfield, UK.

[2] Dowsland, K. A., 1995, Simulated Annealing Solutions for Multi-Objective Scheduling and Timetabling. *Applied Decision Technologies* (ADT '95), London, **205-219**.

[3] Fonseca, C. M, & Fleming, P. J., 1995a,. An Overview of Evolutionary Algorithms in Multiobjective Optimization. *Evolutionary Computation*, **3:1, 1-16**.

[4] Fonseca, C. M, & Fleming, P. J., 1995b, Multiobjective Genetic Algorithms Made Easy: Selection, Sharing and Mating Restriction. *Genetic Algorithms in Engineering Systems: Innovations and Applications*, Sheffield, **45-52**.

[5] Linkens, D. A. & Nyongesa, H. O., 1993. A Distributed Genetic Algorithm for Multivariable Fuzzy Control. *IEE Colloquium on Genetic Algorithms for Control Systems Engineering*, Digest No. 199/130, **9/1 - 9/3**.

[6] Marett, R. & Wright, M., 1995, The Value of Distorting Subcosts When Using Neighbourhood Search Techniques for Multi-objective Combinatorial Problems. *Applied Decision Technologies*, London, **189-202**.

[7] Goldberg, D. E., 1989, *Genetic Algorithms in Search, Optimization & Machine Learning*. Addison-Wesley.

[8] Holland, J. H., 1992, Genetic Algorithms. *Scientific American*, **66-72**.

[9] Horn, J. & Nafpliotis, N., 1993, Multiobjective Optimisation Using the Niched Pareto Genetic Algorithm. *Illinois Genetic Algorithms Laboratory (IlliGAL)*, report no. 93005.

[10] Ryan, C., 1994, Pygmies and Civil Servants. *Advances in Genetic Programming*, MIT Press.

[11] Schaffer, J. D., 1984, *Some experiments in machine learning using vector evaluated genetic algorithms*. PhD dissertation, Vanderbilt University, Nashville, USA.

[12] Schaffer, J. D., 1985, Multiple Objective Optimization with Vector Evaluated Genetic Algorithms. *Genetic Algorithms and Their Applications: Proceedings of the First International Conference on Genetic Algorithms*, **93-100**.

[13] Srinivas, N. & Deb, K., 1995, Multiobjective Optimization Using Nondominated Sorting in Genetic Algorithms. *Evolutionary Computation*, **2:3, 221-248**.

[14] Sun, Y. & Wang, Z., 1992, Interactive Algorithm of Large Scale Multiobjective 0-1 Linear Programming. *Sixth IFAC/IFORS/IMACS Symposium on Large Scale Systems, Theory and Applications*, **83-86**.

[15] Syswerda, G. & Palmucci, J., 1991, The Application of Genetic Algorithms to Resource Scheduling. *Genetic Algorithms: Proceedings of the Fourth International Conference*, Morgan Kaufmann, **502-508**.

Handling Probabilistic Uncertainty in Constraint Logic Programming

Vladislav B.Valkovsky[1], Konstantin O.Savvin, Michael B.Gerasimov

Software Engineering and Computer Application Department
St.Petersburg State Electrotechnical University
Prof. Popov str., 5, St.Petersburg, 197376, Russia
[1]E-mail: vlad@ailab.etu.spb.ru

Keywords: reasoning with uncertainty. Probabilistic Logic Programming. Constraint Logic Programming. network based project planing.

Abstract

The paper describes the approach and prototype of the tool for handling uncertainty in probabilistic interpretation in the framework of Constraint Logic Programming (CLP). The approach is intended to formulate real-life scheduling. planing. manufacturing problems more adequate to real situation with regard to uncertainty of input information. The main features and implementation of the tool integrated in the CLP(R) system [1] are described briefly. Theoretical background of the approach is a probabilistic logic introduced by N.Nilsson [2]. Calculation of probabilities for the goals in probabilistic logic programs is performed by means of Monte-Carlo method as suggested in [3]. The pilot application for dealing with network based project planing is presented in order to demonstrate the technology of the tool usage. We also discuss possible ways of speeding up the Logical Inference with Probability (LIP) by means of the so-called *Success Formula* that allows to skip unproductive trials and increase efficiency.

1. Introduction

A lot of planing. scheduling. manufacturing. management problems that arise in practice can be formulated as exhaustive search or optimization problems in terms of Operations Research (OR) domain. Many effective algorithms were developed to solve certain types of such problems. They are often intended to be implemented on computers by procedural languages. But there is a potential possibility to solve search problems by means of declarative languages as well. particularly in Prolog. This can be done by exploiting system build-in backtracking mechanism to organize search with classical "generate-and-test" approach [4]. Obviously this approach leads to complete exhaustion of variants and is not suitable for complex real-life problems of meaningful dimension.

The situation changed cardinaly with involvement of Constraint technology in Logic Programming. Constraint Logic Programming languages allow researchers to implement searching and optimization algorithms far more efficiently with the aid of different mechanisms of constraints processing [5][6]. This caused the appearance of large amount of CLP-applications that efficiently solve scheduling. manufacturing. planing and other optimization problems. It should be noted that in many cases the CLP implementations of algorithms have at least the same efficiency as specially designed procedural-language implementations [7]. In addition. CLP implementation usually requires significantly less efforts on development and support stages.

CLP conveniently integrates. within the same environment. both the classical LP functionality and constraint technology features. This effective combination could be used more profitably than for solving the problems mentioned above in classical OR statement. LP is commonly used as one of the main tools for implementation of different ideas. methods. techniques of Artificial Intelligence domain. in particular. knowledge representation and uncertainty handling. Having the tools that allow to represent and handle uncertain information in CLP language we could formulate complex real-life problems more adequately to the nature. It could be possible to avoid some simplifications and assumptions used to reduce real problem to classical deterministic formulation. apply more complex and flexible models of the processes and phenomena and thus obtain more valid and reliable results.

All mentioned above made us try to implement CLP-based instrumental system that provides representation and handling of uncertain information in probabilistic interpretation. and to investigate the potential possibility of developing applications for solving real-life problems on the basis of this tool. Further in the paper we give the description of the tool that allows to design probabilistic logic programs and provide logical inference with probability in the CLP(R) environment by means of the Monte-Carlo approach. Pilot application for solving network based project planing problem [8] in extended non-deterministic statement is described to demonstrate possible technology of developing applications on the basis of the tool. Then we discuss a method for increasing the efficiency of the tool implementation by means of applying the so-called *Success Formula* which prevents apriori unsuccessful trials. At the end we discuss preliminary results and perspectives of the approach usage.

2. Probabilistic representation of uncertainty

Dealing with uncertainty is one of the key problems in AI. There are many approaches operating with different kinds of uncertainty in knowledge. Every model of uncertainty representation has its own positive and negative features, is more or less universal, has more or less well defined theoretical basis. However, the decision about which model to apply is determined first of all by its suitability to certain problem and correspondent domain.

We deal with probabilistic representation of uncertainty and our choice is determined mainly by the fact that the uncertainty in problems like planing, scheduling, manufacturing, management usually has frequency (stochastic) nature. Some statistical data related to the problem domain often exists. Certain domain events can be described directly as random or can be under the influence of random factors. It is logical and semantically correct to treat the validity of some assertion in this case as probability of the fact that the assertion is true. Probabilistic representation also gives the possibility to operate directly in terms of probability distribution and probability density function. This, for instance, allows to provide correct and straightforward calculation of risks in planning and decision-making applications. On the contrary, it is not quite adequate to represent and handle such uncertainty by means of other well-known approach - Fuzzy Logic. Besides, there are some arguable points in theoretical background of this approach [9], that causes wide resonance among researchers.

Another reason of our choice is that the probabilistic interpretation of uncertainty is well-founded theoretically. It is mathematically grounded on the theory of probabilistic logic suggested by N.Nilsson [2]. This model has clear semantics, described in the terms of *possible worlds*, operates with probabilities and uses strict methods of theory of probabilities. The principles and semantics of probabilistic logic can be naturally used to generalize LP to Probabilistic Logic Programming (PLP) [10].

At last, the possibility to use the statistical trials (Monte-Carlo) method for calculation of probabilities for goals in probabilistic logic programs allows to avoid some problems, that are common when dealing with uncertainty, in particular handling dependent uncertain sentences. It is also possible to provide some additional features that are very useful for practical applications (see below).

2.1. Computing probabilities with the Monte-Carlo method

The detailed description of probabilistic logic programming and possible techniques for computing probabilities of the goals in uncertain (probabilistic) logic programs can be found in [3][10]. Here we only summarize some key points. Probabilistic logic program consists of the set of certain sentences (ordinary part) and set of uncertain sentences with associated probabilities. The probability of the uncertain sentence can be treated as the probability that the sentence is present in the program. Suppose that at some moment it is known whether each uncertain sentence is present in the program or not. In this case, the uncertain program becomes an ordinary logic program and one can check by running the interpreter whether some goal succeeds in this program or not. In terms of PLP such certain program is considered as *possible instance* of initial probabilistic program. This is a natural transfer of the *possible worlds* semantics from probabilistic logic to PLP.

Different possible combinations of presence of M uncertain sentences in the program implies 2^M different possible instances of probabilistic program. Thus to compute probability of some goal in the uncertain program one can enumerate all possible different program instances and for each instance check whether the goal is succeeded or not. Ratio between the number of programs where the goal succeeds and total number of programs gives the desired probability. Obviously, the method is not suitable for large M. In [3] it is suggested to use Monte-Carlo method in such cases. On each trial of the method, a random instance of probabilistic program is generated and an attempt to prove the goal is performed by a Prolog interpreter. The ratio between the number of successful attempts and total number of trials gives the approximate estimation for the probability of the goal. The amount of trials required to satisfy certain precision and validity of the estimation can be calculated by Hoeffding theorem [11]. So for $N \geq \dfrac{1}{2\varepsilon^2}\ln\dfrac{2}{\delta}$, the inequality $|\rho - \rho'| \leq \varepsilon$ is satisfied with the validity $1 - \delta$, where p - probability estimation of the goal's success, ρ' - its exact value, N - number of iterations, ε - precision.

2.2. Extending CLP(R) in order to handle probabilities

The design of our extension for CLP(R) intended to handle probabilistic uncertainty is based on the approach described above. The extension is implemented as a set of protected predicates of CLP(R). Predicates are organized in several groups. The main groups are **descriptive, generative, start** and **service** predicates. Limited space of the paper does not allow to describe the predicates of the tool in detail, so we only give a summary of features and a brief review of implementation principles. More detailed description of the tool is given in [12]. The features provided by the tool are:

1. assignment of probabilities to the program sentences (facts, rules, alternatives of a rule);
2. description of different schemes of dependency between program clauses:
   - set of sentences forms a complete group of events;
   - one and only one sentence in the set is true;
   - at least one sentence in the set is true;
3. computation of probability estimations of the goal success;
4. assignment of conditional probabilities to the program sentences and calculation of probability estimations of the success of conditional sentences;
5. description of stochastic parameters with given distributions (e.g. normal, exponential, uniform);
6. statistical processing of results.

Let's first consider simple example to demonstrate how probabilistic programs can be represented in CLP(R) on the basis of tool. Suppose that in the graph presented at Fig. 1 the edges **ab, ac, cd, bd** are present with probabilities 0.5, 0.9, 0.5, 0.9 respectively. It is necessary to calculate the probability that the path from **a** to **d** exists. This problem can be represented by probabilistic program shown on Fig 2. The tool provides all the

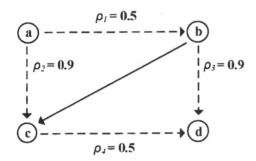

Figure 1. Example

```
% certain sentences
edge(b,c).
edge(X,Y) :- pr_edge(X,Y).
path(X,Y) :- edge(X,Y).
path(X,Y) :- edge(X,Z), path(Z,Y).
% uncertain sentences
pr_edge(a,b).  pr_edge(c,d).      % probability 0.5
pr_edge(a,c).  pr_edge(b,d).      % probability 0.9
% goal
?-path(a,d).              % probability 0.7
```

Figure 2. Probabilistic program

necessary to represent uncertain sentences and provide computation of probability for the goal.

The predicates of **descriptive** group are used to describe uncertain sentences of probabilistic program. In our example, we can use the simplest version that describes independent probabilistic facts and replace uncertain sentences in program at Fig. 2 by the following line:

mcdat_facts(pr_edge, [pr_edge(a,b), pr_edge(c,d), pr_edge(a,c), pr_edge(b,d)], [0.5, 0.9, 0.5, 0.9]).

The parameters are: some unique atom that is used to identify the description, a list of probabilistic facts and a list of the corresponding probabilities. Descriptive predicates only keep information that is required to generate random instances of probabilistic program. Actual generation is performed by calling predicates of **generative** group. Each kind of descriptive predicates has a paired generative predicate. Being called, a generative predicate tries to access the description of the corresponding kind with identifier specified as a parameter of the call. If description is found, the generation is performed according to it and the results are recorded in program database. Thus in the example the call to mcgen_fact(pr_edge) will record some of uncertain facts pr_edge/2 into the program database depending on random chance. This call must be done on each iteration of Monte-Carlo method before the attempt to check the goal, so it is convenient to modify the goal in the following way:

pr_path :- mcgen_fact(pr_edge), path(a,d).

Calculation of goal probability estimation is initiated by **start** predicate: mc_start(pr_path, 200, Prob). Here pr_path is the header of the modified goal, 200 is a specified amount of trials and Prob is variable for storing the result of computations. This predicate runs metainterpreter that in its turn performs a specified amount of trials for the modified goal. During each trial the modified goal generates random instances of uncertain sentences and tries to prove the initial goal on the obtained random instance of probabilistic program. At the end, the probability estimation is computed as ratio between the number of successful trials and total number of trials.

The use of the Monte-Carlo method helps to solve in a relatively simple way the problem of handling different kinds of dependencies between uncertain sentences. This problem is a bottleneck for the most of other methods dealing with uncertainty. In statistical trials technique, only the stage of generation of random instance of probabilistic program is affected by the problem. The randomly generated combination of uncertain sentences additionally must be checked in order to satisfy the dependency restrictions. The tool includes predicates that describe dependencies of types: "one and only one sentence from the set is present in the program", "at least one sentence from the set is present in the program", "complete group of events". In a similar way it is possible to represent conditional events and compute conditional probabilities of goals. The only difference is that on each iteration, not only a goal itself should be checked but also the (sub)goal that represents the condition. The estimation for final conditional probability (P_C) can be computed on the basis of standard formula for conditional probability from Probabilities Theory, as a ratio between the amount of trials where both a goal and condition succeed (N_{GC}), and an amount of trials, where condition succeeds (N_C): $P_C = N_{GC} / N_C$. More over, this technique can be used to check whether certain goals are independent or not.

Statistical trials approach also allows to combine conveniently the LIP with straightforward statistical modeling facility. Thus the descriptive group contains special predicates that describe distribution law of stochastic variables. Some parameters of domain objects or events probabilistic program deals with can be represented by these stochastic variables. The instances of such stochastic variables can be generated and recorded into the program database on each trial in the same way as the instances of uncertain sentences. Afterwards, these instances of stochastic parameters can be accessed inside the body of the goal to calculate some necessary characteristics of domain objects. The set of values calculated and accumulated in all trials can be used to build sampling probabilistic distributions of required characteristics or to perform other statistical processing. This feature could be very useful in considered types of problems for building adequate models of problem domain. The tool includes also several **service** predicates that help with logging of calculated stochastic characteristics, building histograms, statistical processing and exporting this information.

2.3 Abduction problem

The direct problem in uncertainty handling is to calculate the probability of the goal success based on the given probabilities of uncertain sentences in probabilistic logic program. The reverse problem is to estimate the unknown probability of some uncertain sentences in probabilistic logic program based on the known probability of the goal. This problem is known as one of the abduction problems.

The simplest way to obtain the approximate relation between the probability of the goal and the probability of uncertain sentence with the help of the tool is to calculate the probability of the goal for several values of the probability of uncertain sentences with subsequent smoothing.

For a particular case, when uncertain sentences are independent, it was shown that probability of the goal and probabilities of uncertain sentences are bound by the polylinear relation [13]. For instance, for the case with one unknown probability of uncertain sentence this relation is $p = ap_1 + b$, and for the case with two unknown probabilities of uncertain sentences - $p = ap_1 + bp_2 + cp_1p_2 + d$, where p - probability of the goal; p_1, p_2 - probabilities of uncertain sentences; a, b, c, d - some numeric coefficients. The values of the coefficients can be evaluated in the following way. We can prepare the system of linear equations with regard to unknown coefficients by computing the probability of the goal for several (depending of the number of unknown coefficients) sets of different arbitrary selected values of probabilities of uncertain sentences. Afterwards, this system can be solved by exploiting CLP(R) language features automatically.

When the relation between probabilities of goal and uncertain sentences is known, it becomes possible, in particular, to estimate the actual value of probability of some of the uncertain sentences, if probability of the goal is known and some assumption about values of probabilities of other uncertain sentences is made. Of course, the same operation can be performed for intervals of probabilities as well. So the possibility to solve mentioned case of abduction problem is useful for analyzing the state of probabilistic knowledge bases during the design, modifications or maintenance (to estimate how adequately and correct certain situation of problem domain is represented, to determine how sensitive some queries/goals to the probability of some events, etc).

The abduction is very important also in diagnostic systems, especially for analysis of those critical situations or events that must have occured in a given object system but which are not directly observable. It gives an ability to estimate probabilities of such events with the help of the model of the system and results of observations in real situation.

3. Pilot application

In this section we present pilot system for dealing with probabilistic network-based project planing to illustrate the possible application of the approach described above. The detailed review of the network-based project planning problem can be found for example in [14][15]. The input of the problem is the set of tasks of some project to be accomplished. For each task it is known which tasks are to be finished before it and which are to be started directly after it. In the traditional deterministic formulations it is assumed that durations of all tasks are fixed and known from previous experience or expert estimations. It is generally accepted in the project analysis and optimization methods to represent the set of tasks by a network (directed graph) according to their precedence. The main results of solving the problem are the time required to finish the project, set of tasks that belong to so-called *critical path* and set of time parameters for each task that characterize the schedule of its execution. The base tasks' time parameters are early start time, late finish time and free reserve of time. The last one is the time of possible delay in task execution that does not cause the delay of subsequent tasks. Tasks with no free reserve compose critical path of project execution. The length of critical path is the minimal time required to complete the project. Delay of any of the tasks on critical path leads to delay the whole project. Classical method to solve the problem is the Critical Path Method (CPM) [14].

The stochastic formulations of the problem are also known for a long time. The well-known example is the classical PERT model [16]. In [15] it was also suggested to consider the tasks durations as random values with certain predefined distribution. The only difference from deterministic case is that algorithm operates with parameters of distributions (expectation and deviation) of tasks durations instead of the fixed values, and the results are the expectations of required parameters that are valid with some confidence level. However, this approach requires that certain conditions concerning applicability of the central limit theorem are to be satisfied. Particularly, they are: fixed critical path, large enough amount of tasks on it, certain kind of distributions and independence for the tasks times.

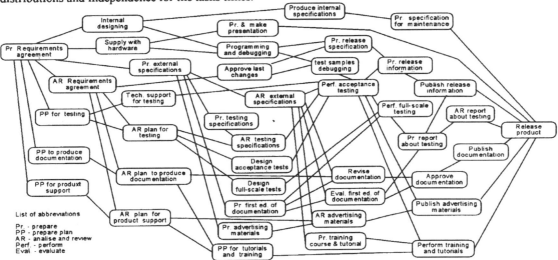

Figure 3. Network representation of the project

The project considered in the application is the software development project. The materials presented in [17] are used to prepare the input data for the system. The network of the project is shown at Fig. 3. The nodes of the graph correspond to the tasks and links represent the order (left to right) of task execution. With the help of the approach for handling uncertainty we can describe the process of project execution in more details and more adequate to real situation. The tool allows to design probabilistic knowledge base (KB) that will keep information about all factors that can affect project execution.

The probabilistic knowledge base is designed on the basis of the network representation. It provides models of tasks execution and computes duration for them. Several tasks (about 10) are described as complex *macro-task*. The model of such tasks describes a few possible outcomes with some probabilities. Some "unexpected" delays are possible during execution of these tasks as a result of 2-4 reasons with certain probabilities. The occurrence of these reasons depends on several random factors with certain probabilities. These reasons and factors can be outcomes of other tasks or some random events that are not present in project network directly. The values of the delay are considered as random variables with certain distribution (exponential, normal, uniform) or composition of them. Thus about 10-15 pseudo-random numbers generators are used to obtain the instance of macro-task duration values. The scheme of description for one of such macro-task ("Programming and debugging") is presented at Fig. 4. Note that the same random factors or outcomes of some tasks can affect execution of several tasks simultaneously and thus the correlation takes place between the values of tasks duration. Other tasks are represented by more simple models using few random parameters.

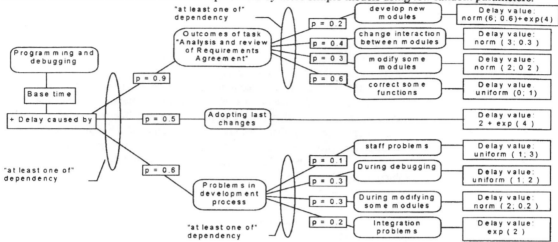

Figure 4. Model of the task "Programming and debugging"

Implementation of CMP-like algorithm for project evaluation widely uses the advantages of CLP(R) operational model. Use of delayed constraints mechanism and min/max constraints allows to compute tasks' time parameters by means of only one network traversal. In fact, the implementation of algorithm is simply a declarative representation of formulas for calculation those parameters and several additional predicates that organize the recursive propagation of formulas' constrains through the project network and binding of correspondent problem domain variables.

The algorithm is integrated with KB via variables that correspond to the durations of the tasks. The computations are organized in following way. On each iteration of LIP the KB produces values of durations for all tasks or, in other words, the possible "instance" of the project execution. Then this instance is evaluated by the algorithm to determine the time required to finish the project and critical path. During the computations all information that is necessary for future statistical processing is logged. When all iterations are completed the required statistical processing can be performed. The time spent to make one iteration of LIP in the system is about 0.8-0.9 sec (PC 486DX2-66). The number of iterations for the different goals and queries varied from 200 to 1000. The time of evaluation of all parameters for all tasks of project with 200 iterations is about 180-190 sec.

Let's consider what are the results that the system can provide comparing to classical methods. First of all, the system allows to obtain sampling probability density functions for duration of project macro-tasks. The examples are presented at Fig. 5. 6. Analysis of these histograms for different variants of tasks models can help to estimate how adequately the model is described in the knowledge base, and one can see that the assumption about any certain predetermined kind of distribution (e.g. normal) for tasks duration (it is assumed in stochastic formulations of the problem) is not quite reliable is many cases.

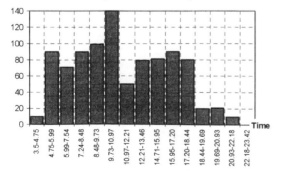

Figure 5. Histogram of probability density function for task "Revise documentation"

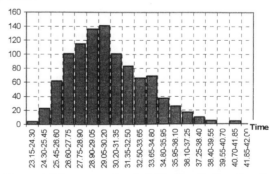

Figure 6. Histogram of probability density function for task "Programming and debugging"

Using the information logged during project evaluation it is possible to build sampling distribution for the project completion time. Analysis of the series of such histograms made for different initial probability(ies) of some event(s) in KB can be useful. For instance, probabilities for outcomes of the task "Analysis and review of the Requirements Agreement" (see Fig.4) can reflect the degree of the company's accommodation level to customer requests. Figure 7 presents the series of histograms for the project completion time having been made for different values of probability of the outcome "necessary to develop new modules". The time (in some abstract units) is along horizontal axis and the probability of that the project finishes until certain time is on vertical one. Thus one can easily estimate the risk of breaking certain deadlines (e.g. Time = 77) if company takes certain strategy of responding to new customer requests. Such analysis for events or factors which can cause delay of the project or some of its stages allows to forecast the situation, make appropriate management decision and estimate the risk of breaking the schedule or the resources limits.

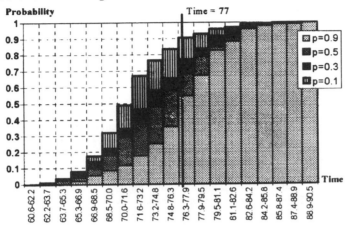

Figure 7. Histogram of probability distribution function for the project completion time

It is also possible to perform some useful analysis with use of abduction. For example, having computed the expectation of project completion time for several different values of probability of certain event one can obtain dependency between the project completion time and the probability of this event. Knowing such dependencies for many random factors or events it is possible to determine the factors which have the most significant influence on the execution of project or some separate stages of it. Such information could be important to analyze project comprehensively and make management decisions like force or delay some tasks, perform additional investments and so on.

The system allows to retrieve useful information also by the user-defined queries. Some examples are:
- what is the probability of event "project can be finished earlier that in e.g. 80 units of time" (1);
- what is the probability of event "the task 'programming and debugging will take less than e.g. 27 units of time" (2);
- what is the probability of the event (1) under condition (2) or vice versa;
- what is the probability of the event "project will not be delayed more than e.g. 85 units of time if it is known that outcomes of the task "Analysis and review of the requirements agreement" are: "necessary to modify some modules of designed software" (3); "necessary to change the interaction between some modules of designed software" (4); "(3) or (4)", "(3) and (4) simultaneously", etc.

Thus designed system allows to obtain significantly more information for project analysis and management than classical standard deterministic methods for network-based project planning. In fact, it is possible to represent the model of the project as detailed as in necessary by means of probabilistic KB. So the system could be used as a suitable instrument for project analysis, supporting management decisions, forecasting of decision consequences, etc. Some more information concerning the system can be found in [18].

4. Efficiency questions

Implementation of logical inference with probability via the Monte-Carlo method rises up the question of efficiency. The reason is that the main way to increase the precision and validity of probabilistic estimations is to increase the amount of trials. But, if $\varepsilon < 0.1$ and $\delta = 0.05$ then approximation $N > 1/\varepsilon^2$ takes place, i.e. when ε is small, the number of iterations is large enough. This fact made us look for the ways of reducing the time required to calculate the probability estimations.

One of the possible ways to increase LIP efficiency is method suggested by E.Dantsin in [10]. This method is based on construction of the so-called *logical success formula* (LSF) for the given goal and logical program being analyzed. LSF is the boolean formula, which defines invariably the truth of the goal, depending only on the uncertain sentences of probabilistic logic program. It assumes that all certain predicates are always inferable and thus they do not affect the goal's inferability. Boolean variables in LSF reflect the presence of uncertain sentences in the program. To be more accurate - variable X_i takes the value "1"(true) if the corresponding uncertain sentence is presented in the program, and the value "0"(false) otherwise. Therefore, having constructed the LSF we have no necessity to run an interpreter on each trial to determine whether the goal succeeds or not. Instead of it, we need only to produce the instances of uncertain sentences, assign correspondent 0/1 values to the variables of LSF and then check the value of the formula. If the LSF value is "1" then the goal succeeded, if "0" - goal failed. Such a formula can be obtained by means of *setof*-like technique.

However, LSF is oriented mainly to pure logical programs. Success of the goals in logical programs that solve practical problems depends not only on uncertain clauses. Thus the assumption that all certain clauses are inferable is not correct for such programs. Real programs contain different non-logical predicates, computations involving stochastic parameters that affect the goal's success, etc. In addition, CLP-programs operate with constraints which bind variables in clauses that also significantly affect the result of inference process. Thus for LSF to be correct in such programs we have to take into account also a certain sentences inferability. This can require difficult analysis of inferability of certain sentences in the program and, in some cases, it can be impossible to determine whether some sentence is inferable or not, without starting interpreter.

This approach is not applicable in straightforward way in our probabilistic extension also because of the fact that real programs often produce some useful information on successful iterations (so-called *side effect*) and such information can be even more important than the fact of the goal success itself. So in order to be able to obtain this information we have to start interpreter anyway on instances of the program where the goal succeeds. Thus reducing of LIP time in our extension can be accomplished at the expense of the trials where the goal failed. Further we present an approach for increasing the efficiency of LIP for our probabilistic extension, which can be applicable to real programs.

Thus it is necessary to run interpreter on iterations where goal succeed anyway in order to obtain possible side effect information. We can simply consider the sentences (including all non-logical ones and constraints), which are not known to be inferable or not, to be inferable (probably except several trivial cases when ground facts are absent). So, the representation of such predicate in the final LSF will be "1"(true) on the corresponding places. LSF can be obtained with a procedure that analyses only the syntactical structure of the program [19] as well as by means *setof*-like method.

Obviously, the application of the described method results in approximate LSF (ALSF), which has the following meaning. If the ALSF is false on some instance of the program, then the goal is uninferable, and we have not to start the interpreter. Otherwise, we have to start an interpreter, because truth of the ALSF does not mean the goal's success and the goal can fail during the inference process.

This approach can be extended for the case when stochastic parameters supported by probabilistic extension are involved in constraints. As in the case with LSF, we want to reduce the task of determining the success of the goal (via starting the interpreter) to a more simple task: in our case - to evaluation of the value of some formula. Such formula should take into account solvability of the constraints which depend on stochastic parameters generated on each iteration. Let's denote such formula *constraint success formula* (CSF). It is possible to build CSF by exploiting feature of delayed constraints of CLP(R) in the following way.

In the program below the instances of the values of stochastic parameters (identified by rndPar1 and rndPar2) retrieved from program database are involved in some constraints:

```
% Description of stochastic parameters by descriptive predicates of the tool.
mcdat_rndExp ( rndPar1, 2.0 ).
mcdat_rndUniform( rndPar2, 0, 10 ).
% goal contains the call to generative predicates of the tool. Generated instances of stochastic parameters are
% recorded into the program database as arg. of the term with the functor defined in correspondent descriptions
goal: -  mcgen_rndExp(rndPar1), mcgen_rndUniform(rndPar2), ...,
         some_constraints(V1, V2, ...), rndPar1(V1), rndPar2(V2),...
```

Suppose that the terms *rndPart1* and *rndPart2*, that are present in the program database, contain unbound variable as argument instead of some actual value. In this case, when the goal succeeds, all constraints related to variables V1 and V2 are in the delayed set and can be retrieved via additional variables included in the header of the clause as it is shown below:

```
goal( rndParams(V1,V2) ): - ...
```

For this modified goal. CSF can be constructed by means of *setof*-like techniques. Having constructed the CSF we can use it in the same way as ALSF. On each iteration, we generate random parameters and compute the value of CSF. If CSF equals to "0" we skip this iteration, otherwise we start an interpreter in order to be able to produce possible side-effect information.

CSF in constraint programs is the natural supplement to LSF in the logic programs. So the integration of these parts seems to be very natural. Let's denote such integrated formula as *full success formula* (FSF). Both ALSF and CSF are constructed by means of same the *setof*-like techniques. During the ALSF or CSF construction. each variant of the goal's success. obtained by *setof*, produces some subformula. These subformulas are linked to each other by disjunction. In the ALSF case such subformula is a conjunction of 0/1 variables which represent uncertain clauses. In the CSF case, subformula is a set of constraints that involve the goal'a arguments which are bound with stochastic parameters used in the program. Thus, in order to obtain FSF, we need to build ALSF and CSF for the given goal and then merge two sets of subformulas. The procedures for obtaining ALSF and CSF guarantee the scheme of the merging presented on Figure 8. Here $ALSF_i$ and CSF_i are subformulas of ALSF and CSF respectively.

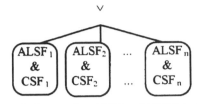

Figure 8. Integrating ALSF and CSF into FSF

The straightforward usage of FSF reduces LIP elapsed time. In the scheme "generate and test". this FSF can be expressed via Prolog sentences; and random vector of variables corresponding to uncertain sentences or stochastic parameters can be checked every time it is generated. If FSF equals to "0", iteration can be skipped. But in some cases FSF can be used even more effectively. We have investigated some useful variants for the case when FSF consist of only one subformula, i.e. represented by conjunction of constraints[1].

One of the variants is the possibility to apply CLP ideology "constrain-and-generate" which is widely used in search algorithms implemented in CLP paradigm. Generation of random vector can be performed with constraints representing FSF being already active. Making so it is possible to obtain "0" even before all the variables in random vector are generated. So one can reduce time spent on generation of the instances for uncertain sentences or stochastic parameters on unsuccessful iterations.

Another possibility is to use CSF-part of formula to generate stochastic parameters so that generated values from the very beginning belong to the admissible domain[2]. In particular, vector of n stochastic variables can be considered as a point in n-dimensional space and CSF defines some area in this space. In this case, if CSF constrains are already in an active set, we can generate each next stochastic parameter using not only information from corresponding **descriptive** predicate but also the information retrieved from current set of constraints. In order to retrieve the information we can dump the current constraints set[3], convert the result into the functor form and try to find terms like:

$$Var < Number_1 \text{ and/or } Var > Number_2$$

These terms represent a closed or a half-closed interval for some variable not generated yet. Here: *Var* is variable not generated yet. and $Number_i$ - some numeric constant. When the interval for some variable is found. we can transform parameters of the distribution so that the generated value will apriory satisfy the constraints.

5. Resume

In this section we make a preliminary analysis of the described approach for handling probabilistic uncertainty.

The key point of the approach is the use of statistical trials method as the main technique. It determines the variety of advantages of our approach. In particular, it provides an ability to manipulate with different types of dependencies between uncertain sentences and allows to use stochastic variables directly in the uncertain programs. These facilities are well-suited for designing adequate and flexible models of objects. events or phenomena of problem domains that in their turn provide more relevant solutions of the problems. It is necessary to note that expenses for computing probabilities by the Monte-Carlo method grow approximately linear with increasing of amount of uncertain sentences involved. This allows to manipulate with uncertain programs and knowledge bases of large size.

---

[1] active constraint set in CLP(R) cannot contains disjunction of constraints.

[2] It is possible when stochastic parameters are used only to produce "side-effect" information on successful iterations and do not affect success of the goal itself

[3] By means of **dump** predicate of CLP(R)

The implementation of the approach in CLP language gives the possibility to integrate conveniently relevant description of problem domain with effective algorithms within the same environment, that allows to develop applications in shorter time. The implementation itself is valuable. It is not specialized for any certain kind of applications, and can be used as generic tool for handling uncertainty represented by probability. The tool is flexible enough and can be easily extended or adopted to be integrated in other constraint languages.

Certainly, the approach is not free of drawbacks. For example, with increasing of the required precision and validity of calculated probability estimations, the amount of trials grows fast and consequently the time of computations increases as well. However, this problem can be partially solved with the use of *Success Formula* described in section 4. Improvements of this method are the subject of future investigation.

The other disadvantage is that the assignment of initial probabilities and preparing the input data for probabilistic models is not trivial task. The problems of such kind are inherent for all methods dealing with uncertainty. In some cases the problem can be solved by comparing several models and choosing the best one. Certainly, the use of abduction (see subsection 2.3) for estimation of unknown probabilities can significantly help to solve this problem. The investigations in this area are also the subject for future work.

Despite the mentioned drawbacks, we think that the described approach in general looks to be perspective and suitable for solving complex real-life problems in such areas as planing, scheduling, manufacturing, decision-making systems, diagnostics, etc.

References

[1] Jaffar J., Michaylov S., Stuckey P.J., Yap R.H.C., 1992, The CLP(R) Language and System, *ACM Transactions on Programming Languages and Systems*, 14(3), 339-395.

[2] Nilsson N.J., 1986, Probabilistic Logic, *Artificial Intelligence*, 18(1), 71-87.

[3] Dantsin E., Kreynovich V., 1989, Probabilistic Inference in Forecasting Systems, *Reports of the Russian Academy of Science (mathematics)*, 307(1), 17-21 (in Russian).

[4] Sterling L., Shapiro E., 1986, *The Art of Prolog*, MIT Press, Cambridge, MA.

[5] Fruhwirth T., Herold A., Kuchenhoff V., Provost T., Lim P., Monfroy E., Wallace M., 1993, Constraint Logic Programming. An Informal Introduction. Tech. Report ECRC-93-5, Munich, Germany.

[6] Van Hentenryck, P., 1989, *Constraint Satisfaction in Logic Programming*, MIT Press, Cambridge, MA.

[7] Aggoun A., Beldiceanu N., 1993, Extending CHIP in order to solve Complex Scheduling and Placement Problems, *Math. Comput. Modelling*, 17(7), 57-73.

[8] Bigelow C.G., 1962, Bibliography on Project Planning and Control by Network Analysis, *Operations Res.*, 10, 728-731.

[9] Elkan C., 1994, The paradoxical success of Fuzzy Logic, *IEEE Expert - Intelligent systems & their applications*, Aug.,

[10] Dantsin E., 1992, Probabilistic Logic Programming and their Semantic, *Lecture Notes in Computer Science*, 592, Springer-Verlag, 152-164.

[11] Hoeffding W., 1963, Probability inequalities for sums of bounded random variables, *J.Amer.Statist.Assoc.*, 58(301), 13-30.

[12] Valkovsky V., Gerasimov M., Evgrafov A., Savvin K., 1996, Probabilistic Constraint Logic Programming. Tech Report TR-1996-8, Dept. of Software Engineering and Computer Application, St.Petersburg State Electrotechnical University, St.Petersburg, Russia.

[13] Dantsin E., Valkovsky V., 1996, Abductive Reasoning in Probabilistic Prolog, *Proceedings of the Second Workshop of INTAS-93-1702 project Efficient Symbolic Computing*, St.Petersburg, Russia, October.

[14] Souder W.E., 1978, Project choice. Project Tasks Scheduling and Project Management in *Handbook of Operational Research*, ed. Moder J.J. and Elmaghraby, Van Nostrand Reinhold Company.

[15] Ventsel E.S., 1972, *Operations Research*, Moscow, Russia (in Russian).

[16] Fazard W., 1962, The origin of PERT, *The Controller*, (December 1962), 598-602, 618-621.

[17] Gunter R.C., 1979, *Management methodology for software product engineering*, A Wiley-Interscience publication, J.Wiley&Sons, NY-Chichester-Brisbane-Toronto.

[18] Valkovsky V.B., Gerasimov M.B., Savvin K.O., 1996, CLP with Probability in Scheduling Problems, *Proceedings of the Second International Conference on the Practical Application of Constraint Technology*, London, UK, April, 299-315.

[19] Gerasimov M.B., 1995, *Investigation of efficiency of logical inference with probability in CLP*, MSc Thesis, St.Petersburg State Electrotechnical University, St.Petersburg, Russia (in Russian).

A Voxel-Based Representation for the Evolutionary Shape Optimisation of a Simplified Beam: A Case-Study of a Problem-Centred Approach to Genetic Operator Design

P. J. Baron[1], R. B. Fisher[1], F. Mill[2], A. Sherlock[2], A. L. Tuson[1]

[1]*Department of Artificial Intelligence, University of Edinburgh, 80 South Bridge, Edinburgh EH1 1HN, {peterba, rbf, andrewt}@dai.ed.ac.uk*

[2]*Manufacturing Planning Group, Department of Mechanical Engineering University of Edinburgh, King's Buildings, Mayfield Road, Edinburgh, EH9 3JL, {F.Mill, A.Sherlock}@ed.ac.uk*

Keywords: Shape Optimisation, Genetic Algorithms, Voxels, Directed Mutation

Abstract

This paper examines a voxel (N-dimensional pixel) based representation for shape optimisation problems, and shows that although a basic genetic algorithm performed poorly on a simplified beam design problem, the use of three domain specific operators improved performance greatly. Additionally, the use of a 'directed smoothing' operator that preferentially adds material to high stress areas was examined and found to assist evolutionary search. This paper demonstrates how domain knowledge and an understanding of how genetic algorithms work can be used to inform the design of suitable operators.

1. Introduction

Shape optimisation within constraints is a hard problem from the field of Mechanical Engineering. The objective is to design a shape that best satisfies some predetermined goal whilst at the same time maintaining some property of the shape within a constraint, or perhaps even a set of constraints.

Previous work has applied genetic algorithms (GAs) [1] to shape optimisation problems with encouraging results. Using engineering software packages to evaluate the suitability of a given design, successful shapes have been evolved — examples include [2, 3, 4]. However, previous work on evolutionary shape optimisation has primarily been concentrated around parametric representations of structural design and shape optimisation problems which presume a family of solution shapes.

This paper focuses upon the use of a voxel (N-dimensional pixel) based representation instead, within which the shapes being optimised are represented as a series of binary 0's and 1's. This approach has the advantage that it can describe any topology, and makes no assumptions about the form of the final solution [5]. Furthermore, a voxel based representation, by virtue of its directness of representation allows domain knowledge to be easily added, and to the level felt appropriate by the designer; the following examples will help illustrate this point.

First, areas of the voxel representation can be fixed to be permanently on or off; this allows the designer to prohibit material from being placed in locations where it is not desired. A second example lies with the ease in which existing designs can be utilised by the system — all that is required is for the initial design to be digitised and the bitmap used to initialise the population of the genetic algorithm.

Finally, taking beam design as an example, holes in the shape of the beam are entirely possible and even probable if some of the mass of the beam is occupying a low-stress area — a parametric representation can only create holes where the user is expecting them to be required and has defined the appropriate parameters.

However, [2] argues that using a binary voxel representation leads to the following problems: a long length of the chromosomes (often greater than 1000 bits); the formation of small holes in the shape; no guarantee that the final shape produced will be smooth; and even if the parents represent a valid shape, the children will not necessarily be valid. On the other hand, these objections appear to be offset somewhat by the work in [3], which has used a voxel-based approach, claiming satisfactory results. However, there appears to be little attempt to include domain knowledge into the genetic algorithm, most probably leading to the very long amount of time required to obtain these results; in this case, more than 23 hours.

This paper describes a case-study of how domain knowledge and an understanding of how genetic algorithms work can be used to inform the design of suitable operators for shape optimisation, as well as a test of the suitability of a voxel-based representation for shape optimisation problems.

2. The Optimisation of the Cross Section of a Beam

One of the problems encountered in shape optimisation, no matter what approach is taken, is interfacing the optimiser with a suitable evaluation package. For example, in wing optimisation, the wing shape has to be smooth, else the CFD (Computational Fluid Dynamics) package will act strangely: either returning negative drag, or resulting in the program crashing. Problems involving Finite Element Analysis (FEA) evaluation packages are somewhat better behaved, but interfacing is still not trivial, and the calculations do take some time.

The shape optimisation of a beam cross-section is the problem considered in this study. Evaluation of the candidate cross-sections was made using bending theory for symmetrical beams, considering only normal stresses [6]. This is a greatly oversimplified model, but sufficient to test the operators devised in this paper, as well as making repeated experimentation feasible.

Each candidate solution can thus be represented as a 2-D grid of voxels, with the optimisation objective being to minimise the mass m of the beam whilst ensuring that at all points (ie. at all voxels) in the beam the normal stress does not exceed a maximum stress (σ_{max}), which is a constant that is determined by the material to be used. This maximum stress constraint is given by Equation (1):

$$\sigma_{max} \geq \frac{-My}{I} \text{ for all voxels} \tag{1}$$

where M is the bending moment, y the perpendicular distance of the voxel from the neutral axis, and I is the second moment of area of the cross-section. The neutral axis of a shape is defined as a line which passes through the centroid of mass of the shape. As the voxels are of uniform size and density, the centroid can be found by taking the average of the positions of all the occupied voxels. For a symmetric beam the neutral axis would be horizontal.

The bending moment M is a constant determined by the loading on the beam. The mass m of the beam is proportional to the area of the cross-section and hence the number of voxels turned on. The second moment of area is given by Equation (2):

$$I = \sum_{i \in \{all \ on \ voxels\}} y_i^2 \, dA \tag{2}$$

where y_i is the distance of the ith voxel from the neutral axis and dA is the area of a voxel (as we are dealing with a cross-section).

The optimum cross-section for a beam using this evaluation can be deduced to being a flange at top and bottom of the design domain. For the experiments described here, the following values of the constants were used: $M = 2 \times 10^6$ Nm, $\sigma_{max} = 100$ MPa, and a beam of dimensions 320×640mm was considered.

In practice, this would correspond to an I-beam, but that also requires a web to connect the two plates of the beam together. In a full calculation with shear stresses, the web would arise so to counteract this additional stress. However as shear stress is not represented in this simplified problem, a connectivity requirement in the form of a repair step was added, whereby all voxels must be connected (by a 4 connect rule) to a seed voxel in the centre top edge of the beam. In addition, a straight web was enforced before the connectivity repair step. This was found, in formative experiments, to prevent the formation of a crooked web (as the "sheer-less" physics model used does not prevent this), and improve the results obtained slightly.

3. A Basic Genetic Algorithm Implementation

This study builds upon the following basic implementation of a genetic algorithm. The encoding digitises the shape as a 32×64 grid and represents this as a 1-dimensional string of 2048 bits. Standard two-point crossover (applied with probability 0.3), and bit-flip mutation (applied with bitwise probability 0.001) were used to operate on this encoding. After the application of each operator, pixels that were not connected to the seed voxel were set to zero. An unstructured, generational population model of size 20 with rank-based selection with selection pressure 1.7 was used.

Strings are initialised to random binary bit strings with a preselected 'density' percentage; the higher this value, the more voxels are initially turned 'on'; for this study, this was set to 70%. All initial population strings must pass the validity checks used by the evaluation function: the number of active voxels must be non-zero; the second moment of area must be greater than 1×10^{-12} (goes to 0 if insufficient active voxels); and the fitness must be greater than 0.0001. The rationale behind this was to ensure that there were few instances where the lack of connected voxels would lead to blank, or sparsely filled initial solutions after the connectivity repair procedure was applied.

NOTE: as the aim of this work was to investigate whether a voxel-based representation was a feasible approach, some possible problem simplifications were not considered. One of these was that the symmetry of the beam was not exploited. Of course, full advantage of such problem features would have been taken in a more realistic and computationally expensive problem,

3.1. The Fitness Function

The fitness function was designed to minimise the area of the beam (number of active voxels) within the maximum allowed stress, with a penalty value being applied to any solution which broke that constraint. A small additional factor, $\frac{1}{S}$, was included in the fitness calculation based upon the maximum stress point in the beam. This had the effect of causing any valid solutions to continue to evolve towards better solutions (in this case a beam which minimises area and minimises the maximum amount of stress present). The fitness, F which is to be maximised, is given by Equation (3) below:

$$F = \frac{1}{V - \frac{1}{S} + k \times max\{(S - \sigma_{max}), 0\}} \qquad (3)$$

where V is the number of active voxels (ie. beam mass), S the highest stress (calculated using Equations (1) and (2)) felt by any of the pixels, σ_{max} the maximum allowed stress, and k a constant which can be adjusted to vary the weight of the penalty associated with the maximum stress constraint (in this study it was set to $k = 5.0 \times 10^{-5}$).

3.2. Results Obtained

The basic genetic algorithm was found to give disappointing results. When allowed to run to convergence (over 2000 generations), the shapes produced, though recognisable as approaching an I-beam shape, were highly irregular and possessed small holes. Figure 1 illustrates this by showing the 4 best solutions from ten genetic algorithm runs. From these results it appears, at least for a simple genetic algorithm implementation, that the potential problems described in [2] do manifest themselves.

4. An Improved Genetic Algorithm Implementation

The performance of the basic genetic algorithm was disappointing, however, many successful applications of the genetic algorithm use domain-specific operators [7]. Knowing the nature of the problems that arise with this approach, would it be possible to design operators to overcome them? With this end in mind, the basic operators were replaced with domain-specific operators in order to improve the following perceived problems:

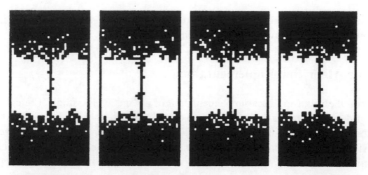

Figure 1. Typical End-of-Run Results Obtained by the Basic GA

- Crossover effectiveness;

- Removal of holes and isolated pixels;

- Removal of rough edges.

The operators are: a 2-dimensional crossover operator, a smoothing mutation operator, and a mutation operator that mutates a 2x2 area of the voxel grid. All of these operators also address one weakness in the basic genetic algorithm implementation: the encoding of a 2-dimensional problem as a one-dimensional string. Each of these operators will now be described in turn, and results of experiments to evaluate their effectiveness will be summarised.

4.1. A Mutation Operator for Smoothing

If a human designer was to examine the results in Figure 1, it would be likely that the designer would modify the design to remove the small holes and rough edges, as the designer's intuition would indicate that this would improve the quality of the beam. So one solution to the poor performance of the basic genetic algorithm would be to introduce an operation that embodied this intuition.

A mutation operator for smoothing was therefore devised. An area of x and y sizes ranging from 2×2 pixels to $\frac{1}{4}$ of the dimensions of grid was randomly selected. The most common value for the pixels in the area selected was then found, and then written to all of the pixels in that area, as illustrated by Figure 2.

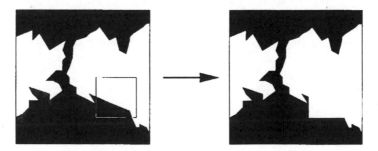

Figure 2. The Smoothing Mutation Operator

Examination of the results obtained, in a few formative runs, confirmed that this operator can remove isolated pixels with great success. With this as the only mutation operator, the shape was optimised to two near perfect horizontal bars at the vertical extremities (an I-beam) after 1000 generations.

4.2. Two-Dimensional Crossover

Another problem encountered in representing what is a 2-dimensional optimisation problem as a one-dimensional chromosome arises with *linkage*. In a 1-D encoding, voxels that correspond to spatially close points on the actual shape can be far apart on the string. Therefore, use of crossover can more easily disrupt building blocks such as one part of the shape being of high fitness, because the bits that correspond to it are spread throughout the string.

Such a situation has been encountered before in the use of a genetic algorithm to solve the source apportionment problem [8, 9]. That investigation found that representing the problem as a 2-D matrix, and devising a crossover operator (UNBLOX) that swapped a 2-dimensional section between solutions led to improved performance (Figure 3). For the purposes of this study, the block size was allowed to vary between a 1×1 grid to $\frac{1}{2}$ the x and y grid dimensions.

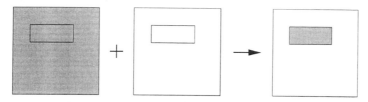

Figure 3. The UNBLOX Crossover Operator

When UNBLOX was implemented for this problem it was able to produce a recognisable I-beam after 500 generations — a large improvement in performance over the one-dimensional two-point crossover operator that shows that linkage is a significant factor in this problem.

4.3. 2×2 Area Mutation Operator

The third operator was devised for three reasons: first, the belief that as we are tackling a 2D problem, the operators should reflect this; second, a 2×2 area was thought sufficient to eliminate many of the loose pixel/wobbly-line problems found with the basic genetic algorithm; third, applying the operator only where at least one pixel is present would give a good chance of a worthwhile modification being found, and would be useful at the end of a genetic algorithm run where only small local changes would be required.

To this end, this operator acts on a 2×2 area of the chromosome array, and only modifies the contents if at least one voxel in the chosen area is turned on and one other voxel is turned off. This is so that this mutation operator would work well on the boundary of the evolved shapes. If so, standard bit-flip mutation is applied to each voxel in the area. Use of this operator increased the rate at which excess material was chopped away, whilst the ability of the genetic algorithm to continue to find improvements after convergence was also improved. In addition, the final form of the I-beam was found to be cleaner.

4.4. Putting it all Together

After the formative evaluations of each of the new operators above, it was necessary to see if these operators would work effectively in combination. Therefore, the genetic algorithm was run with the following settings: $p(UNBLOX) = 0.3$, $p(2 \times 2\ mutation) = 0.125$, and $p(Smoothing\ mutation) = 0.125$ (all probabilities are per-string); where each operator is applied sequentially to the string with these probabilities, and the sizes and locations of the operations were selected with a uniform probability. The probability of the 2×2 mutation operator was increased by 0.0005 per generation to a maximum of 0.4 as this was found to improve the quality of the results obtained slightly. All of the other genetic algorithm settings were left unchanged from the basic genetic algorithm.

Figure 4 shows some typical end-of-run results, which are near-perfect I-beams, with no holes — a noticeable improvement over the basic genetic algorithm (Figure 1). Plots of the fitnesses against generation for the basic and improved genetic algorithms are given in Figure 6 (these figures are an average over 10 runs). As can readily

Figure 4. Typical End-of-Run Results Obtained by the Improved GA

be seen, the new operators have dramatically improved the performance of the genetic algorithm, finding a better quality solution more quickly than the basic genetic algorithm, thus vindicating the approach used.

The only caveat that was found was that the size range permitted for the UNBLOX and smoothing operators does interact with the amount of bending moment applied, and this can affect the results of the optimisation somewhat. This is thought to be because the size of the changes produced by these operator should be comparable to the size of the features of the final shape, which is determined in part by the bending moment. For example, if the final shape was to have thin plates at the top and bottom, then using a smoothing operator with a large size range would be very likely to remove a good section of thin plate. Fortunately, in our experience, finding a sensible range proved straightforward so long as a little common-sense was used.

5. Guiding the Search: Directed Smoothing

At this point, the operators are still applied in a somewhat undirected fashion. One manifestation of this was the observation that the smoothing operator occasionally removed material from areas that were of high stress, and added material to areas where it was not needed. Obviously, this goes against an engineer's intuition which would suggest the opposite. Therefore, a directed smoothing operator was designed to make use of some of the useful information available to the system as a result of the evaluation function's operation. This operator was based upon a 'directed mutation' operator that has been successfully used in timetabling [10].

The evaluation must, as a necessary step to calculating the results, obtain a value for the stress present at every voxel location. The simple physics model we are using here will also calculate a close approximation to the stress present in each of the empty voxel locations as if they were solid. Thus, we can obtain an array of values representing the stress on each voxel and the average stress value for the whole array. In addition, this approach should be extensible to real a FEA package if we use material with a Young's modulus 1×10^{-3} times that of the solid material; this approach has been used successfully in [11].

This information can then be used to guide the smoothing operator in a more intelligent way. The smoothing operator was modified so that after an area has been selected, all voxels in that area with above average stress are switched on, and all voxels with below average stress are turned off. This algorithm will therefore smooth an area of voxels to match the underlying stress values according to the average value for the whole shape.

The experiments in Section 4.4 were repeated with the directed smoothing operator replacing the smoothing operator. Figure 5 shows some typical end-of-run results, which again are near-perfect I-beams, therefore any fears about this operator misleading the search appear unfounded. A plot of the fitness against generation for the directed genetic algorithm, along with the plots for the basic and improved genetic algorithms for easy comparison, are given in Figure 6 (these figures are an average over 10 runs).

It is apparent that this algorithm has been found to operate extremely well, causing much faster convergence to a near-optimal solution when used in combination with the other 'improved' operators, whilst giving a solution of equivalent quality. This will be of great use when experiments using a full FEA package are undertaken, as these packages are very computationally intensive.

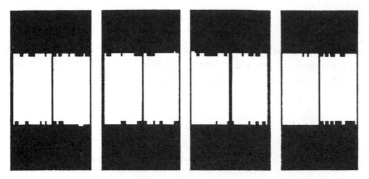

Figure 5. Typical End-of-Run Results Obtained by the GA with Directed Smoothing

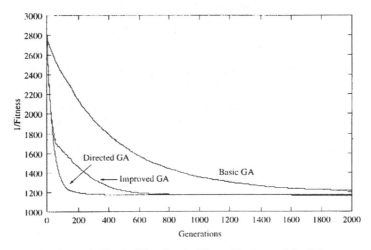

Figure 6. Fitness Plots for the Three Versions of the GA

6. Conclusions

Results presented here show that, for the optimisation of a simplified beam cross-section, the use of a 'standard' genetic algorithm with two-point crossover and bit-flip mutation led to disappointing results — the final shape was highly irregular and possessed small holes. Three domain specific operators were then used: a 2-dimensional crossover operator, a smoothing mutation operator, and a mutation operator that mutated a 2×2 area of the voxel grid. These were designed to overcome perceived problems with schema linkage, irregular shapes and small holes.

With these operators, the genetic algorithm was able to produce shapes that were known to be optimal for this problem — thus providing some vindication of this approach and showing, contrary to previous arguments in the literature, that this approach can be used for evolutionary shape optimisation *when used in conjunction with suitably designed operators*. This is, of course, with the caveat that the problem studied here is somewhat simplified; work applying this approach to real-world shape optimisation problems is currently underway.

Finally, a 'directed smoothing' operator that adds material to high stress areas was examined to ascertain the degree that this form of domain knowledge can assist evolutionary search, and was shown to speed up search significantly.

In summary, this paper describes a case-study of how domain knowledge and an understanding of how genetic algorithms work can be used to inform the design of suitable operators, as well as demonstrating the suitability of a voxel-based representation for a simple shape optimisation problem.

Acknowledgements

Thanks to the Engineering and Physical Sciences Research Council (EPSRC) for their support of Andrew Sherlock and Andrew Tuson via studentships with references 95303677 and 95306458. The authors would also like to thank the reviewers for their relevant and useful comments.

References

[1] Holland, J. H., 1975, *Adaptation in Natural and Artificial Systems*. Ann Arbor: The University of Michigan Press.

[2] Watabe, H., Okino, N., 1993, A Study on Genetic Shape Design, *Proceedings of the Fifth International Conference on Genetic Algorithms, University of Illinois, USA, July*, pp. 445-450.

[3] Chapman, C. D., Saitou, K., Jakiela, M., 1994, Genetic Algorithms as an Approach to Configuration and Topology Design, *Journal of Mechanical Design*, **116**, 1005-1012.

[4] Husbands, P., Jeremy, G., McIlhagga, M., Ives, R., 1996, Two Applications of Genetic Algorithms to Component Design, *AISB Workshop on Evolutionary Computing, Sussex University, UK, April*, pp. 50-61.

[5] Smith, R., 1995, *A First Investigation into a Voxel Based Shape Representation*, Manufacturing Planning Group, Department of Mechanical Engineering, University of Edinburgh, UK.

[6] Gere, J. M., Timoshenko, S. P., 1984, *Mechanics of Materials 2e*, Brooks/Cole Engineering.

[7] Davis, L., 1991, *Handbook of Genetic Algorithms*, New York: Van Nostrand Reinhold.

[8] Cartwright, H. M., Harris, S. P., 1993, Analysis of the distribution of airborne pollution using genetic algorithms, *Atmospheric Environment*, **27**, 1783-1791.

[9] Cartwright, H. M., Harris, S. P., 1993, The Application of the Genetic Algorithm to Two-Dimensional Strings: The Source Apportionment Problem, *Proceedings of the Fifth International Conference on Genetic Algorithms, University of Illinois, USA, July*, pp. 631.

[10] Ross, P., Corne, D., Fang, H.-L., 1994, Improving Evolutionary Timetabling with Delta Evaluation and Directed Mutation, *Parallel Problem-solving from Nature — PPSN III, Jerusalem, Israel, September* pp. 556–565.

[11] Bendsoe, M., Kikuchi N., 1988, Generating Optimal Topologies in Structural Design using a Homogenization Method, *Computer Methods in Applied Mechanics and Engineering*, **71**, 194-224.

Comparison of Crossover Operators for Rural Postman Problem with Time Windows

Kang, Myung-Ju[1] and Han, Chi-Geun[2]

[1]*Dept. of Computer Engineering, Kyung Hee University,*
Ki-Heung, Yong-In, Kyung-Ki do, 449-701, Korea
mjkang@nms.kyunghee.ac.kr

[2]*Dept. of Computer Engineering, Kyung Hee University,*
Ki-Heung, Yong-In, Kyung-Ki do, 449-701, Korea
cghan@nms.kyunghee.ac.kr

Keywords: Genetic Algorithm, MOX, Multiobjective optimization, Pareto-optimal

Abstract

In this paper, we describe a genetic algorithm and compare three crossover operators for Rural Postman Problem with Time Windows (RPPTW). The RPPTW which is a multiobjective optimization problem, is an extension of Rural Postman Problem (RPP) in which some service places (located at edge) require service time windows that consist of earliest time and latest time. To solve the RPPTW which is a multiobjective optimization problem, we obtain a Pareto-optimal set that the superiority of each objective can not be compared. We perform experiments using three crossovers for 12 randomly generated test problems and compare the results. The crossovers using in this paper are Partially Matched Exchange (PMX), Order Exchange (OX), and Modified Order Exchange (MOX) which has been modified from the OX. For each test problem, the results show the efficacy of MOX method for RPPTW.

1 Introduction

The Rural Postman Problem with Time Windows (RPPTW) is considered in this paper. The RPPTW is an extension of the Rural Postman Problem (RPP) in which some service places (located at edge) require service time windows that consist of earliest time and latest time[4]. The RPPTW is defined as follows. If a service man arrives at a service place before the earliest time, he must wait until the service place is ready for the service and the cost of traveling for the traveling service man increases. Also, if a service man arrives at a service place after the latest time, some cost penalty would be given to the service man from the service place. Hence, service man would like to arrive at the service place within the given time windows in order to reduce his total traveling cost and total penalty. So, the RPPTW is a multiobjective optimization problem.

A study of multiobjective optimization using genetic algorithms was proposed by Rosenberg in 1960's and Schaffer tried Vector Evaluated Genetic Algorithm program in 1984 [3]. In recent, the studies of multiobjective optimization have been applied to shortest path problem on acyclic network[1], analysis for water quality[6], vehicle routing problem[8] and so on.

The multiobjective optimization problems are difficult to obtain the optimal solution. Single objective optimization problems have a single optimal point, whereas multiobjective optimization problems have a set of optimal points known as the Pareto-optimal set. Each point in the Pareto-optimal set is optimal in the sense that a component of the cost vector is nondominated by at least one of the remaining components. The Pareto-optimal set is defined as follows [2]:

Definition 1. Inferiority

A vector $u = (u_1, ..., u_n)$ is said to be inferiority to $v = (v_1, ..., v_n)$ iff u is partially less than v, i.e.,

$$\forall i = 1, ..., n, u_i \leq v_i \wedge \exists i = 1, ... n : u_i < v_i).$$

Definition 2. Superiority

A vector $u = (u_1, ..., u_n)$ is said to be superior to $v = (v_1, ..., v_n)$ iff u is superior to v, i.e.,

$$\forall i = 1, ..., n, u_i \geq v_i \wedge \exists i = 1, ... n : u_i > v_i).$$

Definition 3. Non-inferiority

Vector $u = (u_1, ..., u_n)$ and $v = (v_1, ..., v_n)$ are said to be non-inferior to one another if u is neither inferior nor superior to v.

That is, Each element in the Pareto-optimal set constitutes a non-inferior solution to the multiobjective optimization problem.

This paper describes a genetic algorithm for the RPPTW and compare the performances of crossovers. In such RPPTW as Traveling Salesman Problem (TSP), the order of strings in each chromosome is important. Hence, we use partially matched exchage (PMX) proposed by Goldberg and Lingle[3, 7], order exchange (OX) proposed by Davis[3, 7], and modified order exchange (MOX) that we propose.

2 Multiobjective Optimization with Genetic Algorithms

In recent, Genetic Algorithms[3, 7] (GAs) emerged as one of the most effective and robust search algorithms. In a single objective optimization and search, the desired end result is a single solution, whereas, in a multiobjective optimization, the goal is to find a set of solutions distributed all along the Pareto set of the different objective functions[2]. As GAs always work with a population of solutions while progressing from one generation to the other, rather than a single solution at a time, they are particularly attractive for multiobjective optimization which deals with a set of Pareto solutions rather than a single solution. Once the final Pareto set is found, we can choose a suitable solution from this set according to our purpose. In this paper, to solve multiobjective optimization problem, following two objectives are considered to obtain the Pareto-optimal set.

- *Minimizing the total routing cost.*

- *Minimizing the total penalty.*

3 Rural Postman Problem with Time Windows and Objective Function

RPP is to find the shortest traveling path that passes a set of edges of a given graph at least once. Figure 1 shows a traveling path of RPP. In the Figure 1, a-a', b-b', c-c', d-d', and e-e' are the edges in E'($\subseteq$ E) that must be passed at least once in the path. a'~b, b'~c, c'~d, and d'~e are the paths that should be decided in order to find the shortest traveling path.

RPPTW is an extension of the RPP in which some service places (located at edge) require service time windows that consist of earliest and latest time[4]. If a service man arrives at a service place before the earliest time, he must wait until the service place is ready for the service and the cost of traveling for the traveling service man increases. Also, if a service man arrives at a service place after the latest time, some cost penalty would be given to the service man from the service place. Hence, service man would like to arrive at the service place within the given time windows in order to reduce his total traveling cost and total penalty.

The followings are parameters, an objective function, and a fitness function for RPPTW.

Paremeters :

- $d_{e_i^2, e_{i+1}^1}$: Cost calculated by Dijkstra algorithm from the second node of the ith edge ($\in$ E') of a tour to the first node of the $(i + 1)^{st}$ edge of the tour ($\in$ E'), where $e_i = (e_i^1, e_i^2)$ and $e_i^1, e_i^2 \in V$.

- c_{e_i} : Cost in the ith edge of the tour, where $e_i \in$ E'.

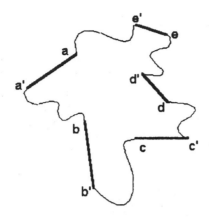

Figure 1: The traveling path

- a_i : Arrival time at the service place in the ith edge of the tour.

- p_i : Penalty at the service place in the ith edge of the tour.

- TW_i : Time Windows at the service place in the ith edge of the tour that consist of followings :

$$TW_i = \begin{cases} e_i & - & earliest\ time \\ l_i & - & latest\ time \end{cases}$$

- C : Total routing cost

- P : Total penalty

- F(C, P) : Objective function

Objective function :

$$min\ F(C, P) = (C \times P)^m,$$

where m is a non-negative integer.
Subject to :

$$C = \sum_{i=1}^{n}(c_{e_i} + d_{c_i^2, c_{i+1}^1}) \quad and \quad P = \sum_{i=1}^{n} p_i$$

where,

$$p_i = \begin{cases} 0, & e_i \leq a_i \leq l_i \\ e_i - a_i, & a_i < e_i \\ a_i - l_i, & a_i > l_i \end{cases}$$

and n is the size of E', and if $i{=}n$, let $i{+}1{=}1$ (we assume that the tour starts at edge 1 and ends at edge 1).
Fitness Function :

$$F^* = \frac{1}{F(C, P)}$$

4 Genetic Algorithm

4.1 The Structure of Chromosome

An undirected graph G = (V, E) comprises a set V of n vertices, $\{v_i\}$, a set E $\subseteq$ V $\times$ V of edges connecting vertices in V and a subset E' ($\subseteq$ E) that is a set that must be passed at least once.

In this paper, the chromosome consists of two kinds of strings. One is for describing the visiting order of the edge in E' and the other is for a set of binary codes (0 or 1) that indicate the decoding information. For example, assume that E' = $\{1, 2, 3, 4, 5\}$, where 1, 2, 3, 4 and 5 denote edges (a, a'), (b, b'), (c, c'), (d, d'), and (e, e'), respectively, and 0 and 1 denote directions of the edges. If the decoding information of an element is 1, the direction of the tour is reverse. For example, assume that the following describes the structure of chromosome.

Edge (E') Information	1	3	2	4	5
Decoding Information	0	1	0	1	0

The 0 of edge 1 means that in the tour we travel from a to a', and the 1 of edge 3 denotes a path from node c' to c, because the decoding information is 1.

4.2 Modified OX

Crossover is an operator that exchanges some strings in two selected chromosomes appropriately and a pair of new chromosomes are produced.

The order of strings are important in our problem. Hence, we use PMX proposed by Goldberg and Lingle[3, 7], OX proposed by Davis[3, 7], and MOX which has been modified from the OX.

This paper will describe MOX method only which we propose. Both PMX and OX were described in [3, 7].

MOX builds an offspring by choosing a subsequence of a tour from one parent and preserving the order and position of as many strings as possible from the other parent. Hence, the children can inherit larger characters from the parents than the other methods (PMX, OX) and the possibility of premature convergence can be reduced. A subsequence of a routing is selected by choosing two random cut points, which serve as boundaries for reordering operations.

For example, the two parents with two cut points marked by '|',

$$p1 = (012 \mid 345 \mid 6789)$$

and

$$p2 = (987 \mid 654 \mid 3210)$$

would produce offsprings in the following way. First, the segments between cut points are copied into offspring :

$$o1 = (* * * \mid 345 \mid * * **)$$

and

$$o2 = (* * * \mid 654 \mid * * **).$$

Next, we remove strings 3, 4, and 5, which are already in the first offspring from the sequence of the strings in the second parent. And we get

$$9 - 8 - 7 - 6 - 2 - 1 - 0.$$

This sequence is place in the '*' positions of the first offspring in order :

$$o1 = (987 \mid 345 \mid 6210).$$

Similarly we get the other offspring :

$$o2 = (012 \mid 654 \mid 3789).$$

4.3 Mutations

In this paper, three mutation methods are applied. The first is that a decoding information of edges is flipped at the selected point marked by '—' (mutation1). The second is that two selected points marked by '—' are swapped (mutation2). The last is that the substring between two cut points marked by '|' along the length of the chromosome is inversed (mutation3). These are :

- mutation1. Reverse

$$p = (1001\underline{1}10011)$$
$$o = (1001\underline{0}10011)$$

- mutation2. Reciprocal exchange

$$p = (9876\underline{5}4321\underline{0})$$
$$o = (9876\underline{1}4325\underline{0})$$

- mutation3. Inversion

$$p = (987 \mid 6543 \mid 210)$$
$$o = (987 \mid 3456 \mid 210).$$

Here, the reverse method is applied to the decoding information, and both the reciprocal exchange and the inversion are applied to the edge information in the chromosome.

5 Experimental Results

The GA was programmed in MSC++ version 4.1 and tested on an IBM PC Pentium Pro for 12 randomly generated problems which was generated by the same method as [4].

Table 1 describes the problems applied to GA. In GA, the size of population is 100, and we evolve the population for 100 generations. The selection scheme in this paper is roulette wheel method according to fitness function. Each crossover (MOX, OX and PMX) rate was 0.6, the mutation1 (Reverse) rate was 0.05, the mutation2 (Reciprocal Exchange) rate was 0.04, and the mutation3 (Inversion) rate was 0.03. A Pareto-optimal set of each generation is maintained and the final Pareto-optimal set of the algorithm is the solution of a problem.

The results are shown in Table 2 and Figure 2 ~ Figure 13. Table 2 describes the size of Pareot-optimal set obtained by GA according to each crossover operator for 12 test problems and Figure 2 ~ Figure 13 describe the results of comparison of crossover operators (PMX, OX and MOX). The results show that the MOX method is more efficient than the existing PMX and OX operators for 10 test problems except problem 1 and 8. This is because the MOX can preserve the order and position of as many strings as possible from parents and the chromosomes in the current generation can inherit larger characters from the chromosomes in the old generation than the existing crossover methods. Hence, in our GA using MOX, the possibility of premature convergence is reduced.

6 Conclusions

In this paper, we introduce a genetic algorithm and compare three crossovers (PMX, OX and MOX) for RPPTW. According to the experimental results, we can know clearly that the proposed MOX crossover method produces more and better Pareto-optimal solutions than the existing PMX and OX methods for our test problems. The comparison made on the basis of the number of Pareto-optimal solutions describes that the proposed MOX method produces more in number and better results.

In the future, we can solve RPPTW by Simulated Annealing and Tabu search algorithms.

Acknowledgements

The authors would like to thank anonymous referees. We also thank Byung-Hyun Jun, Jai-Won Jang, Hyun-Nam Lee, Kyung-Won Park, Sung-Geun Lee and Jung-Eun Lee for their careful proofreading of the manuscript.

References

[1] Azevedo, J. A. & Martins, E. Q. V., 1991, An Algorithm for the Multiobjective Shortest Path Problem on Acyclic Networks, *Investigacao Operacional*, Vol. 11, No. 1.

[2] Fonseca, C. M. and Fleming, P. J., 1993, Genetic Algorithms for Multiobjective Optimization: Formulation, Discussion and Generalization, *Proceedings of the Fifth International Conference.*

[3] Goldberg, D. E., 1989, *GENETIC ALGORITHMS in Search, Optimization & Machine Learning*, Addison Wesley.

[4] Kang, M. J. and Han, C. G., 1996, A Genetic Algorithm for the Rural Postman Problem with Time Windows, *Proceedings of 20th International Conference on Computer & Industrial Engineering*, pp. 321-324.

[5] Lenstra, J. K. and Rinooy Kan, A. H. G., 1976, On General Routing Problems, *Network*, Vol. 6, pp. 273-280.

[6] Makowski, M., Somlyody, L. & Watkins, D., Multiple, 1995, *Criteria Analysis for Regional Water Quality Management: The Nitra River Case*, IIASA.

[7] Michalewicz, Z., 1994, *Genetic Algorithms + Data Structures = Evolution Programs*, 2rd rev. and extended ed. Springer-Verlag, Berlin, Heidelberg.

[8] Thangiah, S. R., 1995, *Vehicle Routing with Time Windows using Genetic Algorithms*, Practical Handbook of GENETIC vol. 2, ALGORITHMS, CRC Press.

Table 1: Test problems

Test Problems	$|V|$	$|E'|$
1	20	17
2	30	13
3	40	21
4	50	16
5	20	14
6	30	14
7	40	14
8	50	38
9	20	19
10	30	12
11	40	15
12	50	47

Table 2: The experiment result

Test Problems	Pareto-optimal set		
	PMX	OX	MOX
1	19	5	20
2	53	32	30
3	9	7	16
4	8	8	13
5	43	20	70
6	11	7	11
7	21	16	31
8	4	4	6
9	54	21	58
10	22	13	27
11	19	12	44
12	6	11	8

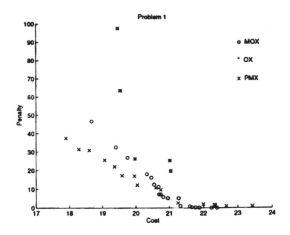

Figure 2: The experimental result of problem 1

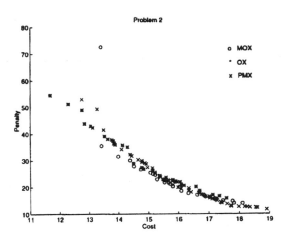

Figure 3: The experimental result of problem 2

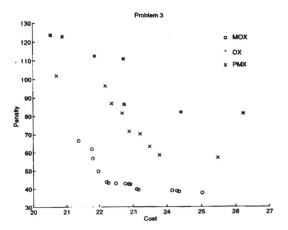

Figure 4: The experimental result of problem 3

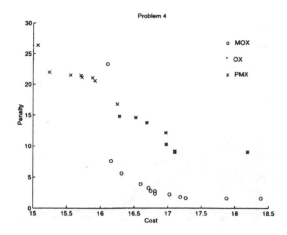

Figure 5: The experimental result of problem 4

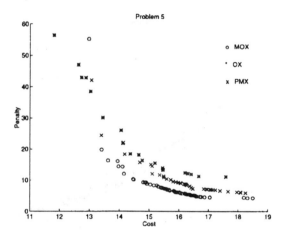

Figure 6: The experimental result of problem 5

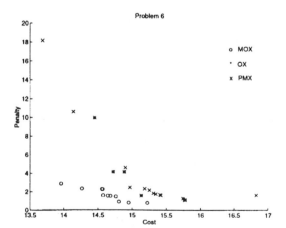

Figure 7: The experimental result of problem 6

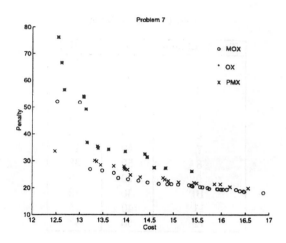

Figure 8: The experimental result of problem 7

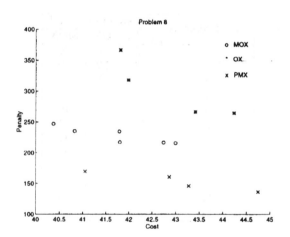

Figure 9: The experimental result of problem 8

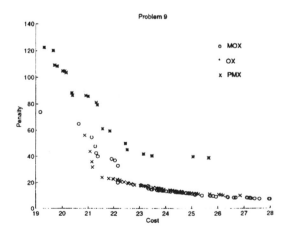

Figure 10: The experimental result of problem 9

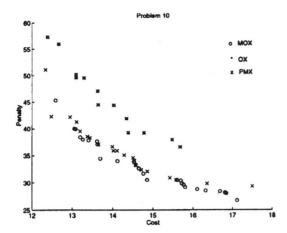

Figure 11: The experimental result of problem 10

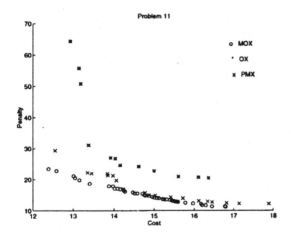

Figure 12: The experimental result of problem 11

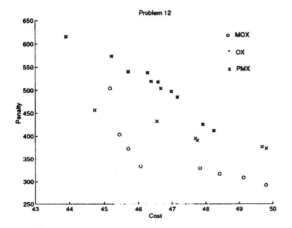

Figure 13: The experimental result of problem 12

On Generating Optimum Configurations of Commuter Aircraft using Stochastic Optimisation

R. Pant[1] and C. M. Kalker-Kalkman[2]

[1]*Aerospace Engineering Department, Indian Institute of Technology, Powai, Mumbai, India 400076, Tel : 00-91-22-5783558, Fax : 00-91-22-5782602, E-Mail : rkpant@aero.iitb.ernet.in*
[2] *Department of Mechanical Engineering, Delft University of Technology, 2628 CD Delft, The Netherlands, Tel : 00-31-15-2783983 Fax ; 00-31-15-2781397, E-Mail : C.M.Kalker-Kalkman@wbmt.tudelft.nl*

Keywords: stochastic optimisation methods, simulated annealing, monte-carlo methods, genetic algorithms, aircraft conceptual design, graphic functions

Abstract

This paper discusses the application of stochastic techniques to obtain optimum configurations of a commuter aircraft by minimising a 3 component composite objective function *OF*. Four cases arise depending on which of the three terms of *OF* are considered. Owing to the highly non-linear and discontinuous behaviour of *OF* and the presence of several closely packed local minima, gradient based techniques were not suitable, hence stochastic optimisation techniques viz. Simulated Annealing (SA), Monte-Carlo method (MC), and Genetic Algorithms (GA) were employed. In the GA and the MC method, completely new populations of designs of better quality could be generated during the run time, with the aid of some graphic functions. The optimum configurations obtained by the three stochastic methods were then compared with those obtained earlier with a modified univariate method (MUV). It was found that the stochastic methods always arrived at better configurations for all the cases compared to MUV, and the best results were achieved by the SA.

1. Introduction

During the conceptual design stage, the optimum configuration of a civil aircraft is generally obtained by minimising a single/mixed objective function (such as the Direct Operating Cost or Gross Weight), within the constraints imposed (such as the field length requirement, one-engine-operative climb gradients). Recently, a new objective function *OF* was proposed by Pant, Gogate and Arora [1], that looks at the optimum aircraft configuration from the passengers' point of view. This objective function is a measure of the net amount spent by the passenger for a multi-segment travel plan, and consists of a summation of the direct spending (in terms of the fare paid), and the indirect spending (in terms of the money equivalent of the time spent) in travel. A methodology for incorporation of *OF* in aircraft conceptual design and optimisation procedure was outlined. As a test case, this methodology was applied to obtain the optimum configuration and fleet size of a commuter aircraft for meeting the needs of a hypothetical network. Owing to some specific features of the problem (which are discussed later), *OF* turned out to be highly non-linear and discontinuous, hence conventional gradient-based methods could not be used for optimisation. This paper discusses the application of three stochastic optimisation techniques, viz. Simulated Annealing (SA), Monte-Carlo methods (MC) and Genetic Algorithms (GA) to minimise *OF*. The optimum configurations obtained by these methods are then compared by the ones obtained by an iterative directional search method.

2. Formulation of the Problem

A passenger pays for travel in direct terms (in the form of the fare F) and in indirect terms (in the form of the money equivalent of time spent in travel C_{time}). Further, if the aircraft is to be operated in a region not yet connected by air, extra investment C_{ap} is needed for development and maintenance of airports and related infrastructure. In conventional studies, this term is included in F as landing charges and/or the airport tax. From the point of view of the airline, it is sufficient to minimise F alone, which is very closely related to the sum of the operating costs incurred. However, if the overall system costs are to be considered, then C_{ap} have to be considered apart from F, which may or may not be charged from the passengers as a part of their fare. From the passengers' perspective, C_{time} is also important, especially for business travellers. The general expression for *OF* can be considered as

$$OF = F + a.C_{ap} + b.C_{time} \tag{1}$$

where a and b are weighting coefficients. The 4 cases that arise from the above definition of *OF* are case 1 with a = b = 0, case 2 with a = 1, b = 0, case 3 with a = 0, b = 1 & case 4 with a = b = 1.

In the absence of specific data, the test cases reported in [1] were carried out for operation over a hypothetical network of 100 contiguous cities separated by 216 Km from each other in a honeycomb pattern. Journeys from each city to any of the 6 neighbouring cities only were considered, and the average daily passenger demand for the entire network was assumed to be 12000 passengers. The average *Value of Time* of passengers was taken to be US $ 60 per hour (1986 base), which is the money equivalent of one hour of the passenger's travel time. These four parameters were termed as *network variables*.

Six variables were chosen to represent the configuration of the aircraft, viz. passenger capacity (N_{pax}), maximum lift coefficient (C_{lmax}), wing loading (*W/S*), wing aspect ratio (*AR*), wing thickness ratio (*t/c*), and the landing gear type (*LGType*). Two variables related to the way in which the aircraft is operated, i.e. the cruising velocity (V_{cr}), and cruising altitude (H_{cr}.) were added to this list, making a set of 8 design variables. Upper and lower limits were assigned to these design variables, based on previous experience, as shown in Table 1.

Table 1. List of Design Variables with their ranges.

Parameter	Symbol	Units	Lower Limit	Upper limit
Passenger Capacity	N_{pax}		6	60
Cruising Velocity	V_{cr}	kmph	325	575
Max. Lift Coefficient	C_{Lmax}		1	3
Wing Loading	*W/S*	kg/m^2	125	775
Wing Aspect Ratio	*AR*		5	12
Cruising Altitude	H_{cr}	km	0.460	3.0
Landing Gear Type	*LGType*		Fixed (0)	Retractable (1)
Wing Thickness Ratio	*t/c*		0.10	0.20

Except N_{pax} and *LGtype*, all design variables were continuous. N_{pax} had certain specific integer values due to the inevitable jumps in the number of passengers seated abreast with increase in capacity (to ensure a suitable fuselage slenderness ratio). *LGtype* was either 0 or 1 (depending on whether the Landing Gear was retractable type, or fixed type). Three constraints were intrinsically imposed in the algorithm to calculate *OF* viz.

1) Cruise power fraction ≤ 75 % , from engine life considerations.
2) Climb angle with one engine inoperative ≥ certain specified value, from safety considerations.
3) Field length required ≤ specified value.

The problem was thus formulated in the form

Minimise *OF*

where *OF* = *OF*(N_{pax} ,C_{lmax}, *W/S*, *AR*, *t/c*, *LGType*, V_{cr}, H_{cr}), subject to the constraints mentioned above.

Expression for F, C_{time} and C_{ap} in terms of the network variables and design variables were developed. These expressions, and the details of the algorithm developed to obtain the value of *OF* for any given set of the design variables are given in [1].

2.1 Peculiarities of the Objective Function *OF*

The algorithm starts with assumed values of 5 parameters called *initiators*, viz. aircraft gross weight, maximum cruising speed at sea-level, propeller efficiency, cruise lift to drag ratio and aircraft power loading.

These are then corrected in an iterative fashion, as and when sufficient information is available to arrive at their values. The iterative nature of the algorithm leads to the limitation that convergence is not guaranteed for all sets of design variables. In other words, for many sets of design variables, the algorithm does not converge; these cases are termed infeasible and a very high value is artificially assigned to *OF*. Further, some design variable sets lead to violation of the constraints imposed and are rejected in a similar fashion. This feature of the algorithm makes *OF* highly non-linear and discontinuous, hence conventional gradient-based optimisation methods were considered unsuitable [2].

2.2 Modified Univariate Method (MUV)

In [1], a modified Univariate method (MUV) was employed, which was essentially an interactive cyclic directional search in all the design variables, but one at a time. The strength of the method was that the step size, direction and extent of search could be altered in keeping with the trend of variation of *OF*, thus avoiding infeasible regions to a great extent. This led to a substantial reduction in the number of calculation cycles required, and also provided a *feel* for the function. The method, however, was quite cumbersome, since the values of the design variables were manually manipulated by the user. Further, there was no way to prove that the global minimum was reached. The details of the algorithm for *OF* and the MUV method are given in [1].

3. Stochastic Optimisation Methods

Stochastic Optimisation methods are the ones whose results depend on multiple evaluations of the objective function for random combinations of the design variables. They are not influenced by the non-linearity and/or discontinuity of the objective function. These methods are more likely to arrive at the global optimum compared to the gradient based methods, due to the presence of the random element inherent in them, which also makes them function independent to some extent. However, this random element also brings with it a couple of problems.

- Two runs of the method from the same starting point may end up with different results, depending on the path taken by the algorithm. These methods usually require a user-specified seed(s) for the pseudo-random number generators that they use. Different seeds correspond to different sequence of numbers being generated, hence the path followed by the algorithm differs. However, this very feature can be utilised to ascertain the *confidence level* on the ``globalness" of the optimisation achieved. For e.g., if several runs with different sequence of random numbers employed to generate trial points end up at the same (or very similar) optimum, it is likely to be a very strong local optimum, if not the global optimum itself.
- The number of function evaluations for stochastic optimisers is usually much larger than that required by calculus based optimisers. For smooth, unimodal and continuous functions, stochastic methods are very much inferior to calculus based methods.

Simulated Annealing (SA), Monte-Carlo Methods (MC) and Genetic Algorithms (GA) the three stochastic methods which have been widely reported in literature of being able to successfully and efficiently tackle such ill-behaved objective functions [2,3,4]. In the next sections we will explain how those three methods have been applied for minimising *OF*. Results obtained for the four cases outlined above by the MUV method developed earlier and these three stochastic optimisation methods are also compared.

3.1 Simulated Annealing (SA)

SA was introduced by Metropolis et al. [5] and is based on the thermodynamical analogy of annealing of metals. When molten metal is allowed to anneal (cool slowly), it eventually arrives at a low energy state. If, on the other hand, it is quenched (cooled suddenly), it assumes a high energy state. SA tries to minimise some analogue of the energy in a manner similar to annealing to achieve the global minima. It was first proposed by Kirkpatrick et al. [6] for optimisation of combinatorial problems (in which the objective function is defined in a discrete domain) and was successfully employed for objective functions involving very large number of variables (even tens of thousands). Corona et al. [2] were one of the many researchers who have extended the applicability of this method for objective functions involving continuous variables.

Details of the SA algorithm

For the present work, the SA algorithm developed by Goffe et al. [7] based on the methodology proposed by Corona et al. [2] was employed. The algorithm starts with a high initial value of temperature T_{init} and a starting set of design variables. Trial sets are then generated using random numbers from the set [-1,1] and initial step length for each design variable v_i. If the function value for the trial set is lower than that for the previous one, the trial set is accepted. Acceptance of a trial set yielding higher function value is random, with a probability decreasing exponentially with the temperature. After N_S steps through all design variables, their step lengths are adjusted, to ensure that roughly half of all the moves are accepted using a varying criterion c_i, in line with the approach followed by Metropolis et al. [5]. A very high acceptance rate implies that the function domain is not being fully explored, while a very low acceptance rate means that the new trial points are being generated too far away from the current optimum. Both of these imply that the algorithm is not progressing efficiently and involves wasting of computational effort. After carrying the above loop N_T times, the temperature is gradually reduced employing a geometric schedule governed by the parameter r_t. The algorithm is stopped when the reduction in the function value in N_e successive cycles is less than a small number e.

In short, SA explores the entire domain of the function and tries to optimise it while moving both uphill and downhill, enabling it to escape from the local optima. It is largely independent of starting values, which are very critical for most conventional methods. The change in the step length as the algorithm proceeds provide an insight into the sensitivity of the design variables with the objective function, since large step lengths indicate that the objective function is quite flat in that design variable and vice versa.

Tuning of the SA parameters

The 8 parameters related to the SA viz. T_{init}, v_i, N_S, c_i, N_T, r_t, N_e and e are to be "tuned" in keeping with the nature of the objective function being optimised. A bad choice for these parameters can make the algorithm extremely inefficient and may even result in failure to arrive at the global optimum. Values of 1.0, 10, 2.0, 4 and 0.001 were assigned for v_i, N_S, c_i, N_e and e respectively, as recommended by Corona et al. [2], and based on experience. The most suitable values of the remaining 3 parameters viz. T_{init}, N_T and r_t were determined by numerical experimentation. Details of some of the runs are listed in Table 2. It may be noted that the configurations termed as the *best* optimum in Table 2 represent the best among the solutions obtained from several trial runs that were carried out for each case.

It was decided to keep an upper limit of 10^4 function evaluations for each case, which corresponds to about one and half hours computational time on the SUNSPARC workstation. As suggested by Goffe et al. [7], trial runs were carried out with high T_{init} to determine the temperature T^* at which the step lengths begin to decline sharply. Reduction in T_{init} leads to a reduction in total number of evaluations required (N_{eval}), however there is a danger of getting stuck in a local minima if one starts with a very low T_{init}. The best way to overcome this problem is to ensure that T_{init} is slightly higher than T^* determined by initial runs, as shown for case 2 and 4. A very high r_t can lead to rapid quenching, as was observed in the first run for case 1; values between 0.25 to 0.4 gave best results. A reduction in N_T leads to much lower N_{eval} (especially when coupled with low T_{init}) as can be seen by comparing the two best results for case 3. Based on these experiments, the values of T_{init} = 10, r_t = 0.3, and N_T = 2 were assigned for all the optimisation cycles. Note that the ratio of number of accepted evaluations (N_{accp}) to N_{eval} is roughly 0.5 for most of the runs ending in .the best solutions.

Table 2. Experimental runs for tuning of SA parameters

	T_{init}	r_t	N_T	T^*	T_{final}	N_{accp}	N_{eval}	*OF*	optimum	Comments
Case	10^6	0.80	4	---	3518	4981	5000	13.455	local	quenched,high r_t
1	10^2	0.30	4	2.7	5.3E-4	2763	4161	13.247	local	high N_{eval}
	10	0.30	2	3.0	5.9E-4	916	1761	12.992	best	lower N_{eval}
Case	10^5	0.25	4	6.0	2.3E-5	3859	5441	22.587	local	high N_{accep} / N_{eval}
2	10	0.25	4	0.6	3.8E-5	1853	3201	22.267	best	lower N_{eval}
Case	10^4	0.30	4	24	5.9E-4	1857	3521	69.632	local	high N_{eval}
3	10^5	0.25	4	1.5	2.3E-5	3850	5441	68.752	best	high N_{eval}
	10	0.30	2	0.9	5.9E-5	876	1761	68.758	best	lower N_{eval}
Case	10^2	0.40	4	9.0	1.6E-5	2219	4481	93.123	local	high N_{eval}
4	10	0.30	2	0.8	1.8E-5	861	1921	92.623	best	lower N_{eval}

Figure 1 shows the convergence history of SA for case 1, along with the variation of the design parameters at the end of each cycle. Convergence criteria is finally met at the end of 16 cycles, each of which corresponds to

272

110 evaluations of *OF*. It can be seen that after the 11th cycle, fluctuations in the value of all design variables are markedly reduced, except for H_{cr}, which continues to fluctuate right till the end. This indicated that *OF* was a weak function of this variable, which was confirmed later by the results obtained by other optimisation methods.

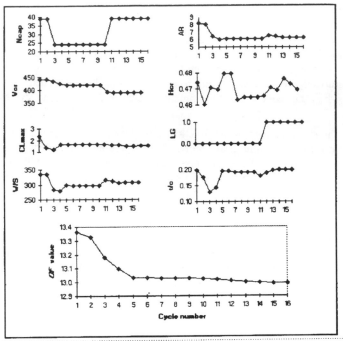

Figure 1. Convergence history of case 1 with Simulated Annealing

3.2 Monte Carlo Method

In [3], [11] and [12] a program GOOD (Generator Of Optimal Designs) has been described that implements a Monte Carlo method. The algorithm for calculation of *OF* was coupled to the program GOOD, with the aid of BAD (Better Analysis and Definitions) which is a pre-processor for GOOD. In BAD, an input file is created that defines the variables involved and the subroutines and formulae to be computed for *OF*.

With the aid of this program, random values for the independent variables are generated. The user has to provide lower and upper bounds for those values. The dependent variables are then computed and if they are not within certain user-specified bounds, the design is rejected as infeasible. For feasible designs, the objective function is computed and if this appears to be a feasible solution, it is compared to the "current optimum". As soon as a new optimum is found, it is plotted on the screen, i.e. a line is drawn from the "previous optimum" to the "new optimum". This way the user gets an insight in the regions where the better solutions are found. The program has been constructed in such a way that the user is able to modify the search intervals for the independent variables. This was possible by coupling a graphical interface system to the program GOOD. The system thus enables the user to interact with the program and to control and direct the process of optimisation, which leads to a much faster convergence to an optimum. An illustration of this grahical interface system is provided in Figure 2 ahead.

3.3 Genetic Algorithms

Genetic Algorithms [4, 8-10] are based on the Darwinian model of "survival of the fittest". They work by maintaining a pool (termed population) of several competing designs which are randomly combined to find improved solutions. One can consider the members of the population as carriers of good qualities that can be inherited by new generations. Each population member is represented by a string of binary coding that encodes the design variables. These are analogous to the genes of the biological individual in an evolutionary chain. Discrete design variables are represented exactly by binary numbers, while the continuos variables can be represented to any required degree of accuracy by discrete numbers defined within the range of design variables.

The probability of survival of each member to the next generation of the population depends on its "fitness", which is directly related to the value of the objective function. The search proceeds towards an optimal solution by iteratively selecting fitter individuals for further reproduction. A mechanism called *crossover* is employed for creating the next generation, in which the strings of two fit parents from the present generation are interchanged at a randomly selected crossover location. To avoid getting stuck in a local minimum, an operator called *mutation* is employed to each bit with a very low probability, in which the bits are flipped. These mechanisms are discussed in detail in [8-10].

The construction of a new population has been organised in such a way, that *child* designs are rejected when they are not feasible or when the objective function is worse than the worst of the current population. This means that they are only admitted to the new population if they are better than the worst member. They are also rejected when an identical member already exists in the population. This ensures a steady improvement in the population as a whole. Better solutions are usually found within a few generations, however, in many cases so-called *premature convergence* occurs, see [11]. This happens when the members do not differ a lot from each other, hence new members will not bring any worthwhile improvement in the population. In GOOD, premature convergence can be avoided, since it is possible to start a completely new population within user-specified intervals, at any stage. The best member of the population is, however, retained, in order to maintain the quality. It is found that in this way the algorithm avoids premature convergence, and finally converges to the global optimum.

For the present problem, the initial population was kept quite small (10), since *OF* was very complicated, with few feasible solutions. As the iterations proceeded, the population size was increased to 20, 40 or even 100, in the interval where feasible or better solutions were found. The probability of crossover was initially set to 0.4 %, which was steadily increased to a maximum value of 10 %. A heuristic cross-over option is also available in GOOD, as discussed in [11]. A Tournament based selection criteria was employed. The number of generations that were required to converge to an optimum solution varied between around 150 to 200. Several runs starting with totally random initial population were carried out, and in most cases, the same or very similar optimum solution was reached. This provided sufficient confidence that the solution so obtained was likely to be the global optimum, or at least a very strong local optimum.

The program GOOD described above for MC method also has a Genetic Algorithm version [11], which was used in the present problem. Here again, it is possible to interact with the code at various stages of program execution, using the same graphical interface. It is also possible for the user to specify new search intervals interactively, in which completely new populations are generated in the regions where better solutions can be expected. Figure 2 is a snapshot of the GA version of the GOOD program running on a PC and it illustrates how the convergence history of the algorithm can be observed as the iterations proceed.

4. Comparison of Results Obtained by Various Methods

A comparison of the results obtained by all the methods for the 4 cases is given in Table 3. For all cases, configurations having a lower *OF* were obtained using SA and GA, compared to the ones obtained by the MC and MUV. The optimum configurations obtained by various method do not radically differ from each other, except for case 1, in which the MC and MUV got stuck in the local minimum, since the optimum aircraft capacity is 24 passengers as against 39 passengers for SA and GA. For all the cases, MUV required a very low number of function evaluations. However, due to its interactive nature, the total time needed to arrive at the optimum solution was more than that needed by the other three methods. The largest number of function evaluations among all the four methods were needed by GA and MC. For all the cases, the fastest and the best results were obtained with SA, however, a few trial runs were required to arrive at the appropriate values for the tuning parameters. It may be noted that several runs of SA, MC & GA were carried out for each case, most of which converged to the optimum configurations reported in Table 3, indicating that the results obtained were fairly reproducible.

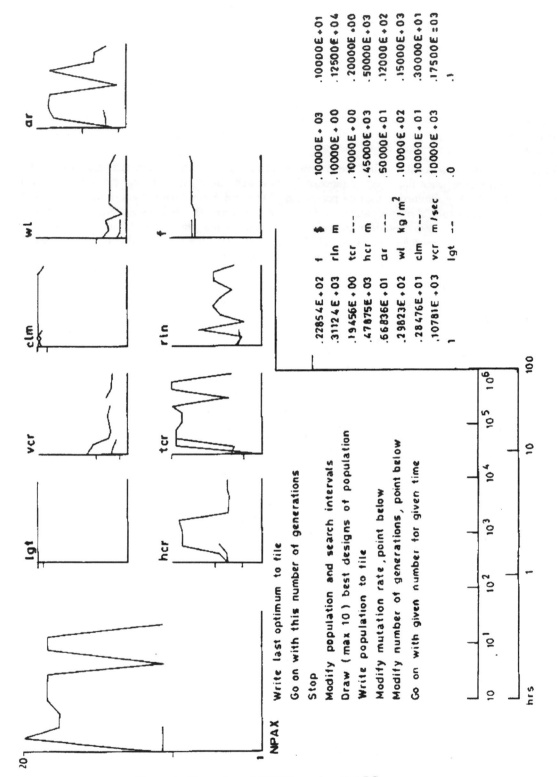

Figure 2. Illustration of GOOD running on a PC

Table 3 Optimum configurations obtained by various methods

	N_{cap}	V_{cr}	C_{Lmax}	W/S	AR	H_{cr}	LGType	t/c	OF value	N_{eval}	Method
units		KMPH		KG/m²		KM			US $/pax/flt		
Case	39	388.98	1.52	306.56	6.24	0.470	FIX	0.20	12.992	1761	SA
1	39	385.45	1.75	225.55	5.41	0.465	FIX	0.20	13.169	14909	GA
	24	396.50	1.70	272.45	6.70	0.498	RET	0.17	13.223	2393	MC
	24	396.00	1.80	262.31	6.40	0.460	FIX	0.18	13.450	281	MUV
Case	24	360.86	2.99	138.46	5.04	0.465	FIX	0.20	22.267	1921	SA
2	24	366.05	3.00	126.39	5.00	0.463	FIX	0.18	22.296	7320	GA
	24	369.29	2.99	136.05	5.00	0.462	RET	0.18	22.311	52685	MC
	24	400.86	3.00	148.96	5.00	0.460	RET	0.16	22.660	320	MUV
Case	6	556.34	1.45	712.43	7.56	0.480	RET	0.10	68.752	1761	SA
3	6	540.94	1.91	728.10	7.39	0.465	RET	0.10	68.830	12806	GA
	6	540.54	2.00	730.54	7.19	0.484	RET	0.10	68.852	18528	MC
	6	550.80	2.40	669.08	7.30	0.460	RET	0.10	69.320	153	MUV
Case	6	487.01	3.00	154.50	5.00	0.470	RET	0.11	92.623	1761	SA
4	6	490.50	2.99	158.75	5.00	0.465	RET	0.10	92.605	3666	GA
	6	496.84	2.99	152.06	5.00	0.461	RET	0.11	92.651	2703	MC
	6	486.00	3.00	153.84	5.00	0.460	RET	0.10	92.970	115	MUV

5. Conclusions and Future Work

In conclusion, this study established the efficacy of both SA and GA in tackling a complicated multi-modal objective function about which no prior knowledge of function behaviour existed. The MC method is a kind of biased random walk, since all the trial configurations are generated randomly, and no previous information is retained, except the details of the current optimum. Despite this, it was able to find reasonably good solutions for the three cases, although it needed a large number of function evaluations, especially for case 2 & 3. The three stochastic optimisation methods were seen to converge to better or similar solutions compared to the gradient based optimiser (MUV).

For the present problem, SA was found to be a very robust and easy to implement optimisation algorithm for, and came up with the best results for all the three cases. It needs several trial runs before the actual optimisation could commence, to tune the various optimisation related parameters. This could be quite cumbersome in some cases. It also requires a very large number of function evaluations compared to most classical methods. The GOOD GA code was seen to have good progress in the initial stages of the optimisation run, but slowed down considerably towards the end. This is because stochastic optimisation codes are very good in exploring the function domain to quickly identify the areas of promise, but then they are very slow in converging to the final solution. Thus, it may be worthwhile to provide a mechanism by which one could switch to a gradient based optimisation method from within the GOOD code. One such method is Stochastic Iterated Hill Climbing (SIGH) [13]; and similar hybrid method for multidisciplinary aircraft design optimisation has been recently used by Bos [14].

Finally, it should be kept in mind that there is no optimisation method which is universally good for all applications. Wolpert & McReady [15] have proposed the *No Free Lunch* theorem of optimisation, according to which any (optimisation) algorithm performs only as well as the knowledge concerning the cost (objective) function is put into use in the cost (objective) evaluation procedure. Further, they have concluded that on an average, all optimisation methods are equally effective, and have suggested that an optimisation method should be tailored to suit the problem at hand, using the information about the function.

References

[1] Pant, R. K., Gogate, S. D. and Arora P. , 1995, Economic parameters in the conceptual design optimisation of an air taxi aircraft. *Journal of Aircraft*, **32** , 696-702.

[2] Corana, A., Marchesi, M., Martini, C. M. and Ridella, S., 1987, Minimising multimodal functions of continuous variables with the Simulated Annealing algorithm. *ACM Transactions on Mathematical Software*, **13**, 262-280.

[3] Kalker-Kalkman, C. M. , 1991, Optimal design with the aid of Randomization methods. *Engineering with Computers*, **7**, 173-183.

[4] Lin C.-Y., Hajela, P., 1992, Genetic Algorithms in Optimisation problems with discrete and integer design variables. *Engineering Optimisation*, **19**, 309-327.

[5] Metropolis, N., Rosenbluth, A., Rosenbluth, M., Teller, A. and Teller, E., 1953, Equation of state calculations by fast computing machines. *Journal of Chemical Physics*, **21**, 1087-1090.

[6] Kirkpatrick, S., C. D. Gelatt, Jr., and M. P. Vecchi., 1983, Optimisation by simulated annealing, *Science*, **220** , 671-680.

[7] Goffe, W. L., G. D. Ferrier, J. Rogers. , 1994, Global Optimisation of Statistical Functions with Simulated Annealing, *Journal of Econometrics*, **60**, 65-100.

[8] Davis, L. , 1991, *Handbook of Genetic Algorithms*, Van Nostrand Reinhold, New York.

[9] Goldberg, D. E., 1989, *Genetic Algorithms in Search, Optimisation and Machine Learning*, Addison-Wesley, Reading.

[10] Zbigniew, M., 1992, *Genetic Algorithms + Data Structures = Evolution Programs*, Springer - Verlag, New York.

[11] Kalker-Kalkman, C. M., Offermans, M. F., 1995, A general design program based on genetic algorithms with applications. *Proceedings of the 21st ASME Design Automation Conference*, Boston, USA, September.

[12] Kalker-Kalkman, C. M., 1994, A design program based on the Monte Carlo Method with Applications, *Advances in Computer Aided Engineering, CAD/CAM research at Delft University of Technology*, report on the VF- project CAD/CAM 1989-1994, Delft University Press, ISBN 90-407-1017-1.

[13] Bramlette, M. F., Cusic, R., 1989, A Comparative Evaluation of Search Methods Applied to Parametric Design of Aircraft, *Proceedings of 3rd International Conference on Genetic Algorithms*, 213-218.

[14] Bos, A. H. W., 1996, Multidisciplinary Design Optimisation of a second generation Supersonic Transport Aircraft using a Hybrid genetic/Gradient-Guided Algorithm, *Ph.D. Thesis*, Faculty of Aerospace Engineering, Delft University of technology, Delft, The Netherlands

[15] Wolpert, D, McReady, W. G., 1995, No Free Lunch Theorems for search, *Working Paper 95-02-010*, Santa Fe Institute, New Mexico, USA.

Part 6: Engineering Design

Papers:

Application of Genetic Algorithms to Packing Problems - A Review

E. Hopper and B. Turton

School of Engineering, University of Wales Cardiff
Newport Road, Cardiff CF2 3TD, UK
phone: +44-1222–874425; fax: +44-1222–874420
HopperE@cf.ac.uk; Turton@cf.ac.uk

Keywords: genetic algorithms, packing, nesting, review

Abstract

This paper reviews two and three-dimensional packing problems and their solution utilising genetic algorithms. The dimensions and the geometry of the figures involved in the allocation process classify packing problems. Since the genetic representation of the problem is vital to the performance of the genetic algorithm, particular emphasis is put on the techniques used for the representation of the problem and the genetic operators. With most of the approaches being hybrid genetic algorithms their decoding strategies are also briefly discussed.

1. Introduction

Cutting and packing problems are concerned with finding a *good* arrangement of multiple *items* in larger containing regions (*objects*). The placement is described by a set of rules or constraints. The objective of the process is to maximise the material utilisation and hence minimise the "wasted" area. This is of particular interest to industries involved with mass-production as small improvements in the layout can result in savings of material and a considerable reduction in production costs. The complexity of the problem and the method of solution depend upon the geometry of the objects and the constraints imposed.

For the solution of small packing problems algorithms have been developed with the aim of producing maximal material utilisation. For larger combinatorial problems these techniques become inefficient due to the vast number of possible solutions and the computation time grows exponentially. These problems are said to be NP-complete [1] [2]. As conventional methods fail to produce an optimal solution heuristic methods have been developed, that can produce *'near-optimal'* solutions in reasonable time. Such techniques include genetic algorithms, simulated annealing, tabu search and artificial neural networks [2] [3]. Reviews and surveys of methods to solve various types of packing problems are given in [4] [5] [6] [7] and [8].

This paper provides an overview of the research activities that approach two and three-dimensional packing problems using genetic algorithms. Readers who are not familiar with the concept of genetic algorithms should refer to the recent literature [9] [10] [11]. Section 2 explains the need for heuristic methods for the solution of packing problems. The research carried out on approaching the packing problems with genetic algorithms is introduced and discussed in section 3. Section 4 concludes with a summary of the major aspects, which are important for the application of genetic algorithms to packing problems.

2. Genetic Algorithms and Packing Problems

The first researcher to apply genetic algorithms to packing problems was Smith [12] in 1985 in a bin-packing problem. At the same time Davis [13] summarised the techniques for the application of genetic algorithms to epistatic domains using the example of two-dimensional bin-packing. Since then various types of packing problems have been approached ranging from regular to arbitrary shapes in two or more dimensions.

Complex epistatic problems are commonly approached by a two-stage procedure, using a hybrid genetic algorithm. In this the genetic algorithm manipulates the encoded solutions, which are then evaluated by a decoding algorithm, which transforms the packing sequence into the corresponding physical layout. The decoding method used can be deterministic or heuristic. Since domain knowledge can be built into the decoding procedure the size of the search space is reduced. However, the need for a decoding heuristic results in a loss of information from one generation to the next. The quality of the solutions is determined by comparison with manual layouts, which have proven to be time consuming but very effective.

3. Overview of Problems

A variety of packing problems have been tackled using genetic algorithms. Since the geometric properties influence the complexity of the problem and the magnitude of the search space these properties are most useful for classification (see Figure 1).

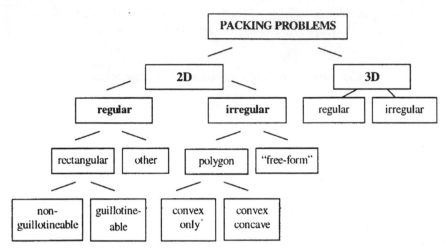

Figure 1: Classification of packing problems approached by GAs

Apart from the spatial dimensions of the object the geometry of the items is a very important criterion. In the following section problems encountered in the literature will be introduced. Since the representation of the problem is the key to an efficient application of genetic algorithms, particular emphasis is put on the description of the encoding technique and the data structure of the resulting genetic operators. Where available the performance of the genetic algorithm in comparison to other solution approaches is briefly stated.

3.1. Two-dimensional Packing Problems

Most of the packing problems found in the literature are two-dimensional. The two main groups are distinguished by the items to be placed, which can be of regular or irregular shape. Regular figures have shapes that are determined by a few parameters [6] (e.g. rectangles, circles), whereas the term irregular applies to asymmetrical convex and concave shapes.

3.1.1. Regular Packing Problems

Packing of Rectangular Shapes:

The packing of rectangular items distinguishes between guillotineable and non-guillotineable patterns [6] [7] [8] [14]. In both cases the aim is to find the arrangement of rectangles producing the least waste. Some bin-packing problems are included in this category. The objective is to minimise the number of bins. For details concerning these algorithms see Table 1.

a) Non-guillotineable Packing Problems

Four approaches to non-guillotineable packing problems will be discussed in the following section.

Genetic algorithms were first applied by Smith [12] to the packing of rectangular figures into a single rectangular bin of fixed dimensions. The algorithm described is a combination of a genetic algorithm and a heuristic procedure decoding the list of items into a packing pattern. For the evaluation of a packing pattern the ratio of the packed to unpacked area of the bin is taken. Two decoding procedures for the conversion of the permutation into the layout were developed. The Slide algorithm places the rectangle in one corner from where it "falls" to the corner furthest away under orthogonal movements resulting in a zigzagging into a stable position. The Skyline procedure tries all stable positions and orientations of an item in a partially packed bin. This takes longer, however, it covers more of the search space. To determine where to place an item the algorithm evaluates the partial pack by relating the distance between the item to be packed and all other items. The comparison with a packing program that is based on heuristics and dynamic programming, shows that the genetic algorithm achieves the same packing density. However, it runs 300 times faster.

The genetic algorithm Jakobs [15] proposed allocates rectangular figures on a rectangular board of a fixed width with the aim of minimising the height of the occupied area. This problem has been extended to polygons involving convex and concave features (see section 3.1.2). In both cases the genetic algorithm is combined with deterministic procedures in order to improve the solutions obtained. Since the height is not sufficient for the comparison of different packing patterns, the fitness function also takes into account the largest resulting contiguous remainder. The decoding of the permutation representing a packing sequence into the physical layout follows a deterministic rule known as Bottom-Left-algorithm (BL-algorithm). Each rectangle is placed as far as possible to the bottom and then as far as possible to the left of the containing rectangular area.

Jakobs' genetic algorithm starts with an initial population, where one of the permutations is a width-sorted sequence of items and the rest is generated randomly. Unlike the classical reproduction process, where the poor individuals are removed by selection, Jakobs identifies the worst individual of the population using the BL-algorithm and replaces it with the offspring according to steady-state replacement. Comparisons show that the performance of the genetic algorithm is better than the deterministic BL-algorithm.

Kröger et al. [16] [17] [18] approach a large packing problem in a similar way to Jakobs [15], again different sized rectangles are to be packed into a single bin of fixed width and open height with the objective of minimising the height. The encoding technique is based on a directed binary tree, where each node represents a rectangle in the packing scheme. Directed edges enter a node leading from one rectangle to its immediate neighbour at the top or at the right. Decoding the genotype into the packing scheme determines the location at which the lower left corner of the rectangle is to be placed, either to the top or to the right of the preceding rectangle. The position is then fixed by the BL-rule placing the rectangle as far as possible to the bottom and the left side of the bin. In order to generate a unique packing scheme each node is assigned a priority value, so that the rectangle with the highest priority is placed next in case of a conflict. The data structure not only encodes the set of rectangles, also contains information about their orientation and priority.

The fitness evaluation of a packing pattern considers the height and the width. The genetic operators have been adapted to the problem with the mutation operators modifying the set of edges, the orientation and the priority values. The crossover consists of taking a sub-tree from the first parent and placing it at the root position of the offspring. The missing rectangles are then taken from the second parent while the orientations are kept and the priority values are modified such that the packing sequence is maintained. Results indicate that the genetic algorithm developed is able to solve large packing problems in reasonable time.

Falkenauer and Delchambre [19] have developed a genetic algorithm for the classical two-dimensional bin-packing problem, where the objective is to minimise the number of bins. Consequently the fitness function has to take into account how efficiently the bin capacity is utilised. The encoding technique uses one gene per item to represent the bin, in which it is packed. This method of encoding does not perform well in combination with the classic crossover and mutation operators. Therefore an alternative data structure was proposed with a chromosome consisting of two parts, an object part and a group part. The object part identifies which items form a bin, the group part contains one gene per bin introduced in the object part.

The genetic operators only work on the group part and have to handle strings of variable length. They make use of two heuristic procedures, the First Fit (FF) and the First Fit Descending heuristic (FFD). The FF places the objects one by one in the first bin with sufficient space and starts a new bin in case the ones used so far are already full. The FFD sorts the objects according to their sizes before applying the FF rule. The crossover procedure ensures that important information corresponding to the items stored in the bins, is transmitted between the generations. After random selection of two crossover points per parent the bins between the crossing sites are inserted into the first offspring. In order to avoid items appearing twice in the solution the relevant bins originating from the first parent are deleted. Since the deleted bins may also contain items that are not present in the bins coming from the second parent, these missing items are reinserted with the FFD procedure creating additional bins if necessary. For the mutation process a few bins of the chromosome are selected randomly and the items they contain are reinserted in random order with the FF algorithm. An inversion operator is also implemented changing the order of the bins in the group part of the chromosome. This can improve the transmission of certain genes to the offspring. Comparison with the FFD algorithm has shown that the performance of the genetic algorithms is superior.

Table 1: Comparison of the genetic algorithms for non-guillotineable 2D packing problems

	Smith [12]	Jakobs [15]	Kröger [16] [17] [18]	Falkenauer et al. [19]
Representation	permutation	permutation	directed binary tree	two part chromosome
Fitness	ratio packed to unpacked	remaining area	height, width	exploitation of bin capacity
Reproduction		proportional selection	proportional, ranking, best, random	
Crossover	order cross-over	order cross-over	problem specific operator	problem specific operator
Mutation	random reordering of string; rotation	inversion, exchange of elements, rotation	variation of set of edges, orientation, priority	deletion of bins, reinsertion of objects with FF
Decoding	Slide Pack algo Skyline Pack algo	BL-algorithm	encoding structure + BL-algorithm	

The single bin-packing problems in [12] and [15] use permutations to encode the sequence of the placement. Order-based chromosomes and modified crossover and mutation operators are suitable for combinatorial problems. They only contain information about the order of packing. The physical layout is determined by decoding procedures, which are similar for both of the genetic algorithms described. In contrast the data structure developed by Kröger et al. [16] [17] [18] not only determines the order of placement, but also contains some information about the position, however, to fix the position a decoding procedure is still needed. The data structure permits the inclusion of some domain knowledge in the genetic algorithm. In the case of [12] and [15] this domain knowledge is only utilised by the decoding algorithm. In order for the genetic algorithm to operate most efficiently as many of the characteristic features of the problem should be inherited by the offspring as possible. If these features are hidden in the decoder, the succeeding generations cannot benefit. From this point of view the data structure developed in [16], [17] and [18] is superior. The encoding technique proposed by Falkenauer and Delchambre [19] is adapted to the problem of bin-packing. It not only has to consider the efficient packing of a single bin, but also the best packing for the combination of several bins. This is possible with the two chromosomes per individual approach. The domain part of the problem is contained in the decoding procedure.

b) Guillotineable Packing Problems

Guillotineable packing problems have been approached with genetic algorithms by two researchers. Both approaches are based on a tree representation.

The representation Kröger [14] proposes ensures that the packing pattern is guillotineable. The relative arrangement of the rectangles is described as a slicing tree structure. In the tree the "leaf"-nodes correspond to the rectangles to be packed, whereas all other nodes represent the hierarchy of guillotine cuts needed for the packing scheme. Apart from guaranteeing a guillotineable solution this representation contains the complete subtrees which can be manipulated separately. The fitness of a string is related to the height of the packing pattern.

In order to preserve the knowledge that is stored in the subtrees a special crossover operator has been developed that does not separate the slicing tree structure and the nodes. Only subtrees with a certain packing density, that have no rectangle in common are transmitted to the offspring. After reducing the first parent to the subtrees to be inherited, the subtrees from the second parent are separately inserted into the new string together with a new cut-line. The offspring is completed by the insertion of single rectangles that are missing. In terms of mutation five different operators are applied randomly involving swapping of adjacent subtrees, inversions of cut-line and rectangle orientation as well as rotation of partial packing patterns by 90°. A hillclimbing strategy is implemented in the genetic algorithm aimed at improving the fitness of a recently mutated or recombined string. The solutions produced by the genetic algorithm are superior to those found by other heuristics including iterative, non-iterative algorithms as well as a random search strategy and a simulated annealing algorithm. Genetic algorithms and simulated annealing achieve significantly better results than the primitive heuristics, with the genetic algorithm being closer to the best known solution.

In order to reduce the complexity of the problem Kröger introduces the concept of meta-rectangles, which describe a group of adjacent, densely packed rectangles, that are summarised to one large rectangle. In this way partial layouts are frozen yet the shape is still flexible enough to be grouped with other rectangles. In terms of recombination the crossover operator has to ensure that the meta-rectangles are transmitted to the offspring. This produces a significant reduction in the run times and leads to an improvement in the average best solutions.

András et al. [21] propose a solution similar to Kröger's algorithm [14] for guillotineable cutting that is based on the tree representation, where each node is either further cut into two pieces or remains non-cut. In order to encode this problem a data structure has been developed with each node containing information about the dimensions of the piece, the position and orientation and if there is a cut. The fitness of the individuals is related to the packing density. A combined crossover - mutation operator exchanges subtrees between two parent strings. After the crossover it may be necessary to modify the offspring to guarantee feasible solutions, which adds a mutational component to the operation. Unfortunately, the quality of the solutions is not measured against another method so that the performance of the genetic algorithm cannot be evaluated.

Since the guillotineable packing problem is more constrained than the non-guillotineable one, the data structures developed in [14] and [21] implement these additional constraints rather than leaving them to the decoder. The solutions generated by the genetic algorithms part of the procedure are all feasible. With the aid of the meta-rectangle strategy proposed by Kröger [14], which allows storage of partial layout information, efficient subtrees can be identified easily and passed on to the offspring.

Packing of Regular Shapes other than Rectangles:

The only research on packing regular shapes other than rectangles by genetic algorithms has been carried out by George et al. [22]. A hybrid genetic algorithm is combined with a heuristic method to pack different-sized circles into a rectangular area. During the packing process, when any given set of circles has already been placed on the rectangle, so-called position numbers are used to indicate possible locations for the remaining circles.

The encoding technique of the genetic algorithm makes use of the position numbers, which are defined with respect to the sides of the object and the circles already placed. Instead of evaluating every possible position of a circle in the packing pattern, only an initial position is allocated to each circle, this serves as a default position and is only modified if it decodes to an infeasible packing configuration. The initial positions of all circles are stored in a string, with the first cell containing the position of the first circle etc. As a measure of fitness the density of the circles in the rectangle is used. The genetic operators applied are proportional selection and one-point crossover and mutation in order to generate new sequences of position numbers. The decoding procedure attempts to place a circle according to its initial position number in the string. If this position is not feasible or not defined, the position number is incremented until a feasible position is found, discarding the circle when no feasible location has been found. In this way all defined positions of a circle are examined.

The genetic algorithm is compared to heuristic methods using the same decoding procedure. The heuristic methods, including the genetic algorithm, only differ in the way the position numbers have been generated in the strings. They are stated in detail in [22]. The comparison also includes a heuristic method, in which the position numbers are generated randomly. Performance comparison for different problems types have shown that the techniques using randomisation, i.e. genetic algorithms and the random technique, outperformed the other heuristics, when a balance must be reached between quality and computational effort. The advantage of the data structure in [22] is that domain information is implemented in the genetic algorithm as part of the procedure. The task of the decoder is to check the feasibility of the layout.

3.2.2 Irregular Packing Problems

The category of irregular packing problems summarises the packing of polygons as well as arbitrary shapes. A number of researchers have approached the packing of polygons some including cavities, no references have been found to genetic algorithms implementing the packing of arbitrary shapes. In packing problems involving polygons it is important to distinguish between convex and concave shapes. A genetic algorithm has been developed for a packing problem involving convex polygons. The packing of sets of polygons, including concave ones, increases the complexity of the problem and has been approached by two researchers. A summary of the algorithms is given in Table 3.

Packing of Convex Polygons

Fujita et al. [23] propose a hybrid approach combining the genetic algorithm with a local minimisation algorithm to solve a nesting problem with convex polygons. The fitness of an individual is related to the waste produced by the corresponding packing pattern. The local minimisation algorithm is used to decode the permutations into the physical layout. This algorithm uses a Quasi-Newton method to manipulate the relative positions between the objects, which are defined by a set of variables. The computational example shows that the genetic algorithm converges on a feasible packing pattern. Since the performance of the hybrid genetic algorithm has not been compared to packing achieved by other methods, it is not possible to judge its efficiency. The computation time for the evaluation of the layout and the calculation of the overlaps is considered to be very large.

Packing Convex and Concave Polygons

Although the approaches of Ismail and Hon [24] [25] and Jakobs [15] involve approximation processes to transform irregular shapes into items with rectangular features, the procedures proposed are different. Ismail and Hon approximate an irregular figure by a set of smaller squares which results in a digitised image with orthogonal borders. Jakobs circumscribes the polygons with rectangles and has extended the genetic algorithm proposed for the packing of rectangles to convex and concave polygons. The polygons are first embedded in rectangles with a minimum area. After the genetic algorithm has produced a solution a shrinking-algorithm is applied to the packing pattern, attempting to move the polygons closer together and hence minimising gaps between them. In order to improve the shrinking-step, reflections of the original polygons are also tested. Comparisons between the BL-algorithm, and the genetic algorithm in combination with the deterministic shrinking algorithm, show that the solution obtained by the genetic algorithm can be improved by the additional shrinking step, since it allows utilisation of the space "wasted" by the embedding process. This shrinking step is especially advantageous in the case of concave features.

The main drawback of this approach is that the dense packing of the layout is left to the decoder incorporating orientation trials and shift procedures. Since only the packing sequence is implemented in the genetic algorithm, the main characteristics of the layout cannot be inherited by the offspring.

Ismail and Hon approximate an irregular figure by a set of smaller squares which result in a digitised image with orthogonal borders. In [24] a method is developed for the pairwise clustering of two identical convex and concave polygons, which is implemented in [25] and used within a genetic algorithm for the packing of a set of polygons. For pairwise clustering two techniques were developed, one involving a genetic algorithm. In both approaches the same technique is used for the description of the two-dimensional shapes. After circumscribing the shape with the minimum rectangular area a grid is superimposed converting the shape into a two-dimensional array, where solid and empty elements are numerically represented by 0 and 1. When clustering two shapes two parameters are used to describe their relative position to each other. Another four parameters are introduced to represent the mirroring of the shapes along the axis. These parameters are combined to a binary multi-parameter string, defining a clustering solution.

The fitness reflects the best orientation for maximising the material utilisation and includes a penalty for overlapping. Subsequent decoding of the string into a layout is straightforward. Comparing the performance to the deterministic method that has been developed by Ismail and Hon [24], the genetic algorithm is found to produce denser packing of figures with concave features due to the limitations of the other method, whereas for other shapes the solutions have been identical.

Ismail and Hon [25] extended the clustering method proposed in [24] to cover the nesting of dissimilar shapes in combination with a heuristic rule. Applying the above technique the individual shapes are first digitised and represented as a two-dimensional grid array. The working area, where they are going to be placed is also represented as a grid, so that the packing problem consists of arranging a number of digitised shapes. Additional to the two parameters, which describe the relative position of a shape to the others, three parameters define mirroring and rotation. The overall genetic string is a sequence of the encodings for each individual shape. This data structure can result on the genetic algorithm producing infeasible solutions.

The fitness function applied does not only reflect the packing density but also introduces a penalty for overlapping. The decoder uses a complex set of parameters and rules to describe the relative positions and the placement of the polygons. The task of the decoder is simply to check the feasibility of the layout imposing penalties for over-lapping. Penalties are a less efficient guide to the search compared to a decoding algorithm that avoids producing constrained results [13].

Table 2: Comparison of the genetic algorithms for 2D irregular packing problems

	Fujita et al. [23]	Jakobs [15]	Ismail et al. [24]	Ismail et al. [25]
Problem	convex polygons	convex, concave polygons	pairwise clustering of convex/ concave polygons	nesting of convex, concave polygons
Representation	permutation	permutation	multi-parameter string, binary	multi-parameter string, binary
Fitness	wastage in packing pattern	remaining area	density of packing, penalty	density of packing, penalty
Reproduction	truncation, linear scaling, elitist plan, expected value plan	proportional selection	proportional selection, elitist	proportional selection, elitist
Cross-over	order cross-over	order cross-over	one point crossover	one point crossover
Mutation	random removal and reinsertion in string	inversion; exchange of items; rotation	classic	classic
Decoding	local minimisation algo	BL-algorithm		set of rules

3.2   Three-dimensional Packing Problems

Unlike two-dimensional packing problems, three-dimensional ones deal mainly with regular and in most cases cuboid objects to be loaded onto a pallet or into a container. In the literature five problems of this type have been approached using genetic algorithms. An overview of the algorithms developed for regular 3D problems is given in Table 3. The complexity increases when irregular objects are to be placed. Only one approach to the packing of arbitrary shaped three-dimensional items by genetic algorithms has been made to-date.

3.2.1. Regular Packing Problems

Prosser [26] has developed two genetic algorithms for a highly constrained pallet loading problem The problem consists of loading stacks of plates onto a minimum number of pallets without violating geometrical and weight constraints. The main drawback of the first genetic algorithm is the evaluation function calculating the number of pallets used. Since it only produces few distinct values for densely packed pallets, there is a lack of guidance for the search process. Apart from that the evaluation is not computationally efficient.

To overcome these disadvantages, a hybrid genetic algorithm was developed in order to deliver one maximally loaded pallet. Loading one pallet after the other up to the weight limit reduces the number of possible loading patterns and results in a faster convergence on a solution. In addition to the weight, another constraint is the number of items on one pallet. In order to load one pallet up to the first violation of the constraints the fitness function has been related to the loaded pallet weight. The genetic algorithm is then applied to the reduced problem iteratively until the allocation of the stacks is concluded. For example if a maximum of six elements of a string can be loaded onto a pallet the inversion operator is modified such that elements are randomly swapped between the first six positions of a string and beyond. The computational effort of the evaluation is less than that required in the first genetic algorithm. The comparison of the performance of the two genetic algorithms and the branch-and-bound method shows that the second genetic algorithm produces better solutions in less time.

Corcoran and Wainwright [29] have proposed a hybrid genetic algorithm for a three-dimensional loading problem, where the bin consists of a single, open-ended bin. The algorithm is based on the so-called level technique, which places items level by level into a single, open-ended bin, whose height is determined by the highest object on that level. In extending this strategy to three dimensions the bin is divided into "slices" along the height and "levels" along the length. In order to evaluate the fitness of the packing pattern two heuristic algorithms are used. The Next Fit algorithm places an item into the next available position on the level without exceeding the width of the bin; if that constraint is violated a new level is started. The First Fit procedure, however, searches from the beginning until a suitable position is found. The performance comparison for various sets of items shows that the genetic algorithms produce better packing patterns than the two heuristics, which have equal packing efficiencies.

A similar approach to the level technique used by Corcoran and Wainwright [29] has been developed by Bortfeldt [28]. He proposes a genetic algorithm for a container-loading problem, where a set of boxes of different sizes is packed maximising space utilisation. Unlike other genetic algorithms that have been designed for packing problems this one does not use a heuristic method for the decoding the genotype during the evaluation stage. A heuristic algorithm is used to generate an initial population and in the crossover and mutation operator. It is based on a layer technique, similar to the heuristic method used by Corcoran and Wainwright. The depth of a layer is determined by the first box (layer defining box = LDB) in the layer, width and height are the same as the container dimension. A further procedure is implemented to fill the layer, where each allocated box generates three spare spaces inside the layer: beside, in front and above. The fill algorithm searches for the best fitting pair of boxes and places them into the layer.

In order to keep the decoding effort low a problem specific data structure has been developed. The chromosome representing a packing pattern includes the number of layers and for every layer the LDB, its rotation and the set of boxes allocated in that layer. The main advantage of this encoding technique is that the genotype of the individuals is sufficient for evaluation of the fitness. It guarantees the feasibility of the individuals generated by the genetic operators without the need for decoding. The concept of the crossover technique is to transfer the best layers of both parents. When no more layers can be taken and the offspring is still incomplete, new layers are generated by the heuristic algorithms described above. During the mutation stage the boxes with the lowest frequency of appearance in the new population are determined. New packing plans are then generated by the heuristic algorithm, which uses these boxes as the LDB of the first layer. The individuals generated in this way are exchanged for the ones with the lowest performance in the population. A performance comparison shows that the genetic algorithm produces a packing pattern with a higher space utilisation.

Lin et al. [27] take a different approach that only considers spatial constraints of the container dimensions. Their algorithm also takes into account additional constraints such as the weight of the boxes that are loaded into a container. A heuristic method enhanced by a simulated annealing algorithm is combined with a genetic algorithm. In order to take account of the weight constraint, i.e. placement of heavier boxes below lighter ones, the loading layout of the container is encoded as a multi-chromosome string, where each relative chromosome represents the loading pattern of one layer. Crossover and mutation only operate on the corresponding relative chromosomes without inter-chromosome exchanges. For the decoding of the strings into the layout a set of heuristic rules is used. In terms of solution quality the genetic algorithm outperforms the heuristic method enhanced with the simulated annealing algorithm the computational cost being dependent on the problem size.

Table 3: Comparison of the genetic algorithms for 3D regular packing problems

	Prosser [26]	Corcoran et al. [29]	Bortfeldt [28]	Lin et al. [27]
Problem	pallet loading with weight constraint	loading of open container	container loading; only spatial constraint	container loading with weight constraint
Representation	permutation	permutation	problem specific	multiple-chromosome
Fitness	weight of pallets	height		set of heuristic rules
Reproduction	truncation, rank, expected value plan		rank selection, elitist	prop. Selection, elitist
Crossover	order-based, adapted to loading constraint	PMX, Cycle, Order2; Rand1	problem specific	partially mapped crossover (PMX)
Mutation	exchanging elements from first 6 positions with rest	elements swapped and rotated;	problem specific	inversion
Decoding	set of constraints	First Fit, Next Fit		set of heuristic rules

3.2.2 Irregular Packing Problems

The concept of packing irregular items into a cylindrical container was introduced by Ikonen et al. [30]. Objects need to be packed utilising cavities of larger objects with no orientation restrictions. In order to represent the order and the orientation, in which the parts are packed a multi-chromosome representation is used. The first chromosome consists of a permutation, which encodes the packing order. The second one contains the orientation of each element. Linear order crossover is applied to rank selected parents. A swapping operator in the mutation stage changes the position of two parts in the chromosome and a mutation operator is applied to the second chromosome changing orientation values randomly. The geometry of the parts is included in the decoding procedure which translates the strings into their packing layout and evaluates their fitness. Packing density is related to the distance of each part from the origin. In order to reduce the probability of infeasible solutions an adaptive penalty function is incorporated in the fitness function. This penalty function is especially advantageous when it is not known in advance how difficult it is to find feasible solutions and how effectively the objective function differentiates between solutions. Only the underlying concepts for this genetic algorithm have been developed so far, results have yet to be published.

4. Conclusions

This review of two and three-dimensional packing problems shows that genetic algorithms have been successfully implemented in the solution of a large variety of optimisation problems. The solution space of combinatorial problems is enormous and increases with the complexity of the problem, in particular with the geometry of the items. Since most packing problems are NP-complete, heuristic algorithms are used. Deterministic methods cannot solve the problem in polynomial time. Among other heuristic methods genetic algorithms offer the ability to search large and complex solution spaces in a systematic and efficient way.

Convergence of high quality solutions depends upon the representation of the packing problem. It is important that the encoding technique, which describes possible packing patterns, allows us to recognise characteristic features of the packing schemes. It is also advantageous to design the data structure such that substructures of layouts are accessible and can be manipulated easily. For packing problems order-based chromosomes can be used to represent packing sequences. An appropriate modification of the data structure allows maintenance of certain efficient substructures of the layout. At the same time the genetic operators need to be adapted to the encoding technique, so that they support the inheritance of important layout features, which are meaningful and effective for the packing objective.

The evaluation function is closely related to the performance of a genetic algorithm, since it is used to guide the search process. It needs to describe the problem in such a way that it allows differentiation between the quality of the solutions. If the evaluation function is not appropriate the search cannot be directed and operates randomly.

The common feature underlying most of the genetic algorithms developed for packing problems is the two-stage approach. A genetic algorithm is used to explore and manipulate the solution space, whereas a second procedure is needed to evaluate the solutions generated. Therefore the phenotype needs to be constructed in order to check quality and feasibility of a packing scheme, this can be done either by a deterministic or a heuristic method.

The decoding method influences the computational effort required and limits the genetic algorithm, since it may not support the inheritance of certain features by the offspring. In order to reduce the genetic algorithm's dependency on the decoding method, it is desirable to find a data structure that allows us to calculate the fitness from the genotype rather than the decoded phenotype. Genetic algorithms perform best when they transmit a maximum amount of information from one generation to the next. Consequently, hybrid genetic algorithms should be used with caution as they often incorporate the knowledge within the decoding method rather than the genetic structure. In approaches that do not involve a decoding algorithm, the geometry of the figures must be considered in the data structure.

Much of the work presented in this review concentrates on the material utilisation as the main criteria for the packing process. However, in reality packing problems are more constrained (Table 4).

Table 4: Future developments for GAs in cutting and packing

Problem	Explanation	Example
Cutting costs	cutting is assumed to be free, this is not the case	steel can be cut on per ton basis or per unit length of cut; a pattern may be split into objects to be cut per ton or per unit length
Irregular objects	packing into irregular objects	e.g. leather industry: shape of object is not regular; packing into objects that have already fixed items within them
Material uniformity	penalise use of low standard sections	e.g. leather industry: packing of non homogenous sections of the object can be penalised depending on the item to be packed
Un/ Packing order	ease of construction	e.g. loading of vehicles: order of packing is decisive for the unpacking of the vehicle, which depends on the future destination of the items
Item movement	pattern might not be physically realisable	particularly the realisation of 3D patterns might no be possible due to path planning constraints
Item location	location can depend on additional criteria	weight and size constraints limit possible locations in the object, e.g. in 3D container loading problems
Split items	split items to improve packing	e.g. metal industry: where feasible, large items can be split and welded after cutting to improve the packing
Complex cost of waste	cost based on use of 'waste' material	trim loss of one object could be sold on; cost will depend on shape and use of 'waste'
Time cost	termination depend on benefit/ time in nesting	due to scheduling problem and deadlines real time may become part of evaluation
variable cost of bin	selection of best 'set' of objects for packing	e.g. ship building industry: some items in the order list can be cut from different objects, i.e. steel thickness, area

The strengths of a genetic algorithm approach lie in its ability to search large and complex solution spaces in a systematic and efficient way. In addition the fact that it is not dependent on a particular problem structure allows the user to utilise different methods for the decoding of the genotype. A key weakness behind much of the work is a lack of comparison with known benchmark problems and algorithms. As work in the area of genetic algorithms and packing advances we can expect more sophisticated constraints to be included and improved comparative studies of the more general problems. Table 4 shows just some of the future possibilities.

Bibliography

[1] Sedgewick R, 1992, Algorithms in C++, Addison-Wesley, Reading.

[2] Reeves C., 1993, Modern Heuristics for Computational Problems, Basil Blackwell, Oxford.

[3] van Laarhoeven P. J. M., Aarts, E. H., 1987, Simulated Annealing: Theory and Applications, D. Reidel Publishing Company, Dordrecht.

[4] Dowsland K. A., Dowsland W. B., 1992, Packing problems, European Journal of OR, vol. 56, pp. 2-14.

[5] Dowsland K. A., Dowsland W. B., 1995, Solution approaches to irregular nesting problems, European Journal of Operational Research, vol. 84, no. 3, pp. 506-521.

[6] Dyckhoff H., 1990, Typology of cutting and packing problems, Eur. J. of OR, vol. 44, no. 2, pp. 145-159.

[7] Hässler R. W., Sweeney, P. E., 1991, Cutting stock problems and solution procedures, European Journal of Operational Research, vol. 54, part 2, pp. 141-150.

[8] Hinxman A. I., 1980, The trim loss and assortment problems, Eur. J. of OR, vol. 5, part 1, pp. 8-18.

[9] Davis L, 1991, Handbook of Genetic Algorithms, Van Nostrand Reinhold, New York.

[10] Goldberg D. E., 1989, Genetic Algorithms in Search, Optimisation and Machine Learning, Addison-Wesley Publishing Company, Reading.

[11] Mitchell M., 1996, An introduction to genetic algorithms, MIT Press, Massachusetts.

[12] Smith D., 1985, Bin-packing with adaptive search, in: Grefenstette (ed.), Proceedings of an International Conference on Genetic Algorithms and their Applications, Lawrence Erlbaum , pp. 202-206.

[13] Davis L, 1985, Applying adaptive search algorithms to epistatic domains, Proceedings of the 9th Int. Joint Conference on Artificial Intelligence, Los Angeles, pp. 162-164.

[14] Kröger B., 1995, Guillontineable bin-packing: A genetic approach, Eur. J. of OR, vol. 84, pp. 645-661.

[15] Jakobs S, 1996, On genetic algorithms for the packing of polygons, Eur. J. of OR, vol. 88, pp. 165-181.

[16] Kröger B., Schwenderling P., Vornberger O., 1991, Parallel genetic packing of rectangles, in: Parallel Problem Solving from Nature 1st Workshop, Springer Verlag, pp. 160-164.

[17] Kröger B., Schwenderling P., Vornberger O., 1991, Genetic packing of rectangles on transputers, in: P. Welch (ed.), Transputing, part 2, IOS Press, Amsterdam pp. 593-608.

[18] Kröger B., Schwenderling P., Vornberger O., 1993, Parallel genetic packing on transputers, in: J. Stender (ed.), Parallel Genetic Algorithms: Theory and Applications, IOS Press, Amsterdam, pp. 151-185.

[19] Falkenauer E., Delchambre A., 1992, A genetic algorithm for bin-packing and line balancing, Proceedings of the 1992 IEEE International Conference on Robotics and Automation, Nice, Fr, vol. 2, 1186-1192.

[20] Syswerda D., 1991, Schedule optimization using genetic algorithms; in: Davis (ed.); Handbook of Genetic Algorithms; Van Nostrand Reinhold, New York.

[21] András P., András, A., Zsuzsa, S., 1996, A genetic solution for the cutting stock problem, Proceedings of the First On-line Workshop on Soft Computing, Aug. 1996, Nagoya University, pp. 87-92.

[22] George J. A., George, J. M., Lamar, B. W., 1995, Packing different-sized circles into a rectangular container, European Journal of Operational Research, vol. 84, pp. 693-712.

[23] Fujita K., Akagji, S., Kirokawa, N., 1993, Hybrid approach for optimal nesting using a genetic algorithm and a local minimisation algorithm, Proceedings of the 19th Annual ASME Design Automation Conference, Part 1 (of 2), Albuquerque, NM, USA, vol. 65, part 1, pp. 477-484.

[24] Ismail H. S., Hon K. K. B., 1992, New approaches for the nesting of two-dimensional shapes for press tool design, International Journal of Production Research, vol. 30, part 4, pp. 825-837.

[25] Ismail H. S., Hon K. K. B., 1995, Nesting of two-dimensional shapes using genetic algorithms, Proceedings of the Institution of Mechanical Engineers, Part B, vol. 209, pp. 115-124.

[26] Prosser P., 1988, A hybrid genetic algorithm for pallet loading, in: B. Radig (ed.), ECAI 88 Proceedings 8th European Conference on Artificial Intelligence, Pitman, London pp. 159-164.

[27] Lin J. L., Foote B., Pulat S., Chang C. H., Cheung, J. Y., 1993, Hybrid genetic algorithm for container packing in three dimensions, Proc. of the 9th Conf. on Artificial Intelligence for Applications, pp. 353-359.

[28] Bortfeldt A., 1994, A genetic algorithm for the container loading problem, in Rayward-Smith (ed.): Proceedings of the Unicom Seminar on Adaptive Computing and Information Processing, pp. 749-757.

[29] Corcoran A. L., Wainwright R. L., 1992, Genetic algorithm for packing in three dimensions, Proceedings of the 1992 ACM/SIGAPP Symposium on Applied Computing SAC '92, Kansas City, pp. 1021-1030.

[30] Ikonen I., Biles W. E., Kumar A., Ragade R. K., 1996, Concept for a genetic algorithm for packing 3D objects of complex shape, Proc. 1st Online Workshop on Soft Computing, Nagoya University, pp. 211-215.

Generic Evolutionary Design

P. J. Bentley[1] and J. P. Wakefield[2]

[1]*Department of Computer Science, University College London,*
Gower St., London WC1E 6BT, UK.
Tel. 0171 391 1329 P.Bentley@cs.ucl.ac.uk (corresponding author)
[2]*Division of Computing and Control Systems, School of Engineering*
University of Huddersfield, Huddersfield HD1 3DH, UK.
Tel. 01484 472107 J.P.Wakefield@hud.ac.uk

Keywords: generic evolutionary design, automated design, genetic algorithms

Abstract

Generic evolutionary design means the creation of a range of different designs by evolution. This paper introduces generic evolutionary design by a computer, describing a system capable of the evolution of a wide range of solid object designs from scratch, using a genetic algorithm.

The paper reviews relevant literature, and outlines a number of advances necessitated by the development of the system, including: a new generic representation of solid objects, a new multiobjective fitness ranking method, and variable-length chromosomes. A library of modular evaluation software is also described, which allows a user to define new design problems quickly and easily by picking combinations of modules to guide the evolution of designs.

Finally, the feasibility of generic evolutionary design by a computer is demonstrated by presenting the successful evolution of both conventional and unconventional designs for a range of different solid-object design tasks, e.g. tables, heatsinks, prisms, boat hulls, aerodynamic cars.

1. Introduction

Evolution is one of the most powerful search processes ever discovered [8]. In the natural world, evolution has created an unimaginably diverse range of designs of greater complexity than mankind could ever hope to achieve [6]. Natural evolution is the ultimate in generic evolutionary design.

In recent years, researchers have begun using computer algorithms that mimic this process of evolution, in order to automate stages of the human design process [10]. Designs have been successfully optimised using *genetic algorithms* (GAs), with some remarkable results [9]. More recently, simple conceptual designs have been generated by evolutionary computation [11]. However, as yet, evolution has not been used to perform the entire design process.

It is evident that natural evolution is eminently capable of generating new conceptual designs from scratch, evaluating these designs and optimising them. We are all living proof of this. Consequently, if evolutionary computation techniques of suitable similarity are used to perform the whole process of design, it seems probable that a computer system could become capable of equally creative and diverse design.

For this work it was decided that a computer system should be created that was capable of evolving the shape of a range of different solid-object designs from scratch. Such a system would demonstrate the feasibility of using evolution to perform the entire design process, and would show the creative potential of generic evolutionary design by a computer.

This paper briefly describes the investigation into and the creation of such a generic evolutionary design system. In addition, a range of designs successfully evolved by the system are presented and discussed.

2. Background

There are five separate but related areas of research relevant to the subject of generic evolutionary design by a computer:

(i)   The optimisation of existing designs.
(ii) The generic optimisation of designs.
(iii) The creation of designs by computers.
(iv) The creation of art by computers.
(v)   Genetic algorithms.

2.1. The Optimisation of Existing Designs

The development of non-generic optimisation systems, capable of optimising selected parts of existing designs, is a common area of research [10]. Numerous examples of the optimisation of designs exist, many using GAs or other adaptive search methods. For example, computers have been used to optimise: oil-pump pipelines, floorplans, structural topologies, finite impulse response digital filters, microwave absorbing materials, hydraulic networks, aircraft landing struts, VLSI layouts, and even spacecraft systems [8,9,10].

Some of these systems have been used with considerable success to optimise real-world problems. For example, as described by Holland, a design of a high-bypass jet engine turbine was typically optimised in eight weeks by an engineer; the genetic algorithm optimised a design in only two days, "with three times the improvements of the manual version" [9].

Whilst the wide variety of applications being tackled shows that computers can be used to successfully evaluate and optimise many different types of design, every one of these optimisation systems, without exception, suffers from two major drawbacks. Firstly, every one can only optimise existing designs - it would be quite impossible to use any of them to create a new design. Secondly, every one is application-specific - they can only optimise the single type of design they were created to optimise, and no others.

2.2. The Generic Optimisation of Designs

Generic design optimisation (i.e. the optimisation of more than one type of design by a single system) is a rare subject for research. As described in detail by Pham, Bouchard's 'Engineer's Associate' provides a limited generic framework to work with systems that can be represented by equations [11]. Alternatively, Culley's 'GPOS' (general purpose optimisation system) consists of a toolbox of optimisation algorithms, capable of optimising a range of different applications (once interfaced appropriately) [5].

However, Tong's 'Engineous' [14] is perhaps the most successful generic system, having been demonstrated on over twenty design optimisation tasks, including the optimisation of: turbine blades, cooling fans, DC motors, power supplies and a nuclear fuel lattice. A large portion of the system is comprised of complex interfacing software to allow the use of existing design evaluation packages. The system relies heavily on expert systems containing much application-specific knowledge to guide the evolution of a GA. Tong claims that "the current version of Engineous has demonstrated the profound impact such a system can have on productivity and performance" [14].

2.3. Creation of Designs by Computers

Research into the subject of design *creation* (typically concentrating only on conceptual designs) is growing. Early work concentrated on cognitive simulations, i.e. attempting to make a computer 'think' in the same way as a human, when designing. Such systems attempted to create descriptions of designs at an abstract level, typically using an expert system to 'design'. For example Dyer's 'EDISON' [7] represented simple mechanical devices such as doors and can-openers symbolically in terms of five components: parts, spatial relationships, connectivity, functionality and processes. A combination of planning and invention using 'generalisation', 'analogy' and 'mutation' attempted to modify these components to fulfil the design specification. Unfortunately, it seems that the abstract level at which reasoning was performed was too low, so the system was unable to handle any problems apart from the simplest cases [11]. Another approach consisted of invention based on 'visualising potential interactions' [15]. This generated descriptions of designs in terms of high-level components and the interactions between them, using qualitative reasoning and quantitative algebra. Again, the proposed system could only deal with highly simplified designs [11].

Many creative design systems reduce the complexity of the problem by presenting the computer with a number of high-level design building blocks which must be ordered correctly to form a design. For example, Pham describes a "preliminary design system" known as TRADES (TRAnsmission DESigner) [11]. When given the type of input (e.g. rotary motion) and the desired output (e.g. perpendicular linear motion), the system generates a suitable transmission system to convert the input into the output. This GA is presented with a set of building blocks such as rack and pinion, worm gear, and belt drive.

Perhaps the work that can most accurately be described as creative design is the recent work of Rosenman, who attempts to evolve new floorplans for houses [12]. Two dimensional plans are 'grown' using a simplified GA to modify 'cells' organised hierarchically using grammar rules. However, the system requires much problem-specific knowledge and the elaborate representation used may actually prevent complex shapes from being formed.

2.4. Creation of Art by Computers

The use of computers to create art (usually with GAs and similar adaptive search algorithms) is growing in popularity amongst some artists. For example, Todd and Latham have successfully evolved many three dimensional 'artistic' images and animations [13]. Their two-part system, consisting of 'Form Grow' and 'Mutator' uses an evolutionary strategy which creates and modifies shapes composed of 'artistic' primitive shapes (e.g. spiral, sphere, torus). John Mount showed his 'Interactive Genetic Art' on the internet (at http://robocop.modmath.cs.cmu.edu.8001/). This work utilised a GA to modify fractal equations that defined two dimensional images. Additionally, the biologist Richard Dawkins has demonstrated the ability of computers to evolve shapes resembling those found in nature [6]. Using a simple evolutionary strategy that modifies shapes arranged in tree-structures, he has produced images resembling the shapes of life-forms, e.g. 'spiders', 'beetles', and 'flowers'.

However, all of these art creation systems require the images being evolved to be evaluated by a human (i.e. artificial selection). Moreover, despite the fact that some of these systems can produce some complex three dimensional shapes [13], none of them have been used to produce anything more than 'pretty pictures'.

The system described in this paper combines for the first time the creative evolutionary techniques pioneered by artists (and biologists) with the more rigorous methods of automatic creative design. This has resulted in a novel generic design system which has the 'creative properties' of the art systems and is capable of the generation of a wide range of useful designs [1]. Furthermore, it is the 'innovative flair' (Goldberg, 1989) of the genetic algorithm that gives the system such capabilities.

2.5. The Genetic Algorithm

Perhaps uniquely for one type of search algorithm, the genetic algorithm has become widely used in all of the areas of research related to generic evolutionary design. In these and many other domains, the GA has been shown repeatedly to be a highly flexible algorithm, capable of finding good solutions to a wide variety of problems [8].

The GA is based upon the process of evolution in nature. Evolution acts through large populations of creatures which individually reproduce to generate new offspring that inherit some features of their parents (because of random *crossover* in the inherited chromosomes) and have some entirely new features (because of random *mutation*). Natural selection (the weakest creatures die, or at least do not reproduce as successfully as the stronger creatures) ensures that more successful creatures are generated each generation than less successful ones. In nature, evolution has produced some astonishingly varied, yet highly successful forms of life. These creatures can be thought of as good 'solutions' to the problem of life. In other words, evolution optimises creatures for the problem of life.

In the same way, within a genetic algorithm a population of solutions to the problem is maintained, with the 'fittest' solutions (those that solve the problem best) being randomly picked for 'reproduction' every generation. 'Offspring' are then generated from these fit parents using random crossover and mutation operators, resulting in a new population of fitter solutions [8]. As in nature, the GA manipulates a coded form of the parameters to be optimised, known as the *genotype*. When decoded, a genotype corresponds to a solution to the problem, known as a *phenotype*.

The *robustness* of GAs [9], combined with the fact that the human design process has been formally compared to the working of a GA, and that the GA is the closest analogy to natural evolution in Computer Science, make the choice of a GA in a generic evolutionary design system seem wholly justified [1].

3. The Generic Evolutionary Design System

When applying a genetic algorithm to any new application, four main elements must be considered. First, the phenotype must be specified, i.e. the allowable solutions to the problem must be defined by the specification and enumeration of a search space. Second, the genotype (or coding of the allowable solutions) must be defined. Third, the type of genetic algorithm most suitable for the problem must be determined. Fourth, the fitness function must be created, in order to allow the evaluation of potential solutions of the problem for the GA.

Since a genetic algorithm was used to form the core of the generic evolutionary design system, these four elements can be identified in the system. Designs are searched for using a multiobjective genetic algorithm as the 'search-engine' to evolve solutions. To achieve this, the GA manipulates hierarchically organised genotypes (or coded solutions). The genotypes are mapped to phenotypes (or designs) defined by a low-parameter spatial-partitioning representation. These phenotypes are analysed by modular evaluation software, which provides the GA with multiple fitness values for each design. Figure 1 illustrates how these four elements are combined to allow the evolution of a range of different solid object designs from scratch.

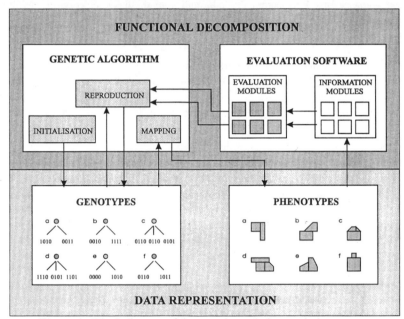

Figure 1. Block diagram of the generic evolutionary design system.

3.1. Phenotypes

Evolving designs, or phenotypes, from scratch rather than optimising existing designs requires a very different approach to the representation of designs. When optimising an existing design, only selected parameters need have their values optimised (e.g. for a jet-turbine blade, such parameters could define the length and cross-sectional area at specific parts of the blade). To allow a GA to create a new design, the GA must be able to modify more than a small selected part of that design - it must be able to modify every part of the design. This means that a design representation is required, which is suitable for manipulation by GAs. Many possible representations exist, and some have been used by the evolutionary-art systems: Todd and Latham [13] used a variant of constructive solid geometry (CSG), others have used fractal equations (e.g. John Mount), and Dawkins used tree-like structures [6]. However, for a system capable of designing a wide variety of different solid object designs, a more generic representation is needed.

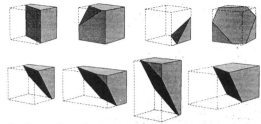

Figure 2. Examples of primitive shapes used to represent designs.

After some investigation, a new variant of spatial-partitioning representation (known as 'Clipped Stretched Cuboids'), was developed for this work. This representation combines methods from CSG and traditional spatial partitioning representations, to allow the definition of a wide range of solid objects using a number of primitive shapes in combination [3]. Primitive shapes consist of a rectangular block or cuboid with variable width, height and depth, and variable three dimensional position. Every cuboid can also be intersected by a plane of variable orientation (see fig. 2), to allow the approximation of curved surfaces. Intersected cuboids, or primitives, require nine parameters to fully define their geometry. Designs are defined by a number of non-overlapping primitives.

This solid-object representation is capable of the definition of a wide range of solid objects using relatively few primitives to partition the space. Significantly, the fewer the primitives in a design, the fewer the number of parameters that need to be considered by the GA. Additionally, this representation enumerates the search-space such that similar designs are placed close to each other, minimising discontinuities, and easing the task of finding an evolutionary path from a poor design to a better design [1].

3.2. Genotypes

The genetic algorithm within the system never directly manipulates phenotypes. Only coded designs, or genotypes are actually modified by the genetic operators of the GA. Every genotype consists of a single chromosome arranged in a hierarchy consisting of multiple blocks of nine genes, each gene being defined by sixteen bits, see fig. 3.4. This arrangement corresponds to the spatial partitioning representation used to define the phenotypes, with each block of genes being a coded primitive shape and each gene being a coded parameter.

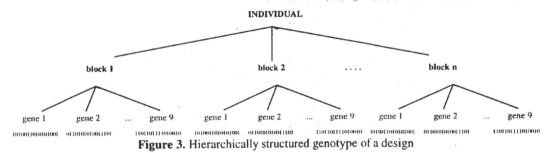

Figure 3. Hierarchically structured genotype of a design

A mutation operator is used within the genetic algorithm to vary the number of primitives in a design by adding or removing new blocks of nine genes from chromosomes. This permits evolution to optimise the number of primitives in addition to the geometries of primitives in designs. (A new primitive is added to a design by splitting a randomly chosen primitive into two. A primitive is removed by simply deleting that primitive from the genotype). However, varying the length of chromosomes in this way can cause the crossover operator to produce meaningless offspring [4]. To overcome this, a new type of crossover operator, known as hierarchical crossover, was developed. This new version of crossover uses the hierarchical arrangement of the chromosomes to find points of similarity between two chromosomes of different sizes. Once such points are found, hierarchical crossover uses the tree-structure of the chromosomes to efficiently generate new offspring without loss of meaning [4].

Hierarchical crossover is used by the GA to generate all offspring (i.e., with a probability of 1.0). Mutation is used to vary the number of primitives in a design with a default probability of 0.01 and a standard mutation operator is used to vary single bits within genes with a default probability of 0.001.

3.3. Genetic Algorithm

The genetic algorithm used within the system is more advanced than Goldberg's simple GA [8]. For example, two populations of solutions are maintained: the main *external population*, and the smaller *internal population*. All new solutions are held in the internal population where they are evaluated. They are then moved into the external population (i.e. 'born' into the 'real world'), replacing only the weakest members of the external population. Other different features include the use of an explicit mapping stage between genotypes and phenotypes, and the use of multiobjective techniques within the GA.

To begin with, the GA has the internal population of solutions initialised with random values to allow the evolution of designs from scratch (i.e., the GA begins with randomly shaped 'blobs'). However, if required, a combination of random values and user-specified values can be used to allow the evolution of pre-defined components of designs, or of selected parts of designs.

The GA then uses an explicit mapping stage to map the genotypes to the phenotypes. This resembles nature, i.e., the DNA of an organism is never 'evaluated' directly; first the phenotype must be grown from the 'instructions' given in the DNA, then the phenotype is evaluated [6]. By performing this process explicitly, the system is able to gain some advantages. For example, should a symmetrical design be required, only half a design needs to be coded in the genotype and hence evolved by the GA. This partial design can then be reflected during the mapping stage to form a complete design, which is then evaluated. This mapping stage is also used to enforce the rules of the solid-object representation, by ensuring that any designs with overlapping primitives are corrected so that their primitives touch rather than overlap [2].

Next, the GA calls user-specified modules of evaluation software to analyse the phenotypes and obtain multiple fitness values for each individual solution (most design problems are multiobjective problems). The GA must then determine from these multiple fitness values which phenotypes are fitter overall than others. In other words, the GA has to be able to place the phenotypes into order of overall fitness, using multiobjective optimisation techniques to handle the many separate fitness values produced by the evaluation software.

After performing comparisons between the performances of existing and new multiobjective ranking techniques, it was found that one of the new methods developed for this work allowed the GA to evolve the best designs most consistently. This multiobjective method automatically scales the separate fitness values of each phenotype, according to the effective ranges of the corresponding functions, in order to make them commensurable [1]. The fitness scores are then simply summed to provide a single, overall fitness value for each phenotype. In addition, by multiplying each scaled fitness value by a user-defined weighting value before aggregation, the new method also incorporates the concept of 'importance', allowing a user to increase or decrease the relative importance of any objective [1].

Once overall fitness values have been calculated for each individual solution, the GA moves the individuals from the internal population where all new individuals are held, into the main external population. However, unlike the simple GA, this GA does not replace an entire population of individuals with new individuals every generation. In a similar way to the steady-state GA, this GA only replaces the weakest (less fit) individuals in the external population with new individuals from the smaller internal population, allowing the fittest individuals to remain in the external population over multiple generations. Unusually, the GA also prevents very fit individuals from becoming immortal by giving every individual in the external population a pre-defined lifespan. Once the individual reaches this lifespan, they become very unfit and thus are quickly 'killed' by new individuals taking their places. This prevents poor individuals with high scores, caused by the random variations of noisy evaluation functions, from corrupting evolution [1].

Finally, the GA favours individuals with higher overall fitnesses when picking 'parents' from the external population. The randomly chosen parent solutions (with fitter solutions preferentially selected) are then used to generate a new internal population of offspring using hierarchical crossover and the mutation operators.

The GA then maps the new genotypes to the phenotypes, evaluates the new phenotypes, and continues the same process as before. This iterative process continues until either a specified number of generations (i.e. loops) have passed, or until an acceptable solution has emerged.

3.4. Evaluation Software

All parts of the system described so far are generic, i.e. they can be applied to a wide range of different solid-object design problems. However, there is an element of the system that must inevitably be specific to individual design applications: the evaluation software. Designs must be evaluated to instruct the GA how fit they are, i.e.

how well they perform the desired function described in the design specification. Hence, the evaluation software is a software version of the design specification, which must be changed for every new design task.

In an attempt to reduce the time needed to create evaluation software for a new design problem, all parts of the various different types of evaluation software created for this work have been implemented as re-usable modules. In other words, it is proposed that many designs can be specified by using a number of existing evaluation modules in combination. Moreover, wholly new design tasks will only require the creation of modules of evaluation software that do not already exist, thus dramatically shortening the time needed to apply the system to a new application. Over time a large library of such modules could be developed, to reduce the future need for new modules. Examples of the existing modules in the library of evaluation software developed as part of this project include: *minimum size, maximum size, specific mass, specific surface area, stability, supportiveness, ray-tracing*, and *particle-flow simulator*.

In addition to a library of different evaluation software modules (or fitness functions), a library of phenotype information modules is maintained. This is necessary because many modules of evaluation software require specific information about a design in order to calculate how fit that design is. Using a distinct information module to calculate, say, the mass of a design, allows all evaluation modules that need this value to share the information generated. Hence, such information on phenotypes need only be generated once, to supply all evaluation modules that require it. Examples of the information modules in the library developed as part of this project include: *vertices, mass, centre of mass, extents, primitive extents* and *surface area*.

Consequently, complete design applications are specified to the evolutionary design system by the selection of a combination of modules of evaluation software, and their corresponding desired parameter values. The system then enables the appropriate information modules which supply all of the evaluation modules with the necessary information on the current phenotype. A number of separate fitness values are generated by the evaluation modules for each design, which are used by the GA to guide evolution to good solutions.

4. Designs Evolved by the System

In total, fifteen different design tasks were presented to the system: tables, sets of steps, heatsinks, optical prisms (right-angle, roof, derotating, rhomboid, penta, abbe, porro), and streamlined designs (train fronts, boat bows, boat hulls, saloon cars, sports cars). Each of these tasks involved the evolution of a design with an entirely different shape, in order to allow that design to perform the desired function. Despite some of these problems being deceptive for the GA, this generic system was able to evolve not only fit, but acceptable designs (as judged by humans) for all fifteen problems [1]. Most designs took around 500 generations to evolve, using internal and external population sizes of 160 and 200 respectively.

The first task was to evolve the design of a table. This was specified by using five evaluation modules: *size, mass, flat upper surface, supportiveness* and *unfragmented*. These defined that good table designs should be an appropriate size and mass, should have a flat upper surface capable of supporting heavy objects without the table toppling over, and that the design should be whole (i.e. no part should 'float free' of the main design). Figure 4 shows an evolved design which uses four legs to provide the required stability. Alternative solutions have used other concepts such as a single wide base or a very low centre of mass to provide stability [2].

The second task was to evolve the design of a small set of steps, specified with similar modules of evaluation software as used for the table problem, except that three flat surfaces at specified heights were desired. Figure 5 shows an intricate evolved design using two side supports for stability, with the top step being further supported by a column at the rear. Behind the steps the design is hollow to reduce the mass.

Figure 4 Evolved Table

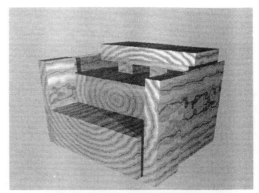

Figure 5. Evolved set of steps

Figure 6 Evolved heatsink

Figure 7. Evolved porro prism

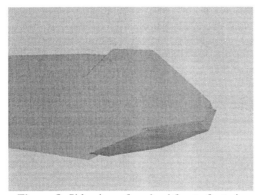

Figure 8. Side view of evolved front of a train

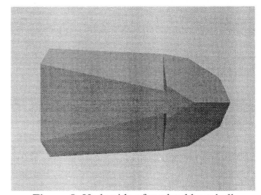

Figure 9. Underside of evolved boat hull

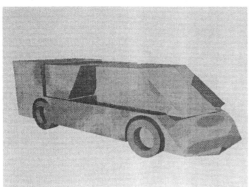

Figure 10. Evolved saloon car (wheels added)

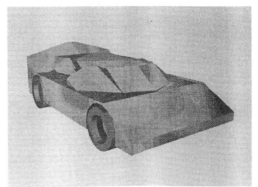

Figure 11. Evolved sports car (wheels added)

The third task was to evolve designs of heatsinks (to dissipate the heat of CPUs). This task was specified using the evaluation modules: *size, mass, unfragmented,* and *surface area.* A very high value for the surface area was desired (to define, in effect, that the surface area, and hence the approximate ability of the heatsink to radiate heat, should be maximized). A number of unusual designs were evolved, with the system often dramatically increasing the number of primitives used to represent each design in order to create detailed and uneven surfaces with large surface areas. Figure 6 shows a more traditional-looking evolved heatsink design.

The fourth type of problem was to evolve a number of different types of optical prism. All of these problems were specified using the evaluation modules: *size, unfragmented,* and *ray-tracing.* The ray-tracing module was used to define the characteristics of light travelling into the prisms, and evaluate how well the characteristics of the emerging light matched the desired characteristics for each type of prism. Figure 7 shows an almost perfect porro prism evolved by the system (for this problem using a collection of initially randomly-positioned, previously-evolved right-angle prisms).

The fifth type of problem was to evolve a number of different types of 'streamlined' designs. These were specified using the evaluation modules *size, unfragmented,* and *particle-flow simulator.* The 'particle-flow simulator' was used to provide an approximation of water and air-flow by firing particles at designs and calculating the forces generated when collisions with the designs occurred. By defining that minimal forces on the front of designs were required, designs with low water or air-resistance were specified. Figures 8, 9, 10, and 11 show successfully evolved 'streamlined' designs of a train, boat hull, saloon car and sports car, respectively. As can be clearly seen, the system has independently discovered a variety of techniques to allow the desired forces to be generated on each design. For example, fig. 8 shows a design of a train with a pointed nose, shaped to minimise wind resistance and generate an overall down-force. Figure 9 shows the underside of a boat hull, angled to cut through water cleanly and provide the required amount of up-force. Figure 10 shows a design of a saloon car, evolved around a fixed chassis, which uses a sloping windscreen and bonnet (windshield and hood) to reduce wind resistance and generate the desired down-force. Finally, fig. 11 shows an evolved sports car that has a sloped bonnet, curved windscreen and large back spoiler which all work in unison to generate the required amount of force pushing down on each wheel, whilst minimising the air-resistance.

5. Conclusions

This paper has presented a new way of using computers in design. It has shown that it is possible, feasible and useful to produce a generic evolutionary design system capable of successfully creating a range of new and original solid-object designs.

This novel computer system has four main components:

- A new low-parameter spatial-partitioning representation, used to define the shape of solid-object designs.
- Hierarchically structured genotypes combined with a new hierarchical crossover operator, which allow child designs to be efficiently generated from parent designs with different sized genotypes without loss of meaning.
- A steady-state multiobjective genetic algorithm, using an explicit mapping stage between genotypes and phenotypes, preferential selection of parents and a life-span operator, which forms the main search-engine at the core of the system.
- Modular evaluation software, which is used to guide evolution to functionally acceptable designs, with new design tasks being quickly specified by the user picking combinations of existing evaluation modules from a library.

The research described in this paper has demonstrated the use of a computer to perform generic evolutionary design by evolving consistently acceptable designs for fifteen different design tasks. Designs evolved by the system were based on sound conceptual ideas, 'discovered' independently by the system. The shapes of designs were optimised in order to ensure that they performed the desired function accurately. The system evolved a range of conventional and unconventional designs for all problems presented to it.

In conclusion, evolutionary design has been performed in nature for millennia. This research has made the first steps towards harnessing the power of natural evolutionary design, by demonstrating that it is possible to use a genetic algorithm to evolve designs from scratch, such that they are optimised to perform a desired function, without any human intervention.

References

[1] Bentley, P. J., 1996, *Generic Evolutionary Design of Solid Objects using a Genetic Algorithm.* Ph.D. Thesis, University of Huddersfield, Huddersfield, UK.

[2] Bentley, P. J. & Wakefield, J. P., 1996a, The Evolution of Solid Object Designs using Genetic Algorithms. *Modern Heuristic Search Methods,* John Wiley & Sons Inc., **Ch 12, 197-211**.

[3] Bentley, P. J. & Wakefield, J. P., 1996b, Generic Representation of Solid Geometry for Genetic Search. *Microcomputers in Civil Engineering* **11:3, 153-161**.

[4] Bentley, P. J. & Wakefield, J. P., 1996c, Hierarchical Crossover in Genetic Algorithms. *Proceedings of the 1st On-line Workshop on Soft Computing* (WSC1), Nagoya University, Japan, **37-42**.

[5] Culley, S. J. and Wallace, A. P., 1994, Optimum Design of Assemblies with Standard Components. *Proc. of Adaptive Computing in Engineering Design and Control - '94*, Plymouth, **163-168**.

[6] Dawkins, R. 1986, *The Blind Watchmaker*, Longman Scientific & Technical Pub.

[7] Dyer, M. Flower, M. and Hodges, J., 1986, 'EDISON': an engineering design invention system operating naively. *Artificial Intelligence* **1, 36-44**.

[8] Goldberg, D. E., 1989, *Genetic Algorithms in Search, Optimization & Machine Learning*, Addison-Wesley.

[9] Holland, J. H., 1992, Genetic Algorithms. *Scientific American*, **66-72**.

[10] Parmee, I C & Denham, M J, 1994, The Integration of Adaptive Search Techniques with Current Engineering Design Practice. *Proc. of Adaptive Computing in Engineering Design and Control -'94*, Plymouth, **1-13**.

[11] Pham, D. T. & Yang, Y., 1993, A genetic algorithm based preliminary design system. *Journal of Automobile Engineers* **v207:D2, 127-133**.

[12] Rosenman, M. A., 1996, A Growth Model for Form Generation Using a Hierarchical Evolutionary Approach. *Microcomputers in Civil Engineering* **11:3, 163-174**.

[13] Todd, S. & Latham, W., 1992, *Evolutionary Art and Computers*, Academic Press.

[14] Tong, S.S., 1992, Integration of symbolic and numerical methods for optimizing complex engineering systems. *IFIP Transactions (Computer Science and Technology)* **vA-2, 3-20**.

[15] Williams, B. C., 1990, Visualising potential interactions: constructing novel devices from first priciples. *Proc. of Eighth National Conference on Artificial Intelligence (AAAI-90)*, Boston, Mass.

Evolving Digital Logic Circuits on Xilinx 6000 Family FPGAs

T. C. Fogarty[1], J. F. Miller[1], P. Thomson[1]

[1]*Department of Computer Studies*
Napier University,
219 Colinton Road, Edinburgh

t.fogarty@dcs.napier.ac.uk
j.miller@dcs.napier.ac.uk
p.thomson@dcs.napier.ac.uk

Keywords: Evolutionary electronics, evolvable hardware

Abstract

This paper describes work which attempts to evolve circuit solutions for combinational logic systems directly onto Xilinx 6000 FPGA parts. The reason for attempting to evolve designs direct onto the device is twofold: (i) every circuit has a known functionality and (ii) every circuit must be able to be placed on the chip and then routed. Using evolutionary techniques allows us to consider these two important aspects of design and implementation as a single problem. The paper describes the basic method adopted, using a network list (netlist) chromosome and genes which represent circuit module function, and then discusses some of the results achieved, plus difficulties encountered, and some of the additional problems which still require to be solved in this new and exciting area of research.

1. Introduction

This paper is inspired by both the authors' own attempts to synthesise digital logic circuits by using genetic algorithms to resolve the variable-ordering problem of decision diagram representations [1][2], and by the work of Adrian Thompson who has previously evolved circuit solutions directly on the 6000 family of fine-grained field-programmable gate array (FPGA) devices produced by Xilinx [3].

Thompson's work describes the direct evolution of circuits on to existing hardware components which exploit the underlying analogue structure of a device that was designed for digital applications. This means that solutions which are evolved in this manner cannot be replicated easily from the existing design on to other identical physical parts. Where our approach differs, is that we are attempting to create only digital solutions, and, in this paper, those which concentrate upon finding combinational solutions to existing combinational design problems. However, this is just as important to the designer, as many of the problems that are encountered when attempting to place a synthesised circuit - by which we usually mean minimised in terms of numbers of gates used - on to an FPGA part are due to the automatic place and route (APR) software's inability to find routes on the actual chip for the design implementation. In this paper we have taken a radical new approach to circuit synthesis which addresses this difficulty. We evolve circuit functionality under the strict conditions of specific routing strategies that we know will be acceptable to the existing structure of the target FPGA part.

We also feel that this approach provide a benefit to engineers as we are effectively encouraging the design of novel circuit solutions which, when studied, could lead to a better understanding of design principles, or may even create new principles. In this paper we will demonstrate, in our results, via a small example of how hitherto unknown designs of existing circuits may emerge. The design discovered by the genetic algorithm for this example was surprising in several ways. Firstly, that it was possible to evolve this type of function in such a simple way, and, secondly, the way in which the actual solution deviated from the conventional design.

Evolvable Hardware is a very new topic for research [4][5][6][7] and at present is largely confined to the use of evolutionary algorithms in the synthesis of digital logic at the systems level [4]. There are very few examples where researchers are actually attempting to evolve the functionality of digital circuits, and even fewer where this is being done with the specific implementation platform in mind. In fact, as far as we are aware, Koza [5] and Higuchi [6] head the only two other groups who have attempted to evolve actual digital circuits at gate level, and only Higuchi mentions implementation issues, and has a particular device in mind - a programmable logic device or PLD. Additionally, the most complex gate-level digital circuit that has been designed using genetic principles, we believe, is Koza's two-bit adder without carry [5].

In the results section of this paper, we will show our evolved designs with 100% functionality for the one-bit, two-bit and three-bit adders all with carry, and the two-bit unsigned multiplier. The latter three designs are by far the most complex digital circuits that have to date been created by purely evolutionary methods with no human design constraints. In the case of the two-bit multiplier, the circuit produced is radically simpler and more efficient than can be obtained by known design methods of minimisation. This would tend to suggest that it is easier for evolution to find simpler solutions than more complex ones.

2. The Evolvable Hardware Method

The method adopted in our technique is to consider, for each potential design, a geometry (of a fixed size array) of uncommitted logic cells that exist between a set of desired inputs and outputs. A *chromosome*, represented as a list of integers, may then be created which forms a set of interconnections and gate level functionality for these cells from output back toward the inputs i.e. a network connection list, or *netlist*.

Each of the uncommitted logic cells is capable of assuming the functionality of any two-input logic gate, or, alternatively a 2-bit *universal logic module* (ULM) with single control line. Therefore, for a 4 by 4 geometry, a typical chromosome may look something like this:-

2 1 6   4 7 -13   0 3 -12   0 8 3   0 5 2   4 2 -1   7 1 9   8 1 -10   9 11 -7   4 8 -8   0 15 -4   11 17 -14   11 18 19   11 18 14   17 14 -1   0 15 -4   22 21

Figure 1. A typical netlist chromosome for the 4 by 4 geometry.

Notice, in this arrangement that we split the chromosome up into groups of three integers. This relates to the connection of the two inputs to the gate or ULM. The third value may either be positive - in which case it is the control connection of a ULM - or negative - in which case it is a two-input gate, and the modulus of the number indicates the function according to Table 1 below. The first input to the cell is called A and the second input called B for convenience. For the logic operations, the C language symbols are used: (i) & for AND, (ii) | for OR, (iii) ^ for exclusive-OR, and (iv) ! for NOT. There are only 12 entries on this table out of a possible 16 as 4 of the combinations: (i) all zeroes, (ii) all ones, (iii) input A passed straight through, and (iv) input B passed straight through are considered trivial - because these are already included amongst the possible input combinations, and they do not affect subsequent network connections in cascade.

Considering Figure 1 it can be seen that the first cell, i.e. number 8, indicated by being first in the chromosome list, has its A input connected to input 2, its B input connected to input 1, and its control input connected to input 6. This means that it is a ULM device, and the control is driven by input 6 because the third number on the list is positive. If we consider the second group of three digits, we see that the third integer is negative, -12 in fact. From Table 1, this indicates that this is an OR gate with both inputs inverted, and so the cell inverts inputs 4 and 7 and then ORs these together.

In this way, the entire circuit is defined. Note that, in this example, the two outputs are driven by the outputs of cells 22 and 21 respectively - these are the final two values on the chromosome, and because they drive individual outputs, they do not appear in a group of three integers like all other connections.

This, of course means, that for every circuit, there will inevitably be cells that are not actually connected to any of the outputs. These are redundant for the circuit they define, and are removed when the chromosome is analysed. This analysis is performed after the algorithm has evolved 100% functionality, and is not done as part of the basic procedure.

Table 1. Cell gate functionality according to negative integer value in chromosome.

Negative Integer	Gate Function
-1	A & B
-2	A & !B
-3	!A & B
-4	A ^ B
-5	A | B
-6	!A & !B
-7	!A ^ B
-8	!A
-9	A | !B
-10	!B
-11	!A | B
-12	!A | !B

It should be noted that the chosen netlist structure for the chromosome ensures that:-

1) all circuits are valid combinational logic circuits, and
2) chromosomes may be manipulated in such a way as to guarantee that all subsequent child chromosomes, after crossover and mutation by a genetic algorithm, will also represent valid combinational logic circuits.

It was considered important, in choosing the representation, that it should fill these two criteria as time would be wasted effecting repair to a less robust representation. It is also important in terms of translation of the evolved solutions directly on to the FPGA devices.

The fitness of each chromosome is defined as the percentage of the correct output bits for every input combination of the complete specification file. We do not attempt to evolve 100% functional circuits using an example set, but instead use the entire circuit specification. This is the only reliable method we have to discovered to date. In our work, we do not attempt to carry out an exercise in machine learning by trying to generalise from a limited example set since we only interested in evolving actual fully-functional, practical circuit solutions.

3. Results

We now present the results of some of the arithmetical circuits that we have thus far been able to evolve using the above method.

The circuits that we are looking at are binary adders which incorporate both input and output carry, and unsigned multipliers. Some of the results are extremely interesting when these are compared with the circuits that have been created by designers. Of course, many of these arithmetical circuits have standard forms, particularly the adders which tend to be comprised of cascaded *full-adders* with the option of incorporating additional circuitry to provide a *carry look-ahead* facility. This latter approach tends to make the circuits much more complex to implement in terms of the numbers of gates used. Evolution is capable of developing solutions for these applications where the carry circuitry is not constrained by considering its relationship to the sum, but is treated rather as an independent output of the entire combinational system. This appears to provide several advantages in that the resulting carry circuitry is greatly simplified when compared to that required by a carry look-ahead strategy. Treating the multiplier as the development of a purely combinational transform also appears to give beneficial results in terms of the resources required. In fact, when we evolved the 2-bit unsigned binary multiplier, and compared it with the circuit that would be achieved by conventional design techniques, the result was staggering.

Figure 2 below shows the evolved design for the 2-bit multiplier. This circuit contains only eight logic gates. Compare this with the design achieved by treating this as a straightforward combinational design and using conventional minimisation methods, which requires twenty gates, then this is a significant result!

302

When one considers that actual Boolean algebra representation of the respective solutions, then one is able to see where the evolutionary approach is scoring over the conventional methods.

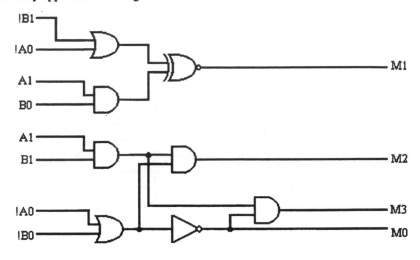

Figure 2. The evolved 2-bit unsigned binary multiplier

$$M0 = \overline{\overline{A0} + \overline{B0}}$$

$$M1 = \overline{(\overline{B1} + \overline{A0})} \oplus A1.B0$$

$$M2 = A1.B1.(\overline{A0} + \overline{B0})$$ (1)

$$M3 = A1.B1.(\overline{\overline{A0} + \overline{B0}})$$

$$M0 = A0.B0$$

$$M1 = A1.\overline{B1}.B0 + A1.\overline{A0}.B0 + \overline{A1}.A0.B1 + A0.B1.\overline{B0}$$ (2)

$$M2 = A1.\overline{A0}.B1 + A1.B1.\overline{B0}$$

$$M3 = A1.A0.B1.B0$$

Equations (1) above show the evolved representation, whereas equations (2) show the solution minimised by *Karnaugh Map* [8]. The K-Map solution conveys a regular sum-of-products representation for the desired output functionality, whereas the evolved solution shows a mixture of both sum-of-product and product-of-sum forms that a designer would find extremely difficult to develop, as there are no rules that tell him/her which parts of the equations to expand and which parts to leave unaltered. This, then, is the difficulty for the conventional minimisation approach: *no such rules exist.* Additionally, in the evolved solution, there is great deal of re-use of the functional sub-blocks that are used to derive particular outputs. Consider, as an example, the derivation of outputs *M0* and *M2*. If *M0* had been merely represented in the AND form i.e. as a single AND gate (from equations 2), then it would not have been possible to re-use this in the derivation of *M2*. This is highly novel. In addition to making the circuit substantially simpler than the conventionally minmised representation, the maximum gate delay is the same i.e. a maximum of three gate delays, which is in itself an improvement upon the conventional cellular design approach to multipliers (using full adders) which incurs a thirteen gate delay.

It is important to note that these evolved designs are created within a fixed geometry of logic cells (where a cell is a two-input gate or 2-1 ULM), and these are pruned down to remove redundancy after the design is completed. In other words, once the circuit has evolved, there are inevitably cells which contribute nothing to the solution - by either not being connected to any output, or by possessing a non-varying gate output. These are removed by a program which analyses for these conditions and performs the removal automatically. We often find that the final circuits require between 33% to 50% of the original gate resource within the applied geometry. This means that for an 8 by 8 geometry, for example, the solution would probably only require between 21 and 32 of the 64 original cells.

Figure 3 and Figure 4 show the evolved designs for the one-bit and two-bit full adders respectively.

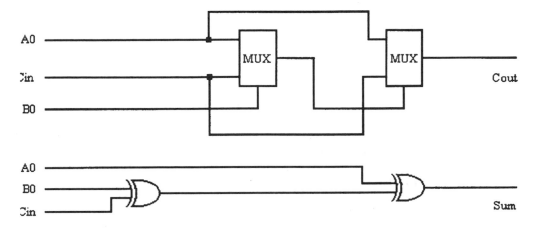

Figure 3. The evolved one-bit full adder with carry

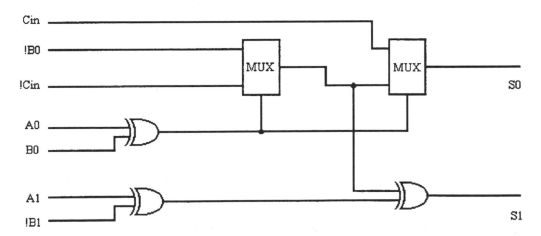

Figure 4(a) The evolved two-bit adder sum

It is interesting to note that the SUM output of the one-bit adder with carry circuit in Figure 3 is identical to that used in designed full adders. However, the carry is radically different. There is no attempt made by evolution to use aspects of the sum derivation - as is done in conventional designs - but rather the carry is developed in an entirely different way. This has the effect of producing a more economical and more elegant circuit solution, and again one which is not intuitively obvious to designers.

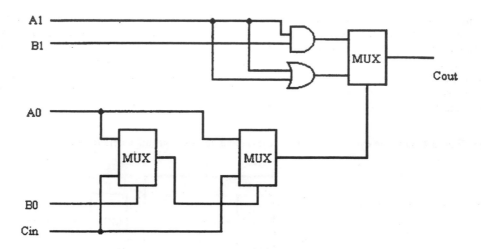

Figure 4(b). The evolved two-bit full adder carry

We were intrigued having evolved the one-bit adder carry to see whether evolution would use this sub-function when evolving a higher adder carry. In other words, we were asking the question as to whether evolution was in fact finding a generic way in which to construct carry circuits for any adder. We found the answer to be in the affirmative as can be seen in Figure 4 where the evolved solution for the two-bit adder with carry is presented. Interestingly, the control for the rightmost ULM ·is in the form of a one-bit carry circuit. This, then, controls whether the final output is to be the AND of the two most significant bits, or else the OR of these same two bits. This, once again, is a very elegant solution. Additionally, because the sum and carry are now both treated as combinational outputs in parallel, the carry does not suffer from the ripple effect of passing through both the full adders of a conventional design. That is, with the evolved design, the carry is available just as quickly as the sum. Using a carry look-ahead approach, even for this small adder, the conventional design would use 14 FPGA cells in the 6000 family, whereas the evolved design uses only 10. We envisage that this type of saving will probably scale-up significantly when we attempt to evolve larger adder designs.

It should be noted, that in evolving both the one-bit and two-bit adders, we have evolved both circuits as single multiple-output entities and, alternatively, as separate sum and carry circuits. We always found that it was much simpler to evolve solutions using the latter approach. We also found that this approach produced consistently more efficient solutions such as those shown the Figures 3 and 4.

4. Conclusions

In this paper we have presented an evolutionary approach for the design of combinational logic circuits. The method uses a genetic algorithm to evolve both a netlist structure and functionality for logic cells as represented on Xilinx's 6000 family of FPGA parts. The fitness of solutions is measure by their degree of correlation with a desired system functionality.

We feel that this approach to combinational circuit design has a distinct advantage over conventional design methods because: (i) we are not concerned about the application of complex rules to the minimisation of logic, and so there is no requirement for possibly abstract and lengthy implementation of these in computer code, (ii) we know for a fact that our solutions will fit on to the target implementation device, because we are designing with these specific architectural constraints built-in, (iii) our solutions are already routed for the target part, and so potential routing bottlenecks are automatically avoided, and (iv) extremely novel solutions may be discovered which may assist our understanding of design principles.

The points (ii) and (iii) above are particularly important as this removes the necessity for place and route software to be used. This can be both time-consuming and non-productive if the minimised design is unable to be routed due to limitations on the part that were previously unanticipated. We have had personal experience of this with designs which were perfectly easily minimised by synthesis software, both proprietary and our own algorithms [9], but could not then be routed on to the Xilinx parts. This new approach takes a much more organic view, with the design effectively being grafted on to the part in question.

Clearly, there are difficulties with the evolutionary design of circuits in scaling these designs up to larger systems. Some of the questions that remain unanswered regarding scalability are:

(i) what is the largest circuit that can be evolved using current computer hardware?
(ii) what is the ideal geometry for any given circuit?
(iii) what is the ideal routing constraint?
(iv) can larger systems be realistically constructed from smaller sub-functions?

We are currently addressing some of these questions, and expect to report our findings in due course.

References

[1] P Thomson and J. F. Miller, 1996, "Symbolic Method for Simplifying AND-EXOR Representations of Boolean Functions using a Binary-Decision Technique and a Genetic Algorithm.", *IEE Proceedings on Computers and Digital Techniques*, Vol. 143, No. 2, pp. 151-155.

[2] J. F. Miller, P. V. G. Bradbeer and P. Thomson, 1996, "Experiences of using Evolutionary Techniques in Logic Minimisation", *1st Online Workshop on Soft Computing*.

[3] A. Thompson, 1996, "Silicon Evolution", in *Genetic Programming 1996: Proceedings of the First Annual Conference*, J. R. Koza, D. E. Goldberg, D. B. Fogel and R . L. Riolo Eds. Cambridge, MA, 1996, pp. 444-452, The MIT Press.

[4] E. Sanchez and M. Tomassini, Eds., 1996, *Towards Evolvable Hardware*, Vol. 1062 *of Lecture Notes in Computer Science, Springer-Verlag*, Heidelberg.

[5] J. R. Koza, "Genetic Programming", 1992, The MIT Press, Cambridge, MA.

[6] H. Iba, M. Iwata and T. Higuchi, 1996, "Machine Learning Approach to Gate-Level Evolvable Hardware.", Vol. 1062 *of Lecture Notes in Computer Science, Springer-Verlag*, Heidelberg.

[7] M. Sipper, E. Sanchez, D. Mange, M. Tomassini, A. Perez-Uribe and A. Stauffer, 1997, "A Phylogenetic, Ontogenetic, and Epigenetic View of Bio-Inspired Hardware Systems.", *IEEE Transactions in Evolutionary Computation*.

[8] A. E. A. Almaini, 1994, "Electronic Logic Systems.", pp. 305-308, *3rd Edition, Prentice-Hall International*.

[9] P. Thomson and J. F. Miller, 1997, "Comparison of AND-XOR Logic Synthesis using a Genetic Algorithm against MISII for Implementation on FPGAs.", submitted to *GALESIA '97*.

A Dialogue Module at the Conceptual Product Design Stage through Genetic Algorithm

H.J. Wu, L.C. Shih & C.H. Ding

Dept. of Industrial Design,Tunghai University, Taichung, Taiwan
hsien@s867.thu.edu.tw

Keyword: Genetic algorithm, conceptual product design, dialogue

Abstract

Conceptual product design involves a series of the designer's creative thinking processes and consecutive communication with other disciplines are required in this stage to reach a final solution. During this stage, for example, different format of dialogues between designers and engineers, salesmen and customers are required to collect information about the product concept. For the product designers, the decision making procedure in this stage is the configuration inside the black box where designers' mental activities are frequently misunderstood. Another problem usually found in the conceptual product design stage is that product designers are ignorant of the product requirements defined by others due to insufficient communication. In order to enhance the performance of the conceptual product design, a novel approach based on genetic algorithm (GA) is presented in this paper. Using GA, a dialogue module is established to reproduce feasible design solutions and the evaluation of these alternatives are then calculated inside the dialogue module. The process of conceptual product design is similar to the idea evolution with implementation of GA. An application of the proposed method is also presented briefly in this paper.

1. Introduction

In terms of the creative thinking process, it is complicated for a product designer to formalize the task of decision making at conceptual design stage. The designers have to collect information about the product concept all the time with other experts in various manner through dialogues in which the designers are able to catch suitable creativity under the limitation of manufacturing technologies and market trends. 2D drawings or 3D prototypes are then utilized to implement this product concept by the designers based on evaluating the result from those dialogues. Therefore, the purpose of those dialogues are providing the designers with various perspectives. It optimize the product design without sacrificing the creativity which is the key issue to develop a product from a designer's viewpoint.

The Genetic algorithm (GA) is a highly effective search algorithm based on evolution and is widely used in solving engineering problems such as model matching, structural design, and cutting stock problem[1,2,3,4]. The concept of evolution in the GA provides a completely different standpoint and methodology of acquiring optimization. Basic description of this algorithm and its applications can be found in [5 & 6]. In this paper, a dialogue module of the conceptual product design stage is established through genetic algorithm. The activity inside this dialogue module is defined as the process of reproduction. The product designers generate design parameters based on the reproduction of various product criteria requested by different factors in the product development cycle. In addition to the reproduction procedure through genetic algorithm, this dialogue module also provides the product designers with simulation ability of constraint satisfaction (CS). As a result, the process of conceptual product design stage is based on the idea evolution with implementation of GA and simulation technique in one approach where genetic algorithm plays an important role in determining product design parameters.

2. Problem definition

The conceptual product design process to be considered in this paper is that of general industrial designers use in their routine design works. An essential characteristic of this process is the integration between designers' creativity and functional requirements requested by other divisions related to product development. During this stage, mistake about decision making should be minimized to prevent from further losses at later stage where errors are difficult and costly for recovery. Therefore, designers tend to take as many design factors (manufacturing technology, market trends, etc.) as possible into consideration when they are working at generating feasible design solutions. In this case, several dialogues are used by designers to communicate with other product divisions for idea evaluation purpose. Since there are difficulty in obtaining designers' mental activity such as aesthetic concerns, this paper focuses on the process of making dialogues and the contents of those dialogues to avoid excessively subjective judgment.

Since the purpose of these dialogues for designers is to catch possible product idea under certain conditions, information collected in these dialogues can be with the patterns of 1)design limitations, 2) special requirements, and 3) associated idea stimulation. Based on these imaginable response received from all dialogues, designers perform idea modification to meet product requirements and perhaps create more unique design alternatives. In addition to idea modification or so-called idea evolution, it can be viewed as the continuous procedure of constraint satisfaction (CS) in these dialogues. It then becomes the motivation of using GA to regenerate and evaluate design alternatives because of the similarity in terms of the reproduction process and different levels of evolution in nature.

3. GA implementation in conceptual product design

As stated previously, the genetic algorithm for the conceptual product design which is presented in this paper is developed for producing more practicable design idea, solving constraint satisfaction problem, and making design decision. To develop GA implementation for this concern, the first step is to acquire an appropriate coding to represent the product idea in stead of 2D drawings. In this paper we used two types of strings (A & B type) to build the dialogues module.

Type A string is a binary string representation of the product idea in which each bit indicates the response from other product divisions about a specific design parameter with value 0 (unsatisfied) and 1(satisfied). Type B string is a string representation of the product idea itself that each bit number stands for one possible design choice of a particular design parameter with value starting with 1 (higher values simply mean different choices). The length of strings is based on the number of design parameters used by the designers, and the sequence of the bits in one string is not restricted. Different bit sequences show different priority of design consideration decided by the designers. Type B strings are then represent all feasible design alternatives and type A strings respond the degree of satisfaction of one individual design solution. Reproduction of strings are performed by crossover operator applied to two B strings randomly selected from the population. One-point crossover is used and the position of crossover is also selected randomly for experiment purpose. Mutation operator is defined as the change of bit sequence in a string (both in A and B strings) to shift priority.

The procedure of establishing the dialogue module (from designer's standpoint) is listed as follows:
1 set design parameters as the label of each bit ,
2 create Type B strings as long as feasible design idea exist,
3 put all Type B strings into the population B,
4 create A strings corresponding to B strings (dialogue),
5 put A strings into population A,
6 select two B strings for one-point crossover,
7 rearrange bit sequence for mutation if required,
8 evaluate new strings and convert to A strings (dialogue),
9 calculate the degree of satisfaction for each A string and choose the A string with maximum value, and the formula of calculating the degree of satisfaction is user-defined (based on the designer's interpretation of different product design requirements),
10 decide whether to add new B string into population B based on the result of step 9,
11 repeat step 6-10 or stop.

In this building procedure, dialogues between product designers and other product divisions occur at step 4 and step 8. Step 4 indicates the response from different perspective according to a particular design and shows the degree of satisfaction under certain priority. Step 8 is used to evaluate the new design idea that shows designer's creativity in different manner. As a result, Type A string is primarily made to solve the constraint satisfaction problem through evaluating the degree of satisfaction, while Type B string is used to generate more product idea with the implementation of GA concept. The framework of the application proposed here is shown in figure 1. With the combination of product concept evolution and design constraint satisfaction, GA here provides a capacity of integrating designer's creativity and practical limitation resulted from real word.

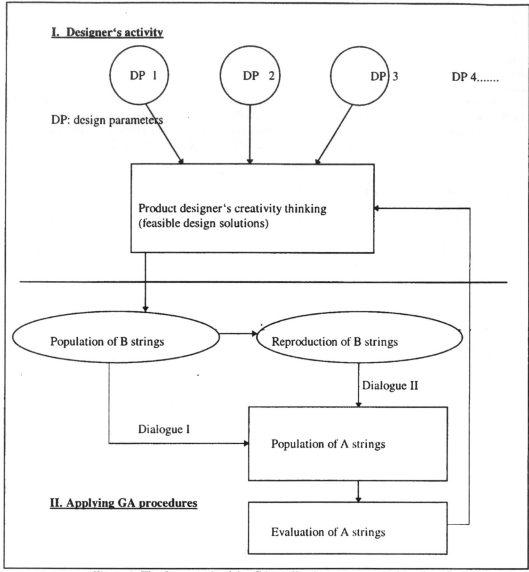

Figure 1. The framework of the GA application on conceptual product design

4. Example

Home appliance product (microwave) is used as the example to show the process step by step to build the designer's dialogue module through GA.

1 set design parameters as the label of each bit,
> => material, size, panel location, door open type, display pattern (from left to right)

2 create Type B strings as long as feasible design idea exist,
> => (B1: 23123 and B2: 22122)
>
> => (B1: plastic, large, upper-right, vertical, digital)
>
> => and (B2: plastic, medium, upper-right, vertical, graphic)

3 put all Type B strings into the population B,

4 create A strings corresponding to B strings (dialogue),
> => dialogue with marketing division (A1:01100 and A2: 11101)

5 put A strings into population A,

6 select two B strings for one-point crossover,

=> one-point at position 3,

    => (B3: 23122 and B4: 22123)

7 rearrange bit sequence for mutation if required,

    => mutation on B3 and B4 (reverse bit sequence)

    => (B5: 22132 and B6: 32122)

8 evaluate new strings and convert to A strings (dialogue),

    => dialogue with marketing division again (A5:01110 and A6:11110)

9 calculate the degree of satisfaction for each A string and choose the A string with maximum value,

    => set formula of degree of satisfaction

    => $E(A1) = 0 \times 5 + 1 \times 4 + 1 \times 3 + 0 \times 2 + 0 \times 1 = 7)$

    => $E(A2) = 1 \times 5 + 1 \times 4 + 1 \times 3 + 0 \times 2 + 1 \times 1 = 13)$

    => $E(A5) = 0 \times 5 + 1 \times 4 + 1 \times 3 + 1 \times 2 + 0 \times 1 = 9)$

    => $E(A6) = 1 \times 5 + 1 \times 4 + 1 \times 3 + 1 \times 2 + 1 \times 0 = 14)$

10 decide whether to add new B string into population B based on the result of step 9,

    => add B6 and give up B5

11 repeat step 6-10 or stop.

At step 9, the evaluation function (user defined) is set to calculate the degree of satisfaction. In this example, degree of satisfaction is evaluated through the arrangement of bit sequence which represents design priority (mutation operator provides sequence rearrangement). The addition of new B strings indicates that the product designers accept the suggestion from other divisions and useful information is collected through dialogues to reproduce new product idea.

5. Conclusions and further work

This work has demonstrated the application of genetic algorithm in the conceptual product design stage where designers' creativity usually contradict product functional requirement or specific limitation defined by other product divisions during the product development cycle. In this paper, a dialogue module is established from designer's viewpoint through genetic algorithm whose reproduction process is similar to the process of generating product design idea. With the reproduction procedure using GA, more creativity thinking can be created through these dialogue modules. Genetic algorithm plays an important role in determining product design parameters with implementation of product idea evolution. Constraint satisfaction problem is also discussed in this paper by evaluating the degree of satisfaction based on the dialogue with other product divisions. As a result, designers are able to achieve the concept of product design without sacrificing their creativity.

One extension of this work would be the automatic coding system of converting 2D design into strings. Rules of priority rearrangement and the development of new evaluation function are ongoing research topics.

References

[1] Ravichandran, B. & Sanderson, A.C., Model-based matching using a hybrid genetic algorithm, *Proc. of 1994 IEEE International Conference on Robotica and Automation*, Sad Diego, 1994, pp.2064-2069.

[2] Zarubin, V.A., Genetic algorithm in the role of a shell for structural evolution simulation at the conceptual design stage, *Proceedings of the first online workshop on soft computing*, Nagoya University, Japan, 1996, pp. 205-210.

[3] Moss, R. & Gabriel, S., Implementation of a genetic algorithm to spacecraft structural stability design, *Proceedings of the first online workshop on soft computing*, Nagoya University, Japan, 1996, pp.83-86.

[4] Peter, A. et al., A genetic solution for the cutting stock problem, *Proceedings of the first online workshop on soft computing*, Nagoya University, Japan, 1996, pp.87-92.

[5] Goldberg, D.E., Genetic algorithms in search, optimization, and machine learning, Addison-Wesley, 1989.

[6] Davis, L., Handbook of Genetic Algorithms, Van Nostrand Reinhold, 1991.

Soft vs. Hard Computational Issues in Configuration Design

Stephen Potter, Pravir K. Chawdhry, Stephen J. Culley

University of Bath, Bath, BA2 7AY, UK.
E-mail: {S.E.Potter, P.K.Chawdhry, S.J.Culley}@bath.ac.uk

Keywords: configuration design, neural networks

Abstract

The engineering task of configuration design, the combination of pre-defined domain entities into a system that meets some specified requirements, is ill-defined: there is no computationally expressible algorithm available for consistently producing adequate designs.

This suggests that Artificial Intelligence (AI) techniques must be applied to produce an automated design tool. However, past attempts at construction have relied on hard computing techniques, usually in the form of 'hard-wiring' design rules into a knowledge base, with the obvious necessity for all these rules to be available in an explicit form. This is rarely the case outside simple domains.

This design 'knowledge', then, is the crucial factor in such systems, and the difficulties involved in its acquisition and expression are persistent obstacles to their construction. In this paper, we show how a soft computing approach (artificial neural networks) can be applied to the problem of capturing and expressing some of the more nebulous elements of this knowledge, which is then incorporated within a conventional hard computing framework to provide a useful design tool. The use of neural networks would seem to be particularly agreeable in a design context, as they display some emergent properties associated with aspects of design creativity.

1 Introduction

The configuration design process is the task of arranging and specifying parameter values for pre-existing domain components, each having some known individual behaviour, in such a way as to construct a system (or systems) which provides the functionality required of it by some set of (customer-supplied) functional requirements, while not violating any of the supplied constraints placed on solutions.

According to Simon and Newell [1], a well-structured problem conforms to the following three criteria;

1. It can be described in terms of numerical values, scalar and vector quantities.

2. The goals to be attained can be specified in terms of a well-defined objective function.

3. There exist computational routines (algorithms) which permit the solution to be found and stated in actual numerical terms.

Configuration design problems are described in a mixture of qualitative and quantitative terms, as are the solutions to such problems and there exists no objective algorithm for producing solutions to given problems. Hence, configuration design is ill-structured and so, as Simon and Newell go on to say, any automation of the process, of necessity, must eschew traditional algorithmic hard computing approaches and rely on some application of artificial intelligence (AI) techniques, that is, techniques which expressly endeavour to mimic the flexible and adaptive sort of intelligent behaviour shown by humans.

Previous approaches to this automation (for examples, see [2, 3]) have involved initiating a dialogue with a human design expert, in an attempt to access the knowledge which is then to be placed at the core of an expert system shell. However, the assumption made is that this knowledge is expressible in clear, lucid terms by the expert. This is seldom so, and any knowledge gained is likely to be simplified and incomplete at best and widely inaccurate at worst due to the highly abstract and compressed nature of the experiential information which constitutes expertise (see [4] for a deeper discussion of the problems involved). This breakdown in knowledge transfer has been termed the 'knowledge bottleneck' [5] of expert systems production. Clearly, another approach is called for. This paper advocates the use of 'soft' AI techniques in tandem with hard to surmount these problems. A soft AI approach to a problem is one for which no explicit algorithm exists beforehand, but rather, a meta-algorithm exists, that is, an algorithm for finding the algorithm. Conversely, a hard approach consists of applying an explicit, transparent algorithm to the problem.

The following section describes configuration design in a useful light, that of a search through a space of design abstractions. Subsequently, the categories of knowledge in this process are then discussed, which allows a description of the process in more formal terms, and a consideration of the computational issues in representing the knowledge. The final sections describe a configuration tool, built according to the precepts developed and present some of the conclusions that have been drawn.

2   Configuration Design as Search

Another useful concept postulated by Newell and Simon [6] is the idea that all human goal-oriented symbolic activity is a search problem and it is useful to cast the configuration design problem in this light.

The configuration design process may be seen as a search through the space of design abstractions, from a totally abstract initial state (the requirements, in terms of functionality and constraints) to a totally concrete state(s), in which a complete design(s) is specified in terms of actual physical components and valid types of connections between them, by way of a series of intermediary states of decreasing levels of abstraction. The task, then, becomes one of recognising a plausible path through this space; this path must consist of a sequence of known operations for transforming one state into the next. There may be more than one valid path through the space, of which any one may be chosen, or a number, so as to produce several, equally valid, candidate solutions. (It should be noted that this treatment of a problem-solving search differs conceptually from that introduced by Newell and Simon; they envisaged of it as proceeding through states described at the same level of abstraction.)

For such a problem, Newell and Simon then go on to enumerate several approaches to finding a solution to a search problem. These include generate and test, hill climbing, means-end analysis and operator subgoaling, none of which would seem to be suitable for providing efficient designs, as they could involve examining a large proportion of the possible states in an extremely large state space. For such a high-level intelligent task as configuration design there would seem to be a need to explicitly apply heuristic 'knowledge' of how to perform the task (search the abstraction space). This knowledge would allow any reasonable description of the requirements for the solution to be understood and then related to elements of a more concrete abstraction, in domain terms, and, in so doing, focus the search upon a manageable (and, hopefully, the correct) area of the abstraction space. From this area, the next transformation can then be made, and so on, until, finally, the search alights upon a small number of plausible concrete solutions.

3   The Role of Knowledge in Design

'Knowledge' is a highly ambiguous term; there are many different types of skills and abilities to which this term is applied. In the following discussion, there is no intention to enter into a discussion of what it means when we say we 'know' something; it is purposely left as a rather loosely defined generic expression to encompass the wide range of abilities which must be exploited within any successful design episode. In these terms, design knowledge is that resource which allows the search through the abstraction space to proceed in an efficient and (mostly) correct manner, to produce solutions to the posed problem.

Within the KADS project [7], the knowledge which is applied in any problem-solving episode can be categorised as belonging to one of the following epistemological groups, described here as they would apply to the problem of design;[1]

---

[1] The subsequent description is a simplified account of the categories of knowledge defined in the CommonKADS project rather than that defined earlier in KADS-I; in the course of the evolution of the project, an additional category of knowledge

- *domain knowledge* — this category contains knowledge of the entities which constitute the domain under consideration. These entities are of necessity at different levels of abstraction and complexity. In terms of configuration design, this group includes knowledge of the physical elements (and their behaviour) which may constitute a solution, knowledge of the relationships that may be used for combining these elements, knowledge of how groups of related elements (up to and including the system level) behave, how to provide parameter values, and so on.

- *inference knowledge* — that is, knowledge of what conclusions (i.e. design decisions) may be made, given the possession of some piece of information. In other words, how an abstract element of a design may be made 'more concrete' according to the requirements specified, the intermediate abstractions already chosen, design choices made elsewhere, etc. Hence, this type of knowledge governs the transformations that can occur whilst performing the design search.

- *strategic knowledge* — this is knowledge of how elements of inference knowledge can be arranged and controlled so as to provide a complete strategy for producing a design. This amounts to a set of high level methodologies for controlling the search through the abstraction space from requirements to solutions.

To these, we have felt it necessary to add a further category, that of *working knowledge*. This category is unique for each separate design episode. It contains the specific requirements for the particular design, the design choices made, knowledge of the reasons for modifications to a design, even feedback from the customer, etc. In such, this category represents the dynamic knowledge accumulated during the episode; from this dynamic knowledge, it is possible for the designer to learn, that is, to modify her beliefs and prejudices in the other categories in response to new information. This reflects the nature of expertise as being closely related to and dependent upon experience; there is no such thing as the 'complete' designer.

However, while these categories are beneficial in distinguishing the various levels of knowledge from one another, they are not sufficient. The omission is of a basic-level knowledge which may be loosely termed 'common-sense'. This is knowledge which may have no explicit bearing upon the process, but which, as every designer possesses it (although it can never be the same in any two individuals) plays an implicit role. *A priori* knowledge[2] necessarily falls into this category, allowing *modus ponens*, inductive principles, mathematical axioms, etc. With this common-sense knowledge, flexible, adaptive, intelligent behaviour becomes possible; for example, the infinite range of possible functional requirements and constraints can be understood, and the implications for a particular design to be internally re-described in terms of the effects on the domain (less abstractly, the designer's knowledge about the physical properties of the 'real world' may allow a solution to be constructed which is constrained to fit into a limited area). However, the use of this knowledge goes much further; one major aspect of creativity in design is the formation of solutions by making some analogous link to structures, properties, descriptions of a second, quite separate domain.[3] Obviously, this creativity comes about only because there is knowledge of the second domain (and of the manner in which analogical references can be formed). It is debatable whether common-sense knowledge *per se* is sufficient for intelligent behaviour, but it is a necessary component of it.

This knowledge category, its content, scope and representations even less certain than the others, is the most difficult to express computationally, and may simply be impossible for computers built according to current theories and technologies. Nevertheless, leaving aside the philosophical issues of ontology and epistemology, until it is incorporated, automated design tools (and indeed, all tools purporting 'intelligent' behaviour) will fall short of the human ideal (similar arguments have been put forward in [8], and elsewhere). However, even with this limitation, it is possible to construct useful systems, incorporating some of the other knowledge or, at least, a practicable approximation to it. The following section contains a formal treatment of the design process to enable the construction of such a system for configuration design, but omits consideration of this currently unobtainable 'common sense' knowledge.

4 A Formal Treatment of the Configuration Design Process

This section draws heavily on the work of Wielinga et al. [9], the major difference being the omission of all consideration of knowledge in the earlier work. Before the process itself can be considered, it is

---

was found to have been rendered obsolete.

[2] Leaving aside germane questions of whether such knowledge exists, and if so, what it consists of.

[3] A second form of creativity, that of combining known domain concepts in novel ways, *is* possible using soft computing techniques, as shall be seen.

necessary to provide some definitions of the entities and concepts which represent information within the process. Hence, the design problem for a particular domain has the following constituents;

- a set of possible functional requirements, F: that is, a set including all the functional requirements that could possibly be specified for a solution to meet;

- a set of possible constraints (non-functional requirements), C: likewise, the set of all the constraints to which a solution may be expected to conform;

- a set of physical elements of the domain, from which a design can be constructed, E;

- a set of relationships between the elements of E: that is, the possible methods by which the domain elements may be connected to one another;

- a set of design structures, that is, a set of the sets of possible combinations of relationships and elements that can occur in the domain, D;

- a set of parameters which elements may have, H;

- a set of combinations of particular elements and the corresponding parameters they possess, Q;

- a set of values which parameters can assume, V;

- a set of assignments of values to parameters of particular elements, A.

In addition, the three types of design knowledge, that is, domain knowledge (K_D), inference knowledge (K_I) and strategic knowledge (K_S) are assumed. It should be noted that;

$$E \cup L \cup D \cup H \cup Q \cup V \subset K_D$$

which provides a definition of some of the constituents of the domain knowledge.

In addition, the concept of working knowledge (K_W) is introduced. This includes information and concepts provided to and inferences made during a particular configuration design episode. K_T, the total design knowledge, includes all the various categories of knowledge. (These terms are summarised, with definitions were applicable, in table 1).

Now, these definitions allow the formalisation of the configuration design process in the following manner. A particular problem, P, is defined in terms of some customer supplied functional requirements, F_P and constraints, C_P (note that all terms denoted with a subscript P refer to a particular problem). Hence;

$$P = < F_P, C_P >$$

where $F_P \subset F$ and $C_P \subset C$ (also, F_P and C_P represent the first elements of K_W). Similarly the problem space, that is, the space of all possible configuration design problems may be defined as;

$$PS = \{< f, c > | f \subset F \wedge c \subset C\}$$

The initial stage of the process involves the rigorisation and formalisation of the given functional requirements and constraints sets in order to establish and express implicit functional requirements and constraints respectively: this is achieved using the domain knowledge;

$$F_P \cup K_D \vdash F_D$$
$$C_P \cup K_D \vdash C_D$$

Now, this full expression of the constraints allows the determination of the total space of designs which may fall under consideration for this problem, X;

$$C_D \cup K_D \vdash \exists X \bullet X = \{x_i \in E \times D \times A | feasible(x_i) \wedge satisfies(x_i, C_D)\}$$

Table 1. Summary of terms

functional requirements set	$F = \{f_i\}$
constraints (non-functional requirements) set	$C = \{c_i\}$
elements set	$E = \{e_i\}$
relationships names set	$L = \{l_i\}$
design structures set	$D = \{d_{ki\ldots n}\}\{\{< l_k, e_i, \ldots, e_n >\}\}$
parameters set	$H = \{h_i\}$
element-parameters set	$Q = \{q_i\} = \{< e_i, h_{ij} >\}$
element-parameter values set	$V = \forall q_i \bullet q_i \in Q : \{v_{ijk}\}$
assignments set	$A = \{a_{ijk}\} = \{< e_i, h_{i,j}, v_{ijk} >\}$
domain knowledge	K_D
inference knowledge	K_I
strategic knowledge	K_S
working knowledge	K_W
total design knowledge	K_T

where $feasible(x)$ is a function which determines whether or not x represents a valid design within the domain, that is, whether x is complete, constructable, all elements are connected in some manner and so on, and $satisfies(x, y)$ is a function which determines whether (a design) x satisfies (some set of constraints) y. Note also that the set, X, of candidate solutions may well be empty if the constraints prove too stringent.

The actual design process may then be defined as the process of inferring some set, S, consisting of designs from the space of candidates which also provide the necessary functionality, from the functional requirements and constraints using the total design knowledge;

$$F_D \cup C_D \cup K_T \vdash \exists S \bullet S = \{x_i \in E \times D \times A | feasible(x_i) \wedge satisfies(x_i, C_D) \wedge (x_i \cup K_D) \models F_D\}$$

where each element of the solution set, S, is both feasible and does not violate the constraints, as before, and which semantically provides the desired functionality (and, if the requirements prove too taxing for the domain, the set may be empty).

It should be noted that, along with the domain, inference and strategic knowledge, the working knowledge is necessary to the inference of a set of solutions; this is expressing the concept that previous design choices inform and influence subsequent decisions. As these decisions are made, they are added to K_W (and thence to K_T). Assuming that K_W remains associated with a particular design episode, and so information may be added at a later date through external testing, customer feedback, etc. then it represents a source of augmentation and modification of the other categories of knowledge (e.g. learning new relationships between functional relationships and domain elements, new high-level design strategies).

From this treatment, the configuration design process within any particular domain may be expressed as;

$$CDP = < P, K_D, K_I, K_S, S >$$

That is, given some definition P of a problem, the requisite types of knowledge allow the creation of a set S of design solutions to P. (It should be noted that no consideration of the relative merits of solutions is implied within this treatment, all solutions being regarded as equally valid.)

5   Soft Computing and the Configuration Design Process

In the above formalisation of the configuration design process, and as stated earlier, the knowledge is the essential element for performing the task. The automation of the task, therefore, must involve an integration of each of the different types of knowledge. The domain knowledge, in the terms in which it was described in the formalisation, is relatively easy to access; it exists in textbooks, manufacturer's catalogues and manuals, etc. It can, on the whole, be expressed through traditional hard computing, as, almost by definition, knowledge of procedures and routines at this level does exist in a well-structured form.

However, the inference and strategic knowledge are both harder to define and to capture, both being of a form susceptible to all the problems associated with the knowledge bottleneck. A further problem occurs in trying to represent the imprecise and probably incorrect knowledge gained; 'harder' Artificial Intelligence techniques, such as rule-bases, do not embody the flexibility which a human could apply, and either the rules are grossly simplified in an attempt to make the system as general as possible or else there follows an explosion in the number of rules leading to a unmaintainable and inscrutable system. These problems would seem to be an insurmountable obstacle to the automation of the process.

Nevertheless, it is possible to gain an impression or image of some of this knowledge using soft computing techniques. Artificial neural networks have been seen to 'learn' complicated pattern-matching techniques. As the inference knowledge may be viewed in terms of a series of pattern-matching inferences, from a description of (part of the) problem at one level of abstraction, to a description at a lower level (that is, closer to the design solution), it would seem that this type of knowledge can be extracted and represented using soft computing, given training data of the required format (artificial neural networks, in shunning representations at the symbolic level, reason with data at a sub-symbolic level; often, this entails more complex representations than seem most 'natural', especially in problems in which there is a mixture of qualitative and quantitative information, as is the case with configuration design). Additionally, their non-deterministic outputs can be used to produce design alternatives, by recording and reasoning with concepts that, although not adjudged the best, are still rated as sufficiently good by the network to merit further consideration. A further property of neural networks is that given rich enough raw training material, can produce acceptable generalisations of the domain concepts so as to cope with previously unseen cases in an 'intelligent' manner. This latter emergent ability is particularly appealing in the context of design, as it represents a form of creativity, in that existing concepts may be arranged or related in a novel way. (This aspect of creativity in neural networks has also been noted by Boden.[10])

There remains the problem of acquiring the strategic knowledge, particularly problematic as machine learning techniques require examples from which to learn, examples which don't exist for this type of knowledge. The only solution at the current time would be to hypothesise some approximation of how a particular design task can be performed (in terms of starting with a definition of the functional requirements and constraints, and finishing with a complete solution or solutions, by way of a series of abstractions and inferences to move from one abstraction to the next). If this can be done, then an automated system for producing designs can be constructed. To convey these ideas more completely, an example of such a system is described in the following section.

6 An Automated Configuration Design Tool

To assess the capabilities of soft computing techniques when applied to the extraction and expression of design inference knowledge, an implementation of a prototype tool to design fluid power systems was undertaken. (The task was simplified to a certain extent by the omission of certain domain concepts and the stage of proposing values for parameters of domain entities.)

The initial step in doing this is to provide some strategic knowledge (that is, the highest level knowledge of how the design search is to be performed); this is comprised of the following five-stage methodology. This methodology is considered a hypothesis about the task in this particular domain and would be 'hard' knowledge within the system.

1. Prompt user to provide a set of requirements and constraints. This stage represents an interface between the users, who provide their specifications for a design, and the design process itself.

2. Select most suitable circuit template based on the stated requirements. Analysis of examples of designs had revealed that there exists a set of basic circuit configurations (or templates - e.g. see figure 1(a)), each of which was found to be common across a number of examples in the archive. Each templates include a number of labelled locations (or slots) into which components or groups of components can be inserted in order to provide the functionality particular to that circuit. A template, although it is expressed in convenient terms of actual domain components, can also be seen as a partial abstraction; in that it implicitly contains a description of the high-level functionality that must be embodied in every hydraulic circuit; that is, there must be a power source, connected to a method of directing this power, connected to some actuation of this power.

3. Recognition of some set of *basic functions* **which the requirements indicate ought to be achieved by candidate solutions.** This stage is that of determining the functionality which solutions should display to satisfy the requirements. These basic functions are domain-dependent abstractions, examples being, say, *control the speed of actuator extension* and *hold the load stationary in the event of system failure.*

4. Select components to achieve each of these basic functions, and determine their positions within the chosen template. Once a set of the required basic functions has been recognised for the given requirements, it is necessary to map these onto a set of physical components, and determine the position of these components within the chosen template. The abstract form of the basic functions are thereby translated into more concrete aspects of the system design.

5. Incorporate the chosen components within the template. Finally, the components must be inserted into their prescribed positions within the circuit template, to give a number of valid configurations that would meet the stated requirements.

By way of an example, a given system specification may indicate that the template shown in figure 1(a) is that required and that the function control the speed of actuator extension must be satisfied in any solution. It can then be inferred that the meter-out component arrangement, figure 1(b), in slot B of the template is the most apt for achieving this, giving the complete solution (figure 2). Note that if no component is chosen for a particular slot, then that slot is considered connected by pipe-work to complete the circuit.

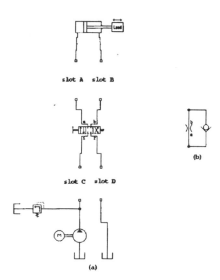

Figure 1. (a) An example circuit template, and (b) a meter-out component

Once the strategic knowledge has been defined, the areas in which inference knowledge is to be applied may then be identified; these are in the selection of a suitable template and a set of basic functions given a set of requirements and in the selection of components and their circuit positions from a combination of requirements, template and basic function information. For the previously stated reasons, neural networks were selected as the mechanisms for the learning and subsequent application of this inference knowledge. Figure 3 shows an architecture for a configuration design system implementing the task methodology and indicating the positions at which the neural networks are utilised.

The next stage is to identify and collect the sources of the inference knowledge. This knowledge is assumed to be implicitly contained within domain design examples. However, before this knowledge can be extracted, these examples must be collated, structured and processed so as to reflect the information that is embedded in them. An archive of design cases is maintained; this archive, in its raw form, contains a number of design cases, each consisting of some set of requirements, which may be numerical, textual or graphical in nature, and a corresponding design that has been produced to meet these requirements, in the form of the design of a fluid power system in this instance. In addition to merely storing the examples in this manner, a knowledge acquisition formalism is followed for each so as to provide a perspective onto the raw data that will emphasise the implicit information contained in the example, providing the training data for the neural networks at different abstraction levels within the system. For example, the template selection network is trained on data consisting of requirements and the label of the template upon which the corresponding solution is based.

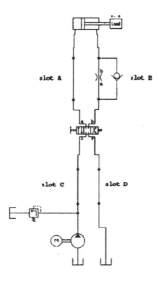

Figure 2. A solution circuit

Domain knowledge of entities and their relationships, in this instance, is expressed in terms of knowledge of the templates which exist, of components and the template positions in which they may appear and of the basic functions. Working knowledge in the system consists of recording the specified requirements, the design choices and inferences made, solutions formulated, and so on. It is important in this case, as for example, the choice of a particular component is dependent on the earlier choice of basic functions. Although it is not discussed here, this knowledge is intended to form the basis of future modifications to the system, particularly in the event of design failures occurring. This reaction to failure is an obvious intelligent faculty which is often overlooked in such systems.

For the purposes of demonstration, it was decided to implement a reduced version of the system, reasoning with a single template. Hence, task 2 above is unnecessary, but it was felt that this, of all the tasks assigned to neural networks, represented that most akin to the pattern recognition problems (selecting one of many) dealt with successfully by networks in the past. Additionally, implementation of a subset of the networks for component-position selection (task 4) was felt adequate to prove the whole. To build the system, a basic set of requirements, to be specified in a particular format, was introduced and the neural networks were trained accordingly (using a modification of the TDNN architecture [11] due to its amenability to the handling of time-dependent variables, by which the requirements are described), and incorporated within a software shell which provided the necessary interface routines and internal flow of information, forming the complete automated configuration design tool for fluid power systems.

A problem encountered was the scarcity of relevant examples: it seems to be a feature of the domain that designs are not stored along with the requirements which dictated them. This dearth of examples is countered somewhat by the deconstruction of the problem into sub-tasks, each of which is performed by a smaller neural network than would probably be needed for the whole task, and the subsequent re-use of examples for training different networks, but it remains a very real problem. As it is, sixteen examples from the archive have been used to train the prototype configuration design system.

The system was tested with sets of requirements distinct from those used during the training: for example, the following requirement was chosen as a test:

318

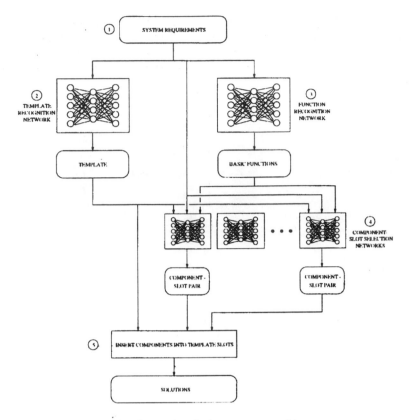

Figure 3. Configuration design system architecture

Extend a load of 1000 tonnes a distance of 0.7 metres vertically in 100s, and then retract the load, now increased to 1005 tonnes, back through the same distance in 200s.

After this information had been expressed in the appropriate format and has been processed by the system, four solutions where proposed (an example is shown in figure 4). These were all felt to represent valid candidate designs. (This prototype system is described in greater detail in [12].)

7 Conclusions

This paper has described the computational issues at the heart of the configuration design process. Central to the automation of this process is the need to incorporate knowledge. However this knowledge is not of a uniform type, nor is it completely and easily available. Upon closer consideration of this knowledge, and what it must embody, an approach using different but complementary AI approaches, both hard and soft computationally, would seem to offer the greatest opportunities for building an automated design tool. To provide an example of how such a combined strategy would appear in practice, the constituents of a system to configure fluid power systems is described; the success of the system formed of these constituents would seem to add force to the argument for adopting a more eclectic approach. In this context, artificial neural networks, as the soft computing paradigm, also offer some features of design creativity in that their recognised generalisation abilities imitate, to a certain extent, the ability of humans to combine domain elements in novel formations. This behaviour has yet to be modelled through hard techniques.

Acknowledgements

This research has been funded by the UK EPSRC, grant number GR/K22273 for the Engineering Design Centre at the University of Bath.

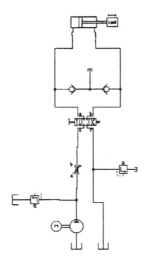

Figure 4. A sample solution

References

[1] Simon, H. A. and Newell, A., 1958, Heuristic problem solving: the next advance in operations research, *Operations Research*, **6(1)**, 1–10.

[2] McDermott, J., 1982, R1: A rule-based configurer of computer systems, *Artificial Intelligence*, **19(1)**, 39–88.

[3] Marcus, S., Stout, J. and McDermott, J., 1988, VT: An expert elevator designer that uses knowledge-based backtracking, *AI Magazine*, **9(1)**, 95–112.

[4] Berry, D., C., 1987, The problem of implicit knowledge, *Expert Systems*, **4(3)**, 144–151.

[5] Lenat, D. B., 1983, The role of heuristics in learning by discovery: three case studies. In *Machine Learning: An Artificial Intelligence Approach*, Michalski, J, K, Carbonell, J. G. and Mitchell, T. M. (Eds.), Tioga Pub. Co., Palo Alto, CA., USA, 243–306.

[6] Newell, A., 1980, Reasoning, problem solving and decision processes: the problem space as a fundamental category. Reprinted in P. S. Rosenbloom, J. E. Laird and A. Newell (Eds.), 1992, *The Soar Papers: research on integrated intelligence*, Volume 1, MIT Press, Cambridge, Mass., USA, 55–80.

[7] Wielinga, B.,Van de Velde, W., Schreiber, G. and Akkermans, H., 1992, The KADS knowledge modelling approach. In R. Mizoguchi et al. (Eds.), *Proceedings of the 2nd Japanese Knowledge Acquisition for Knowledge-Based Systems Workshop, Hitachi Advanced Research Laboratory, Hatoyama, Saitama, Japan*, 23–42.

[8] Hayes, P. J., 1985, The second naive physics manifesto. Reprinted in R. J. Brachman and H. J. Levesque, 1985, *Readings in Knowledge Representation*, Morgan Kaufmann Publishers Inc., CA., USA, 468–485.

[9] Wielinga, B., Akkermans, J. M. and Schreiber, A. T., 1995, A formal analysis of parametric design problem solving, In B. R. Gaines and M. A. Musen (Eds.), *Proceedings of the 9th Banff Knowledge Acquisition for Knowledge-Based Systems Workshop, Alberta, Canada*, SRDG Publications, University of Calgary, Canada.

[10] Boden, M. A., 1991. *The Creative Mind: Myths and Mechanisms*, Abacus, London, UK.

[11] Waibel, A., Hanazawa, T., Hinton, G., Shikano, K. and Lang, K. J., 1989, Phoneme recognition using time-delay neural networks, *IEEE Transactions on Acoustics, Speech and Signal Processing*, **37**, 328–339.

[12] Potter, S., 1997, *The development of machine learning architectures for engineering design*, Report Number 16/97, School of Mechanical Engineering, University of Bath, Bath, UK.

Airframe optimization based on structural evolution simulation by means of the GA and sequence of numerical models

V.A. Zarubin*, A.V. Chernov**,
E.F. Filatov**, A.V. Teplykh**

Professor of Samara State Aerospace University
Moskovskoe Shosse, 34, Samara, 443086, Russia
Telephone: +7-8462-587340
Fax: +7-8462-351836
E-mail: zarubin@mb.ssau.samara.ru

**Graduate student of Samara State Aerospace University*

Keywords

Sturctural optimization, Aircraft structure, Design variables, Conceptual design, Genetic algorithms.

Abstract

The peculiarities of aircraft structural optimization are discussed. It is shown that structural mass is very important indicator which must be minimized for successful airframe. However the accurate optimal problem formulation and solution are possible if some of most influential parameters are defined in advance and fixed during search of structural mass minimum. Therefore to include these influential parameters in the set of design variables (DVs) more general function, relative expenses, is chosen as a goal function and structural optimization problem is formulated as a task of airframe relative expenses minimization subject to broad spectrum of constraints. Vector of design variables {x} is divided into three subvectors {s}, {r} and {m}. Vector {s} describes principal parameters which define aerodynamic layout, structural scheme and structural material type. Vector {r} consists of shape and skeleton parameters, for example position of spare in wing structure or angle of stringers orientation. Material distribution parameters like thickens or area of cross section are described in {m} vector. Optimization problem solution is presented as evolution of the conception, therefore the design process is an iterative one, where the sequence of models are used and values of design variables are verified. GA is used in external loop of the algorithm for search of the best set of s-type of DVs which are treated as structural genes. In inner loops the genotype string of each individual is decoded into phenotype and the embryo is developed as the sequence of numerical models. Skeleton (r-type) and muscular (m-type) parameters are optimized by ordinary methods of structural mass minimization. Conceptual design of short-range passenger aircraft is presented as a numerical test. The test demonstrates the workability of the method, algorithm and software created in this research.

Introduction

New aircraft usually is conceived after firm confirmation of the demand in such machine at aviatransportation market what must follow from detailed analysis of this market and evaluation of sufficient probability to sell successfully these new aircraft after their production. This analysis is complicated process, but the result is formulated clearly, for example, there is need for 700 passenger jets for 100 - 120 passengers, range is equal to 1500 - 2000 kilometers and so on with definition of cruise speed, runway length, landing speed, etc., etc.. This description of main ideas what the new aircraft must be is developed into more detailed aircraft presentation at different design stages, where the idea transforms to preliminary design, then detailed design, experimental production and natural testing, then mass production and exploitation. Aircraft computer models are used at all of these stages and the later stage we are at, the more detailed models are used. Computer models are implemented not only for checking the ideas but for efficient search of the better variants and optimization of the design variables. Computer models simulate all processes of aircraft production, maintenance and modification. The decision of aircraft writing off is made on the results of computer analysis too. These processes resemble the life cycle of living organism which is developed from embryo (preliminary design) into living being (detailed design and production), its growing up (maintenance, repairs and modification) and death (writing off).

Success of new aircraft depends of many factors. Some of them are well defined, for example, availability of new engines, new avionics, equipment, advanced maintenance technologies, etc., but some are pure pieces of luck, which is difficult to predict. In this paper we will concentrate on airframe structural optimization, therefore it is very important to understand the factors which define the success in aviation structural design. From this point of view we attract the attention to several peculiarities, which differ aviation structural design from other load bearing structures design. The main of them is that structural mass is very clear criteria of structural success and it is obvious that aircraft which has minimum of structural mass and all constraints fulfilled will have benefits in comparison with heavier one. Structural mass is a function of aerodynamic and inertia loads which are functions of structural mass itself. The only way to solve the problem of accurate load definition is usage of sequence of verified structural and aerodynamic models step by step. Again this process resembles an embryo development.

Aircraft production and maintenance phases are characterized with the important role of CAD/CAM/CAE too. The parallel engineering technology, which is used at all phases and processes of structural life, starting from the initial design stage, then in detailed design, production, operation and up to decision making about quit from operation, uses a virtual object model, which is a totality of different structural models, including geometrical, elastic, inertia, aerodynamic, technological, economical, etc. Virtual model gives a chance to simulate the whole life cycle of a structure and accumulate data for its improvement. The improvement can be achieved with the help of different methods of optimization and the genetic algorithms (GAs) in particular.

The modern structural analysis is based on the methods of numerical simulation. Finite element (FE) analysis is widely used for the purpose, but only optimization makes this tool a real mean for structural design.

History of structural optimization application to aviation structure design has several milestones. Discrete methods of structural analysis and finite element method appeared in design practice in the 60's and gave a good chance to describe the structural material distribution parameters or design variables (DVs) in a natural way as cross section areas or thickness of structural elements presented by finite elements. Fully Stressed Design (FSD) algorithm was and continues to be an effective technique to find rational material distribution with stress and fabrication (minimum or maximum structural element size usually) constraints.

Structure is fully stressed if each structural element has maximum allowable stress at least in one of load cases. The recurrence relation of FSD is very efficient in structural sizing with stress constraints only. Several iterations (5-7 usually) are sufficient for convergence to FSD. This is very important feature, because each iteration includes the full procedure of FE- analysis.

The FSD approach has many known drawbacks like inability to find the global optimum, impossibility to work when different materials are used in structure and some others. But in the RIPAK package [1] the FSD realization includes some special techniques, which make this algorithm very useful for real design practice. For example, DVs can be combined in groups and the sizes will have changes during optimization with keeping of the relations prescribed to the group.

In the 70's Optimal Criteria (OC) and Mathematical Programming (MP) methods were developed and applied for structural mass minimization subject to broad spectrum of constraints: stresses, displacements, frequencies, dynamic response, aeroelasticity, etc. Then in 80's shape and skeleton parameters were included in practical structural optimization. The optimization tools in structural design became powerful means which made it possible to find the effective solutions and to create competitive airframe structures. The optimization subsystems are part of all commercially available finite element packages now [2,3].

Optimization subsystems support search of shape, skeleton and material distribution parameters. Structural mass is a goal function as usual. However some of very influential parameters, for example type of structural material, type of structural load bearing scheme (monocock, semimonocock, truss, etc.), topology of this scheme, some geometrical parameters like wing aspect ratio or angle of sweepback are not included into set of Design Variables (DVs). These parameters are defined by chief designer and his close team from the preliminary design department. They use different models of aircraft efficiency analysis and define the main parameters which describe aircraft appearance, performance and major features. The OC and MP methods are not useable for multiextremal functions with discrete DVs. New methods and approaches should be implemented for the problem solution.

The GAs are the means for this purpose. Robust performances of the structure are defined at this stage and the GA [4,5] is the best tool to provide such feature for the airframe.

The accuracy and quality of conceptual design predetermine the success of the whole project in a great amount. The more detailed simulation, numerical testing and optimization are done at this stage, the fewer mistakes from this stage will penetrate into following ones. To provide such kind of simulation for each possible variant of conceptual design we have to supply the GA with the control functions over all these processes. As a result the GA becomes a shell which envelops all these processes of structural simulation in the computer net combining distributed DBs and applications.

The aim of this paper is to present the multidisciplinary approach to aviation structure optimization based on complex structural and efficiency analysis and combination of the Genetic Algorithms (GA) with ordinary optimization methods [6,7]. The GA is used for forming the idea (conceptual design) which is then developed into detailed design with the help of the totality of computer models and Sequential Linearization in the Optimality Criteria Methods (SLOCM) [2] optimization.

The tasks which have to be solved to reach this aim are.

1. Analysis of the optimality criteria and optimal problem formulation.
2. Methods, algorithms and software creation.
3. Numerical testing.

The remainder parts of this paper contain the results, achieved by the authors on the way of these problems solution. The conclusion and plans for future works are formulated at the end.

2. Optimal problem formulation

2.1. Design variables

Parameters which are varied during different design stages can be presented as a vector $\{x\}$ which consists of three subvectors: $\{m\}$, $\{r\}$ and $\{s\}$.

$\{m\}$ is a vector of structural material distribution parameters, for example, areas or thickness of structural elements cross sections. Usually these DVs are defined in FE-model as geometrical parameters of finite elements and they don't influence FE-mesh.

$\{r\}$ is a vector of shape and skeleton parameters. In FE-model these parameters are presented as height, radius, angle, position, etc.. These parameters define FE-mesh geometry, but don't influence topology. For example, the position of spar in wing chord percentage or stringers orientation in wing panel don't influence finite element topology description but they change the node coordinates if varied.

$\{s\}$ is a vector of main parameters which describes aircraft appearance, structural type, material, topology, etc.. The variation of these parameters may lead to a new FE-mesh with different type and number of elements and node coordinates. For example, in [6] the helicopter tail-boom structure is presented by two load bearing scheme types, truss and semimonocock, which led to principally different FE-models for each type of structural scheme.

2.2. Constraints

During structural design we have to satisfy many demands and constraints imposed upon the structure. Models of FE-analysis allow us to calculate values of many of constraint functions like stresses, displacements, frequencies, critical loads, critical velocities, etc. Preventing of violation of these constraints results in structural mass increase. Therefore minimum of structural mass is a very important demand for successful structure. However it must be achieved by variations of m- and r-type parameters only. The parameters of s-type, as it was said before, have to be defined on the basis of more general criteria, like aircraft efficiencies.

2.3. Goal function

Choice of the criteria for aircraft structure evaluation during the design stage is a very important question. These criteria influence not only the aircraft performances but its future life and fate.

If main parameters {s} are defined it is a common practice to use structural mass as a goal function of {m} and {r} DVs, but for {s} parameters this function doesn't work properly. For example, wing aspect ratio λ has a positive sensitivity coefficient respect to the structural mass, what means that λ increase leads to structural mass increase. But there is a desire to make λ bigger with the purpose to reduce the induced drag. Structural mass in the role of goal function doesn't allow to do this and λ will be placed to its minimum value in such optimization.

There are many other main parameters from s-type group which in the case of structural mass minimization take its boundary values. Wing profile thickness will try to attain its maximum, angle of sweepback - minimum, fuselage length aspect ratio - minimum too. But these three parameters have influence to drag, velocity, fuel and transportation efficiency with opposite effect in comparison with their influence to structural mass.

Take-off mass seems to be more general criteria which includes links between these conceptual DVs and fuel efficiency. But from the numerical tests [11] it follows that structural mass is more sensible to changes in comparison with the mass of fuel. Therefore after all these improvements of aerodynamics the take-off mass becomes bigger due to structural mass increase. Some reduction of fuel mass is not considerable.

It is obvious that we must take more general criteria of structural efficiency than structural or even take-off mass.

Two criteria of transportation efficiency E_{trRove} and E_{trkv} were taken into consideration [8,9]. They have a form:

$$E_{trRove} = 1 / G_{com} K_{cruis} , \tag{1}$$

and

$$E_{trkv} = k_v \, m_{off} \, v_{cruise} / q_L \, m_{empt} \tag{2}$$

where G_{com} is relative payload mass (absolute payload related to take-off mass); K_{cruis} is a coefficient of aircraft aerodynamic performance in cruise regime; k_v is a coefficient of relations between flight and cruise speed; m_{off} is a take-off mass; v_{cruise} is a cruise speed; q_L is a fuel consumption per kilometer; m_{empt} is operation empty mass of aircraft.

Numerical tests [11] of these criteria revealed that they influence upon some parameters as it could be predicted, for example they resist the reduction of wing aspect ratio. But for many other parameters, for example angle of wing sweepback or material type they don't work.

It became obvious, that for s-type of DVs determination we must use more general criterion or set of criteria. Therefore we chose a_{rel}, a relative expenses [9], as more advanced one.

$$a_{rel} = a + a_{inv} , \tag{3}$$

where a_{rel} is a relative expenses; a is a prime cost of transportation; a_{inv} is a relative investment. And

$$a = A / K_{com} G_{com} V_{fl} , \tag{4}$$

where A is the expenses for maintenance and exploitation of the aircraft in one flight hour; K_{com} is the coefficient of commercial payload (not full loading due to season changes in transportation); G_{com} is a mass of payload; V_{fl} is a flight velocity.

Structural mass is presented indirectly in a and directly in a_{inv}. The expenses becomes smaller with structural mass decrease.

This goal function (3) takes into consideration most of the parameters which are interested for us at early design stages. But of course this criterion should be developed in the future works.

2.4. Optimal problem formulation

Optimal problem is formulated as following.

Minimize $a_{rel}(x)$ subject to $G_j(x) \leq 0$ and $x^l_i \leq x_i \leq x^u_i$. (5)

where vector $\{x\} = \{s, r, m\}$; $i = 1,...,n$; n is number of DVs; $j = 1,...,l$; l is number of constraints.
Expressions (5) mean that we want to find such parameters of aircraft appearance, shape, skeleton and material distribution which minimize relative expenses and satisfy the constraints imposed upon the aircraft structure.

3. Method of problem solution

Method is based on the aircraft models adaptation during design process.

In accordance with [4] the mechanisms of adaptation provide a progressive modification of some parameters which in GA approach are considered as a set of genes combined in a chromosome.

In structural design this approach suggests that all DVs are encoded as genes and combined in a string or chromosome. In our research only principal DVs are treated like genes (in accordance with [6] they are called the DVs of s-type), other DVs are the parameters which describe not the set of chromosomes, but the external sizes of the object. It is very natural to propose that s-type of the DVs are the principal parameters, which have to be encoded and treated as genes. And other DVs are the parameters of structural "skeleton" (r-type) and "muscular" systems (m-type) which have to be treated by correction and training accordingly [6].

Another distinctive peculiarity of our approach is the presence of special procedures of structural development, interpreted in [7] as training or life simulation and correction of body (not genes) parameters in the mechanisms of adaptation.

4. Algorithm of problem solution.

Following steps and actions lead to problem solution.

0. Preliminary research. Formulation of main demands. Prototype design. Initial data forming.
1. Optimal problem formulation.
2. The initial population of several possible variants of load bearing structure is created. s-type parameters are encoded in a chromosome (string) for each individual.
3. The life of each individual is simulated in accordance with the constraints and demands imposed.
3.1. Genotype is decoded into phenotype.
3.2. Structural mass is calculated in accordance with the simplest empirical model of aircraft existence. Prototype parameters and statistics data are used.
3.3. Calculation of the take-off mass of the first approximation.
3.4. Load definition.
3.5. Creation of the FE-model of the first level.
3.6. Structural mass of FSD calculation.
3.7. Calculation of the take-off mass of the second approximation.
3.8. Conceptual design.
3.9. Creation of the FE-model of the second level.

3.10. Creation of the aerodynamic, inertia and other computer models of totality of discrete models of the second level.

3.11. Minimum structural mass problem formulation.

3.12. Structural optimization by SLOCM.

3.13. Economical model tuning.

3.14. Goal function calculation.

4. The exit conditions are checked. If the number of iterations is exhausted or the process converged, for example the difference between best and worst individuals is smaller than prescribed one, then go to 6, else go to 5.

5. The GA's selection, crossover and mutation procedures from [5] are used. The new generation is created. Go to 3.

6. Stop.

5. Description of the software.

Goldberg's programs from [5] are taken as a basis for the GA realization. The POLINA FE-package [10] is used for structural simulation and structural mass evaluation. The package consists of following blocks and stand-alone programs.

1. Main block provides initial data (like size of population, number of iterations, chromosome structure, details of interactions with FE-package) input, program tuning and control.

2. Block of encoding-decoding transformations of genotype into phenotype and back.

3. Block of structural mass of the first approximation level evaluation.

4. Block of take-off mass of the first approximation level evaluation.

5. Generator of FE-model of the first level.

6. Generator of loads for FE-model of the first level.

7. Block of structural mass of the second approximation level evaluation.

8. Block of take-off mass of the second approximation level evaluation.

9. Block of goal function value calculation.

10. Block of GA selection.

11. Block of GA crossover and mutation.

POLINA FE-element package [10] is used for structural behavior simulation and structural mass of FSD estimation.

6. Numerical example.

The short air route passenger aircraft was taken as a subject for numerical demonstration.

Preliminary research [11] revealed that the 100 seats plane with the range of 1500-2000 km will be in demand for next 20 years.

The main requirements to the aircraft under designing formulated in the research [11] prescribe different variants of cabin layouts, crew staff of 2 pilots and 3 stewards, 800 km/h of cruise speed, 10000 m of cruise flight height, 1800 m of runway length, 220 km/g of landing speed, 2000 km of range with 110000 kg of payload. Other demands define the levels of reliability, safe damages, safety, comfort. Technological, maintenance and environmental restrictions are formulated too.

Optimal problem was formulated as relative expenses (3) minimization subject to minimum structural mass and behavior constraints.

Vector $\{s\}$ consists of the following parameters: λ is wing aspect ratio; λ_f is fuselage aspect ratio; c_0 is relative aerofoil thickness; η is wing's narrowing; χ is backsweep angle; H_{cr} is cruise height; V_{cr} is cruise speed; mat is type of material. Genotype-phenotype relations are presented in Tables 1 and 2.

The initial population of 50 individuals was created by random way. The best individual in this generation has the parameters shown in Table 3. The parameters of the worst one are placed in Table4.

Table 1. Genotype-phenotype relations, 2 bits strings

DV:	c_0	η	λ_f	mat
String:	--	--	--	--
00	0.10	2	6.5	D16
01	0.11	2.5	7.5	B95
10	0.12	3	8.5	AlLi1
11	0.13	3.5	10	AlLi2

Table 2. Genotype-phenotype relations, 3 bits strings

DV:	λ	H_{cr}	V_{cr}	χ
String:	---	---	---	---
000	7	3	500	0
001	8	6	600	5
010	9	7	650	10
011	10	7.5	700	15
100	10.5	8	750	20
101	11	8.5	800	25
110	11.5	9	850	30
111	12	10	900	35

Table 3. The best individual of the initial population

λ	λ_f	V_{cr}	χ	c_0
10	8.5	700	35	0.11

η	H_{cr}	mat	fitness
2.0	10	D16	11.12535

Then algorithm of the fourth section is implemented. In this research the algorithm is reduced and steps 3.8-3.12 are excluded, what means that structural optimization is done by FSD approach only without full structural simulation and SLOCM optimization implementation. Each genotype is decoded into phenotype and simple empirical models [8,9] are used for aircraft appearance forming and take-off mass evaluation. Finite element model and loads are determined for each individual. Several examples of FE-models are presented at fig. 1.

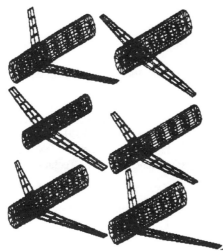

Fig. 1. Examples of FE-models of six individuals from initial population

Then each individual is optimized by FSD approach, its structural and take-off masses of second approximation are calculated and goal function (relative expenses) is evaluated too. The diagram of fitness values are presented at fig. 2.

Table 4. The worst individual of the initial population

λ	λ_f	V_{cr}	χ	c_0
10.5	6.5	900	35	0.13

η	H_{cr}	mat	fitness
3.0	8	AlLi1	29.65755

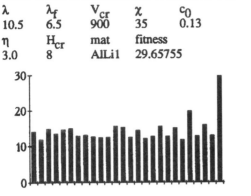

Fig. 2. Fitness diagram of the initial population

Thirty iterations with the probability of crossover equal to 75% and mutation rate equal to 0.01% were done. The process of fitness changes during iterations is presented at fig. 3.

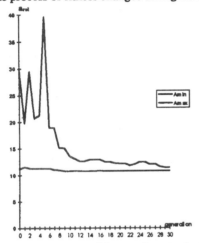

Fig. 3. The process of fitness changes during iterations

In the initial population maximum value of goal function was equal to 29.65755, minimum was equal to 11.12535. During first 7 generations the maximum goal function value had big alterations. After this the tendency for maximum value decrease appeared. Minimum had the reduction tendency from the beginning. The best individual appeared in the ninth generation. Then it disappeared in the 11th generation and appeared again in the 12th one. The 30th generation has the worst individual with 11.39483 and the best one with 10.73364 value of goal function. There are 26 individuals in the population with such value of goal function. The fitness diagram on the 30th generation is presented at fig.4.

Fig. 4. Fitness diagram of the thirtieth generation

The best and the worst individuals in this generation are presented in Tables 5 and 6.

Table 5. The best individual of the thirties population

λ	λ_f	V_{cr}	χ	c_0
10.5	6.5	750	35	0.10
η	H_{cr}	mat	fitness	
2.0	10	D16	10.73364	

Table 6. The worst individual of the thirties population

λ	λ_f	V_{cr}	χ	c_0
9	6.5	750	35	0.10
η	H_{cr}	mat	fitness	
2.0	8.5	D16	11.39483	

Some additional tests [11] showed that for population size equal to 50 and string's length of 20 bits the program reaches the stable optimum after 12-15 iterations.

7. Conclusions and further works

The paper demonstrates the ability of GA to be implemented for aircraft structural optimization. The process of conceptual design based of simulation of airframe model evolution is presented. There are at least two distinctive features in our approach to GA usage in structural optimization.

1. Fitness is a function of the set of DVs which consists of genes and ordinary DVs. For the airframe structure the division of the whole vector {x} of DVs into subvectors {s}, {r} and {m} is natural and well reasoned by their influence upon the structural model. Structural material distribution parameters or m-type of DVs don't change FE-mesh. These parameters or sizes of element cross sections play role of structural muscles. Shape and skeleton parameters or r-type of DVs change FE-mesh geometry. For example, wing ribs may be perpendicular to the forward or rear spar. The orientation influences coordinates of rib's nodes. But it doesn't change mesh topology, element types, material features. And only s-type or conceptual DVs influence airframe's and its model's appearance: is it canard, T-tail or normal stabilizer; does it have an inner beam or not in wing structure; what structural type is used in rear part of fuselage, truss of semimonocock; what type of structural material is implemented and so on. These parameters may change structural appearance and its FE-mesh radically. And only these DVs play role of structural genes.

2. Phenotype has morpho- or embryo-genesis phase during which each individual is developing before taking part in competition for place in mating pool and for chance to give birth for the offspring. Aviation structures are defined by aerodynamic and other loads they create and bear. These aerodynamic, landing, taxi and other loads are defined by the structure itself. Therefore in aviation structural design its natural to have a sequence of structural models used in iteration process with more clear definition of structural parameters first and loads then. In this paper the phenotype development process is presented in step 3 of the algorithm of section 4. The life of each individual is simulated in accordance with the constraints and demands imposed.

The future works of our team are planned in four directions.

1. Modification of POLINA package [10] with the purpose to increase the efficiency of the analysis and optimization codes to be acceptable for execution in an iterative process of structural optimization in the framework of a population of structures in which each individual is represented by a finite element model.

2. Usage of the totality of discrete models for accurate structural behavior simulation. The totality of structural models makes it possible to solve such complicated problems as the flutter speed evaluation, load distribution determination, aircraft balancing during the maneuver, dynamic response during landing and taxi, etc..

3. Analyzing of other GA tools and looking for or creation of the more effective methods, algorithms and software for structural optimization.

4. Creation of the Net Shell for Designing Control (NSDC) which provides a chief designer with the possibilities to present the design process as the set of subprocesses and tasks distributed through departments, teams, specialists and their desks; to prescribe the technology, data bases, data input, processing by particular software and output at each desk; to organize these data flow; to define time-table and responsibility, etc. The NSCD must be a fully portable program which can be tuned for operation in any kind of LAN at major modern computer platforms, available in design offices.

Acknowledgments

The research leading to this paper was supported by "ODA integrated systems" company.

References

[1] Komarov, V.A., Zarubin, V.A., et al, 1984, Automatization of aviation structure design based on FEM. RIPAK FEM package, Samara Aviation Institute, Dep. VINITI, 6.04.84, #3709

[2] Zarubin, V.A., 1994, Structural sensitivity analysis and optimization in the RIPAK package. Part I. Structural Optimization, Vol.8, No. 2/3, Springer-Verlag

[3] Zarubin, V.A., 1995, Structural sensitivity analysis and optimization in the RIPAK package. Part II. Structural Optimization, Vol.9, No. 1, Springer-Verlag

[4] Holland, J.H., 1975, Adaptation in natural and artificial systems. Univ. of Michigan

[5] Goldberg, D.E.,1989, Genetic algorithms in search, optimization and machine learning, Addison-Wesly

[6] Zarubin, V.A., 1996, Multidisciplinary large-scale structural optimization and the place of genetic algorithms in it. Proceedings of 1st Int. Conf. on Evolutionary Computation and Its Application "EvCA'96", Presidium of the Russian Academy if Science, Moscow

[7] V.A. Zarubin, V.A., 1996, Genetic algorithm in the role of a shell for structural evolution simulation at the conceptual deign stage. Proc. of the 1st on-line workshop on soft computing, WSC1, Nagoya University, Nagoya, August 19-30

[8] Badyagin, A.A., Eger, E.M., Mishin, and V.F., Fomin, N.A. 1984, Aircraft design (in Russian). Mashinostroenie, Moscow

[9] Katyrev, I.Ya., Neimark, M.S., and Sheinin, V.M., 1991, Designing of civil aircraft (in Russian). Mashinostroenie, Moscow

[10] POLINA package user's guide, 1995, ODA Integrated Systems, Samara

[11] Chernov, A.V., Filatov, E.F., and Teplykh, A.V., 1997, Genetic algorithm usage in aircraft preliminary design (in Russian). Graduate project, Samara State Aerospace University, Samara

Simulated Evolution and Adaptive Search in Engineering Design - Experiences at the University of Cape Town

John Greene

Department of Electrical Engineering, University of Cape Town,
Private Bag, Rondebosch, South Africa 7701 . jrgreene@eleceng.uct.ac.za

Abstract

This paper reports on experiences in the Department of Electrical Engineering at the University of Cape Town in applying simulated evolution and adaptive search techniques to engineering design. It argues that there is potential, largely unexploited, for the widespread application of evolutionary and adaptive search methods in the routine design of engineering systems and artefacts. The flexibility that these techniques afford in constraint-handling and expression of complex design goals in mixed-mode optimization problems can best be exploited and advanced through their use by designers skilled and experienced in the task-domain. This in turn requires the techniques to be packaged in a form accessible and attractive to the design community at large. We have attempted to address this need by developing a compact package of robust, user-friendly black-box optimizers for use both by designers in industry and within our own graduate and undergraduate curricula.

1. Introduction

Research at the University of Cape Town's Electrical Engineering department encompasses Image Processing, Remote Sensing, Telecommunications and Control, with smaller groups working on Machine Design, Power System Stability studies, Instrumentation and Acoustics. In support of this work, modest activity has been mounted in recent years [1] to develop techniques of optimization by simulated evolution and adaptive search, which have been successfully applied in a number of these research areas [2], [3], [4], [5]. The transfer process has been facilitated by a short postgraduate course on "Adaptive and Evolving Systems" which culminates in a project in which students apply these techniques to their own research activity. A brief version of the material is also taught in a core elective final-year undergraduate course, and many undergraduate 'thesis projects' have been based on it. The success of this approach, and the ingenuity and fruitfulness of some of the projects, have encouraged us to address directly the question of making these tools more accessible to the non-specialist, resulting in the (ongoing) work reported here.

2. Engineering Design, Optimization and Search

Formal synthesis procedures exist in many areas of engineering design, but their use is rare in everyday design praxis, for they tend to be highly complex and limited in scope, and are often too brittle to cope with the complexities of the real world. In reality a good deal of conventional, routine engineering design amounts to a process akin to informal parameter optimization. Forms and structures are selected from a catalogue, explicit or implicit, of possibilities determined by established practice or the experience of the individual designer. This leaves a large number of free variables which are adjusted by a combination of intuition and trial-and-error, simulation and experiment, until a particular range of goals and constraints has been satisfactorily met. In the process, elements of solutions that have worked in other, similar, contexts are reused and combined; occasionally random variants are tried, and even errors may lead to serendipitous discovery. There are potential benefits in formalising and (partially) automating this process through the methods of simulated evolution and adaptive search, and sometimes such an approach opens up altogether new design possibilities.

To consider but one example: a study of the literature indicates many areas in which the possibilities opened up by recent dramatic technological developments have yet to be fully exploited due to the lack of an adequate theoretical basis and design methodology, and one of these is discussed in a recent paper by Avedon and Francis [6]. The authors point to the latent possibilities in digital control due to the availability of the modern digital signal processor which allows the synthesis of high-order controllers at very high data throughputs. They apply methods of Convex Optimization to address the otherwise intractable problems that arise in the design of such controllers, and demonstrate remarkable improvement in both performance and robustness as the order of the controller increases (up to the point where numerical instability in the optimization process sets in). This problem could also be addressed by evolutionary approaches, with a likely improvement in numerical robustness, as well as a relaxation of the constraints which they were forced to place on the transfer function of the controller in order to permit the use of convex optimization. Such instances could be multiplied.

Evolutionary and adaptive search is by no means *limited* to parameter optimization — more exciting and wide-ranging possibilities arise in the search for structures, topologies and algorithms. Designers are constrained not only by the limitations of materials and information, but also more subtly by habits of thought, conventional wisdom, and limitations of human imagination and comprehensibility. We still operate largely under assumptions of linearity, and consider only modular and hierarchically-structured systems, though it is clear in many cases that the resulting performance is inferior to that attainable if we were able to transcend these limitations and exploit the vastly augmented design space and emergent properties attendant upon less constrained and more holistic conceptions. Automated design may ultimately offer possibilities of escaping these constraints and accessing this richer world of possibilities, though for the present there is much useful work to be done to a more modest agenda.

Work at the Department of Electrical Engineering at the University of Cape Town has focused on the application of Genetic Algorithms and other Adaptive search procedures to routine design activity in electrical and electronic engineering, placing particular emphasis on the evolution of *robust* designs. It has been our experience that evolutionary stochastic techniques offer significant advantages over other approaches in terms of the ease and flexibility with which complex goals and constraints can be handled. We believe that the full exploitation and development of this potential requires the active participation of researchers experienced in the subject domain, and to this end we have attempted to make evolutionary and adaptive search techniques accessible and attractive to the design community at large. This paper outlines the approach adopted and discusses some issues that arise.

3. Particular Aspects of the Design Space

Is it possible to characterise engineering design problems in a way which sets them apart from the more general applications of evolutionary and adaptive search? Clearly no hard-and-fast distinction is possible but some general observations may be helpful:

- Many of the routine engineering design problems on which we have focused can be cast as parameter optimization tasks involving from 5-20 variables, each of which can be expressed with a precision on the order of 1% . Thus the configuration space is modest by the standards of, say, typical machine-learning or image-processing tasks, though still far too large for enumerative or random search to be a viable strategy.

- The evaluation of trial solutions is often *costly*. Objective functions are rarely explicitly available in algebraic form, and evaluation is likely to entail finite-element calculations, dynamic simulations or even multiple monte-carlo runs. Computer time will nearly always be dominated by the evaluation process, and overheads imposed by the mechanics of the search process are of relatively little consequence..

- The search space is often *highly constrained*, with the global optimum at the intersection of the performance surface with a constraint boundary. The proliferation of hard constraints may seriously impede the search dynamics, making it difficult for the algorithm to find feasible solutions in the early stages.

- Engineering design problems are often 'mixed mode' optimization problems involving a mix of variables including enumerative, Boolean, and integer data-types and 'real' variables of varying precision. This is perhaps one of the strongest reasons for the promise of evolutionary and adaptive search methods in this field; indeed in this respect they are unique and there is little in the way of competition.

It is rare in engineering design that the goal can be adequately expressed by a scalar objective value; nearly all such problems entail multiple (and often conflicting) criteria. Even in those rare cases where a particular performance criterion seems overwhelmingly dominant, a closer look will reveal a conflicting parallel goal - that of *robustness*. Needle-like optima are of little practical interest in engineering, and frequently performance and robustness criteria must be considered as competing components in a (possibly multi-component) fitness vector.

4. Designer Algorithms or Black-Box Optimizers?

Engineering design was the primary motivation of one strand of early research into simulated evolution [7] and work on Evolution Strategies has preserved this emphasis [8]. Holland's investigations on the Genetic Algorithm [9], motivated by an abstract understanding of the processes of natural evolution, was followed in 1983 by Goldberg's work applying it to an essentially engineering problem [10]. In the late 80s and early 90's the GA rapidly achieved prominence and was applied to a wide range of problems with an engineering flavour.

Current GA-related activity contains at least two distinct strands (although some individual workers encompass both) – research whose prime aim is to understand and refine the mechanism of the GA, and effort directed primarily at developing powerful optimization techniques for practical use. Although introducing many powerful extensions and developments, the former school has tended to remain close to the original conception, using binary encoding and placing great reliance on classical schema theory in directing development of the algorithms, notwithstanding occasional extensions of schema theory to higher-cardinality alphabets and real intervals. It has also tended to emphasise the universality of the GA as a weak method and eschewed *ad-hoc* problem-dependent modifications or the use of problem-domain knowledge in the search (though recent debate on the formal limitations of search seems to have modified this position; we shall say more on this below).

By contrast, the 'practical school' has from the beginning tended to take drastic liberties with the original form, limiting crossover to parameter boundaries, using real-valued vectors as chromosomes, inventing *ad-hoc* problem-domain-specific recombination and mutation operators, introducing heuristics for the global variation of control parameters, employing 'Lamarckian' learning, and even hybridising GA search with hill-climbing, simulated annealing, taboo-search and so on; it is rare today in the application literature to find applications which closely resemble the canonical "Simple Genetic Algorithm".

The case for an *ad-hoc* approach to GA-application has been persuasively argued by two influential books in the field: Davis [11] and Michalewicz [12], both on grounds of performance and also, in the case of Davis, as a matter of strategy. Davis argues that the community of practitioners with optimization problems to solve is much more likely to accept the hybridisation of existing algorithms with the GA (with clear payoff in terms of performance) than to adopt a wholly new and arcane-seeming approach. His book contains many case studies showing how effective this approach can be. Mitchell (who might be said to belong to the other school), in a review of Davis' book [13] points out that it is a valuable starting point for anyone who wishes to begin applying GAs to real problems. But, she warns, it is only a 'starting point' since a great deal of "tinkering with parameters, algorithms and forms of representation will still be required..." to develop an effective application.

However, while this is appropriate in the situation Davis describes (where there is a collaborative relationship between a design expert and GA specialist), our experience suggests that this need for 'tinkering' is precisely what is likely to deter the ordinary solo practitioner or group engaged in routine engineering design tasks from seriously considering evolutionary approaches. If it is true, as we stated in the introduction, that the full exploitation of the potential of evolutionary search in design requires above all its application by designers skilled and experienced in the subject domain (and sensitive to the possibilities and pitfalls therein) we must ask how this can be facilitated and, indeed, whether is a reasonable goal.

Our experience (which as well as application work in an industrial context includes the introduction of evolutionary design methods into the undergraduate and postgraduate engineering curriculum) suggests a clear affirmative answer to the latter question. We have found that there are many areas in which a design practitioner (subject-domain expert or sometimes student), provided with appropriate user-friendly tools for evolutionary optimization, can quickly produce excellent results inaccessible to classical approaches and not otherwise achievable, and occasionally may even advance the state of the art in the application of evolutionary methods to his or her field of expertise. We have therefore addressed the question of how evolutionary and adaptive search techniques can be presented in an accessible way which also helps to overcome a latent prejudice in the design community against a method which from their perspective seems to be something of a 'fiddler's paradise' and an intolerable distraction from the substantive design task in hand.

It is possible that a partial answer to this lies in the development of a GA which self-adapts to the task, and in which the control parameters are evolved along with the search variables. Some excellent work has been done along these lines [14], and the results seem promising, especially for tackling very large and very difficult optimization problems. Nevertheless we believe that there is an important niche at the other end of the complexity spectrum, and that black-box optimizers exist which are adequate to an important range of real-world engineering design tasks without any need for parameter adjustment Freed from the distraction of optimizing the algorithm itself and seeking task-specific representations of trial solutions, the designer is able to devote full attention to the problem in hand, and to exploiting the rich and flexible goal-setting and constraint-handling possibilities which are the *forte* of evolutionary methods. Thus, we would argue, there is a role for simple, robust black-box optimizers in the engineer's tool-kit, and the design of an effective and parsimonious tool-kit is the route we have chosen to pursue.

5. Is There Such a Thing as a Black-Box Optimizer?

We know now that, 'averaged over all search tasks' no search algorithm is better than any other [15]; all search methods have biases, and where these happen to be well matched to the problem, the search will be successful. But how do the theorems on the formal limitations of search relate to the overwhelming empirical evidence that, for example, over a vast range of search problems of practical interest and relevance, Random Search and (phenotypic) Gradient Ascent are hopeless, Simplex search rather better but limited, and Simulated Annealing, Evolutionary Strategies and Genetic Algorithms remarkably powerful. What exactly are the common biases in these problems which lead to the nearly universal superiority of the latter algorithms over the former in a domain which includes combinatorial and parameter-optimization problems as well as much more generalised search tasks? Like Schema Theory, the theorems on the formal limitations of search seem to be powerful general theorems whose practical relevance and applicability are problematic, and there is a need for further theoretical development if they are do be depended upon to guide search practice.

In practical terms, the most urgent need is for clear guidelines relating the biases in particular classes of problem to the specific capabilities of the many available search techniques, but here the state of our knowledge is woefully inadequate. In few areas is there any consensus. We know that in certain classes of problems involving Boolean circuit synthesis the absence of local minima makes hill-climbing a natural choice and it is generally believed that the hill-climber is optimal in certain geometric template-matching problems (partly because they lend themselves to economic incremental fitness evaluation techniques). The effectiveness of genetic methods in certain image-processing tasks has been called into question, and Simulated Annealing claimed to be a better option. More generally, Schaffer, Mathias and Eshelman [16] make a distinction between algorithms using 'pool-based' and 'pair-based' generation, and argue convincingly that the latter can be expected to be superior in optimizing functions in which there is significant inter-variable linkage (to which the biological term 'epistasis' is sometimes applied). By and large, however, there is little clarity and techniques seem to be chosen on the basis of individual preference and familiarity rather than any rational basis. This raises serious problems when choosing benchmarks to evaluate candidate algorithms as potential black-box optimizers.

6. Benchmarking

With the strictures of the No Free Lunch theorem in mind, and having experience of the capriciousness of benchmark testing, we have tried to identify a set of benchmark tests which, as far as possible, is representative of the types of problems to which we hope to apply evolutionary optimization. In view of the extent of the potential problem domain and the lack of an effective functional classification of problem types, much of this remains conjectural, but performance over such a set of tests certainly inspires greater confidence in the algorithm on the part of the in the target design community than tests focused primarily on numeric functions.

In evaluating candidate search methods we have addressed the following design tasks:

- linear circuit synthesis: filters, equalisers etc, including parasitic effects.
- linear controllers: coefficient-matching, frequency-shaping, time-response
- discontinuous controllers: bang-bang-, variable structure sliding mode controllers.
- distributed systems: microwave stripline amplifiers; yagi antennas, array synthesis
- time-discrete systems: FIR and IIR filters
- non-linear regression, system identification and time-series prediction
- automatic calibration and non-linear compensation of instruments.

In addition, numeric function optimization is needed for the preliminary screening of algorithms and to facilitate inter-comparison with published results. It is scarcely possible to ignore the very widely-used 'De Jong' test suite, though many of the functions have been criticised as benchmarks on the grounds of separability (lack of linkage)[17] and susceptibility to simple bit-based hill-climbing [18]. We augment them with the 10-

dimensional Griewank and Rastrigin's functions and the Michalewicz and Langerman functions. Keane's 'bump' function [19] (a highly multimodal 20-dimensional function with a global optimum situated on a constraint boundary) was designed to capture some of the features of engineering design problems and indeed the performance of search algorithms on it appears to correlate well with performance on our benchmark engineering design tasks.

7. The Basic Optimization Toolkit

After a fairly extensive investigation of a wide range of published algorithms on the above benchmark problems we have selected three basic optimization algorithms: a bit-based hillclimber [18], the CHC Genetic Algorithm [20] and an adaptive stochastic search algorithm called Population-Based Incremental Learning (PBIL) [21]. The criteria of choice were as follows: we sought a minimal set of algorithms which, collectively, would be capable of rapidly finding good solutions to problems in all the above-listed classes of engineering design tasks. Equally important they should be attractive to the design community at large, and this was interpreted as being easy to use with a minimal need for experimentation with alternative representations or adjustment of control parameters. Both CHC and PBIL were used with an unvarying set of control parameters (as detailed below) and with constant population sizes of 50 and 100 respectively.

Although testing is still under way, it can already be stated that the selected algorithms appear to meet these requirements: straightforward problems from many of the categories listed above are solvable using simple concatenated binary encoding (Gray coding in the case of CHC) and there is every expectation that satisfactory solutions will be found in all the categories. No clear pattern has yet emerged regarding the specific suitability of a particular algorithm for a particular type of problem – indeed many of the problems yield easily to any one of the algorithms. We have encountered some surprising and potentially useful results in the process: for example, in a search to design a high-order butterworth analogue filter, the algorithm discovered not only the standard textbook solution but a number of alternate realisations which do not seem to be accessible to the classical synthetic approaches (and some of which may offer practical implementation advantages over the well-known solution). In the course of non-linear designs such surprise discoveries are common, but we had not expected it in the course of seeking a solution to a straightforward *linear* design problem.

I will now discuss the algorithms and the reasons for their selection in more detail.

8. Bit-based Hill-Climbing

The hill-climber is not usually thought of as a powerful optimizer, but we have included one in our basic 'tool-kit' for a number of reasons:

- There are problem domains in which it is known to perform well, and may even be optimal. There is a certain complementarity between the GA and the bit-based hill-climber in that there are some problems which are easy for the one and difficult for the other.

- Multiple exploratory runs of a greedy hill-climber from a dispersion of starting points can give useful insights into the nature of the search terrain.

- There is scope for hybridising the hill-climber with the GA in various ways: a hill-climbing step can be incorporated into each generation to improve the 'best.' result through a Lamarckian learning process, and the superior peak-finding ability of the hill-climber in some situations can be used to augment the GA search at the end of a run. The hill-climber can also be useful as part of a clustering operation in which diversity is improved by replacing clusters of highly-performing trial solutions by their mean, thereby avoiding 'wasted' function evaluations.

- A simple adaptation of the bit-based hill-climber suggested by Dekker and Kingdon [22] yields the "morphic hill-climber" [23] which appears to be a remarkably powerful optimizer in its own right with (for our purposes) the outstanding merit of being virtually control-parameter-free. Instead of restarting the search when a local optimum is encountered, the space around the current search point is re-mapped by simply changing the radix in a normal concatenated-string encoding (bases between 2 and 20 are chosen at random). We have as yet conducted only rudimentary testing on the morphic hill-climber but the interim results are encouraging and we intend to pursue this further.

9. The CHC Algorithm

The CHC algorithm, introduced in 1991 by Eshelman [20], is relatively well-known and has a well-deserved reputation as a powerful black-box optimizer. It has been extensively tested in competition with other GAs in a wide range of applications and has consistently performed well (one of the few exceptions being the geometric pattern-matching task at which the hill-climber excels) [17].

CHC uses cross-generational elitist selection, 'heterogenous' half-uniform crossover with 'incest prevention' and no mutation in the ordinary sense; instead, whenever the search converges there is a 'cataclysmic mutation' phase in which the population is regenerated from the current best individual and a set of variants of it in which 35% of the bits are randomly mutated. We have used it in exactly the original form and have not succeeded in 'improving' it in any way by tuning its parameters or varying its population size. Interestingly, Whitley *et al* make a similar comment on the robustness and seeming (at least local) optimality of this algorithm [17].

Recent results published by Eshelman and Schaffer [16] as well as Whitley *et al* [17] suggest that for numerical optimization a real-coded version of CHC would be even better than (or at least complementary to) the Gray-coded version, but we have not yet explored this possibility.

10. Population-Based Incremental Learning (PBIL)

We have found PBIL to be a very powerful and effective search algorithm which has incomprehensibly (we believe) been almost entirely ignored in the literature. It was introduced by Baluja [24] as an abstraction of the GA, in which recombination was dispensed with and the statistics normally maintained in the population represented explicitly in a "probability (or prototype) vector" which is updated in a competitive learning process. As in the standard GA, trial solutions are represented by binary strings, though in the case of PBIL the concatenation of the binary-encoded variables is purely a matter of trivial computational convenience, and the ordering of the parameters in the string is of no consequence whatsoever. This is an advantage from the point of view of ease of use and consistency between differing implementations, but of course it also means that linkage between variables is ignored. This does not seem to result in the limitation in performance which we expected, but it is the reason why we have retained in addition a more conventional GA in our toolkit; it seems reasonable to expect systematic and interpretable differences in performance between the "pool-based" and "pairwise" generative mechanisms, though we have not thus far managed to isolate them.

PBIL is extremely easy to understand, implement and (most importantly for our purposes) to use. A population of typically 100 trial solutions, each in the form of a K-bit string B, are generated randomly, with the expectation of 1s in particular bit-positions determined by the real values in the range $(0,1)$ specified by the corresponding elements of the 'probability vector' PV.

The generation of the test solutions is a simple random sampling of the probability vector:

$$B_i = 1 \text{ if } Random > PV_i \quad \text{else } B_i = 0$$

where $Random$ is a number generated with uniform PDF in the range $(0, 1)$.

In the initial absence of information about the location of the optimum, all the PV elements are initialised to 0.5. The trial solutions are evaluated and only the fittest individual (B^*) is retained. At the end of each generation a small adjustment is made to the elements of the PV to ensure that in the next generation the expected average Hamming distance between the trial solutions and B^* will be reduced. This is achieved by moving the elements of PV 'toward' B, in a manner reminiscent of competitive learning, by a learning factor L (typically 0.1). To maintain diversity and prevent premature convergence, Baluja proposes a small background mutation operation on PV, which we have implemented as a 'forgetting factor' F (typically 0.005). In each generation all the elements of PV are moved a fraction F in the direction of the 'neutral' value 0.5.

i.e.   update: $PV(t+1) = (1-L).PV(t) + L.B^*$ where $L = 0.1$

           forget: $PV(t+1) = PV(t) - F.(PV(t) - 0.5)$ where $F = 0.005$.

Baluja has published an extensive set of tests on a wide range of benchmark problems [25] and we [26], and others [27], have carefully repeated and extended those tests in respect of numeric function optimization. On the usual benchmark functions (especially the more difficult ones) PBIL significantly outperforms simple implementations of the standard GA. Baluja has also published details of its application to large-scale neural-net training [28] and three-dimensional location tasks in surgery [29]. The algorithm can be extended and modified in various ways (e.g. updating PV away from the least-fit trial solution as well as toward the fittest one) but in most cases the improvement in performance is marginal and somewhat task-dependent. In the interests of simplicity and robustness we normally use the simple standard version, invariably with the same population size (100) and control parameters ($L = 0.1$; $F = 0.005$).

PBIL demonstrates a remarkably consistent pattern of convergence; with the individual elements of the PV diverging from 0.5 roughly in sequence from left to right and converging on 0 or 1. The rms deviation of the elements from 0.5 follows a rather predictable curve, and can be used as a useful indication of the state of convergence of the search process. Sometimes examining the pattern of convergence can shed light on characteristics of the search space; as a simple example, failure to converge on the part of the PV element representing a particular variable can be a useful indication of insensitivity of the design to that variable. Early in the search a speedup can be obtained by aborting a generation whenever an absolute fitness increase is detected, and toward the end of a search a good deal of time can be saved by maintaining a lexicographically-ordered list of evaluated trial solutions to avoid needless repetition of the evaluation process. Apart from these small changes we use the algorithm substantially as Baluja describes it.

Even if PBIL were to match or excel CHC in search effectiveness there are good reasons for retaining the latter. Apart from the issue of linkage, another way in which PBIL differs from a GA is in the 'winner-takes-all' nature of its fitness competition. This means that it is not possible to initialise the search by seeding the population with known good solutions in the hope that that a better solution will emerge through the combination of high-evaluating schemata; all but the best-performing trial solution are immediately eliminated and the following generation stochastically created. For similar reasons, it is not easy to see how niche-induction can be implemented, which is essential for multi-criteria optimization. These are very serious shortcomings in the context of engineering design. A possible way forward has however been indicated in a recent publication by Baluja, in the form of a parallel implementation of PBIL, which Baluja calls pPBIL [30].

The pPBIL algorithm consists of a number (Baluja uses 10) of PBIL implementations running effectively in parallel and, most of the time, independently. Each population is created by sampling its own probability vector, and each PV is updated according to the best-performing individual of its own population. However, each of the 10 PVs is actually a *pair* of PVs, *PVa* and *PVb*, and individuals are created by sampling both parts using uniform crossover. Every 100 generations there is a cyclic swap of the *PVb* components between the populations, giving an effect akin to migration in island-model GAs. According to Baluja's published results this has a markedly beneficial effect on PBILs performance as an optimizer; it also opens the way to possibilities of selective initialisation, niche induction and multi-criteria optimization. We are only beginning to experiment along these lines.

PBIL also lends itself to extension to variable length representations, which is of considerable interest in evolving adaptive structures such as neural networks and electronic circuits of varying size and topology. Two colleagues, Jakoet and Schoonees (the latter now at the University of Auckland) have developed a form of Genetic programming which they call PBAP (Population-Based Automatic Programming) [31]. In PBAP a Probability Vector of varying size controls the growth of tree-like structures.

11. Implementation

For exploratory work in engineering design, Matlab ® is in many ways an ideal platform. It permits rapid and relatively error-free prototyping, there is excellent graphics support and seamless integration with a powerful block-diagram oriented simulator (Simulink ®). A wide range of toolboxes is available, including a powerful optimization toolbox with a range of classical optimisers which can complement genotypic approaches. We have therefore implemented CHC, PBIL and the morphic hill-climbers as Matlab functions. In order to permit computer-intensive testing and facilitate interfacing to general-purpose simulators such as SPICE we have also implemented the algorithms in C++.

12. An Illustration: Simple Minimax Strategies for Robust Design

As an illustration of the kind of exploratory work which we hope will be undertaken by the design community using these tools, consider the problem of evolving, say, a filter to meet a certain frequency-response profile. A straightforward search will easily find a set of component values which minimises the squared error between the frequency response and that expressed by a target template. However this is solving an idealisation of the problem rather than the true engineering problem, which could rather be stated: Given a filter topology, find a nominal set of component values such that, for all component values falling within a specified neighbourhood of the nominal values, the mean-squared error (or other appropriate error norm) of the mismatch between performance and template remains within specified bounds.

Of course one can always apply a sensitivity test *post hoc*, but if it fails, the search must be repeated and there is no guarantee of finding the solution (even if one exists), especially since the search may well consistently rediscover the unrobust global (or near-global) maximum. Sequential niching [32] may be helpful in this situation but it also has its problems [33] and we have had little success with it. Instead, it seems more sensible to incorporate robustness directly into the search criteria. For each trial solution a local neighbourhood is examined (corresponding to the tolerance of the components) and the minimum performance of this cluster of perturbed points is returned as the effective fitness of that trial solution. This can be rather computationally intensive, but recourse can be had to the methods of experimental design and response-surface analysis [34].

In a simple version of this we have used a Taguchi orthogonal matrix [35] to fit a linear local model from which the worst-case set of perturbations is calculated. The first order model ignores local interaction between variables so an 'additivity' test, as proposed in [36] carried out at each point to validate the interaction-free model. In the tasks to which we have thus far applied the method the simple model has proved adequate. Where it does not we could use a second-order model (including first-order interactions), or even carry out a small stochastic search-within-a-search at each point to estimate the worst-case combination of perturbations. This is of course computationally intensive but not excessively so in relation to the payoff. It is also easily carried out in parallel. In this way we have been able to evolve filters of near-zero macroscopic sensitivity.

Most of this work has been done using PBIL, and it is uncertain that the results will be the same when a GA is used. There are interesting implications of the negative selection pressure caused by assigning to each trial solution a fitness determined from the 'worst' perturbation in its neighbourhood. This can in fact be thought of as a kind of 'negative Baldwinian learning'! It remains to be seen how this will impact on the search dynamics.

In a similar way we have evolved robust linear controllers by testing each candidate controller on an ensemble of perturbed 'plants', calculating the fitness in terms of the lowest-performing variant in each case. An initial optimization run followed by analysis of the local shape of the response surface in parts of the design space where local maxima tend to cluster, following Parmee et al [37], is helpful in the rational choice of distribution of the perturbed plants. This is a simple strategy of very wide applicability in engineering design.

13. Future Directions

As stated above, the primary reason for seeking a compact simple algorithmic toolkit is to encourage experimentation in the design community, especially in relation to domain-specific issues like goal expression and constraint handling. When the validation of the toolkit algorithms has been completed we will attempt to assemble some case studies along these lines to provide suggestive starting points.

At some point we will have to grasp the nettle of vector-valued objective functions to replace the rather arbitrary weighting strategies we have been using. Despite much useful work in this area, many problems remain. Pareto-ranked evaluation seems a promising direction, possibly coupled with preference expression as proposed by Fonseca and Fleming [38] but this is not sufficient – a single-population GA is not capable of sustaining multiple maxima due to genetic drift, and some method of niche induction is essential. Again, a great deal of work has been done but there is little consensus as to the most promising approach. We are currently examining the possibility of applying deterministic niching as proposed by Mahfoud [33] to CHC and pPBIL.

14. Conclusion

I have described a small, carefully selected set of stochastic optimization tools with an emphasis on robustness and simplicity of use. CHC, PBIL and simple bit-based hill-climbing have emerged as strong contenders for inclusion in this set, the former in 'standard' form, and the latter two possibly in modified (parallel and 'morphic' respectively), forms. Although results on really difficult optimization problems will inevitably be sub-optimal with respect to algorithms tailored more specifically to the task domain, we believe (and our

338

experiences suggest) that the availability of such a package will do much to facilitate and encourage experimentation within the design community on the routine design of engineering artefacts and systems through simulated evolution and adaptive search., hopefully leading to a fuller exploitation of the unique potential of these approaches.

References

[1] Greene, J. R., 1996, Population-based Incremental Learning as a Simple Versatile Tool for Engineering Optimisation, *Proceedings of the First International Conference on Evolutionary Computation and its Applications (EVcA 96), Moscow, Russia, June*, pp. 258-269.

[2] Pagliari, G., and Greene, J. R., 1996, Image Registration, Parameter Tuning and Approximate Function Evaluation, using the Genetic Algorithm and Digital Image Warping, *Proceedings of 4th I.E.E.E. Africon Conference in Africa (Africon '96), Stellenbosch, South Africa*, September pp. 536-541.

[3] Thithi, I., Control System Parameter Identification Using the Population Based Incremental Learning (PBIL), *I.E.E. Conference Publication No. 427, UKACC International Conference on Control (CONTROL '96), London, U.K., September*, pp. 1309-1314.

[4] Wing, M., 1996, *Design of an Energy-Efficient Permanent Magnet DC Brushless Motor*, PhD Thesis, University of Cape Town, Cape Town, South Africa.

[5] Horrell, J., and du Toit, L., 1996, Antenna Array Design using PBIL, *Proceedings of 4th I.E.E.E. Africon Conference in Africa (Africon '96), Stellenbosch, South Africa*, September pp. 276-281.

[6] Avedon, R. E. and Francis, B. A., 1993 Digital Control Design Via Convex Optimization, *Journal of Dynamic Systems, Measurement and Control*, **115**, 579-586.

[7] Rechenberg, I., 1965, Cybernetic solution path of an experimental problem, *Library Translation 1122*, Royal Aircraft Establishment, Farnborough, Hants., U.K.

[8] Schwefel, H-P., 1994, *Evolution and Optimum Seeking*, Wiley, New York.

[9] Holland, J. H., 1992, *Adaptation in Natural and Artificial Systems*, 2nd ed., MIT Press, Cambridge, MA.

[10] Goldberg, D. E., 1983, *Computer-Aided Gas Pipeline Operation using Genetic Algorithms and Rule Learning*, PhD Thesis, University of Michigan, Ann Arbor, U.S.A. Davis, L. (Editor), 1991, *Handbook of Genetic Algorithms*, Van Nostrand Reinhold, New York.

[11] Michalewicz, Z., 1994, *Genetic Algorithms + Data Structures = Evolution Programs*, 3rd ed., Springer, Berlin.

[12] Mitchell, M., 1995, *Review of Davis, L. (Editor), 1991, Handbook of Genetic Algorithms*, Unpublished manuscript.

[13] Wang, G., Goodman, E. D., Punch W. F., 1996, Simultaneous Multi-Level Evolution, Technical Report 96-03-01, GARAGe, Michigan State University, Lansing, U.S.A.

[14] Wolpert, D. and Macready W., 1995, No Free Lunch Theorems for Search, Technical Report, SFI-TR-95-02-010, Santa Fe Institute, Santa Fe, NM, U.S.A.

[15] Schaffer J. D., Mathias, K. E., and Eshelman L. J., 1996, Convergence-Controlled Variation, in Belew,

[16] K. and Vose M. (Editors), *Foundations of Genetic Algorithms (FOGA4)*, Morgan Kaufmann, San Mateo, CA. 203-224.

[17] Whitley, D., Beveridge, R., Graves, C. and Mathias, K., 1996, Test Driving Three 1995 Genetic Algorithms: New Test Functions and Geometric Matching. *Journal on Heuristics*, **1**, 77-104.

[18] Davis, L., 1991, Bit-Climbing, Representational Bias, and Test-Suite Design, in Belew, R. K. and Booker, L. B. (Editors), *Proceedings of the Fourth International Conference on Genetic Algorithms*, Morgan Kaufmann, San Mateo, CA, pp. 18-23.

[19] Keane A., 1994, Experiences with Optimisers, *Proceedings of Adaptive Computing in Engineering Design and Control, Plymouth, U.K., September*, pp. 14-27.

[20] Eshelman, L. J., 1991, The CHC Adaptive Search Algorithm: How to have safe search when engaging in nontraditional genetic recombination, in Rawlins G. (Editor), *Foundations of Genetic Algorithms (FOGA2)*, Morgan Kaufmann, San Mateo, CA, pp. 265-283.

[21] Baluja, S., 1994, Population-Based Incremental Learning: A Method for Integrating Genetic Search Based Function Optimization and Competitive Learning, Technical Report CMU-CS-95-163, School of Computer Science, Carnegie Mellon University, Pittsburgh, PA, U.S.A.

[22] Kingdon, J. and Dekker, L., 1995, The Shape of Space, *Proceedings of the First IEE/IEEE International Conference on Genetic Algorithms in Engineering Systems: Innovations and Applications (GALESIA 95)*, London, U.K., September, pp. 543-548.

[23] Kingdon, J. and Dekker, L., 1996, Morphic Search Strategies, *Proceedings of the 1996 IEEE International Conference on Evolutionary Computation, New Jersey, U.S.A.*, pp. 837-841.

[24] Baluja, S., 1995, Removing the Genetics from the Standard Genetic Algorithm, in Prieditis, A. , Russel, S. (Editors), *The International Conference on Machine Learning*, Morgan Kaufmann, San Mateo, CA, pp. 38-46.

[25] Baluja, S., 1995, An Empirical Comparison of Seven Iterative and Evolutionary Function Optimization Heuristics, Technical Report CMU-CS-95-193, School of Computer Science, Carnegie Mellon University, Pittsburgh, PA, U.S.A.

[26] Seaman, B., 1995, *An Empirical Comparison of PBIL and the Genetic Algorithm*, Unpublished undergraduate thesis report, University of Cape Town, Cape Town, South Africa.

[27] Kvasnicka, V., Pelikan, M. and Pospichal, J., 1995, Hill Climbing with Learning (An Abstraction of Genetic Algorithm), *WEC1 Unpublished Internet On-Line Seminar, Nagoya, Japan.*

[28] Baluja, S., 1996, Evolution of an Artificial Neural Network Based Autonomous Land Vehicle Controller. *IEEE Transactions on Systems, Man, and Cybernetics - Part B: Cybernetics*, 26(3), 450-463.

[29] Baluja, S. and Simon, D. A., 1996, Evolution-Based Methods for Selecting Point Data for Object Localization: Applications to Computer Assisted Surgery, Technical Report CMU-CS-96-183, School of Computer Science, Carnegie Mellon University, Pittsburgh, PA, U.S.A.

[30] Baluja, S., 1996, Genetic Algorithms and Explicit Search Statistics, presented at the *Conference on Neural Information Processing Systems (NIPS-96)* Proceedings to be published by MIT Press.

[31] Schoonees, J. and Jakoet, E., 1997, Population-Based Automatic programming - Part 1: Description of the Algorithm, submitted to *The Fourth International Conference on Neural Information Processing (ICONIP '97)*, Dunedin, New Zealand. (Proceedings to be published by *IEEE Computer Society Press*.)

[32] Beasley, D., Bull, D. R. and Martin, R. R., 1993, A Sequential Niche Technique for Multimodal Function Optimization, *Evolutionary Computation* 1(2), 101-125.

[33] Mahfoud, S. W., 1995, *Niching Methods for Genetic Algorithms*, PhD Thesis, University of Illinois, Urbana-Champaign, IL, U.S.A.

[34] Box, G. P. and Draper, N. R., 1987, *Empirical Model-Building and Response Surfaces*, Wiley, New York.

[35] Phadke, M. S., 1989, *Quality Engineering using Robust Design*, Prentice-Hall International, London.

[36] Roy, R., Parmee, I. C. and Purchase, G., 1995, Sensitivity Analysis of Engineering Designs using Taguchi's Methodology, Research Report PEDC-05-95, Plymouth Engineering Design Centre, Plymouth, U.K.

[37] Parmee, I. C., 1996, The Maintenance of Search Diversity for Effective Design Space Decomposition using Cluster-Oriented Genetic Algorithms (COGAs) and Multi-Agent Strategies (GAANT), *Proceedings of Adaptive Computing in Engineering Design and Control, Plymouth, U.K., September*, pp. 128-138.

[38] Fonseca, C. M. and Fleming, P. J., 1993, Genetic Algorithms for Multiobjective Optimization: Formulation, Discussion and Generalization, in Forrest, S. (Editor), *Proceedings of the Fifth International Conference on Genetic Algorithms*, Morgan Kaufmann, San Mateo, CA, pp. 416-423.

Part 7: Scheduling, Manufacturing and Robotics

Papers:

A Genetic Algorithm Based Hybrid Channel Allocation Scheme

B.Visweswaran[1] and D.K.Anvekar[2]

[1] *Dept. Electrical Communication Engineering, Indian Institute of Science, Bangalore 560 012, India,easwar@protocol.ece.iisc.ernet.in*

[2] *Dept. Electrical Communication Engineering, Indian Institute of Science, Bangalore 560 012, India,dka@ece.iisc.ernet.in*

Keywords: Genetic Algorithms, Mobile cellular communication, Channel Allocation.

Abstract

This paper investigates the application of a Genetic Algorithm (GA) to the hybrid Channel Allocation (HCA) scheme in mobile cellular communication systems. The HCA scheme is a combination of the Fixed and Dynamic Channel assignment schemes. In HCA, the available set of channels is divided into fixed and dynamic sets. The ratio of fixed to dyanamic channels is a significant parameter which defines the performance of the system. This being an optimization problem, it presents an ideal opportunity to apply GAs. The problem of determining the ratio of fixed to dynamic channels by using a GA is approached in two ways. First, the total number of channels available is constrained and the best set is determined. Second, the minimal set of channels which provides optimal performance is determined. In both cases, the GA converged to the optimal solution. A comparison is made between this approach and the corresponding Fixed Channel Assignment scheme.

1.Introduction

The tremendous growth of the mobile user population, coupled with their bandwidth requirements, requires efficient reuse of the radio spectrum allocated to mobile communications. Efficient use of the radio spectrum is also important from a cost-of-service point of view, where the number of base stations required for a certain geographical area is an important factor. A reduction in the number of base stations, and hence in the cost of service, can be achieved by more efficient reuse of the radio spectrum. The basic prohibiting factor in radio spectrum reuse is interference caused by the environment or other modules. Interference can be reduced by making use of channel assignment techniques.

Channel assignment strategies can be classified as a) fixed b) dynamic and c) flexible [1].In all fixed channel assignment strategies, a set of channels is permanently assigned to each cell. The same set of channels is reused by another cell at some distance. The minimum distance at which radio frequencies can be reused with no interference is called the 'cochannel reuse distance'. Due to short term temporal and spatial variations of traffic in cellular systems, Fixed Channel Allocation (FCA) schemes are not able to attain high channel efficiency. To overcome this, Dynamic Channel Allocation (DCA) schemes have been proposed. In contrast to FCA, there is no fixed relationship between channels and cells in DCA. All channels are kept in a central pool and assigned dynamically to radio cells as new calls arrive. After a call is completed, its channel is returned to the central pool. In DCA, a channel is eligible for use in a particular cell if it satisfies the interference constraints. In general, more than one channel may be available for selection.

Hybrid Channel Assignment (HCA) schemes are a mixture of FCA and DCA techniques . In HCA, the total number of channels available for service is divided into fixed and dynamic sets. The fixed set contains a number of nominal channels that are assigned to cells as in the FCA schemes and are preferred for use in their respective cells. When a call requires service from a cell and all its nominal channels are busy, a channel from the dynamic set is assigned to the call. The dynamic set, therefore, results in increased flexibility of the system. The channel assignment procedure from the dynamic set follows any of the DCA strategies. Call blocking probability is defined as the probability that a call finds both the fixed and dynamic sets busy. The possibility that a dynamic channel is free but cannot be assigned due to interference constraints is also taken into consideration here.

The ratio of fixed to dynamic channels is a significant parameter which defines the performance of the system. A measure of the performance of a mobile system is its blocking probability. In general, the ratio of fixed to dynamic channels is a function of the traffic load. When the load is high FCA schemes perform better since they are able to maintain the minimum reuse distance [2]. On the other hand when the load is low DCA schemes are better since they have greater flexibility. In HCA, which combines FCA and DCA, the manner in which the division into fixed and dynamic channels is made assumes importance. This division must be done in such a way the the system is able to adapt to varying traffic loads and give acceptable performance. This paper investigates how a Genetic Algorithm (GA) may be used to efficiently determine this ratio. The GA is applied to a HCA scheme employed in a Highway microcellular environment.

2. Genetic Algorithms

Genetic Algorithms (GAs),[3], are iterative search algorithms based on an analogy with the process of natural selection and evolutionary genetics. GAs ensure the proliferation of quality solutions while investigating new solutions via a systematic information exchange that utilizes probabilistic decisions. It is this combination which allows GAs to exploit the search space with expected improved performance.

The search aims to optimize some user-defined function called the fitness function. To perform this task, GA maintains a 'population' of candidate points over the entire search space. At each iteration, called a 'generation' a new population is created. This new generation generally consists of individuals which are more fit, than the previous ones, as represented by the fitness function. As the population iterates through successive generations, the individuals will in general tend towards the optimum of the fitness function. To generate a new population on the basis of a previous one, a GA performs three steps:

1. Evaluate the fitness score of each individual of the old population.

2. Selects individuals on the basis of their fitness score

3. Recombines selected individuals using 'genetic operators' such as mutation and crossover.

GAs differ from classical optimization methods like steepest descent. simplex, etc., in the following manner:

1. GA works in parallel on a number of parts and not on a unique solution

2. GA requires only a fitness function and no other information nor assumption such as derivatives.

3. Selection and combination are performed by using probabilistic rules rather than deterministic ones.

3. Mechanics of a GA

A simple GA is composed of three operators: 1) reproduction 2) crossover and 3) mutation. Each of these operators are implemented by performing the basic tasks of copying strings, exchanging portions of strings and generating random numbers. The GA begins by randomly generating a population of N strings, each of length m. Each string represents one possible solution to the problem. These strings are evaluated with some objective function and assigned some fitness value. Fitness values are then used to produce a new population of strings. The new strings are again decoded, evaluated and transformed until convergence is achieved or a suitable solution is found.

Reproduction is simply a process by which strings with large fitness values receive correspondingly large number of copies in the new population. Several reproduction schemes like the Roulette wheel ,Elitist strategies etc., may be used. The scheme used, however, is not critical to the performance of the GA.

Crossover provides a mechanism for strings to mix and match their desirable qualities through a random process. The crossover site is selected uniformly at random. When combined with reproduction it is an effective means of exchanging information and combining portions of high quality solutions.

The third operator, mutation, enhances the ability of the GA to find near optimal solutions. Mutation is the occasional alteration of a value at a particular string position. Mutation helps in those situations where neither reproduction nor crossover will ever produce the optimal solution. Any information that may have been lost in previous generations may be reinjected by this operator.

4. Highway Microcellular Structure

A basic cellular system consists of three parts: a mobile unit, a cell site and a mobile switching office (MTSO), with connections to link the three systems. A mobile telephone unit contains a control unit, a transiever and an antenna system. The cell site provides the interface between the MTSO and the mobile units. The MTSO is the heart of the cellular mobile system and provides central co-ordination and cellular administration.

Microcellular digital mobile radio systems have been proposed where the cell shapes are tailored to the physical environment [4]. These cells may vary from the size of a room to a few hundred meters for hand held portables. Microcells for vehicular mobiles, on the other hand, would typically be longer than half a mile. As highways have many mobiles restricted to narrow confines for long periods of time the potential for relatively high teletraffic rates encourages us to consider a relatively complex cellular structure .

Highways may be considered to be composed of segments having different radii of curvature ρ, where ρ is infinity for straight stretches and which may be quite small for mountainous hairpin bends. A cell may be conceptually viewed as a parallelopiped of length L along, and D across, the highway, respectively, where $L \gg D$. The length L, however, might change if the highway has a sudden decrease in curvature, meets an intersection, tunnel etc. In general, the cochannel interference will be greatest in straight highways, as a mobile will experience interference from a transreceiver radiating directly into the mobiles cell.

5. Application of a GA to HCA

As previously mentioned, in the HCA scheme, the ratio in which the available channals are split is of crucial importance. This problem of dividing a group of channels into two can, in general, be approached in two ways:

1. Constrained approach: Here, the number of channels available to the system is constrained and the optimal manner in which these may be divided is determined, i.e., if there are N channels available, the combination (D, S) such that $N = D + S$ where D is the number of dynamic channels and S is the number of static or fixed channels is determined.

2. Unconstrained approach: The number of channels available is assumed to be unlimited and the optimal set is determined. There are again two approaches to this unconstrained optimization problem.

   - To find a subset of channels that results in the lowest blocking probability
   - To determine that subset of channels for which the blocking probability is less than a given threshold.

   In the first case the blocking probability and the total number of channels feature in the fitness function. In the second case it is only the blocking probability that serves as a measure of fitness.

In this paper on the application of GAs to the HCA scheme, the constrained approach and the unconstrained one with the blocking probability below a certain threshold are considered. In most practical cases the emphasis is on a certain grade of service and not on the global minimum that can be achieved. The threshold constrained optimization problem is therefore a more practical scenario.

Population size	Probability of crossover	Probability of Mutation
6	1	.05

Table 1: GA Parameters

6.Constrained Channel Approach

Here, the total number of channels is fixed and the aim is to determine that set which which gives optimal performance. Since the blocking probability is a measure of the performance of a mobile system we are looking for a subset which results in the minimum blocking probability. Since the number of channels is constrained it is sufficient if either the static or dynamic channels is modelled. The other set can be obtained as the difference. This implies that the size of the search space is equal to the number of available channels. Further, since the number of channels is not variable it does not feature in the fitness function.

The aim of this approach is to determine a channel set which performs well for different traffic loads. The fitness function chosen, therefore, is the sum of the blocking probabilities obtained by simulating various call inter-arrival times, i.e.,

$$f(a) = \sum_{r_j} P_b \qquad \forall r_j \in r \tag{1}$$

where $a = (b_1 b_2 b_3 \cdots b_n)$ is a bit string representing a possible solution and $r = (r_1, r_2, r_3, \cdots, r_n)$ is the set of call arrival times.

Considering the FCA scheme, the blocking probability is generally found to be a monotonically decreasing function with increasing inter-arrival times. This implies that when the time interval between call arrivals increases the blocking probability decreases. Large inter-arrival rates translates to low load. For such kind of monotonic functions, DeJong [5] suggests that the elitist strategy produces the best results. In the elitist strategy it is made sure that atleast the most fit string is carried over to the next generation. Since in this investigation the blocking probabilities themselves serve as an estimate of the fitness function, the elitist strategy is followed.

7.Simulation

The system was modelled and the simulations were performed using the Simscript language. The following assumptions were made in this simulation:

1. Call arrivals are Poisson distributed which means that their inter-arrival times are exponentially distributed.

2. Each call requires only one channel.

3. The call holding time is exponentially distributed with a mean of $120 sec$.

4. Blocked calls are dropped.

5. The highway is assumed to be circular, i.e., mobiles leaving the last cell circle back to the first. This assumption was made to eliminate end effects seen in models which consider a segment of a highway.

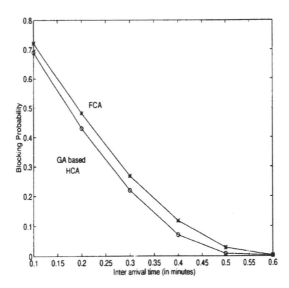

Figure 1: Blocking Probabilities for the constrained channel system with 10 cells and 31 channels.

The GA parameters are given in Table 1. Since the solution space is not very large a modest population size was chosen. Furthur the number of channels available was assumed to be a power of two to simplify the coding scheme.

A 10-cell highway environment was simulated for cases with 31 and 15 channels. Inter- arrival times ranging from 0.1 minutes to 0.6 minutes in steps of 0.1 minutes was considered. The strings representing possible solutions were binary coded into 5 and 4 bits respectively. In the first case the GA converged to 27 fixed and 4 dynamic channels while in the second case it converged to 12 fixed and 3 dynamic channels. In both cases the average number of generations before convergence was achieved was 12. Both configurations were compared with the corresponding FCA implementation. If the ratio of dynamic to static channels is denoted by $D : S$ and $D + S = NC$ where NC is the total number of channels, then while the FCA scheme would employ NC channels per cell the HCA scheme would allocate S channels per cell and D channels would be given to the entire system.

In the highway environment simulated a reuse distance of 2 is considered. Reuse distance refers to the minimum distance at which channel sets of the same frequency are reused. Therefore, in the case considered a minimum of two channel sets are required. When the total number of channels is divided into D and S, a total of $2 * S$ channels are made available to the entire system and each cell is alloted S channels. For example, if the division is 27 fixed and 4 dynamic channels, 8 channels are available to the entire system and 27 channels are alloted to each cell as nominal channels.

The results of the simulation are shown plotted in Figures 1 and 2. The graphs are a plot of the blocking probability against the inter-arrival times in minutes. It can be seen that in both cases the configuration chosen by the GA performs better than the corresponding FCA implementation. While the 31 channel case shows better performance over all the inter-arrival times considered, the 15 channel configuration shows improved performance for high and low loads. For moderate loads it is nearly equivalent to the FCA scheme.

8.Unconstrained Channel Approach

In this approach the constraint that the sum of the dynamic and fixed channels be equal to a constant is removed. This implies that each can take any value bound by an upper limit. In this application the upper limit is assumed to be 31. Since each parameter can take a maximum value of 31 the ssize of the solution space is 961. This problem is similar to the feature selection problem in pattern classification,

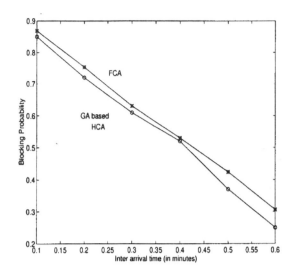

Figure 2: Blocking Probabilities for the constrained channel system with 10 cells and 15 channels.

in which the minimal set of features providing maximum classification accuracy is to be found [6]. Since the subset of channels for which the blocking probability is below a certain threshold is to be determined the fitness function involves both the blocking probability as well as the number of channels. First the set of channels which satisfy the threshold condition is determined and then the total number of channels is used to determine the optimal subset. The threshold is nothing but the blocking probability that is desired for a given traffic load. Once the threshold is satisfied the fitness assigned is given by

$$f(a) = exp^{(sum/62)^2} - 1 \qquad (2)$$

where sum is the total number of channels. If the threshold condition is not satisfied then the blocking probability is itself assigned as the fitness. A function of the type given above is chosen so that among those channels sets satisfying the threshold condition the ones with lesser number of channels is rewarded with a lower fitness. Since this is a function minimization problem these are selected with a greater probability.

9. Simulation

The assumptions made in this case are the same as those in the previous approach . Initially a base load of 5 Erlangs or 150 calls/hour was chosen. A 20 cell highway system was simulated with the threshold fixed at 20%. The GA parameters were the same as those used in the previous approach and are given in Table 1. On simulating the system the GA was found to converge to a configuration of 10 fixed and 3 dynamic channels. Convergence in this case was achieved in an average of 10 generations. With this configuration other interarrival rates were also simulated. The performance of this system was compared to the corresponding FCA implementation with 13 channels. This comparison is shown in Figure 3. It can be seen that the GA based implementation performs significantly better than the FCA scheme over all the inter-arrival times considered. Furthur, the improvement in blocking probability is much greater in the unconstrained optimization case when compared to the constrained cases. This may be due to the freedom that the GA has to select the channel set which provides optimal performance.

10. Conclusion

This paper investigates the application of a GA to the problem of dividing a given set of channels into fixed and dynamic sets for use in a HCA scheme. Two approaches were considered and results obtained

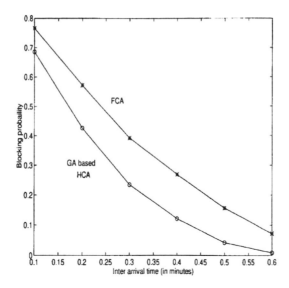

Figure 3: Blocking Probabilities for the unconstrained channel system with 20 cells.

show that the configurations obtained using the GA show better performance when compared to the corresponding FCA implementation. Since the performance is better over a range of inter-arrival times, the process need not be repeated when the traffic load changes. There is, however, the question of scaling. For the small channel configurations considered here the GA performs more optimally than an exhaustive search over all the possible channel combinations. Whether the GA can perform optimally when the size of the problem is increased is to be investigated. Further, the channel constraints have been chosen to be powers of two for simplicity. If this is not the case a suitable coding scheme has to be considered to take care of the illegal strings which may arise during the GA processing. Further study is also being done on how the fitness function may be tuned for the unconstrained optimization problem. A weighted sum of the blocking probability and the number of channels is currently being studied. The optimal weights themselves may be obtained in the same run.

References

[1] Tomson Joe Kahwa, Nicolas D. Georganas, 1978, A Hybrid Channel Assignment Scheme in Large -Scale Cellular Structured Mobile Communication System , *IEEE Trans. Commn.*, **4**. 432-438 .

[2] Katzela, J., Nagahsineh, M., 1996, Channel assignment schemes for cellular mobile telecommunication systems: A comprehensive survey, *IEEE Personal Communications*,**6**, 10-36 .

[3] John, H. Holland, 1992, Genetic Algorithms, *Scientific American*, **7**, 44-50 .

[4] Wong, K.H.H., Steele, R., 1986, Digital Communications in Highway Microcellular Structures, *Proceedings of IEEE Globecom,Houston,TX, December* , 31.2.1-31.2.5.

[5] Goldberg, D., 1989, *Genetic Algorithms in Search,Optimization and Machine Learning*, Addison-Wesley, USA.

[6] Siedlecki, W., Sklansky, J., 1989, A note on genetic algorithms for large-scale feature selection,*Pattern Recognition Letters*, **10**, 335-347 .

Solving Scheduling Problems via Evolutionary Methods for Rule Sequence Optimization

Igor P. Norenkov[1], Erik D. Goodman[2]

[1]Moscow State Bauman Technical University, Russia, norenkov@aicad.isrir.msk.su
[2]GA Res. & Applic. Gp., 2325 Engineering, Michigan State Univ., East Lansing, MI 48824, goodman@egr.msu.edu

Keywords: JSSP, Flow Shop Scheduling, Genetic Algorithms, Heuristic Combination Method, Hybrid

Abstract

The paper is devoted to solution of multistage scheduling problems by genetic algorithms. The Heuristics Combination Method (HCM) is described. The idea of HCM is to optimize the choice and sequence of application of a set of heuristic rules for schedule synthesis, from among a given initial set. Several approaches to improvement of the quality of the schedules synthesized are discussed. One of the approaches is a hybrid evolutionary-genetic method. Some experimental numerical results are given.

1. Introduction

Scheduling problems are found in many areas. As a rule, these problems are NP-hard tasks. Hence, practical problems which are sufficiently large are typically solved only approximately, using various heuristic rules. Unfortunately the degree of optimality of the solutions thus obtained is often not very good, and this quality can not be predicted *a priori*. Evolutionary methods (including genetic algorithms) comprise an important group of approximate methods. There are many publications devoted to genetic algorithm (GA) applications to scheduling problems -- see, for example, review work [1]. Some papers include information about GA applications to scheduling in aircraft design [2], resource distribution [3], manufacturing processes [4], frequency assignment in radio channels of networks [5], *etc.*

One of the main tasks in scheduling by GA is the formulation of the chromosome structure, i.e., the representation of the data about work distribution in time and between servers to be used by the GA. It is possible to utilize representations and operators which require "repair" of the genome after crossover operations, for example, to guarantee that the offspring represents a feasible schedule. However, such repair operations may break some useful building blocks and decrease scheduling efficiency. An interesting alternative is to use an implicit form of schedule representation on the chromosome, in which the gene at the i-th locus does not represent the number of a job or server, but instead the number of a heuristic to be used as the i-th step of generating the schedule. This is the defining feature for the Heuristics Combination Method - HCM [7] described here. The idea embodied in HCM was published simultaneously and independently in [7] and by Fang [8], both in 1994.

This paper describes evolutionary and genetic algorithms created on the basis of the HCM and examples of their applications to solving multistage Flow-Shop Scheduling Problems (FSSP). Section 1 presents the typical multistage JSSP. The HCM is described in Section 2. Section 3 is devoted to algorithmic questions about HCM realization. The experimental numerical results and efficiency evaluation of HCM are given in Section 4.

2. Multistage Scheduling Problems

The HCM may be applied to a wide class of discrete optimization and structure synthesis problems. The features of this class are the following:

1. The synthesis is a multistep process.
2. A fitness value may be computed only when the total schedule is compiled.
3. There are some heuristics that can be used at every step of synthesis.

A typical problem for solving by HCM is the following multistage Flow Shop Scheduling Problem [9]:

- there are N jobs; every job A_i is treated in q stages of service successively from the first stage $(k=1)$ to the last stage $(k=q)$, $i = 1..N$;
- there are M_k machines (servers) at the k-th stage, k = 1..q, the total number of servers is

$$M = \sum_{k=1}^{q} M_k$$

- only one job at a time may be served on a server; once started, the service cannot be interrupted;
- every job belongs to one family, and the j-th server has to undergo a setup change (with associated family-specific setup time) if two consecutive jobs assigned to the j-th server belong to different families;
- matrix P of service times is given; element P_{ij} is the time required to serve the i-th job on the j-th server;
- the matrices E_j of setup times for each server are given; element E_{jir} is the setup time when the i-th job precedes the r-th job on the j-th server;
- costs per unit service and setup time, C_i and R_j, respectively, are known;
- each job has both soft and hard due dates (completion times), D_i and T_i, respectively, and $T_i > D_i$;
- penalties $G1$ and $G2$ are assigned for violation of soft and hard due dates, respectively.

The problem is to find the minimum cost schedule (cost is the anti-fitness function to be minimized), including costs of processing times, setup times, and penalties for violating soft and hard due dates. The most complex problem used in our experiments has been described in [9]. Its initial data are N=105, q=4, M=15. This problem is called N105 in the discussion below.

3. Heuristics Combination Method

A schedule synthesis consists of two parts. The first is choice of order of consideration of jobs. The second part is assignment of jobs to servers. Each part may be determined by a variety of heuristic and/or genetic methods.

Here are some rules for job ordering sometimes used in the first part of scheduling. Rule A1 requires the ordering of jobs in accordance with increasing parameter $GAP = X_i$, where X_i is the time when i-th job service ends on the previous stage. Parameter $GAP = D_i$ is rule A2, and $GAP = (a_1*D_i+a_2)*(a_3*D_i+a_4)$ is rule A3, where $a_1..a_4$ are real coefficients. The set of rules is extended if we take into account only jobs in the interval $[t, t+T_{tr}]$, where t is current time, i.e., minimal X_i, for i in I, and I is the set of numbers of free (not yet assigned) jobs; T_{tr} is an additional parameter changed from one rule to another.

The rules for job assignment to servers are as follows. Rule B1 chooses the server providing the cheapest service. Rule B2 chooses the server providing the quickest service of the job. The set of assignment rules may be extended if we take into account preferences for successive service of same-family jobs.

The composition of rules from the first and second parts generates the set of heuristics. The solution of practical scheduling problem has shown that using single concrete heuristics does not lead to acceptable results. On the other hand, application of different heuristics at different steps of schedule synthesis generates a set of schedules containing significantly better variants. Hence, the task is search for the optimal sequence of application of heuristics.

The HCM is a method for addressing this optimization problem. The chromosome in the HCM includes V = N*q genes. The allele at the i-th locus is the name (i.e., number) of the heuristic to be applied at the i-th step of schedule synthesis. Therefore, the value at every locus need not be unique, and "repair" of offspring chromosomes is not required. The other advantage is that the simple substitution of a set of heuristics leads to adaptation of the GA to a new class of synthesis problems.

A chromosome in the multistage JSSP may be represented as a matrix CR of size $q*N$, where element CR_{ik} refers to the k-th step of synthesis at the i-th stage. Such a representation shows that two directions (horizontal and vertical) of crossover are suitable. The two directions may be explained by epistasis. Epistasis means that the fitness function depends on not only independently on the alleles at the i-th and k-th loci, but also on their joint presence on the chromosome. Alleles at loci with nearby values of index i and index k form building blocks. Disruption of a building block worsens convergence to optimal solution. The natural one-point crossover is good for schemata in one row of CR (they correspond to one stage of the schedule), but it is not effective with respect to schemata formed by alleles in loci with different values of index i under the same value of index k (they correspond to one job at different stages). Such crossover might be termed *horizontal*, and schemata including the alleles in loci CR_{ik} and $CR_{i,k+1}$ are horizontal building blocks.

Schemata including alleles in loci CR_{ik} and $CR_{i+1,k}$ are the vertical building blocks. Better reproduction of vertical building blocks requires a vertical crossover, which is a q-point crossover, and the distance between adjacent crossover points is N. It appears desirable to use both horizontal and vertical crossovers. Coefficient $A = Q_v/Q_h$, where Q_v and Q_h are probabilities of vertical and horizontal crossover, respectively, is one of the parameters for control of the solution process.

4. Hybrid Scheduling Algorithms

The trajectory of a GA toward optimal solution can be characterized by the speed S_f of improvement of the fitness function (FF) during the initial generations and by the level L of stagnation or convergence when FF is almost unchanging. There are some applications in dynamic control when the number of FF evaluations cannot be large and parameter S_f is very important. On the other hand, L is the main index of GA effectiveness in most cases. Therefore it is important to find ways for improvement of L.

It is evident that the quality of the heuristics influences L. The degree of this influence is defined by experiment (see the next section). The proper choice of control parameters such as A is the other natural way. A further way of improvement of solution accuracy is application of a multipopulation (coarse-grain parallel) GA [9] and/or hybrid algorithms. Our multipopulation algorithms were realized using the GALOPPS [9] software.

The hybrid algorithms [6] are a combination of GA and evolutionary algorithms (EA). Although both the GA and EA are evolutionary, EA as used here will mean that a single chromosome or a population of chromosomes is used without crossover. The principal procedures used for our hybrid algorithms (HA) are described as follows:

Procedure SH1 is evolutionary stochastic hillclimbing:

```
FF = fitfun(CC); /*fitfun is procedure of FF evaluation, CC is chromosome*/
i = 0;
while (i<K) /*K is equal to (0.2..0.5)*V, where V is the number of genes*/
{i++;
    Choose at random t loci and new random alleles at them; /* t = 3..8 */
    CC1 = current chromosome;
    z = fitfun(CC1);
    if (z < FF) {CC = CC1; FF = z; i=0;}
}
```

Procedure SH2 is similar, but with choice of t adjacent loci, with random choice of first position.

Procedure SH3 is similar to SH2, but distance between chosen loci is equal to N positions and $t = (1...2)*q$. In other words, the chosen loci correspond to one or two columns of the matrix CR.

Procedure SH4 is like SH1, but simple hillclimbing is used in SH4, with mutations of all genes in turn.

Two alternative procedures may be used for generation of the initial population. In the first procedure, IP1, the population is formed from random chromosomes under the condition

$$FF < TS, \qquad (1)$$

where TS is a threshold value. The second procedure IP2 includes SH1 additionally, where SH1 is applied to chromosomes chosen under condition (1).

The crossover may be horizontal, vertical or mixed with ratio A in the interval [0, 1]. The members of the new generation are selected under the condition

$$FF < FF_{min}+\delta,$$

where FF_{min} is the minimal value of FF at the current time, and δ is a threshold value which depends on the solution method. Every offspring chosen by condition (2) is treated in one of procedures SH1..SH4.

5. Experimental Results

The aim of the experiments was to explore the influence of various factors on the efficiency of solving of several flow shop scheduling problems. We used some test tasks, including the complex task N105 and tests N21A, N21B and N21C, which have different values of matrices P for an example with $N=21$. P is set to P_a for N21A, $P_b=1.3*P_a$ for N21B and $P_c=1.4*P_a$ for N21C. Increasing P (service times) means that the tasks are harder to complete before their soft or hard due dates, which means that the role of the tardiness penalties in the fitness function is increased.

The influence of the set of heuristics chosen from is very large. Table 1 includes the values of the FF in test tasks in which only one heuristic from the set S1...S8 is used. The values of the fitness function without taking into account the penalties are shown the columns labeled "no penalties". For example, no single heuristic is able to satisfy the hard due dates in task N21C.

Problem/ Heuristic	N21A	N21A (no penalties)	N21B	N21B (no penalties)	N21C	N21C (no penalties)	N105	N105 (no penalties)
S1	5629	5529	14376	7196	13841	7691	24583	22417
S2	6702	5602	13376	7226	13961	7801	31391	22822
S3	6635	5565	11406	7216	14927	7737	23851	22470
S4	7694	5604	13391	7271	14941	7781	39423	22759
S5	6699	5579	9484	7304	13008	7818	23084	22879
S6	5696	5636	7413	7253	8928	7758	23283	23113
S7	8720	5590	13318	7148	17935	7765	69813	21979
S8	6521	5381	16093	6913	18583	7413	36602	21863
All	5474	5405	7205	7055	8793	7613	22520	
Pick 3	5451	5351	7165	6995	10728	7568	22059	

Table 1: Performance on four tasks using each heuristic alone and using either 3 best or all of them at random.

The row "All" shows the results of the HCM when all heuristics S1...S8 were used with equal probabilities.

The data from the Table 1 show that no heuristic applied separately leads to satisfactory results. Clearly, some method for selecting and sequencing the heuristics is required to achieve reasonable performance. We might try ordering the heuristics in accordance with the above values of FF or FF_a and by choice of some of the better of them. Row "Pick 3" shows the results when the 3 best heuristics for each problem are used, as determined according to FF_a. They were S8,S1,S3 for N21A and N21C, and S8,S7,S1 for N21B and N105. The hybrid algorithm (HA) with procedure SH2 was used with A = 0.5, t = 6, K= 20, the size of population N_p = 25 (for N21) or N_p = 13 (for N105). Computations were terminated after 10 generations. The last row of Table 1 shows that ordering by FF_a may give good results if due date constraints are not very difficult to satisfy (service times are short relative to due dates specified). In the other case, the better heuristics from the list ordered by FF must be added to the subset of heuristics to be applied. This recommendation was used for N21C. Adding rule S6 led to FF=7703 instead 10728.

The second significant factor that influences solution efficiency is the type of algorithm used. The comparison of a single-population algorithm (SPA) with a hybrid evolutionary-genetic algorithm (HA) was performed for task N21C. The values of FF received after 10,000 evaluations are shown in Table 2. The advantages of coarse-grain parallel genetic algorithms (cgPGA) in comparison with the SPA (single population GA) were described in [10]. Our experiments showed that HA performance was not inferior to cgPGA for this problem. Table 3 presents a sampling of the values of FF obtained in runs on problem N105 by the parallel GA program GALOPPS. Eight parallel subpopulations were used, each with population size N_p. Migration of some members between populations in a ring was performed every 5 generations. Columns show the results after n generations.

N_p	25	37	75	100	115
SPA	8666	7755	7756	7771	7725
HA with δ=4	7747	7707	7711	7730	----
HA with δ=10	7740	7712	7699	----	----

Table 2: Results with various algorithms and population sizes.

Run #	N_p	Gen 4	Gen 8	Gen 12	Gen 16	Gen 24	Gen 48
3	27	22182	22162	22134	22100	22064	-----
4	31	22161	22090	22070	22054	22031	22021
7	39	22181	22132	22055	22025	22013	-----
10	49	22140	22098	22089	22076	22076	-----
12	60	22179	22141	22102	22077	22059	-----

Table 3: Results of 8-subpopulation parallel GA runs on N105, various subpopulation sizes.

Problem	N105	N105
N_p	13	13
Heuristic Proportions	S1:S7:S8 = 4:3:1	S1:S7:S8 = 3:2:3
Population Initialization	IP2, K=40	IP2, K=60
Algorithm	SH1, K=120	SH2, K=60
Parameters	t=3; A=0	t=6; A=0.5
Range of Fitnesses	21956-22053	21887-22058
Average Final Fitness	22005	21980

Table 4: Fitnesses and fitness ranges, parameter settings for SH1, SH2 runs of hybrid algorithm (HA)

The results of the hybrid algorithm (HA) on task N105 after about 30,000 fitness evaluations are presented in Table 4. Here, the notation S1:S7:S8 = 4:3:1 means that a choice of heuristic S1 is made with probability 0.5, heuristic S7 with 0.375 and heuristic S8 with 0.125. Epistasis may be taken into account by selecting one procedure from SH1 (or SH4), SH2, SH3 and by the choice of coefficient A (ratio of vertical to horizontal crossover). Table 5 shows the fitnesses in problem N105 for various values of coefficient A (after 10 generations, N_p=13, K=20).

The table indicates a preference for algorithm SH2, although this conclusion is not strongly supported. The additional reason favoring SH2 is its quicker performance on test N21.

A	0.0	0.2	0.4	0.6	0.8	1.0	Average
SH1	22167	22222	22170	22103	22211	22166	22173
SH2	22150	22102	22129	22121	22123	22142	22128
SH3	22201	22124	22114	22176	22134	22148	22150
Average	22173	22149	22138	22133	22156	22152	----

Table 5: Fitnesses versus algorithm, coefficient A, on N105.

6. Conclusion

An approach based on evolutionary principles of selection of optimal combinations of heuristics is a promising way to seek near-optimal solutions to complex scheduling problems. It may readily be realized using evolutionary-genetic algorithms. The quality of the solutions obtained depends on the subset of heuristics applied. It is advisable to select the set of eligible heuristics based on the peculiarities of the particular problems to be addressed.

Acknowledgements

The authors thank the members of the MSU Manufacturing Research Consortium, and particularly Dr. James Maley of Ford Motor Company, for suggesting this problem and partially supporting the initial phases of this work.

References

[1] Bruns, R., 1993, Direct chromosome representation and advanced genetic operators for production scheduling, *Proc. of fifth int. conf. on GA*, pp. 352-359.

[2] Bramlette, M., Bouchard, E., 1991, Genetic algorithms in parametric design of aircraft, in *Handbook of genetic algorithms*, L. Davis, ed., Van Nostrand Reinhold, New York, pp. 109-123.

[3] Syswerda, G., Palmucci J., 1991, The application of genetic algorithms to resource scheduling, *Proc. of fourth int. conf. on GA*, pp. 502-508.

[4] Yasinovsky, S.,1996, Genetic algorithms based hybrid system for job-shop scheduling, *Proc. EvCA'96*, Moscow, Russ. Acad. Sci., pp. 316-320.

[5] Kapsalis, A., Chardaire, P., Rayward-Smith, V., Smith, G., 1995, The radio-link frequency assignment problem: a case study using GA, *AISB Workshop on Evolutionary Computing*, Springer, pp. 117-131.

[6] Goldberg, D.,1989, *Genetic algorithms in search, optimization, and machine learning*, Addison-Wesley, New York.

[7] Norenkov, I. P., 1994, Scheduling and allocation for simulation and synthesis of CAD system hardware, *Proc. EWITD 94, East-West internat. conf.*, Moscow, ICSTI, pp. 20-24.

[8] Fang, H., Ross, P., Corne, D., 1994, A promising hybrid GA/heuristic approach for open-shop scheduling problems, *ECAI 94, 11th European Conf. on AI*, John Wiley & Sons, New York.

[9] Tcheprasov, V., Punch, W., Goodman, E., Ragatz G., Norenkov, I., 1996, A genetic algorithm to generate a pro-active scheduler for printed circuit board assembly, in *Proc. EvCA'96*, Russ. Acad. Sci., Moscow, pp. 232-244.

[10] Goodman, E., Punch, W., 1995, New techniques to improve coarse-grain parallel GA performance, *Proc. CAD'95*, Yalta, pp. 7-15.

Estimation of Geometrical Parameters of Drill Point by Combining Genetic Algorithm and Gradient Method

Lakshman HAZRA*, Hideo KATO*, Takaharu KURODA†

*Department of Mechanical Engineering, Faculty of Engineering, Chiba University
1-33 Yayoi-cho, Inage-ku, Chiba-shi, 263 Japan
Tel +81-43-290-3184, Fax +81-43-290-3185, E-mail hazra@syshp.tm.chiba-u.ac.jp
†Kisarazu National College of Technology, Japan

Keywords: Drill point geometry, Grinding cone, Computer aided inspection, Steepest gradient method, Evolutionary algorithm

Abstract

The flank surfaces or the point of the twist drill are reground to sharpen the cutting edges by using a drill pointer. In order to maintain the cutting performance of the drill, the inspection of the geometry is very important. This paper deals with a new evaluation method of conically ground drill point geometry. In this proposed method, five geometrical parameters for the evaluation are derived from the kinematic relationship between the drill point and the grinding wheel when regrinding. The parameters are determined by mathematical model matching with measured coordinates of many points on the flank surface. The results show that the combined method of Steepest Gradient Method and Genetic Algorithm is suitable for this matching from the viewpoints of calculation speed and accuracy.

1 Introduction:

Drill is one of the most basic cutting tool in machining process. The geometry of drill point affects productivity, quality and accuracy of the hole making process and the processes following. The point of worn twist drill, or the flank surfaces are reground to sharpen the cutting edges by using drill point grinder. However it is difficult to shape the surface into the target geometry precisely, because of inevitable positioning error, wear of grinding wheel and so on. Accordingly, in order to maintain the cutting performance of the drill, the inspection of the geometry is very important. But conventional drill point geometry inspection methods using metrological instruments, such as simple gauges, profile projector and tool maker's microscope, call for operator's skill and time consumption[1]. Ramamoorthy et.al.[2] have developed a method by using image processing via a noncontact coordinate measuring machine which is simple but it is only for some representative values such as lip clearance angle and chisel edge angle. This paper deals with a new evaluation method for whole geometry of conically ground drill point or flank surface, aiming to develop a computer aided inspection system (CAI) of it. In this method, from the kinematic relationship between the drill point and grinding wheel when regrinding, five geometrical parameters have been developed for evaluation, the brief description of the model is available in next section. Subsequently the experimental set up for drill geometry measurement has been described. Different methods of Evolutionary Algorithm [5][6] have been compared with the synthetic data, which has been discussed in section 4.1. Lastly a combined method of Genetic Algorithm and Steepest Gradient Method has been applied in real data of the drill flank surfaces which have been collected by the experimental setup. The result of this combined method has been compared with true value of the parameters in Table 1.

2 Geometrical Model of Drill Grinding:

Drill point which has ground by conical grinding method is the subject of this investigation. Fig.1 shows the relative position between grinding wheel and drill point. The X-Y-Z system's origin is the center of chisel edge of drill, and Z-axis identifies with drill axis. The x-y-z system's origin is the top of the grinding cone and z-axis identifies with grinding cone axis. Denoting the drill point angle by α and the drill diameter by D, x-y-z system equals X-Y-Z system which turns ζ around Z-axis, translates aD along new X-axis, eD along new Y-axis and pD along new Z-axis, and turn δ around new X-axis. Therefore coordinates (Xi, Yi, Zi) are transformed into coordi-

$$
\begin{Bmatrix} x_i \\ y_i \\ z_i \\ 1 \end{Bmatrix} = \begin{bmatrix} 1 & 0 & 0 & 0 \\ 0 & \cos\delta & -\sin\delta & 0 \\ 0 & \sin\delta & \cos\delta & 0 \\ 0 & 0 & 0 & 1 \end{bmatrix} \begin{bmatrix} 1 & 0 & 0 & aD \\ 0 & 1 & 0 & -eD \\ 0 & 0 & 1 & -pD \\ 0 & 0 & 0 & 1 \end{bmatrix} \begin{bmatrix} \cos\zeta & -\sin\zeta & 0 & 0 \\ \sin\zeta & \cos\zeta & 0 & 0 \\ 0 & 0 & 1 & 0 \\ 0 & 0 & 0 & 1 \end{bmatrix} \begin{Bmatrix} X_i \\ Y_i \\ Z_i \\ 1 \end{Bmatrix} \quad (1)
$$

$$p = (-k_2 - \sqrt{k_2^2 - k_1 k_3}) / k_1 \tag{2}$$

$$k_1 = \sin^2\delta - \cos^2\delta \ \tan^2(\alpha/2-\delta) \tag{3}$$

$$k_2 = -e \cos\delta \sin\delta \left\{1 + \tan^2(\alpha/2-\delta)\right\} \tag{4}$$

$$k_3 = a^2 + e^2 \left\{\cos^2\delta - \sin^2\delta \ \tan^2(\alpha/2-\delta)\right\} \tag{5}$$

$$x^2 + y^2 = z^2 \ \tan^2(\alpha/2-\delta) \tag{6}$$

nates (x_i, y_i, z_i) by Eq.1 to Eq.5. In x-y-z system ,the ideal cone is expressed by Eq.6. Therefore, if we know the diameter of the ground drill (D) and some(n) coordinates (X_i, Y_i, Z_i) of the flank surfaces of the drill, then by minimizing the Eq.7, which is the mean square of relative error in z-axis direction between the coordinates and the grinding cone, the

$$f(a,e,\zeta,\delta,\alpha) = \frac{1}{n} \sum_{i=0}^{n} \left(z_i - \sqrt{\frac{x_i^2 + y_i^2}{\tan(\alpha/2-\delta)}}\right)^2 \tag{7}$$

values of parameters (a, e, ζ, δ, α) can be estimated and this five parameters reasonably describe the drill point geometry.

3 Experimental Setup:

To measure the drill flank surface coordinates , the experimental setup has been developed shown in Fig.2. Here we use LVDT probe as a comparator which is connected to PC via A/D converter. The stepping motor controller of R-θ-Z table is connected to the same PC via I/O port. The minimum measurement possible of the R-θ-Z axis are as follows:
R and Z axis $\rightarrow$ 0.004mm , θ axis $\rightarrow$ 0.012 degree (°).

4 Evaluation of the Parameters :

For evaluation of the parameters, some techniques of Evolutionary Algorithms have been examined and lots of simulation runs have been carried out with an objective to find a best method. For each simulation cases the following range of parameter has been used.:
$a \rightarrow [-1,1]$; $e \rightarrow [-1,1]$; $\delta \rightarrow [15°, 25°]$; $\zeta \rightarrow [0, 10°]$; $\alpha \rightarrow [110°, 125°]$,
where true values of the parameter are as follows:
a =0.01; e =0.05; δ= 20°; ζ= 5°; α=118°.

4.1 Simulation by Evolutionary Algorithm:
Here the problem is same as standard nonlinear Least Square estimation of model parameters, the objective function is not linear with respect to the parameters ,so the nature of it might be Multimodal[6]. To get Global optimum of the parameters , we did the following approach by using synthetic data.
By Steepest Gradient only:
Firstly we used Steepest Gradient Method as optimization method for parameter search. In Fig. 3 the change of parameter α by Steepest Gradient Method against parameter modification number has been shown. In this figure the values of parameter α has been shown by taking two type of initial values of all parameters up to 2000 parameter modification. In this case the value of parameter α goes to 150°, which is far away from true value 118°. Steepest Gradient Method sometimes makes the parameter values go far away from the true value, unless the initial values are near the true values. It is clear that this methods provide local optimum values only and this values depend on the selection of the starting point. Meanwhile it is difficult to find proper initial values in general.
By Genetic Algorithm only:
Then we used Genetic Algorithm(GA) for this parameter estimation problem. Genetic Algorithm is one of best well known Evolutionary Algorithm, inspired by natural evaluation. The flowchart of Evolutionary Algorithm used here is shown in Fig.4. Population number 250 and mutation rate 10% are adopted on the basis of preliminary numerical experiments. The crossover probability is used as 50% i.e. before crossover 124 chromosomes were selected for breeding. Here the floating point coding was used with classical GA. Change of parameter α by GA against generation number has been shown in Fig.5. In this case after 250 generation number the improvement of the value of α does not obtain. It seems that GA makes the parameter converge at infinite number of generations, estimated parameter's line becoming parallel with true parameter's line. Actually the value hardly improves after 200-th generation.
By Evolution Strategies (ES) with Classical Genetic Algorithm:
Evolution Strategies perform very well in numerical domains since they were developed initially to function optimization problem[5]. But here we used for the time being some concept of ES in selection steps with a hope

that we may reach the Global solution or at least Pareto optimal solution. In case of selection the following steps has been used.

i) Arranging the chromosome (strings) from minimum value to maximum value of objective function or performance index f.

ii) Then the classification of population has been done as 3 groups as follows:

Breeding Parents --- approximately first 25% of the population ,which will produce offspring.

Dying Parents --- approximately last 25% of the population, which will die before Mutation.

Neutral Parents --- remaining of the population which will not go Crossover steps.

In Crossover, the random one point crossover has been used between the chromosome lying side by side i.e. first with second, third with forth and so on. The offspring produced is equal in number as the breeding parent, which is also equal into the number of dying parents. So the population before Mutation consists of Breeding parents, Neutral Parents and offspring. Here we use one point random Mutation. Change of minimum value of f with generation number has been shown in Fig.6. Here even using 2% mutation probabilities and population size 50, the similar nature of convergence has been found as before with less number of generation. But some parameters values are not improving or converging to the true values with allowable error[7].

4.2 Estimation by Combining GA and Steepest Gradient Method with measured data:

As Holland suggested that the GA should be used as preprocessor to perform the initial search, before turning the search process over to a system that can employ domain knowledge to guide the local search[5]. We tried Steepest Gradient Method after applying GA to some extent with simulation data as well as data measured by the experimental setup. In Fig.7 and Fig.8 the result of this combine method has been shown. In Fig. 8 the change of performance index f by using the Combined Method of GA until the 100'th generation and subsequently Steepest Gradient Method has been shown. In this case the convergence is best[7]. After using this method the values of the parameters are estimated which has been shown in table 1 with true values. Here the performance index f is equal to the variance in Z-axis direction, which has reached a value 2.6×10^{-6} mm^2 , yet the parameters a, e and ζ reach some distance from true values. This result is caused of exertion unimportance influence on f.

5 Conclusions:

By analysing the relative motion of the grinding wheel and the drill point a method to assess the point geometry by using five parameters has been developed. The results show that the combined method of GA with Steepest Gradient Method may be suitable for the identification of these parameters from the viewpoints of calculation speed and accuracy. But still we are searching more economic approach to get Global optimum solutions for this problem.

Acknowledgements

The first author thanked *Japanese Ministry of Education* (MONBUSHO) for providing him scholarship and to *Jadavpur University*, Calcutta, India for providing him study leave for this research.

References

[1] D. B. Dallas : Tool and Manufacturing Engineers Handbook -- 3rd Edition Mcgraw-Hill (1976) pp.3-23.

[2] B. Ramamoorthy and V. Radhakrishnan : Computer Aided Inspection of Cutting Tool Geometry. *Precision Engineering*, **14**, 1 (1992) pp.28-35.

[3] D. F. Galloway : Some Experiments on the Influence of Various Factors on Drill Performance, *Trans. ASME, J. Engg. Ind.* **79**, 2 (1957), pp191-231.

[4] D. E. Goldberg : Genetic Algorithm in Search, Optimization and Machine Learing, Addison Wesley, (1989) pp.3-14.

[5] Z. Michalewicz : Genetic Algorithm + Data Structure = Evolution Program, Spring-Verlag Berlin Heidelberg, (1992).

[6] T. Back : Evolutionary Algorithms in Theory and Practice, Oxford University Press, (1996).

[7] T. Kuroda, H. Kato, R. Nakamura, Y. Tsuchiya, I. Sakuma and L. Hazra : Computer Aided Inspection of Drill Point Geometry, *Journal of the Japan Society for Precision Engineering (in Japanese)* **63**, 4 April, 1997 pp.575-579.

[8] V. A. Saraswat : Concurrent Constraint Programming, The MIT Press, ENGLAND (1993) pp.85-86.

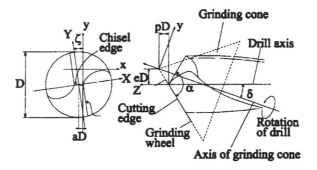

Figure 1. Geometrical Model of Drill Grinding

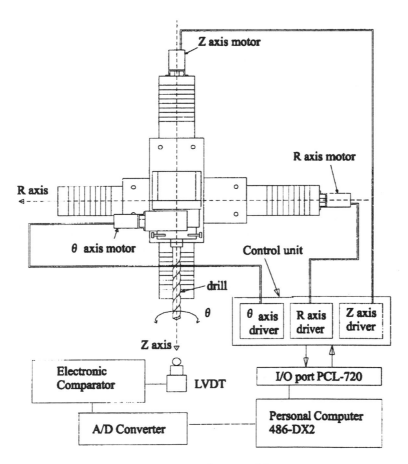

Figure 2. Experimental Setup for Drill Geometry Measurement

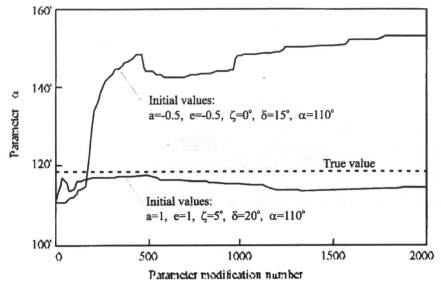

Figure 3. Change of parameter *a* by steepest gradient method

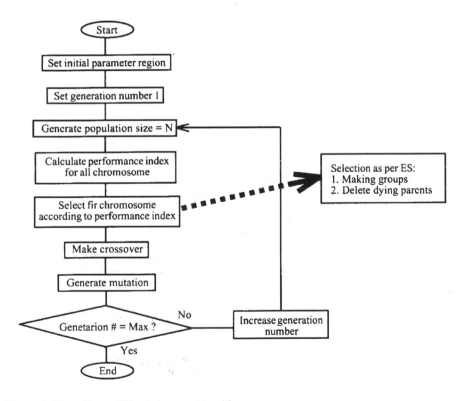

Figure 4. Flow Chart of Evolutionary Algorithm

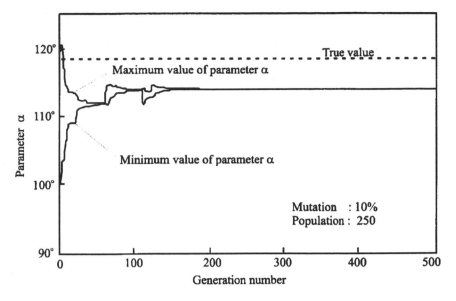

Figure 5. Change of Parameter a by GA

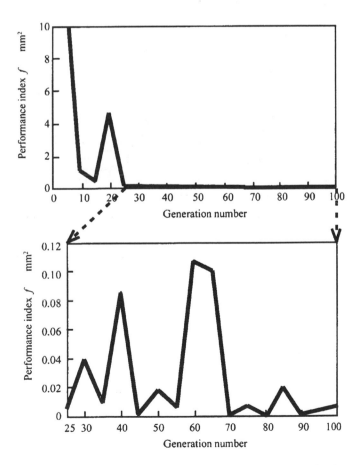

Figure 6. Variation of Objective Function with Generation Number in case of ES with classical GA[7]

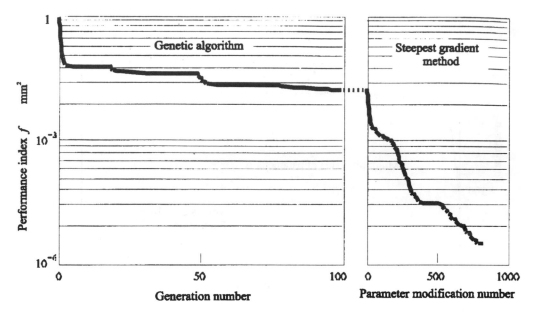

Figure 7. Change of Performance Index f by GA and Steepest Gradient Method (Synthetic Data)

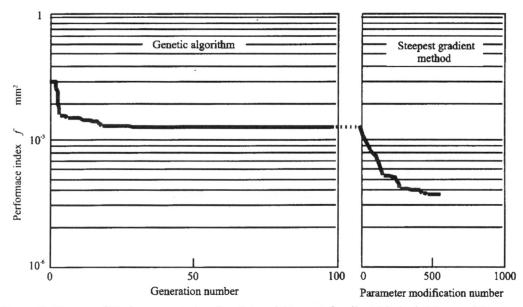

Figure 8. Change of Performance Index f by GA and Steepest Gradient Method (Measured Data)

Table 1. Comparison between initial and estimated values[7][8]

Parameter	a	e	ζ	δ	α	f
Initial region	-1 - 1	-1 - 1	-10° - 10°	10° - 20°	100° - 120°	None
True value*	0.01	0.05	5°	20°	118°	0
Estimated value	0.01131	0.05247	5.89°	20.03°	117.418°	2.6×10^{-6} mm^2

* True value : Measured value by motion measurement

Adaptability by Behavior Selection and Observation for Mobile Robots

François Michaud

Department of Electrical and Computer Engineering
Université de Sherbrooke
Sherbrooke (Québec CANADA) J1K 2R1
Tel.: (819) 821-8000 ext. 2107 Fax: (819) 821-7937
email: michaudf@gel.usherb.ca
http://www.gel.usherb.ca/pers/prof/michaudf

Keywords: Self-Observation, Behavior-Based Mobile Robots, Adaptability, Learning from Interactions, Control Architecture for Intelligent Systems

Abstract

Behavior-based systems are very useful in making robots adapt to the dynamics of real-world environments. To make these systems more adaptable to various situations and goals to pursue in the world, an interesting idea is to dynamically select behaviors that control the actions of the system. One factor that can give a lot of information about the world is the observation of use of behaviors (or *Behavior Exploitation*). Using this factor, the system can self-evaluate the proper working of its behaviors, giving it more adaptability. The paper describes how *Behavior Exploitation* can be used to influence motives using a simulated environment for mobile robots, and to acquire knowledge about the world using a Pioneer 1 mobile robot. It also exposes the repercussion on programming the system, and how it can help in designing a general control architecture for intelligent systems.

1. Introduction

Behavior-based systems [1], i.e., systems that use behaviors as a way of decomposing the control policy needed for accomplishing a task, are very useful in making robots adapt to the dynamics of real-world environments. By giving them basic competences, these systems have shown that some kind of intelligence is able to emerge from reacting to what can be perceived in the environment [2]. However, the behaviors in these systems are usually organized in a fixed and pre-defined manner, giving less flexibility for managing a high number of goals and situations.

Reactivity is only one factor associated with intelligent behavior, and one way to improve behavior-based systems is to make them more adaptive by dynamically selecting behaviors that can control the actions of the system. Other than selecting behaviors by what can be perceived in the environment, control architectures exist that try to combine reactivity with planning (like [3, 4, 5, 6]). The idea is to influence the selection of behaviors with knowledge on the environment (for example by using a model), usually by creating an abstraction on top of behaviors (or control components of different types). Internal variables like motivations [7] can also be used to affect the selection of behaviors.

The topic of this paper is to propose an additional factor that can be used in different ways to affect the selection of behaviors. This factor is the observation of use of behaviors, referred to as *Behavior Exploitation*. It is an important factor because it gives indirect information on the interactions between the system and its environment. Using this factor, the system can self-evaluate the proper working of its behaviors, giving it more adaptability. This paper gives an overview of what kind of interesting influences can be derived from *BehaviorExploitation*. It is organized as follows. Section 2 presents a control architecture proposed for dynamic selection of behaviors and where *Behavior Exploitation* is defined. Section 3 and Section 4 describes two beneficial influences of *Behavior Exploitation*, respectively its effect on motives and its effect for learning about the world. A brief discussion about this factor is then presented in Section 5, and Section 6 concludes the paper.

2. Control Architecture Based on Intentional Selection of Behaviors

This control architecture was designed with the primary objective of proposing a control architecture capable of combining the interesting properties associated with "intelligence", like reactivity, planning, deliberation and motivation, while preserving their underlying principles [8, 9, 10].

This architecture, illustrated by Figure 1, is built on top of behaviors connecting sensory information to actuation. It sets the agent's skills for responding to the situations encountered in the environment. These behaviors all run in parallel and their resulting commands are combined according to an arbitration mechanism to give the resulting control actions.

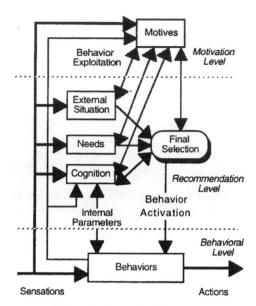

Figure 1. Control architecture

Using the behavior repertoire, the idea is to select these behaviors and to change their respective priority to make the system behave appropriately according to the situations it encounters in the world. Psychology indicates that human behavior is affected by the environment, the needs of the individual and the knowledge acquired or innate about the world [11]. Based on this decomposition, the *Recommendation Level* uses three recommendation modules to affect the selection of behaviors. The *External Situation* module evaluates special external conditions in the environment that can affect behavior selection. The *Needs* module selects behaviors according to the needs and goals of the agent. The third recommendation module, called the *Cognition* module, is for cognitive recommendation. This module learns things about the external environment and how the agent operates in it by observing its reactions, behavior selections and from information sensed from the environment. It can then exploit this acquired knowledge or some innate knowledge to plan or to prepare the use of behaviors. Cognitive recommendations can be influenced by behaviors via the *Internal Parameters* link, as they can influence behavior reactivity using the same link. These three recommendation modules suggest the use of different behaviors to the *Final Selection* module, which combines them appropriately to establish the activation of the behaviors (*Behavior Activation*). The influences of each of these modules can result in a manifested behavior more reactive, egoistic, or rational.

On the top level, the *Motivation Level* is responsible for coordinating and supervising the agent's goals according to its actual experiences in the environment. It is used to examine and to coordinate the proper working of the other modules. Motives are influenced by the environment, the internal drives of the agent, its knowledge and by observing the effective use of the behaviors.

Overall, the agent is then able to adapt its emerging behavior to its perception of the environment, its needs, its knowledge and its ability to satisfy its intentions. Different AI methodologies can be used to implement the modules of this control architecture. It is also important to note that the usefulness of the modules or of the links between them depends on the requirements of the task.

Finally, Figure 1 shows the link called *Behavior Exploitation*, which refers to the observation of use of behaviors. This is different than *Behavior Activation* in the sense that an active behavior is allowed to participate to the control of the agent, and it is said to be exploited if it is used to control the agent (by reacting to the sensations associated with the purpose of the behavior). However, to be exploited, a behavior must first be activated. This paper focuses on the influence of the *Behavior Exploitation* link in this proposed control architecture, and the following sections explain how this factor can be useful for motives and for acquiring knowledge about the world.

3. Behavior Exploitation for Motives

To validate the properties of the control architecture proposed, the first set of experiments used a simulated environment for mobile robots called *BugWorld* [12]. In these experiments, the agent must be able to efficiently reach the targets and must survive by recharging itself when needed. The agent knows nothing about the environment it is in, but has a limited memory to acquire knowledge that can be helpful in its task. For sensing, the agent has at its disposal eight directed proximity sensors for obstacle, each separated by 45° starting from its nose. There are also two undirected target sensors, one on each side. Note that the proximity sensors and the target sensors do not model any existing physical sensing devices. One target in the room is used as a charging station. The agent can also read the amount of energy available, its speed and its rotation. For actions, the agent can control the speed, the rotation and a variable for the color of the agent.

For these experiments, fuzzy logic was used for behaviors as in Saffiotti *et al.* [6]. It was also used by the *External Situation* module and the *Needs* module for recommending behaviors, and by the *Final Selection* module for combining these recommendations. A topological graph, having some similarities with the work of Mataric [13], was used by the *Cognition* module to construct an internal representation of the environment based on experiences of the agent. Finally, activation levels as in Maes [7] were used for motives. More details on these experiments can be found in [8, 9, 10].

In these experiments, the main influence from *Behavior Exploitation* is on motives, especially for detecting improper use of its behavior according to the intentions of the agent. Because behaviors are fuzzy, *Behavior Exploitation* is a fuzzy measure defined in relation (1), approximating the contribution or the importance of the behavior to the fuzzy control actions formulated before defuzzification. It combines the activation of a behavior with its reactivity to the environment.

$$\mu_{exp}(j) = \mu_{act}(j) \otimes \left(\oplus \left[\mu_{B_{rj}} (Action) \right] \right)$$
(1)

Two motives are influenced this way. First, the motive DISTRESS is used to monitor the proper working of behaviors like EMERGENCY (for moving the agent when immediate danger is detected in its front), AVOID (to move away from obstacles), and SPEED (to maintain a constant cruising velocity). These first two behaviors must normally be exploited very briefly to move the agent away from trouble areas. However, if their μ_{exp} remains approximately constant for a long period of time, this may be a sign of conflict between the behaviors used. For the SPEED behavior, a full exploitation for a long period of time is also a sign of trouble indicating that the agent is not able to reach its desired velocity. The SPEED behavior is composed of only two rules indicating when to increase or to decrease the speed of the agent. These rules use two fuzzy linguistic variables overlapping by 50% at the desired velocity, so a constant behavior exploitation of 0.5 indicates normal working condition. DISTRESS influences the use of the BACKUP behavior via the *Needs* module.

The other motive, called DECEPTION, increases when the agent is moving away from a target or a charging station. This is detected by a decrease in the exploitation of the TARGET (to align the robot toward a sensed target) or the RECHARGE (to search for a charging station and to energize the agent) behaviors, respectively. This motive influences the use of the TURN180 to make a U-turn (using the sensors and not a command of 180°) behavior via the *Cognition* module.

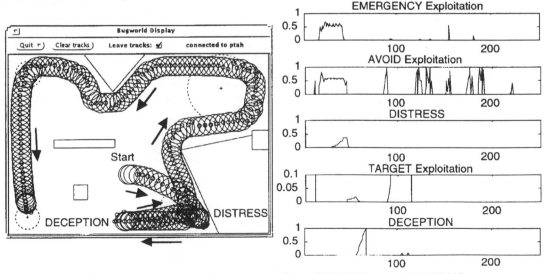

Figure 2: Trace and conditions for the motives DISTRESS and DECEPTION

These two motives help the agent manage the proper use of these behaviors, as shown in the following figures. In Figure 2, the agent starts from a point in the environment leading directly into the lower right corner. Because of a conflict between EMERGENCY, AVOID and ALIGN behaviors, the agent gets stuck in this corner. The simultaneous constant exploitation of EMERGENCY and AVOID excites the motive DISTRESS from which the BACKING behavior is recommended by the *Needs* module. This example indicates that observing the exploitation of behaviors in time may be very useful in managing conflicts between behaviors.

After having backed up away from the corner, the agent starts moving towards the charging station, but observes a decrease in the exploitation of the TARGET behavior. This indicates that it is moving away from a target which, in this case, is the upper right target. The motive DECEPTION is then increased and the agent makes a U-turn by using the TURN180 behavior. The agent continues its path by following boundaries until it reaches the charging station. This shows how the observation of behavior exploitation can serve as a useful indication of the progress of the agent in satisfying its intention of reaching a target.

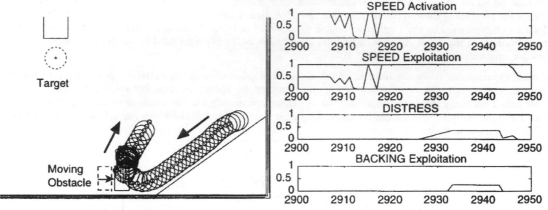

Figure 3: DISTRESS motive when a mobile obstacle came toward the agent

Finally, a similar situation of dissatisfied intention happened when the agent was placed in another room having a moving obstacle in it. Figure 3 shows the obstacle moving toward the agent. The agent tries to move away from it but could not do so because its back side collides with the moving obstacle. The agent does not have any behaviors to avoid obstacles from its back. But, by observing that the SPEED behavior is fully exploited for a long period of time (because the agent wants to move ahead but simply cannot get moving), the motive DISTRESS is excited so that the BACKING behavior can be used.

4. Behavior Exploitation for Learning about the World

The second application of *Behavior Exploitation* involves taking into account the history of behavior use to learn specific dynamics from the robot interactions with its environment, and to use this to select behaviors. Because the algorithm is based on observing behavior use, the robot learns by evaluating its own performance based on the exploitation of its resources and on its own history of past experiences. To do so, the algorithm implemented has four important characteristics: a classification of behaviors according to their respective purpose, a representation mechanism to encode history of behavior exploitation, an efficiency measure to evaluate its interactions with the world, and a recommendation mechanism. To better understand these characteristics, the experimentation setup is first explained in the following section. This research is currently in progress, and only a brief description is given in this paper. More information can be found in [14, 15]. Related work includes Ram and Santamaria [16] who consider history from inputs and from behavior activation for setting future behavior activation, and McCallum [17] in which a reinforcement learning approach based on history is proposed.

Figure 4: Pioneer robot at home with a block

4.1. Experimentation Setup

For the experiments, the general goal of the robot is to search for a block and bring it to the home region. The robot used for learning is a Real World Interface Pioneer I mobile robot equipped with seven sonars and a Fast Track Vision System (Figure 4). This robot is programmed using MARS (Multiple Agency Reactivity System) [18], a language for programming multiple concurrent processes and behaviors. The experiments are conducted in an enclosed rectangular "corral" which contains a block and a home which are distinct in color and detected by the robot's vision system. Other obstacles (like boxes or other robots) and walls are detected with the sonar system.

4.2. Classification of Behaviors

The Pioneer is initially programmed with a set of behaviors classified according to their respective purposes, as follows. First, the robot has a behavior for accomplishing the task of finding and grasping a block, such as *searching-block*, and behaviors for the task of bringing them home, such as *homing* and *drop-block*. These behaviors are referred to as task-behaviors (*tb*). Conditions for activating task-behaviors are given according to the presence or absence of a block in front of the robot and to the location of the robot relative to home (both sensed with the vision system). Second, behaviors such as *avoidance* are given to the Pioneer for handling harmful situations and interference experienced while accomplishing its task. These behaviors are referred to as maintenance-behaviors (*mb*). The designer also provides a third set of behaviors referred to as alternative-behaviors (*ab*), which may be useful for changing the way a task is achieved. No *a priori* conditions are given for their activation: the objective is to let the robot learn when to activate these alternative-behaviors in accordance with its own internal evaluation of efficiency and its own experiences in the world. For example, the Pioneer may decide that is is better to do *wall-following*, *rest* or *turn-randomly* for some time when experiencing difficulties in accomplishing its task. This can result in interesting behavior strategies learned from the point of view of the robot, whose exact conditions need not be specified or even anticipated by the designer.

4.3. Representation Mechanism

The history of behavior use is encoded in graph form and separate graphs are used for each task, i.e., one for searching for a block and one for bringing a block home. Figure 5 shows a typical graph in which nodes represent the type of behaviors used for controlling the robot while accomplishing its task, and links store the number of times a transition between two types of behaviors was observed. The efficiency at the end of the task is memorized in the last node of a path. The sequence of nodes in a path characterizes the sequence of interactions experienced by the robot in the world, and situates the robot relative to those interactions if some repeatable patterns can be found. Initially, the graph for a particular task is empty and is incrementally constructed as the robot goes about its task. When an alternative-behavior is applied, its name becomes the node type to characterize the choice made by the robot at that point in the graph. Similar paths can be followed according to the sequence of behavior exploitation and the nodes stored in the graph. If a complete path is reproduced by the robot, the average of the stored and current efficiencies is computed and the stored value is updated.

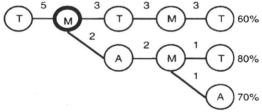

Figure 5: Graph representation of sequences of behavior exploitation (*T* is for task-behaviors, *M* is for maintenance-behaviors, and *A* is for alternative-behaviors)

Finally, it is important to forget paths as the robot builds up its experiences in order to avoid getting stuck in believing that one option is better than the others. The policy used for deleting paths is to delete the oldest path in the graph. The parameter that determines the number of paths to keep is influenced by the number of alternative-behaviors used and the flexibility that the designer wants to give to the robot to adapt to changes in the environment.

4.4. Efficiency Measure

To decide whether or not it may be profitable to use an alternative-behavior, the robot needs an efficiency measure evaluating its performance in accomplishing the task. For the purpose of these experiments, the efficiency measure used evaluates performance also according to the exploitation of behaviors in time. Expressed by relation (2), the efficiency here is characterized by the amount of time required to accomplish the task and the interference experienced during that process:

$$eff(t) = \frac{t_{tb} - t_{mb}}{t_{tb}} - \max\left(0, \frac{t - TT}{TT}\right)$$

(2)

where t represents the current time of evaluation, and TT represents the average total time it took to accomplish the task in the past. The second term in this equation penalizes the efficiency if the robot is taking more time than it normally requires to do the task, based on past experience. A time unit corresponds to one cycle of evaluation of all currently active behaviors.

4.5. Recommendation Mechanism

While the robot is accomplishing a task, the graph representation is used to establish the decision criteria by anticipating the future outcome of its choices. To do this, the robot evaluates the expected efficiency from the current position in the graph. This evaluation occurs only at a maintenance-behavior node, the location where $eff(t)$ decreases. The expected efficiency from any point in the graph is the sum of the memorized efficiencies in connected paths multiplied by their probability of occurrence. These probabilities are derived from the frequency of reproduction of the paths relative to the current position in the graph. For example, if the current position in Figure 5 is the bold node, then the expected efficiency for the subpath starting with the task-behavior is (60*3/5)% while the expected efficiency for the subpath using an alternative-behavior is ((80+70)*1/5)%. The total expected efficiency for this node is the sum of the expected efficiencies or these subpaths, i.e., 66%.

From these measurements, the robot has three options: to decide whether to continue without using an alternative-behavior, to use an untried alternative-behavior, or to use an alternative-behavior that was previously tried. Different selection criteria can be used, based on the expected efficiency and/or the current efficiency at the maintenance-node in the graph.

4.6. Results

Experimentation consists of using the Pioneer robot in different environment configurations (without changing the behavior repertoire or the control parameters of the behaviors) to see what can be learned from the interactions of the robot in these environments. The first tests were done using a static environment to verify that the algorithm can learn stationary conditions from the interactions. Next an environment with multiple robots (up to 3 R1-type robots plus another Pioneer also using the learning algorithm) was used to see if something can be learned from multi-robot group dynamics.

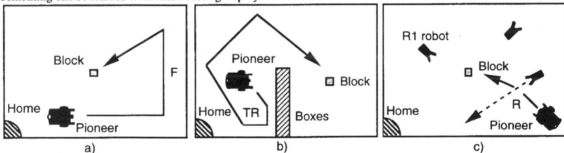

Figure 6: Environment configurations and possible strategies learned

Figure 6 shows the different environment configurations used for experimentation, along with one possible strategy learned for each case when searching a block. In environment a), the block is placed in the center of the "corral". Before doing any experiments, it was assumed that the preferred option would be to use *turn-randomly* in order to make the robot move away from boundaries and go in the center. However, the most commonly found strategy is to move to the first wall (either at the upper left corner or the bottom right corner), and follow it to the upper right corner. From this corner the Pioneer is able to see the block and can easily reach it. The Pioneer is also able to see the block from the other corners but not as reliably, so the option of using *wall-following* is usually preferred.

In environment b), an obstacle made up of boxes divides the "corral" in two. The Pioneer is able to see the block from the upper left corner of the environment, but cannot reach it because it looses sight of the block when avoiding the obstacle at its top part. To get the block, *turn-randomly* used at the start of the searching task can make the robot go toward the right side of the environment and get the block correctly. Changing the environment conditions from putting the block in the center to adding an obstacle like in environment b) was also experimented to see that the algorithm can rapidly adapt to changes in the environment. In this situation, the bias toward the preferred option from the previous environment conditions influences the strategy learned by the robot.

Finally, in environment c), other robots are put in the "corral" to see if some kind of strategy can be learned from group interaction dynamics. This task is much more difficult because it is very unstationary, the R1 robots causing different interactions with the Pioneer over time. It is then difficult to learn a stable strategy to select the different options. However, this causes the algorithm to look for longer strategies to get the task done, and also stabilizes for short period of times in selecting the appropriate options until a decrease of efficiency is observed. One of the possible strategies is to use *rest* when an obstacle (like another robot) is encountered.

A more complete description of the results and a more thorough analysis of the algorithm could have been presented in this paper, but we are still in the process of doing that [15]. However, it is possible to say that the algorithm has shown to give adequate flexibility to the robot by learning the most efficient options to take according to past experiences and from its own evaluation of its resources. In addition, no optimal solution was found for each case because it depends on the actual history of experiences of the robot, this being affected by a lot of factors like noise and imprecision (mostly affecting the vision system (caused by the variation of lighting condition around the block, modifying its color)) and processing constraints (like garbage collection). By taking into consideration the perspective of the robot, these experiments show that we can actually learn from making the robot evaluate its own performance from its interaction with its environment.

5. Discussion

One advantage of using *Behavior Exploitation* to affect the selection of behaviors is that it decreases the burden of the designer in knowing everything about the environment, imagining every possible conflict situation or activation condition, and tuning behaviors for different applications. The fundamental idea is that a purpose is associated with each behavior, and this can be used to reason about the way the agent is behaving in its environment.

Related to this is the question of how to design behaviors, because the influences coming from *Behavior Exploitation* depend on how behaviors are designed. This might depend on the amount of determinism that the designer can have about the tasks that the robot needs to accomplish and the steps required to accomplishing them. For example, well-defined situations can use exclusively-activated behaviors or behaviors that subsumed others, while reactivity and learned conditions from the dynamics of the world might better handle situations that cannot be anticipated precisely during design. Finding a compromise between these types of behaviors is an important issue if we want to make robots autonomously do specific things in unknown environments.

6. Conclusion

Overall, one important contribution from these research projects is that behavior-based systems can be made more adaptive by dynamically changing the configuration of behavior set by factors other than sensed conditions in the environment, thus making a move toward increasingly cognitive means for generating complex behavior. This is done in accordance to the belief that intelligent behavior is related to the dynamic of interactions between the system and its environment [19, 20, 21, 22, 23]. Furthermore, these interactions are established based on the system's ability to adapt to its environment, as well as to its own abilities of interacting with it. In fact, they are affected by its sensing and actuating capabilities, its processing and memorizing capabilities, and its deciding and reasoning capabilities. For any system, these capabilities have limitations. Therefore, it is important to give to the system the ability to cope efficiently with them. *Behavior Exploitation* shows very interesting properties for doing so.

Finally, in spite of the significant progress achieved, intelligence is still a difficult notion to define and to entirely reproduce in artificial systems. No consensus exists on what is intelligence [24], what are the basic and essential characteristics needed by a system to behave intelligently and what could be a general framework of intelligence [25]. At the present time, there are too many things unknown about intelligence to discriminate between the methodologies and control architectures used to reproduce intelligent capacities into artificial systems. Instead, there is an actual need to tie these aspects into a unified framework [26, 27]. It is not claimed here that the proposed architecture fully explains intelligence, but the hope is that it will help establish a better understanding of the nature of intelligence, stimulate further discussions and research on the subject (in accordance with the goal of the current workshop), and how to integrate its aspects into artificial systems.

Acknowledgments

Support from the Natural Sciences and Engineering Research Council of Canada (NSERC) was highly appreciated. The research project involving learning from history of behavior use was done at the Interaction Laboratory of Brandeis University, under the supervision of Maja J. Mataric. Special thanks to Prem Melville for his participation in the experiments with the R1 robots. I would also like to acknowledge the participation of Gérard Lachiver and Chon Tam Le Dinh with the part involving the simulated environment for mobile robots, and thank Nikolaus Almàssy for making *BugWorld* available for experimentation.

370

References

[1] Brooks, R.A., 1986, A robust layered control system for a mobile robot, *IEEE Journal of Robotics and Automation*, **RA-2** (1), 14-23.

[2] Brooks, R.A., 1991, Challenges for complete creature architectures, in *From Animals to Animats. Proc. First Int'l Conf. on Simulation of Adaptive Behavior*, The MIT Press, pp. 434-443.

[3] Bonasso, R.P., Kortenkamp, D., Miller, D.P., and Slack, M., 1995, *Experiences with an architecture for intelligent reactive agents*, Internal Report Metrica Robotics and Automation Group, NASA Johnson Space Center.

[4] Ferguson, I.A., 1992, Toward an architecture for adaptive, rational, mobile agents, in *Decentralized A.I.-3. Proc. Third European Workshop on Modelling Autonomous Agents in a Multi-Agent World*, Werner, E. and Demazeau, Y. (ed.), Elsevier Science, pp. 249-261.

[5] Firby, R.J., 1989, *Adaptive execution in complex dynamic worlds*, Ph.D. Thesis, Dept. Computer Science, Yale.

[6] Saffiotti, A., Ruspini, E., and Konolige, K., 1993, *A fuzzy controller for Flakey, an autonomous mobile robot*, Technical Note 529, SRI International.

[7] Maes, P., 1991, A bottom-up mechanism for behavior selection in an artificial creature, in *From Animals to Animats. Proc. First Int'l Conf. on Simulation of Adaptive Behavior*, The MIT Press, pp. 238-246.

[8] Michaud, F., 1996, *Nouvelle architecture unifiée de contrôle intelligent par sélection intentionnelle de comportements*, Ph.D. Thesis, Université de Sherbrooke, Department of Electrical and Computer Engineering.

[9] Michaud, F., Lachiver, G., and Dinh, C.T.L., 1996, A new control architecture combining reactivity, deliberation and motivation for situated autonomous agent, in *Proc. Fourth Int'l Conf. on Simulation of Adaptive Behavior*, Cape Cod, September.

[10] Michaud, F., Lachiver, G., and Dinh, C.T.L., 1996, Fuzzy selection and blending of behaviors for situated autonomous agent, in *Proc. IEEE Int'l Conf. on Fuzzy Systems*, New Orleans, September.

[11] Dolan, S.L. and Lamoureux, G., 1990, *Initiation à la Psychologie du Travail*, Gaetan Morin Ed.

[12] Almassy, N., 1993, *BugWorld: A distributed environment for the development of control architectures in multi-agent worlds*, Tech. Report 93.32, Dept. of Computer Science, University Zurich-Irchel.

[13] Mataric, M.J., 1992, Integration of representation into goal-driven behavior-based robots, *IEEE Trans. on Robotics and Automation*, **8** (3), 304-312.

[14] Michaud, F. and Mataric, M.J., 1997, Behavior evaluation and learning from an internal point of view, in *Proc. of FLAIRS (Florida AI International Conference)*, Daytona, Florida, May.

[15] Michaud, F. and Mataric, M.J., 1997, *A history-based learning approach for adaptive robot behavior selection*, Tech. report CS-97-192, Computer Science Department, Brandeis University.

[16] Ram, A. and Santamaria, J.C., 1993, Multistrategy learning in reactive control systems for autonomous robotic navigation, *Informatica*, **17** (4), 347-369.

[17] McCallum, A.K., 1996, Learning to use selective attention and short-term memory in sequential tasks, in *Proc. of the Fourth Int'l Conf. on Simulation of Adaptive Behavior*, Cape Cod, September, pp. 315-324.

[18] Brooks, R.A., 1996, MARS: Multiple Agency Reactivity System, *IS Robotics Documentation*.

[19] McFarland, D. and Bosser, T., 1993, *Intelligent Behavior in Animals and Robots*, Bradford Book, The MIT Press.

[20] Pfeifer, R., 1995, Cognition—Perspectives from autonomous agents, *Robotics and Autonom. Syst.*, **15**, 47-70.

[21] Smithers, T., 1994, On why better robots make it harder, in *From Animals to Animats 3. Proc. Third Int'l Conf. on Simulation of Adaptive Behaviors*, The MIT Press, pp. 64-72.

[22] Smithers, T., 1995, Are autonomous agents information processing systems, in *The Artificial Life Route to Artificial Intelligence: Building Embodied, Situated Agents*, Steels, L. and Brooks, R. (ed.), Lawrence Erlbaum Associates, chap. 4, pp. 123-162.

[23] Steels, L., 1995, Intelligence—Dynamics and representations, in *The Biology and Technology of Intelligent Autonomous Agents*, Springer Verlag, Berlin.

[24] Antsaklis, P., 1994, Defining intelligent control. Report of the task force on intelligent control, *IEEE Control Systems*, 4-5 & 58-66.

[25] Steels, L., 1995, The Homo Cyber Sapiens, the robot Homonidus Intelligens, and the 'artificial life' approach to artificial intelligence, in *Burda Symposium on Brain-Computer Interface*.

[26] Albus, J.S., 1991, Outline for a theory of intelligence, *IEEE Trans. on Syst., Man, and Cybern.*, **21** (3), 473-509.

[27] Corfield, S.J., Fraser, R.J.C., and Harris, C.J., 1991, Architecture for real-time control of autonomous vehicles, *Comput. Control. Eng. J.*, **2** (6), 254-262.

Application-Based Time Tabling by Genetic Algorithm

Masahiro Tanaka, Mari Yamada and Osamu Matsuo

Department of Information Technology, Okayama University
3-1-1 Tsushima-naka, Okayama 700, JAPAN
tanaka@mathpro.it.okayama-u.ac.jp

Keywords: Time tabling, Genetic algorithm, Application-based method, Scheduling

Abstract

As is well known, time tabling problem is a combinatorial optimization problem with many hard and/or soft constraints. In most cases the problem includes nonlinear objective function and/or constraints. Several papers on this problem have been published using the genetic algorithm, where the problem is mainly to favor the organization which makes the time table. In this paper, we address the problem using the application slips, with which the clients express their desirable time slots for each lecture. This method is especially effective if the clients are very busy, or if the problem is for commercial use and hence the clients' wishes have high priorities. To make the clients more satisfied even if the first desire is not accepted, the clients are supposed to write several desires of time slots for allocation. Genetic algorithm is used to optimize the order of the application slips to be used.

1 Introduction

In school time tabling, the following constraints occur.

- A tutor must not have two lectures at a same time slot.
- A classroom is not to be used for two lectures at a same time.
- The expected number of audience should not exceed the capacity of the classroom.

These are considered to be hard constraints.

Genetic algorithm has been applied to the time tabling problems, e.g. [1, 2, 3, 4, 5]. Colorni *et al.* [1] took the problem as the assignment of the classroom number into two-dimensional array (tutors × time slots), and coded the problem as an application of the genetic algorithm. Abramson and Abela [2] attached a label to the vector (class number, tutor, classroom), and deduced the problem to assign this label to the time slot by applying the genetic algorithm. The number of lectures that can be held at one time slot is not single, i.e. there are several classrooms available for each time slot, and the number of available classrooms are not the same for all the time slots. Thus, the problem has the chromosome of two-dimensional array with non rectangular form. Burke *et al.* [3] dealt with the examination time tabling.

Paechter *et al.* developed a method where a 4-dimensional vector

$$(\text{time slot, time slot direction, next, next direction})$$

is used as the GA code [6]. With this code, it is possible to avoid the hard constraint problem because, if the first assignment is impossible, it will be tried to another slot. Thus it seems applicable for the time tabling problem. However, it has not been clearly shown whether the GA efficiently produced the complicated code for each lecture.

There are other problems in the above methods. Although it is usually difficult to accept and realize the desire of the clients by those methods.

Several studies have been done on this problem using GA, where the problem is mainly to favor the organization which makes the time table.

In this paper, we address the problem using the application slips, with which the clients express their desirable time slots for each lecture. "Lecture" can be something else for other application areas. This method is especially effective if the clients are very busy, or if the problem is for commercial use and hence the clients' wishes have high priorities. To make the clients more satisfied even if the first desire is not accepted, the clients are supposed to write several (typically three) desires of time slots for allocation.

In this paper, genetic algorithm is used to optimize the order of the application slips to be used. Allocation module is employed to satisfy the hard constraints, i.e. two lectures cannot be assigned to the same time slot, one client cannot deliver two lectures at the same time, "must" lectures for a same student cannot be held at the same time, and so on. All the other efficiencies of the time table can be taken into account by embedding into the fitness function. Various comparisons will be made to justify the superiority of the proposed method.

Creating a time table of a week is a very popular problem that arise at school or "Juku" which is a private place to study after school in Japan. At schools, the time table must be made every year, but the working load for making that is extremely heavy. The problem also exists in various kinds of making programs such as the allocation of sessions in a conference.

The hard constraints are satisfied by the "assignment module" which is used in each iteration of GA. Moreover, GA is good in global performance but it often fails to go into the local optimum. Thus in the approach proposed here, a minor tuning module is applied in the final stage which assigns a lecture by shifting an already assigned lecture.

2 Problem Formulation

2.1 Problem Description

Suppose there are N working days, and there are n_j lecture time slots for the day j. Then the total lecture hour is

$$K = \sum_j n_j.$$

Integers $1, \cdots, K$ are assigned for the time slots, and let the set of the integers be expressed as $\mathcal{K}$. Define the time slot "0" which is the virtual number of time for the lecture that was not assigned. Also, let $\mathcal{K}_0 = \mathcal{K} \cup \{0\}$. Further assume that there are L lectures totally, which are labelled as $1, \cdots, L$ and their set as $\mathcal{L}$. The final problem of time tabling is to define the map $\Phi : \mathcal{L} \rightarrow \mathcal{K}_0$. But, since the application slip is used here, it is necessary to include the point which expresses the degree of the acceptance of the desire in the fitness function.

2.2 Coding

The clients are supposed to submit an application slip that specifies the first, second, $\cdots$, K−th desire of time slot for the lecture to be assigned for each lecture, that are $k_j(i); j = 1, \cdots, K$. It also includes the description of the label $i \in \{1, \cdots, n\}$ and the applicant's number $n(i)$ as the attributes.

The hard constraints that must be kept are (1) a tutor must not have two lectures at a same time slot, (2) a classroom is not to be used for two lectures at a same time, and (3) the expected number of audience should not exceed the capacity of the classroom.

Some other constraints will exist, and this kind of constraints usually vary from organization to organization.

The soft constraints, on the other hand, are such that (1) lectures which some students are better to take are not to be held at the same time, (2) tutors don't like to deliver lectures continuously for a long time.

There are several possible ways of encoding. First, as the simple but not an efficient way, the lecture numbers are lined up vertically, and the time slots are lined up horizontally. The table is to give 1 or 0 which indicates whether such an assignment is taken or not. In this method, there arise hard constraints that must be strictly kept. Each column must be 1 or 0, and each row must be 1. Such a problem can be dealt with by the Hopfield neural net, but it seems practically impossible to carry out because the dimension of the problem is very large.

Next, it is natural to encode the problem where K-dimensional chromosome is prepared to accept the lecture number. With this code, the chromosome is short, but it is very hard to apply GA because, in this case, the hard constraints again arise.

Here, an approach is taken where the length of the chromosome is L (the number of the application slips), and the application slip number is assigned to the locus. The assignment modules tries to assign the application slip in the order of the slip number that appears from the left, and if the time slot of the first desire is already assigned by some other slip, the second desire is tried, and so on. In this way, the hard constraint that the same application slip should not be used is automatically satisfied. Basically, the problem is thus reduced to similar to traveling salesman problem. Thus we apply the GA operations developed for the traveling salesman problem [7].

2.3 An Example of the Problem

This is an example of "juku" time table. A juku is a private supplementary school which crams school children outside school hours. In the juku considered here, each student is taught by a tutor at each time slot. Several tutors come to work each day, and when they come to teach, they teach a couple of hours on the day.

Looking at the time table with the tutors names and subjects, students hand in the application slips which specifies

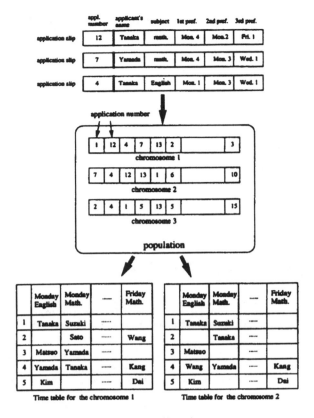

Figure 1. Time allocation stages

the 3 most desirable time slots for each subject. Suppose there are N tutors and M students. The problem here is to assign the application slips to the time table for one week. Suppose that there are m time slots for each working day. Then totally the time table is of the size $5 \times KN$ matrix.

2.3.1 Objective Function and Constraints

The objective function must evaluate the following items as a whole.

- How much the higher desires have been achieved and how little the application slips have no assignment.
- How much the tutors desire have been accepted.

The hard constraints that must be kept are (1) the students cannot take more than 2 lectures at the same time, (2) the tutors cannot deliver more than 2 lectures at the same time.

3 Algorithm

In this method, two modules exist.

1. assignment module
2. minor tuning module

The second module is used at the final stage, where the application slip which was not assigned is tried to be assigned by moving the already assigned application up to once. With this simple modification, it is expected to smooth the assignment a little. Since this modification is similar to the one which humans often do, it is also used as the traditional method for comparison. Note that the assignment does not change when there are as many application slips as the time slots and there is no space for move.

3.1 Assignment Module

The assignment module is used in the GA iteration.

The algorithm is defined as follows.

Step1. Generate the initial population.

Step2. Assignment and evaluation.

Step3. Selection and reproduction.

Step4. Crossover.

Step5. Mutation.

Step6. Assignment and evaluation.

Step7. If the end condition is not satisfied, go to Step3, else End.

Assignment Take out the application slip according to the number of the chromosome. For each application slip, look into the time slot beginning with the first desire time slot. If it is occupied, the next desire will be taken into account. If there is no place of assignment in the three desires, it is kept unassigned.

Selection and Reproduction Ranking by Baker [8] selection is adopted for the reproduction. Individuals of the lowest evaluation by the proportion of Pr is deleted and the same amount of individuals of the highest evaluation of the individuals are reproduced. Moreover, the elitist strategy is applied where the best one remains with no modification.

Crossover In cycle crossover (CX) by Oliver *et al.* [10], some genes are kept at the same place in the next generation. In the ordered crossover (OX1) by Davis [11], relative positions of parent genes are kept in the next generation. In the ordered crossover (OX2) by Syswerda [12], not like OX1, key points are selected randomly and exchange. Also by Syswerda [12], position based crossover (PBX) has been proposed, where the positional information is kept in the procedure of gene exchange. Here we use these operations in parallel.

Mutation Mutation is different from the standard GA because it is necessary that no invalid code may occur by the mutation. Here we define the mutation as the stochastic operation on a single individual, not alike the crossover on two individuals.

1) Exchange A position is selected randomly first. Then select another position of the same individual randomly, and exchange them.

2) Inversion A position is selected randomly first. Then select another position of the same individual randomly, and invert the order of the genes between these two positions.

3) Insert A position (A) is selected randomly first. Then select a position (B) of the same individual randomly, and insert A before B (if A is behind B) or insert B before A (if B is behind A).

3.2 Minor Tuning Module

This module is used on the time table, not on the chromosomes. This is applied once on the time table which the best chromosome created.

Algorithm of Minor Tuning

Step1. Make a list $L_j (j = 1, \cdots, h)$ of the application slips which were not assigned in the best time table created by GA.

Step2. $j := 1$

Step3. $k := 1$

Step4. If the time slot corresponding to the k-th desire of the application L_j is already assigned by the application of himself, go to Step6.

Step5. Check the time table corresponding to the k-th desire of the application L_j. If it is already assigned by another person's application and it can be moved to its second or third desire, and the objective function value is improved, actually reassign it and go to Step7.

Step.6 If $k < 3$, let $k := k + 1$ and go to Step4.

Step.7 If $j < h$, let $j := j + 1$ and go to Step3, else stop.

4 Numeral Example

The table of the used data is shown in the appendix.

Let the number of tutors be 5, the students 20 and the number of application slips $L = 75$.

Here it is obvious that the upper limit of assignment is 100.

The objective function is given by the following rule.

$$\sum_{i=1}^{n} f(r(i)) + \sum_{j=1}^{5N} g(s(j))$$

where $r(i)$ is the rank ($\in \{0, \cdots, 6\}$) for the application slip i, $f(r)$ is the value for the rank r and $f(r(i))$ is the evaluation for the application i.

The rank for i is defined as follows.

- $r := 0$ (default)

- if it is not assigned, $r := r + 3$;

- let t be the time the application was not assigned due to other person's assignment. Then $r := r + t$;

In the best case $r = 0$ and in the worst case $r = 6$. The value for the ranks have been defined as Table 1.

Table 1. Evaluation of the rank of application

r	0	1	2	3	4	5	6
$f(r)$	0	5	15	20	25	50	100

Also, $s(j)$ is the rank of stress for a tutor for one day j, $g(s)$ is the value for the stress s. Here the stress of the tutor has been defined as follows. The rank of the stress is $s \in \{0, 1, 2\}$, where $s = 1$ is the case he or she delivers lectures for three hours. $s = 2$ is the case for 4 hours continuously. The values for the ranks $g(s)$ have been defined as Table 2.

Table 2. Evaluation of the rank of tutor's stress

s	0	1	2
$g(s)$	0	5	10

Let the size of GA be

$$\text{(population } P, \text{ generation } T) = (150, \ 400)$$

This corresponds to evaluate 60000 times of individuals in a GA trial.

For the GA internal parameters, the selection probability Pr, the crossover probability Pc, and the mutation probability Pm. These are defined as

$$(Pr, \ Pc, \ Pm) = (0.05, \ 0.98, \ 0.01)$$

Although epistasy can happen for multiple using the selection, crossover and mutation operations, we first use the elitist selection to obtain the relatively fast convergence as the selection operator.

Next we consider crossover operations. By OX2, the algorithm yielded near optimal solution without a very sensitive effect to it. With CX, the algorithm was very sensitive to the parameter value and it often converged to bad points. With OX1, the algorithm did not converge, which seemed to be a bad operation for this problem by the computer simulation. The effect of PBX and OX2 did not show a significant difference, and there was no special subside effect by using together. Thus we use OX2. A simple example is illustrated below.

```
parent1:   (  1  2  3  4  5  6  7  8  9  )
parent2:   (  4  5  2  1  8  7  6  9  3  )
key point:    *  *              *  *
```

Suppose that the selected key points are 1, 2, 8, 9. The elements of the corresponding positions of the parent 2 are 4, 5, 9, 3, which are kept in the child.

```
child 1:   (  x  x  4  5  9  x  x  x  3  )
```

The remaining elements of the parent 1 remain in the child 1. Child 2 is generated in the same way.

```
child 1:   (  1  2  4  5  9  6  7  8  3  )
child 2:   (  4  5  1  2  8  7  6  9  3  )
```

In our experiment, there was no significant effect by using the mutation operations. Moreover, the inversion yielded a negative result. This means that, in this kind of GA problems, there is no fundamental differences in the crossover and mutation.

To confirm the ability of GA, the result is compared to the ones by the following methods.

Table 3. Best time table by GA

lecture #	1				2			
time slot	1	2	3	4	1	2	3	4
Mon	17	13	18	7	15	16	-	6
Tue	18	17	17	19	3	13	10	5
Wed	-	-	8	18	19	-	6	10
Thu	7	4	-	2	5	11	3	11
Fri	14	13	-	13	10	7	16	3
lecture #	3				4			
time slot	1	2	3	4	1	2	3	4
Mon	7	10	13	17	18	-	-	-
Tue	17	-	3	17	10	-	-	-
Wed	10	17	2	-	-	14	-	6
Thu	8	8	12	10	-	-	18	-
Fri	1	3	19	7	-	4	13	18
lecture #	5							
time slot	1	2	3	4				
Mon	5	18	-	-				
Tue	14	10	7	7				
Wed	7	18	3	4				
Thu	10	10	11	-				
Fri	-	18	-	4				

Random Method Permutation is generated randomly and take the best one among them.

Order Assignment + Minor Tuning The application slips are assigned by the order of acceptance. After that, the minor tuning module is applied.

In the random method, 60000 individuals are generated randomly, which is the same as the number of individual evaluation in GA. The objective function value of the best individual was 315.

Table 4. Best time table by random method

lecture #	1				2			
time slot	1	2	3	4	1	2	3	4
Mon	19	13	-	17	16	11	15	6
Tue	8	2	18	17	-	13	10	3
Wed	7	-	18		19	-	6	10
Thu	7	4	17	18	5	-	3	11
Fri	14	13	-	13	10	7	16	3

Lecture #	3				4			
time slot	1	2	3	4	1	2	3	4
Mon	17	17	13	2	18	18	-	-
Tue	12	-	3	19	18	-	-	-
Wed	10	17	-	-	-	-	10	6
Thu	8	8	7	10	-	-	14	-
Fri	1	3	10	7	-	4	13	-

lecture #	5			
time slot	1	2	3	4
Mon	10	-	-	-
Tue	14	10	7	7
Wed	18	18	3	4
Thu	10	11	5	7
Fri	-	18	-	4

In the second method, the first assignment was with the value 740. After the minor tuning, the value decreased to 575. On the other hand, the objective function by GA was as low as 230. The corresponding time table of the best individually GA is shown in Table 3, and the best one by the random method is shown in Table 4.

It can be seen that the best time table by GA only assigns all the students' desire. In the time table by the random method, two application slips were not assigned. With the order assignment and minor tuning, 5 application slips were not assigned. These results show the usefulness of the genetic algorithm based method. Table 5 shows the result for various numbers of application slips for the same time table. In all the experiments, GA based method yielded the best results.

Table 5. Comparison of evaluates of the best time table by various methods

# of appl.	GA	(genera- tions)	random	order assignment +minor tuning
50	80	(200)	85	130
75	230	(400)	315	575
100	755	(800)	1040	1170
120	1420	(1000)	1945	2125
125	1700	(1200)	2390	2905
150	2380	(1600)	3510	4050
175	3730	(2000)	5000	6140
200	4935	(2400)	6280	7310

When $L = 75$, all the application slips were assigned, thus it was not necessary to use the minor tuning module. When $L = 85$ there were 9 applications that were not assigned by GA, one of which was successfully assigned by using the minor tuning module.

378

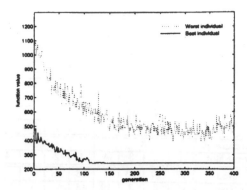

Figure 2. Evaluation of the best and worst individuals by OX2

5 Conclusions

In this paper, the time tabling problem based on applications were studied by using GA. Since the order of assignment of the applications was optimized by GA, it was shown to be quite easy to satisfy the hard constraints, and easily applied to various different problems.

It was also shown that the final result by GA was also locally near optimal. This was assured by seeing that there was not much change by using the minor tuning module.

References

[1] Colorni, A., Dorigo, M. and Maniezzo, V., 1990, Genetic algorithms and highly constrained problems: The time-table case; *Proceedings of the First International Workshop on Parallel Problem Solving from Nature*, Springer-Verlag, Dortmund, FRG, October, pp. 55-59.

[2] Abramson, D. and Abela, J., 1991, A parallel genetic algorithm for solving the school timetabling problem; *Technical Report TR-91-02*, C.S.I.R.O., Division of Information Technology, Department of Communication and Electronic Engineering, Melbourne.

[3] Burke, E. K., Ellman, D. G. and Weare, R. F., 1995, A hybrid genetic algorithm for highly constrained timetabling problems; *Proceedings of the 6th International Conference on Genetic Algorithms*, pp. 605-610, Morgan Kaufmann.

[4] Rich, D. C., 1996, A smart genetic algorithm for university timetabling; *Proc. First International Conf. Practice and Theory of Automated Timetabling: Selected Papers*, pp. 181-97, Springer-Verlag.

[5] Erben, W., 1996, A genetic algorithm solving a weekly course-timetabling problem; *Proc. First International Conf. Practice and Theory of Automated Timetabling: Selected Papers*, pp. 181-97, Springer-Verlag.

[6] Paechter, B., Luchian, H. and Petriuc, M., 1994, Two solutions to the general time table problem using evolutionary methods; *Proceedings of the First IEEE Conference on Evolutionary Computation*, pp. 300-305.

[7] Syswerda, G. and Palmucci, J., 1991, The Application of Genetic Algorithms to Resource Schedling, *Proceedings of the Fifth International Conference on Genetic Algorithms*, Morgan Kaufmann, pp. 502-508.

[8] Baker, J. E., 1985, Adaptive selection methods for genetic algorithms, *Proceedings of the First International Conference on Genetic Algorithms*, Lawrence Erlbaum Associates, Hillsdale, NJ, pp. 101-111.

[9] Strkweather, T., McDaniel, S., Mathias, K. and Whitley, D., 1991, A Comparison of Genetic Sequencing Operators, *Proceedings of the Fifth International Conference on Genetic Algorithms*, pp. 69-76.

[10] Oliver, I. M., Smith, D. J. and Holland, J. R. C., 1987, A study of permutation crossover operators on the traveling salesman problem, *Proceedings of the Second International Conference on Genetic Algorithms*, Lawrence Erlbaum Associates, Hillsdale, NJ, pp. 224-230.

[11] Davis, L., 1985, Job shop scheduling with genetic algorithms, *Proceedings of the First International Conference on Genetic Algorithms*, pp.136-140.

[12] Syswerda, G., 1990, Schedule Optimization Using Genetic Algorithms, *Handbook of Genetic Algorithms*, Davis, L. ed, Van Nostrand Reinhold, New York.

Appendix

Table 6. Application list

appl. #	student #	subject #	1st pref.	2nd pref.	3rd pref.
1	17	3	Mon 2	Tue 1	Mon 1
2	13	4	Fri 3	Fri 2	Thu 3
3	5	2	Thu 1	Tue 4	Thu 1
4	4	1	Thu 1	Thu 2	Fri 1
5	15	2	Mon 1	Mon 3	Thu 4
6	17	1	Mon 4	Tue 3	Tue 1
7	3	3	Mon 3	Fri 2	Wed 3
8	19	3	Fri 3	Tue 4	Fri 1
9	11	2	Wed 3	Thu 2	Mon 2
10	8	3	Thu 2	Thu 1	Thu 4
11	13	1	Fri 2	Mon 2	Fri 3
12	5	5	Mon 1	Thu 3	Thu 1
13	1	3	Fri 1	Fri 3	Wed 1
14	6	4	Wed 4	Fri 4	Fri 1
15	18	5	Wed 2	Wed 2	Fri 2
16	10	2	Wed 4	Tue 3	Wed 3
17	18	1	Wed 3	Mon 3	Fri 4
18	10	5	Mon 1	Thu 2	Thu 1
19	7	1	Thu 1	Mon 1	Tue 3
20	18	4	Tue 1	Fri 4	Tue 3
21	3	5	Wed 3	Wed 4	Wed 1
22	7	5	Thu 3	Tue 4	Mon 4
23	11	5	Thu 2	Thu 3	Wed 1
24	7	5	Wed 1	Tue 2	Thu 4
25	18	1	Thu 4	Tue 1	Thu 4
26	10	3	Mon 2	Thu 4	Thu 2
27	13	2	Tue 2	Thu 1	Tue 1
28	17	3	Tue 4	Mon 1	Fri 4
29	6	2	Mon 4	Wed 4	Wed 4
30	5	2	Thu 1	Thu 1	Mon 4
31	17	3	Wed 2	Thu 2	Tue 1
32	8	3	Thu 1	Tue 3	Fri 1
33	19	1	Tue 4	Mon 1	Fri 2
34	3	2	Fri 4	Wed 3	Fri 4
35	7	3	Thu 3	Tue 3	Mon 1
36	3	2	Tue 4	Tue 1	Tue 1
37	10	5	Thu 1	Tue 2	Tue 2
38	17	1	Mon 1	Tue 4	Thu 3
39	13	1	Mon 2	Fri 2	Tue 1
40	13	3	Mon 3	Mon 4	Wed 2

(to be continued)

appl. #	student #	subject #	1st pref.	2nd pref.	3rd pref.
41	17	3	Mon 4	Thu 1	Tue 4
42	4	5	Wed 4	Wed 2	Tue 4
43	7	3	Fri 4	Mon 1	Fri 3
44	3	2	Thu 1	Fri 2	Thu 3
45	2	3	Thu 3	Mon 4	Wed 3
46	18	4	Mon 1	Fri 3	Tue 4
47	13	1	Mon 4	Fri 4	Tue 4
48	16	2	Fri 3	Wed 2	Wed 3
49	7	1	Mon 4	Tue 2	Wed 1
50	19	2	Wed 1	Thu 1	Thu 3
51	~~10~~	2	Tue 3	Fri 4	Fri 3
52	10	2	Thu 1	Thu 4	Fri 1
53	18	4	Thu 3	Mon 2	Tue 1
54	10	3	Fri 3	Thu 1	Mon 2
55	10	5	Tue 2	Mon 1	Tue 2
56	14	5	Tue 1	Wed 2	Mon 4
57	18	1	Tue 3	Wed 4	Fri 1
58	14	4	Thu 3	Wed 2	Tue 3
59	18	5	Thu 2	Wed 2	Mon 2
60	17	1	Tue 4	Tue 2	Thu 1
61	8	1	Wed 3	Mon 4	Tue 1
62	7	5	Wed 3	Tue 3	Tue 3
63	10	3	Tue 2	Mon 2	Wed 1
64	14	1	Fri 1	Mon 2	Thu 3
65	2	1	Tue 2	Thu 4	Wed 2
66	6	2	Fri 4	Wed 3	Tue 3
67	18	5	Wed 1	Mon 1	Mon 2
68	10	4	Tue 1	Wed 3	Thu 3
69	4	4	Fri 2	Fri 2	Tue 1
70	4	5	Fri 4	Mon 2	Mon 1
71	16	2	Mon 2	Mon 1	Thu 1
72	12	3	Thu 3	Thu 3	Tue 1
73	3	3	Tue 3	Tue 1	Tue 2
74	7	2	Thu 1	Fri 2	Mon 2
75	11	2	Thu 3	Thu 4	Tue 1

A Novel Self-organising Neural Network for Control Chart Pattern Recognition

D. T. Pham and A. B. Chan

Intelligent Systems Laboratory,
Manufacturing Engineering Centre,
School of Engineering,
University of Wales Cardiff,
P.O. Box 688,
Cardiff CF2 3TE, UK
PhamDT@cf.ac.uk   ChanAB@cf.ac.uk

Keywords: self-organising neural networks, control chart patterns, stability-and-plasticity dilemma, firing rule, pattern recognition

Abstract

This paper describes a novel self-organising neural network. The network uses a new type of firing criterion that takes into account the individual components of the patterns to be clustered. To demonstrate the properties of the network, the results of applying it to the classification of control chart patterns are presented. The training and test data sets for the network cover the 6 most common categories of control chart patterns, which are: Normal, Cyclic, Increasing Trend, Decreasing Trend, Upward Shift and Downward Shift. The classification results obtained show that the firing criterion adopted is superior to the usual Euclidean criterion.

1. Introduction

Control charts are employed in Statistical Process Control (SPC) to assess whether a process is functioning correctly. A control chart can exhibit six main types of patterns: *Normal, Cyclic, Increasing Trend, Decreasing Trend, Upward Shift* and *Downward Shift*. Figures 1a-f depict these 6 pattern types. Correct identification of these patterns is important to achieving early detection of potential problems and maintaining the quality of the process under observation.

Supervised neural networks [1] have been employed for automatic identification of control chart patterns [2-6]. However, a disadvantage of using this type of neural networks is that they cannot be easily re-trained to identify new patterns. Due to the inherent randomness in control charts, it is possible that there are patterns not covered by the original data employed for neural network training. In this case, a supervised network would have to be trained again from the beginning using both the original training data and the new patterns to be learnt. This is the classical *stability-and-plasticity* dilemma [7]. This problem can be overcome by self-organising neural networks having an expandable layer of class neurons which enables incremental learning.

This paper presents a self-organising neural network developed for control chart pattern recognition. As with other types of self-organising neural networks, no desired output is provided with any input in the training phase. The network evolves a classification structure to segregate the input data set into clusters as training proceeds. The clusters identify the different classes present in the input data set.

The paper comprises four main sections. Section 2 explains the structure and operation of the proposed neural network. Section 3 gives the results of using the network to classify a small set of control chart patterns to demonstrate the effects of the different network parameters. The results of classifying data sets of more realistic sizes are then presented in section 4 which concludes with guidelines for the design of networks for this application.

382

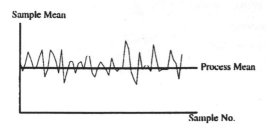

(a) Normal control chart pattern.

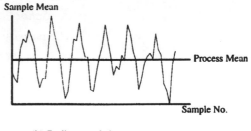

(b) Cyclic control chart pattern.

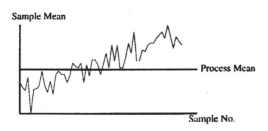

(c) Increasing Trend control chart pattern.

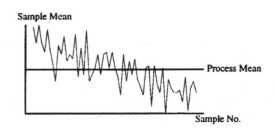

(d) Decreasing Trend control chart pattern.

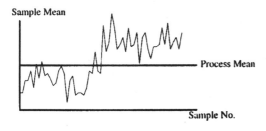

(e) Upward Shift control chart pattern.

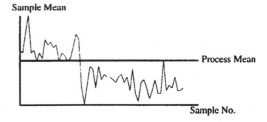

(f) Downward Shift control chart pattern

Figure 1   Main types of control chart patterns

2. Proposed neural network

2.1 Network architecture and firing rule

The proposed network is illustrated in Figure 2.

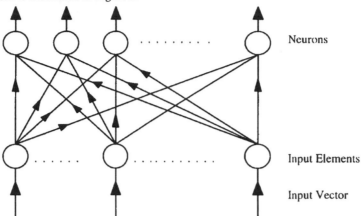

Figure 2 Structure of proposed neural network

The network consists of a layer of input elements followed by a single layer of output neurons. The input elements do not possess any computational capability; they simply distribute the input vectors, as received, to the output neurons. Each output neuron is fully connected to all the input elements and each contains a template pattern for a class. When an input vector is presented, it is compared with the templates in all the output neurons. The neuron containing the class template that best matches the input vector is selected for firing, that is its output is accepted. The firing rule adopted is explained in detail in the following paragraphs.

When an input vector $\underline{x} = [x1, x2, x3\ldots\ldots\ldots, xn]^{\mathrm{T}}$ is presented to the network, it is compared with the template in the first neuron, namely, $\underline{t} = [t1, t2, t3, \ldots\ldots\ldots, tn]^{\mathrm{T}}$. First, all the components in $\underline{x}$ are tested against their counterparts in $\underline{t}$ one by one. Whether or not a component in $\underline{x}$ passes this component test depends on the outcome of the following procedure. Assume that xm is a component of $\underline{x}$ and its counterpart in $\underline{t}$ is tm. If $\dfrac{|xm - tm|}{tm} \leq \alpha$, where α is a user-defined parameter, then xm is regarded as having passed the test. Parameter α determines the range within which xm is allowed to deviate from tm. It can be any real value between 0 and 1, inclusive. For example, given $tm = 1.00$ and $\alpha = 0.2$, then if xm lies in the range 1.2 to 0.8, it will pass the component test.

After every component of $\underline{x}$ has undergone the test, a count is taken of the number of components in $\underline{x}$ that have failed it. The proportion of failed components over the total number of components is compared with another user-defined parameter β (between 0 and 1, inclusive). If it is smaller than or equal to β, the neuron that contains $\underline{t}$ is considered potentially able to represent $\underline{x}$ and therefore to fire. Otherwise, the neuron is disqualified from the subsequent competition for firing.

After $\underline{x}$ has been compared against all the neurons in the network, a competition takes place among the qualified neurons. The neuron that contains the template matching the largest number of components of $\underline{x}$ wins the competition to become the firing neuron.

To illustrate this method of selecting the firing neuron, consider the following example. Suppose that a self-organising neural network has only two output neurons and their associated templates, $\underline{t1}$ and $\underline{t2}$, are:

$$\underline{t1} = [\,0.6, 1.2, 0.2, 1.0, 0.2, 0.2, 0.5, 0.4, 0.3, 0.2\,]^{\mathrm{T}}$$

and

$$\underline{t2} = [\,0.4, 0.4, 0.4, 0.4, 0.4, 0.4, 0.4, 0.4, 0.4, 0.4\,]^{\mathrm{T}}.$$

A vector $\underline{x}$ is provided as the input, where:

$$\underline{x} = [\ 0.5,\ 0.7,\ 0.3,\ 0.6,\ 0.3,\ 0.3,\ 0.6,\ 0.5,\ 0.4,\ 0.3]^{\mathrm{T}}.$$

Assume that $\alpha = 0.15$ and $\beta = 0.35$. Two components in $\underline{x}$ (c2 and c4) fail the component test against $\underline{t1}$ and three components in $\underline{x}$ (c2, c4 and c7) fail the component test against $\underline{t2}$. As the proportion of failed components over the total number of components is less that β in both cases, neuron 1 containing $\underline{t1}$ and neuron 2 containing $\underline{t2}$ both qualify for the firing competition, which neuron 1 wins because $\underline{t1}$ has the higher number of components matching those of the input vector $\underline{x}$. This is the correct result as can be seen in Figure 3 which shows $\underline{x}$ to resemble $\underline{t1}$ more than $\underline{t2}$.

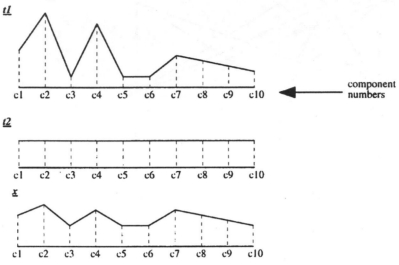

Figure 3 Vectors $\underline{t1}$, $\underline{t2}$ and $\underline{x}$

If the minimum Euclidean firing rule was employed in this problem, neuron 2 would be incorrectly selected for firing because the distance between $\underline{t2}$ and $\underline{x}$ (0.48) is smaller than that between $\underline{t1}$ and $\underline{x}$ (0.70).

2.2 Training and test strategies

Before training the network, the user has to select the values of five parameters: m, n, d, α and β.

m is the number of input elements in the network and is equal to the dimension of the vectors to be classified by the network as well as that of the class templates contained in the neurons.

n is the estimated number of output neurons. It need not be equal to the number of classes in the input data set. If it is higher than the number of actual classes, only part of the estimated number of neurons will be established as output neurons. If n is too high, many trivial class templates will be created at the beginning of the training due to high initial neuron sensitivity as explained later. In practice, the maximum value that n can take, k, is limited only by the amount of available memory. If n is too low, the network will expand beyond the provided set of neurons until a representative classification structure is formed. However, this will increase the overall training time required.

d is a parameter between 0 and 1 that determines the amount by which a template is updated during training. The value of d remains constant throughout the training cycle.

α is the component similarity factor defined previously.

β is the neuron qualification parameter also defined previously. The user has to specify two values for β, referred to as $\beta 1$ and $\beta 2$ ($\beta 1 < \beta 2$). At the beginning of the training, β is equal to $\beta 1$. As the training progresses, β gradually increases to $\beta 2$ in small steps $\Delta\beta$. The step size $\Delta\beta$ is decided by the user.

After the user has set the above parameters, a network with a basic profile will be formed (see Figure 4). Each neuron contains a m-dimensional template. Initially, all these templates are set as null vectors. The first n neurons are made available at the beginning of the training phase since n is the estimated number of classes in the data to be classified.

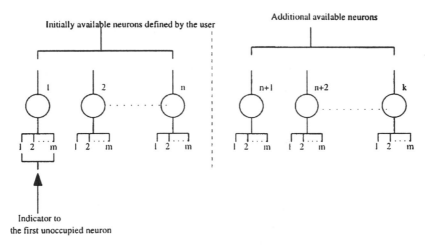

Figure 4   Initial network profile

During training, an indicator is employed to point at the next available neuron with a null template vector. Initially, this indicator points at the first neuron in the network (as shown in Figure 4). A "raw updating" process then takes place which simply copies the input vector into the template of the first neuron. Next, the indicator points to the second neuron in the network as it is now the next available neuron. The first neuron is considered to have been "committed" to a class. This is depicted in Figure 5.

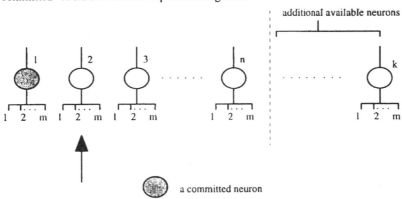

Figure 5   Network profile after presentation of the first input pattern

When the second input vector is applied to the network, it is first tested against the template in neuron 1. If the proportion of the components failing the component test is less than β1, it will be classified into neuron 1. A "subtle updating" process then takes effect on the template in neuron 1 using the following equation to take it closer to the input vector:

$$\text{new } \underline{t1} = \underline{t1} + d[\underline{x} - \underline{t1}] \tag{1}$$

where $\underline{t1}$ is the template in neuron 1 and $\underline{x}$ is the second input vector. However, if the second input vector fails the test, then it will be classified into the next available neuron (which is neuron 2) and the template in that neuron is updated following the "raw updating" procedure. The indicator then points to the next unoccupied neuron. This situation is shown in Figure 6. The training of the network proceeds in the same way for other input vectors in the training set. A training iteration is completed when all the vectors in the training set have been presented once.

When the training becomes advanced, there are two possible outcomes. First, all the input vectors in the training set are classified in a stable manner, that is the same input patterns fire the same neurons in consecutive iterations before all the n neurons are engaged. Second, all the n neurons have been committed but there remain input vectors that cannot be classified into the committed neurons.

386

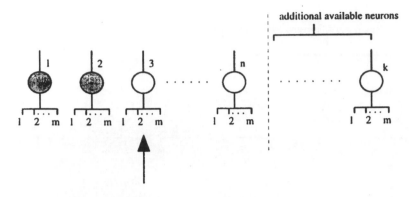

Figure 6   Network profile when 2 classes have been established

In the first case, a stable classification structure has been attained and training can terminate. In the second case, the β value for neuron 1 to neuron n is increased by Δβ to relax the qualification criterion slightly. The training proceeds with the n neurons until either all the input patterns in the data set have been classified or the β value for those neurons has become β2.

In the second case, the (n+1)th neuron is committed (see Figure 7). Parameter β for that neuron initially takes the value β1. β gradually increases to β2 as training progresses. When β equals β2 and if the input patterns in the data set have not yet been completely classified, the next neuron in line will be engaged. This process is repeated until either all the input patterns are classified or the capacity of the network is exhausted (that is all k neurons have been committed).

In the test phase following training, the firing rule employed is as described in section 2.1. β is set equal to β2 for all neurons. Only those neurons committed during the training phase are used. No new neurons are created for new input vectors that cannot be accommodated in the existing neurons. Such vectors are considered unclassifiable by the network.

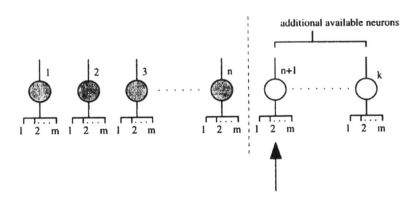

Figure 7   Expansion of the network into the reserved region

3.   Classification of a small set of control chart patterns

This section describes how the proposed network was used to classify a small set of control chart patterns to illustrate the operation of the network and the effects of different combinations of parameters. The set comprised 18 patterns, 3 of each type. Each pattern was a 60-dimensional vector with real numbers between 0 and 1 as components. The correct grouping of the 18 patterns is illustrated in Figure 8.

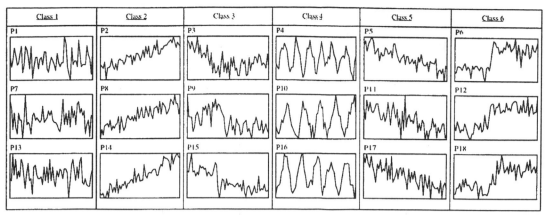

Figure 8 Correct grouping for the 18 control chart patterns

First, a network was created using the following parameters:

α=0.40;   β1=0.15;   β2=0.40;   d=0.30;   Δβ=0.01;   n=6;   m=60.

After 15 iterations, the learning stabilised and 6 neurons were committed. The grouping structure at this stage is illustrated in Figure 9.

There were 6 misclassifications and thus the classification accuracy achieved was 66.7%. As can be observed in Figure 9, the network failed to differentiate between *Cyclic* patterns and *Increasing Trend* patterns completely. They were all assigned to neuron 2. Also, *Decreasing Trend* and *Downward Shift* patterns were accommodated in neuron 4.

A second experiment was carried out with a different set of parameters to attempt to raise the classification accuracy. The parameters used were:

α=0.21;   β1=0.15;   β2=0.18;   d=0.3;   Δβ=0.01;   n=6;   m=60.

The resulting grouping structure is illustrated in Figure 10. The training stabilised after 19 iterations. Altogether, 12 neurons were committed. This means that 6 neurons were created in addition to the original six to accommodate all the 18 control chart patterns. There were no misclassifications, as shown in Figure 10. *Normal*, *Increasing Trend* and *Cyclic* patterns were perfectly grouped under neuron 1, neuron 2 and neuron 4, respectively. The remaining 9 neurons were used to accommodate *Decreasing Trend*, *Upward Shift* and *Downward Shift* patterns. The larger number of neurons required in these cases may be attributed to the fact that the intra-class standard deviation for those pattern types is larger than that for the other types. This means, for example, that the three *Downward Shift* patterns (P3, P9 and P15) are more dissimilar than the three *Cyclic* patterns (P4, P10 and P16).

From these preliminary results, the following observations can be made about the properties of the proposed network. Classification accuracy can be increased by reducing the value of α, adopting a small value for β1 and narrowing the range between β1 and β2. This is because, at the beginning of the training, the neurons have to be very sensitive to subtle differences between patterns in order to create distinct and representative templates for various classes. This high vigilance for each individual neuron is gradually relaxed in the process of training to increase the generalisation ability of the network. However, raising the sensitivity of each individual neuron causes more neurons to be used to classify the data set of 18 control chart patterns, as can be seen in Figure 10. A corollary of this is that when the sensitivity of each individual neuron increases, some neurons would be devoted to classifying very few patterns in the data set. This is not an economical way of using a neuron and can be problematical when there is a large set of patterns to be classified.

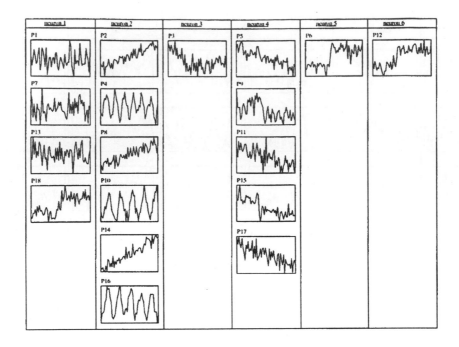

Figure 9   Grouping established with the first set of network parameters

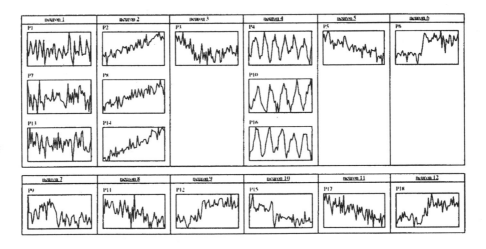

Figure 10   Grouping established with the second set of network parameters

4. Further experiments

A series of 10 experiments were carried out with large data sets. A training set of 366 patterns (61 of each type) and a test set of 132 patterns (22 of each type) were used. The patterns were generated as described in [4]. In each of the 10 experiments, a network was trained. After the training had stabilised, the trained network was tested using the test set. Table 1 summarises the results obtained.

The results for Networks 1, 2 and 3 confirm that the classification accuracy is increased by reducing the value for α and the range between $\beta1$ and $\beta2$ while keeping $\beta1$ at a small value. However, as noted before, the numbers of neurons and training iterations required increased accordingly.

This is particularly evident with Network 4 where, although high training and test accuracies (both above 90%) were achieved, 135 neurons were used and 775 iterations were required before the training stabilised . On average, when considering the training set, each neuron in the IVCA network employed in Network 4 classified fewer than 3 vectors. Neurons in the network were not efficiently used.

Part 8: Dynamic Systems, Identification and Control

Papers:

Neuro-Fuzzy Control Based on the NEFCON-Model Under MATLAB/SIMULINK

Andreas Nürnberger[1], Detlef Nauck[2] and Rudolf Kruse[3]

Faculty of Computer Science, University of Magdeburg
Institute for Information and Communication Systems, Neural Networks and Fuzzy Systems
Universitaetsplatz 2, D-39106 Magdeburg, Germany
Fax : +49.391.67.12018
[1]Phone : +49.391.67.11358, E-Mail: a.nuernberger@iik.cs.uni-magdeburg.de
[2]Phone : +49.391.67.12700, E-Mail: d.nauck@iik.cs.uni-magdeburg.de
[3]Phone : +49.391.67.18706, E-Mail: r.kruse@iik.cs.uni-magdeburg.de

Keywords: hybrid methods, neuro-fuzzy system, system control, neural network, fuzzy system

Abstract

A first prototype of a fuzzy controller can be designed rapidly in most cases. The optimization process is usually more time consuming since the system must be tuned by 'trial-and-error' methods. To simplify the design and optimization process learning techniques derived from neural networks (so called neuro-fuzzy approaches) can be used. In this paper we describe an updated version of the neuro-fuzzy model NEFCON. This model is able to learn and to optimize the rulebase of a Mamdani-like fuzzy controller online by a reinforcement learning algorithm that uses a fuzzy error measure. Therefore we also describe some methods to determine a fuzzy error measure of a dynamic system. Besides we present an implementation of the model and an application example under the MATLAB/SIMULINK development environment. The optimized fuzzy controller can be detached from the development environment and can be used in realtime environments. The tool is available via the Internet.

1 Introduction

The main problems in fuzzy controller design are the construction of an initial rulebase and in particular the optimization of an existing rulebase. The methods presented in this paper have been developed to support a user in both of these cases.

One of the main objectives of our project is to develop algorithms that are able to determine online an appropriate and interpretable rulebase within a small number of simulation runs. Besides it must be possible to use prior knowledge to initialize the learning process. This is a contrast to 'pure' reinforcement strategies [3] or methods based on dynamic programming [1; 15], which try to find an optimal solution using neural network structures. These methods need many runs to find even an approximate solution for a given control problem. On the other hand, they have the advantage that they need less information about the error of the current system state. However, in many cases a simple error description can be achieved with little effort. In this paper we present some methods to determine a fuzzy error measure of a dynamic system.

The first prototype implementation of the described algorithms and the development of a user friendly interface was done in cooperation with the Daimler-Benz Aerospace Airbus GmbH, Hamburg [14]. The tool can be obtained free of charge for non-commercial purposes from our Internet Web-Server (http://fuzzy.cs.uni-magdeburg.de/nefcon or ftp://fuzzy.cs.uni-magdeburg.de/nefcon/nefconma).

2 The NEFCON-Model

The NEFCON-Model is based on a generic fuzzy perceptron [9; 11; 12]. An example, which describes the structure of a fuzzy controller with 5 rules, 2 inputs, and one output is shown in Figure 1. The inner nodes R_1, ..., R_5 represent the rules, the nodes ξ_1, ξ_1, and η the input and output values, and $\mu_r^{(i)}$, ν_r the fuzzy sets describing the antecedents $A_r^{(i)}$ and consequents B_r. Rules with the same antecedent use so-called shared weights, which are represented by ellipses (see Figure 1). They ensure the integrity of the rulebase. The node R_1 for example represents the rule: R_1: if ξ_1 is $A_1^{(1)}$ and ξ_2 is $A_1^{(2)}$ then η is B_1.

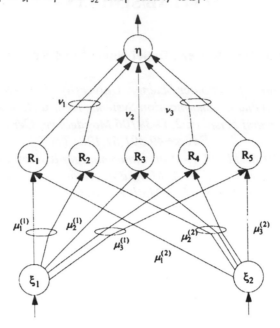

Figure 1. A NEFCON System with two inputs, 5 rules and one output

3 The Learning Algorithms

The learning process of the NEFCON model can be divided into two main phases. The first phase is designed to learn an initial rulebase, if no prior knowledge about the system is available. Furthermore it can be used to complete a manually defined rulebase. The second phase optimizes the rules by shifting or modifying the fuzzy sets of the rules. Both phases use a fuzzy error E, which describes the quality of the current system state, to learn or to optimize the rulebase.

The fuzzy error plays the role of the critic element in reinforcement learning models (e.g. [3; 2]). In addition the sign of the optimal output value η_{opt} must be known. So the extended fuzzy error E* is defined as

$$E^*(x_1, ..., x_n) = sgn(\eta_{opt}) \cdot E(x_1, ..., x_n),$$

with the crisp input $(x_1, ..., x_n)$.

The updated NEFCON learning algorithm learns and optimizes the rulebase of a Mamdani like fuzzy controller [8]. The fuzzy sets of the antecedents and consequents can be represented by any symmetric membership function. Triangular, trapezoidal, and Gaussian functions are supported by the presented implementation.

3.1 Learning a Rulebase

Methods to learn an initial rulebase can be divided into three classes: Methods starting with an empty rulebase [13; 17], methods starting with a 'full' rulebase (combination of every fuzzy set in the antecedents with every consequent) [10] and methods starting with a random rulebase [7]. We implemented algorithms of the first two classes.

3.1.1 Modified Algorithm NEFCON I

The modified algorithm NEFCON I is based on the original NEFCON model [10; 12]. It starts with a 'full' rulebase. The algorithm can be divided into two phases which are executed during a fixed period of time or a fixed number of iteration steps. During the first phase, rules with an output sign different from that of the optimal output value η_{opt} are removed. During the second phase, a rulebase is constructed for each control action by selecting randomly one rule from every group of rules with identical antecedents. The error of each rule (the output error of the whole network weighted by the activation of the individual rule) is accumulated. At the end of the second phase from each group of rule nodes with identical antecedents the rule with the least error value remains in the rulebase. All other rule nodes are deleted. In addition, rules used very rarely are removed from the rulebase. The original algorithm used triangular membership functions, while the improved implementation also supports trapezoidal and Gaussian membership functions. Besides, the algorithm was enhanced for dynamic systems which need a static offset.

3.1.2 The 'Bottom-Up'-Algorithm

The 'Bottom-Up'-Algorithm starts with an empty rulebase. An initial fuzzy partitioning of the input and output intervals must be given. The algorithm can be divided into two phases. During the first phase, the rules' antecedents are determined by classifying the input values, i.e. finding that membership function for each variable that yields the highest membership value for the respective input value. Then the algorithm tries to 'guess' the output value by deriving it from the current fuzzy error. During the second phase the rulebase is optimized by changing the consequent to an adjacent membership function, if this is necessary. The improved implementation supports trapezoidal and Gaussian membership functions, too.

The 'Bottom-Up'-Algorithm is much faster than NEFCON I in case of a large number of input variables and a fine initial fuzzy partitioning. This is caused by the huge initial rulebase used by the NEFCON I algorithm. Nevertheless the 'Bottom-Up'-Algorithm should not be used for complex dynamic systems up to now, because of the heuristic approach of finding the consequents.

3.2 Optimization of a Rulebase

The algorithms presented in this section are designed to optimize a rulebase of a fuzzy controller by shifting and/or modifying the support of the fuzzy sets. They do not modify the rules or the structure of a given network.

3.2.1 The Algorithm NEFCON I

The algorithm NEFCON I [12] is motivated by the backpropagation algorithm for the multilayer perceptron. The extended fuzzy error E^* is used to optimize the rulebase by 'reward and punishment'. A rule is 'rewarded' by shifting its consequent to a higher value and by widening the support of the antecedents, if its current output has the same sign as the optimal output η_{opt}. Otherwise the rule is 'punished' by shifting its consequent to a lower value and by reducing the support of the antecedents.

The original model used monotonic membership functions [18] in the consequents to make it possible to use an backpropagation algorithm. In the current implementation this restriction was removed by storing the activation of every rule during the inference mechanism. Thus it is possible to use symmetric fuzzy sets in the consequents and the antecedents.

3.2.2 The Algorithm NEFCON II

In contrast to the algorithm NEFCON I, which uses only the current fuzzy error E^*, the algorithm NEFCON II also makes use of the change of the fuzzy error E^* to optimize the rulebase [13]. This is a heuristic approach to include the dynamics of the system into the optimization process. Let E^* be the extended fuzzy error at time t and $E^{*'}$ the extended fuzzy error at time t+1, then the error tendency τ is defined as

$$\tau = \begin{cases} 1, & if & (|E^{*'}| \geq |E^*|) \wedge (E^{*'} \cdot E^*) \geq 0, \\ 0, & if & (|E^{*'}| < |E^*|) \wedge (E^{*'} \cdot E^*) \geq 0, \\ -1, & otherwise \end{cases}$$

If $\tau = 0$, the system moves to an optimal state. In this case the rulebase will not be modified. If $\tau = 1$, the error is rising without changing its sign. The output of each rule is increased by shifting its consequents. The antecedents of rules with consequents increasing the output will be 'rewarded', while those with consequents decreasing the output will be 'punished'. If $\tau = -1$, the system has overshot. The output is decreased and the antecedents 'punished' or 'rewarded' accordingly.

396

3.3 Description of System Error

In case of a simple dynamic system the error can be described sufficiently well by simply using the difference between the reference signal and the system response. In case of more complex and sensitive systems the error must be described more exactly to obtain a satisfying rulebase with the presented algorithms.

3.3.1 A Linguistic Error Description

The optimal state of a dynamic system can be described by a vector of system state variable values. Usually the state can not be described exactly, or we are content, if the system variables have roughly taken these values. Thus the quality of a current state can be described by fuzzy rules. With an error definition that uses a linguistic error description with fuzzy rules it is also easily possible to describe compensatory situations [12]. These are situations in which the dynamic system is driven towards its optimal state. In Figure 2 an error description is shown, which is part of the implementation presented in 5. This rulebase also describes an overshoot situation (rules 7 and 8).

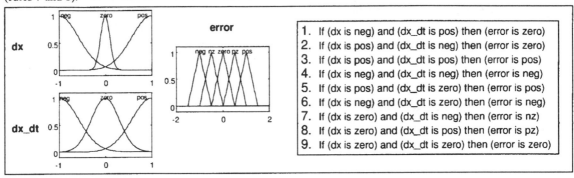

Figure 2. Sample Rulebase for Fuzzy Error Description

3.3.2 An Error Description with 'Fuzzy Intervals'

The error description with 'fuzzy intervals' has been developed for the presented implementation. It makes it possible to describe a 'soft' region for the system response which satisfies our request to the system behavior in a simple and intuitive way.

Figure 3 presents an error description for a simple switch signal. The error signal remains zero, if the response signal of the dynamic system stays in the defined interval between the reference signal and the bounds (thick lines). If the signal leaves the bounds of the interval, the fuzzy error is determined using a linguistic error definition as described above.

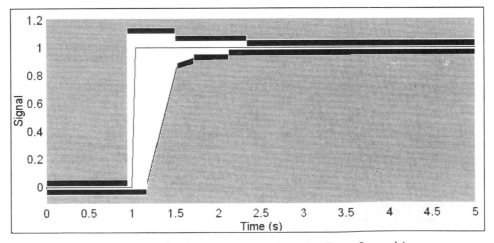

Figure 3. Sample of an Error Description using 'Fuzzy Intervals'

4 Implementation

The aim of the implementation under MATLAB/SIMULINK was to develop an interactive tool for the construction and optimization of a fuzzy controller. This frees the user of programming and supports him to concentrate on controller design. It is possible to include prior knowledge into the system, to stop and to resume the learning process at any time, and to modify the rulebase and the optimization parameters interactively. Besides, a graphical user interface was designed to support the user during the development process of the fuzzy controller.

Figure 4 presents the simulation environment of a sample application during the optimization phase of the algorithm. This example was created under Microsoft Windows NT 4.0.

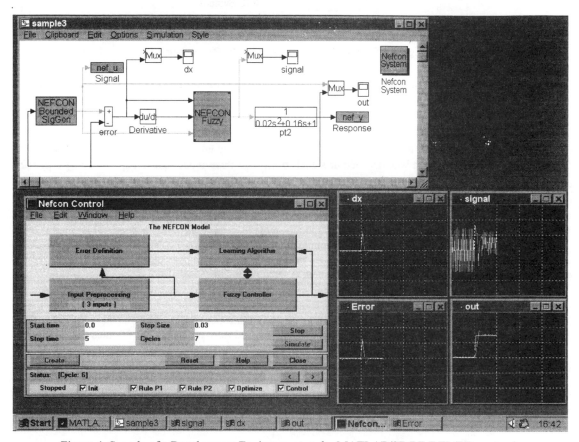

Figure 4. Sample of a Development Environment under MATLAB/SIMULINK (PT_2 system)

5 Example

As an example for the usability of the presented algorithms and error descriptions in practice, we present the simulation results concerning a conventional PT_2 system. Simulation results concerning the classical inverted pendulum problem are comparable to the results obtained by the 'original' NEFCON algorithms presented in prior publications [10; 13; 12].

A PT_2-system models the behavior of a two-mass system, for example a spring-damper combination or a revolution control for an electric motor (see [6; 16]). In classical control theory a PT_2 system is controlled by a PI or a PID controller. A comparison to fuzzy controllers for this problem is considered e.g. in [4].

For the presented example we used a PT_2 system that is given by the following differential equation:

$$\frac{1}{\omega_0^2} \cdot \ddot{x} + \frac{2D}{\omega_0} \cdot \dot{x} + x = V \cdot y \; .$$

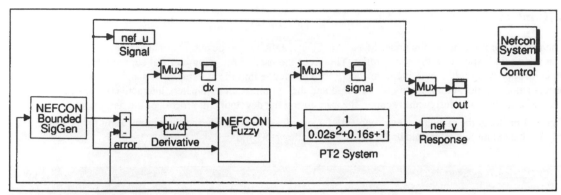

Figure 5. Simulation Environment for the PT$_2$ System

For the constants we chose $\omega_0 = \sqrt{50}$, $D = \frac{2}{5}\sqrt{2}$ and $V = 1$. The transfer function of this system is defined as:

$$T(s) = \frac{1}{0.02s^2 + 0.16s + 1}.$$

The reference signal y' and the fuzzy error were described using 'fuzzy intervals' (see Figure 3). To control the PT$_2$ system a NEFCON system with one output signal $\eta = x$ and three input signals ($\xi_1 = e = y' - y$, $\xi_2 = \dot{e}$, $\xi_3 = y$) was used. The algorithm NEFCON I was selected for the rule learning process, since the heuristic approach of finding the conclusions used by the algorithm NEFCON II would have resulted in an inappropriate rulebase. The reason for this is the integral part, that is needed for control (in a stable state (dy = 0) the dynamic system will probably need an input value y ≠ 0 to remain stable; e.g. an electric motor needs an amperage unequal to zero to maintain a constant number of revolutions). The algorithm NEFCON II was selected for optimization. The input interval of each input variable was partitioned by three trapezoidal fuzzy sets and the output interval by five fuzzy sets. The simulation environment is shown in Figure 5.

The algorithm used a noisy reference signal during rule learning to improve the coverage of the system state space (see cycle 1-3 in Figure 8). The noise was produced by a signal generator included in the implementation. The learning algorithm was applied for 3 rule learning and 3 optimization cycles (with 167 iteration steps each cycle, where each cycle takes 5 seconds system time). The optimized rulebase consists of 25 rules (see Figure 7). The resulting fuzzy sets are shown in Figure 6.

After the learning process was finished, the controller was able to drive the system quite nicely along the desired course (see simulation cycle 7 in Figure 8). However, the control behavior is a little bit 'fidget'. This was implicitly tolerated by the error boundaries defined with the 'fuzzy intervals' and so it could not be improved during optimization.

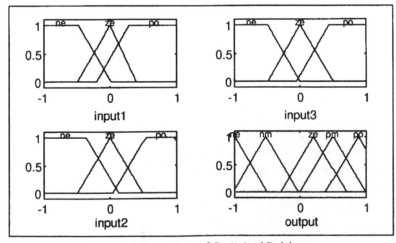

Figure 6. Fuzzy Sets of Optimized Rulebase

```
1.  If (input1 is ne) and (input2 is ne) and (input3 is ne) then (output is ne)
2.  If (input1 is ne) and (input2 is ne) and (input3 is ze) then (output is nm)
3.  If (input1 is ne) and (input2 is ne) and (input3 is po) then (output is ne)
4.  If (input1 is ne) and (input2 is ze) and (input3 is ne) then (output is ne)
5.  If (input1 is ne) and (input2 is ze) and (input3 is ze) then (output is po)
6.  If (input1 is ne) and (input2 is ze) and (input3 is po) then (output is nm)
7.  If (input1 is ne) and (input2 is po) and (input3 is ne) then (output is nm)
8.  If (input1 is ne) and (input2 is po) and (input3 is ze) then (output is nm)
9.  If (input1 is ne) and (input2 is po) and (input3 is po) then (output is nm)
10. If (input1 is ze) and (input2 is ne) and (input3 is ne) then (output is po)
11. If (input1 is ze) and (input2 is ne) and (input3 is ze) then (output is nm)
12. If (input1 is ze) and (input2 is ne) and (input3 is po) then (output is ze)
13. If (input1 is ze) and (input2 is ze) and (input3 is ne) then (output is pm)
14. If (input1 is ze) and (input2 is ze) and (input3 is ze) then (output is ze)
15. If (input1 is ze) and (input2 is ze) and (input3 is po) then (output is pm)
16. If (input1 is ze) and (input2 is po) and (input3 is ne) then (output is pm)
17. If (input1 is ze) and (input2 is po) and (input3 is ze) then (output is po)
18. If (input1 is ze) and (input2 is po) and (input3 is po) then (output is po)
19. If (input1 is po) and (input2 is ne) and (input3 is ne) then (output is nm)
20. If (input1 is po) and (input2 is ne) and (input3 is ze) then (output is nm)
21. If (input1 is po) and (input2 is ne) and (input3 is po) then (output is ze)
22. If (input1 is po) and (input2 is ze) and (input3 is ze) then (output is ze)
23. If (input1 is po) and (input2 is ze) and (input3 is po) then (output is po)
24. If (input1 is po) and (input2 is po) and (input3 is ze) then (output is po)
25. If (input1 is po) and (input2 is po) and (input3 is po) then (output is ze)
```

Figure 7. Learned Rulebase

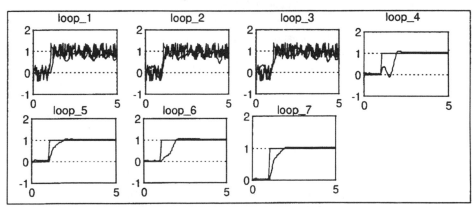

Figure 8. Simulation Results for a PT_2 System

6 Conclusion

Since the updated NEFCON model was implemented under MATLAB/SIMULINK, it is possible to use the model conveniently for the design of fuzzy controllers for different dynamic systems. Additionally, the system configuration can be changed easily. The implementation was designed to be used as an interactive development tool.

In case of dynamic systems with little temporal dependence the rules of the controller will be learned and optimized within a small number of runs. The obtained fuzzy controller will be able to control the dynamic system appropriately. In addition the rulebase is always interpretable. In case of complex systems the quality of the results greatly depends on the definition of the error measure. This is caused by the fact that the NEFCON algorithms use only a simple approach to include the dynamics of the controlled system in the optimization process (see the credit assignment problem [3]). Some variations of reinforcement strategies [1; 7] have to be analyzed in order to determine if it is possible to integrate them into the optimization phase of the presented algorithms. It has to be checked whether they improve the quality of the controller without increasing the number of learning runs significantly.

Remark: MATLAB/SIMULINK is a simulation tool developed and distributed by 'The Mathworks' Inc., 24 Prime Park Way, Natick, Mass.01760; WWW: http://www.mathworks.com.

References

[1] Barto, A. G., Bradtke, S. J., and Singh, S. P., 1995, Learning to act using real-time dynamic programming, *Artificial Intelligence, Special Volume: Computational Research on Interaction and Agency*, **72**(1), 81-138.

[2] Barto, A.G. , 1992, *Reinforcement Learning and Adaptive Critic Methods*, In [19].

[3] Barto, A.G., Sutton R. S., and Anderson, C. W. 1983, Neuronlike adaptive elements that can solve difficult learning control problems, *IEEE Transactions on Systems, Man and Cybernetics*, **13**, 834-846.

[4] Knappe, H., 1994, *Comparison of Conventional and Fuzzy-Control of Non-Linear Systems*, In [5].

[5] Kruse, R., Gebhardt, J., and Palm, R. (Eds.), 1994, *Fuzzy Systems in Computer Science*, Friedr. Vieweg & Sohn Verlagsgesellschaft mbH, Braunschweig, Wiesbaden.

[6] Leonhard, W., 1992, *Einführung in die Regelungstechnik*, Friedr. Vieweg & Sohn Verlagsgesellschaft mbH, Braunschweig, Wiesbaden.

[7] Lin, C.T., 1994, *Neural Fuzzy Control Systems with structure and Parameter Learning*, World Scientific Publishing, Singapore.

[8] Mamdani, E. H., and Assilian S., 1973, An Experiment in Linguistic Synthesis with a Fuzzy Logic Controller, *International Journal of Man-Machine Studies*, **7**, 1-13.

[9] Nauck, D., 1994, A Fuzzy Perceptron as a Generic Model for Neuro-Fuzzy Approaches, *Proceedings of the 2nd German GI-Workshop Fuzzy-Systeme '94*, München, Germany, October.

[10] Nauck, D., and Kruse, R., 1993, A Fuzzy Neural Network Learning Fuzzy Control Rules and Membership Functions by Fuzzy Error Backpropagation, *Proceedings of IEEE International Conference on Neural Networks 1993*, San Francisco, United States, March, pp. 1022-1027.

[11] Nauck, D. and Kruse, R., 1996, Designing neuro-fuzzy systems through backpropagation, In: Pedryz, W. (Ed.), *Fuzzy Modelling: Paradigms and Practice*, Kluwer Academic Publishers, Boston, Dordrecht, London, pp. 203-228.

[12] Nauck, D., Klawonn, F., and Kruse, R., 1997, *Foundations of Neuro-Fuzzy Systems*, John Wiley & Sons, Inc., New York, Chichester, et.al. (to appear).

[13] Nauck, D., Kruse, R., and Stellmach, R., 1995, New Learning Algorithms for the Neuro-Fuzzy Environment NEFCON-I, *Proceedings of the 3rd German GI-Workshop Fuzzy-Neuro-Systeme '95*, Darmstadt, Germany, November, pp. 357-364.

[14] Nürnberger, A., 1996, *Entwurf und Implementierung des Neuro-Fuzzy-Modells NEFCON zur Realisierung Neuronaler Fuzzy-Regler unter MATLAB/SIMULINK*, MSc Thesis, Technische Universität Braunschweig, Braunschweig, Germany.

[15] Riedmiller, M., and Janusz, B., 1995, Using Neural Reinforcement Controllers in Robotics, *Proceedings of the 8th Australian Conference on Artificial Intelligence*, Canberra, Australia.

[16] Tou, J. T., 1964, *Modern Control Theory*, McGraw Hill, New York.

[17] Tschichold-Gürman, N., 1995, *RuleNet - A new Knowledge-based Artificial Neural Network Model with Application Examples in Robotics*, PhD Thesis, ETH Zürich, Zürich, Switzerland.

[18] Tsukamoto, Y., 1979, An Approach to Fuzzy Reasoning Method, In: Gupta, M., Ragade, R., and Yager, R. (Eds.): *Advances in Fuzzy Set Theory*, North-Holland, Amsterdam.

[19] White, D. A., Sofge, D. A., 1992, *Handbook of Intelligent Control. Neural, Fuzzy and Adaptive Approaches*, Van Nostrand Reinhold, New York.

An Optimal COG Defuzzification Method for A Fuzzy Logic Controller

Daijin Kim[1], In-Hyun Cho[1]

[1] *Department of Computer Engineering, DongA University,*
840, Hadan Dong, Saha Ku, Pusan, 604-714, Korea
E-mail : dkim@vlsi.donga.ac.kr and ihcho@xtra.donga.ac.kr

Keywords: Fuzzy Logic Controller, COG Defuzzifier, Genetic Algorithm, Truck Backer-Upper Control.

Abstract

This paper proposes an optimal COG (Center Of Gravity) defuzzification method that improves the control performance of a fuzzy logic controller. The defuzzification method incorporates the membership values and the effective widths of membership functions in calculating a crisp value. An optimal effective width is determined automatically by the genetic algorithm through the training of some typical examples. Simulation results over the truck backer-upper control problem show that the proposed optimal COG defuzzifier reduces the average tracing distance by 23.8% compared with the conventional COG defuzzifier.

1. Introduction

Fuzzy logic controllers (FLCs) have been widely applied to many consumer products and many industrial process controls. In particular, FLCs are very effective techniques for complex and imprecise processes for which either no mathematical model exists or the mathematical model is severely non-linear because they can easily approximate what human experts perform well under such ill-defined environments.

A typical organization of the fuzzy logic controller is shown in Fig. 1. It consists of four principal units: 1) fuzzifier which converts a crisp input to a fuzzy term set; 2) fuzzy rule base which stores fuzzy rules describing how the fuzzy logic controller performs; 3) fuzzy inference engine which performs an approximate reasoning by associating input variables with fuzzy rules; 4) defuzzifier which converts the FLC's fuzzy output to a crisp value for the actual control input over the plant. The control performance is influenced by the selection of the fuzzy sets of the linguistic variables, the shapes of membership functions, the fuzzy rule base, the inference mechanism, and the defuzzification method.

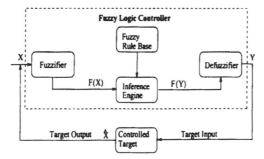

Figure 1. A block diagram of typical fuzzy logic controller.

Many researchers have been performed for the optimal constructions of the above system units. However, little attention has been given to the defuzzification method due to its simple operation and the designer's prejudice about its role. Some researches about optimal defuzzification method are described in the following. R. Yager and D. Filev [1] introduced a parameterized family of defuzzification operators called the SLIDE (Semi LInear Defuzzification) method.

This work was supported by a grant from the university fundamental research support of IITA(96039-BT-12).

The determination method was based on a simple linear transformation of the fuzzy output set of the controller and the optimal values of SLIDE's two parameters by the recursive least square solution of a set of linear equations based on the concept of Kalman filters. But, due to the nonlinearity of the used algorithm, learning of parameters may converge to some local extremum. A. Bastian [2] proposed a modified COG defuzzification method that could control the nonlinearity at the transition between fuzzy logic rules. The defuzzification weights to the overlapping area of the rule consequences were determined by using the backpropagation learning algorithm in a three-layer feedforward neural network. But, the approach was performed on the specific case where all membership functions of an output variable had the same widths and were located at the same intervals. A. Ruiz, J. Gutiérrez, and J. Fernández [3] proposed a trapezoidal fuzzy set representation that improved information processing speed by avoiding division. But, the calculation of area was time-consuming since the area was calculated by an iteration of a recurrence equation.

The motivation of this work is that an appropriate selection of the defuzzification method can improve the control performance of fuzzy logic controller greatly. To realize this motivation, this paper proposes a new COG defuzzification method that incorporates the membership values and the optimal effective widths of membership functions in calculating the crisp value. Furthermore, the optimal values of the effective widths are determined by a genetic algorithm in order to reduce the possibility of trapping to the local solutions. From the practical considerations and experiences in applying the fuzzy logic to control systems, the membership functions of an output variable will have the following characteristics that (1) their overlapping degree is limited to two and (2) their shape is symmetric, and (3) each membership function is allowed to have different width.

This paper is organized as follows. Section 2 describes the conventional COG defuzzifier. Section 3 presents the proposed COG defuzzifiers. Section 4 describes how to determine the optimal defuzzification weight constants. Section 5 applies the proposed COG defuzzifiers to the truck backer-upper control problem and compares the control performance with the conventional one in terms of the average tracing distance. Finally, a conclusion is drawn.

2. The Conventional COG Defuzzifier

The defuzzification process is a mapping from fuzzy control actions defined over an output space into a crisp control action. The aim of defuzzification strategy is to produce a crisp control action that best represents the possibility distribution of an inferred fuzzy control action. The commonly used methods are the max criterion (Max) method, the mean of maximum (MOM) method, and the center of area method. According to the comprehensive studies, a better performance has been reported in the order of COG, MOM, and Max defuzzifier [4]. In the hardware realization, the most widely used defuzzification method is the COG strategy because it yields a better steady-state performance than the other two methods. Recently, more rational defuzzification methods were presented for general-purpose applications [5].

The center of area method gives a crisp value y_c as the center of gravity of the output fuzzy membership functions inferred by an min-max inference engine:

$$y_c = \frac{\int_y y \cdot \mu_Y(y)\, dy}{\int_y \mu_Y(y)\, dy} \tag{1}$$

where y is the support value of the inferred membership functions and $\mu_Y(y)$ is the maximum value of the inferred membership functions at a given support value. So, the numerator and the denominator mean the moment of $\mu_Y(y)$ and the area of $\mu_Y(y)$, respectively.

It has been shown that the COG defuzzifier produces satisfactory control results even if the inferred membership function is replaced with the inferred membership value of the singleton corresponding to the maximum membership value in the original membership function [6]. In this case, the crisp value y_c can be computed by

$$y_c = \frac{\sum_{i=1}^{n} y_i \cdot \mu_Y(y_i)}{\sum_{i=1}^{n} \mu_Y(y_i)} \tag{2}$$

where n is the number of discrete fuzzy terms, y_i is the singleton's support value of the ith fuzzy term in the fuzzy consequence, and $\mu_Y(y_i)$ is the membership value at the singleton's support value y_i.

However, replacing Eq. (1) with Eq. (2) is valid only when (1) the membership functions of fuzzy terms in the output variable are symmetric. (2) they have the same widths, and (3) they are located at the same intervals. In the practical applications, these constraints are not appropriate for an accurate control. Fig. 2 illustrates the misuse of the conventional COG defuzzifier when the third constraint is not satisfied. As can be seen in Fig. 2, both cases have the same singleton's support values but have different types of widths, i.e., the left one with the same widths for two membership functions and the right one with the different widths for two membership functions. Since the conventional COG defuzzifier considers only the membership values at the singletons in calculating the crisp value, the crisp values of both cases are exactly same when Eq. (2) is used to calculate the crisp value. This is far from our instinct that the center of gravity is near to the membership function having the larger area. In other word, the conventional COG defuzzifier fails to reflect the different widths of membership functions. To overcome this failing. a new defuzzification methods is proposed in the next section.

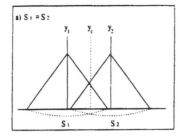 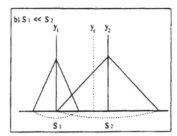

Figure 2. An illustration of misusing conventional COG defuzzifier.

3. The Proposed COG Defuzzifiers

The numerator and denominator of Eq. (1) represents the area of $y \cdot \mu_Y(y)$ and $\mu_Y(y)$ over the output universe of discourse y, respectively. However, since Eq. (2) does not consider the area of $y \cdot \mu_Y(y)$ and $\mu_Y(y)$ but the singleton's support value and a specific membership value, a considerable difference exists between the crisp values obtained from Eq. (1) and Eq. (2). This difference can be reduced by modifying Eq. (2) into

$$y_c = \frac{\sum_{i=1}^{n} A_Y(y_i) \cdot y_i}{\sum_{i=1}^{n} A_Y(y_i)} \tag{3}$$

where n is the number of discrete fuzzy terms, y_i is the singleton's support value of the ith fuzzy term in the fuzzy consequence, and $A_Y(y_i)$ is the area under the clipped membership value of the ith inferred membership function.

Next, we consider how to calculate the area $A_Y(y_i)$ in the case of membership function having the generalized shape $\mu_Y^r(y)$, where the membership function can be either concentrated($r > 1$) or dilated ($r < 1$). Fig. 3 shows the graphical representation of the membership function ($r < 1$). The area $A_Y(y_i)$ under the clipped membership function of $\mu_Y^n(y)$ is given as

$$\begin{aligned} A_Y(y_i) &= \mu_Y(y_i) \cdot s_i - \frac{r}{r+1} \mu_Y^{\frac{r+1}{r}}(y_i) \cdot s_i \\ &= \mu_Y(y_i) \cdot s_i \cdot (1 - \frac{r}{r+1} \mu_Y^{\frac{1}{r}}(y_i)). \end{aligned} \tag{4}$$

The derivation of Eq. (4) is given in the Appendix. It is noted that $n \to 0$, $A_Y(y_i) \to \mu_Y(y_i) \cdot s_i$ and $r \to \infty$, $A_Y(y_i) \to 0$.

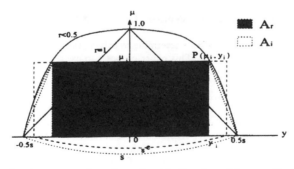

Figure 3. A graphical representation of membership function with generalized shape $\mu_Y^r(y_i)$.

Consider the rectangle whose area is equivalent to the area under the clipped membership value of the membership function $\mu_Y^r(y)$ and the height is $\mu_Y(y_i)$. Then, the rectangle has an effective area $A_Y^e(y_i) = \mu_Y(y_i) \cdot s_i^e$, where s_i^e is an effective width given by

$$s_i^e = s_i \cdot (1 - \frac{r}{r+1} \mu_Y^{\frac{1}{r}}(y_i)).$$ (5)

In this case, the effective crisp value y_c^e is determined by

$$
\begin{aligned}
y_c^e &= \frac{\sum_{i=1}^n A_Y^e(y_i) \cdot y_i}{\sum_{i=1}^n A_Y^e(y_i)} \\
&= \frac{\sum_{i=1}^n \mu_Y(y_i) \cdot s_i \cdot (1 - \frac{r}{r+1} \cdot \mu_Y^{\frac{1}{r}}(y_i)) \cdot y_i}{\sum_{i=1}^n \mu_Y(y_i) \cdot s_i \cdot (1 - \frac{r}{r+1} \cdot \mu_Y^{\frac{1}{r}}(y_i))}.
\end{aligned}
$$ (6)

From the simulation studies, the effective width s_i^e is known to be problem-specific and dependent on the shape of the used membership function. So, we replace the constant $\frac{r}{r+1}$ with a defuzzification weight constant $\omega_i (0 \le \omega_i \le 1)$ in order to find an optimal effective width about the specific problem. Then, the rectangle has an optimal effective area $A_Y^o(y_i) = \mu_Y(y_i) \cdot s_i^o$, where s_i^o is an optimal effective width given by

$$s_i^o = s_i \cdot (1 - \omega_i \cdot \mu_Y^{\frac{1}{r}}(y_i)).$$ (7)

In this case, the optimal crisp value y_c^o is determined by

$$
\begin{aligned}
y_c^e &= \frac{\sum_{i=1}^n A_Y^o(y_i) \cdot y_i}{\sum_{i=1}^n A_Y^o(y_i)} \\
&= \frac{\sum_{i=1}^n \mu_Y(y_i) \cdot s_i \cdot (1 - \omega_i \cdot \mu_Y^{\frac{1}{r}}(y_i)) \cdot y_i}{\sum_{i=1}^n \mu_Y(y_i) \cdot s_i \cdot (1 - \omega_i \cdot \mu_Y^{\frac{1}{r}}(y_i))}.
\end{aligned}
$$ (8)

When the membership value $\mu_Y(y_i)$ is assumed to have an uniform distribution U(0,1), the expectation values of $\mu_Y^{\frac{1}{r}}(y_i)$ is $\frac{r}{r+1}$. When $r \le 2$, $\frac{r}{r+1} \cdot E[\mu_Y^{\frac{1}{r}}(y_i)] = (\frac{r}{r+1})^2 \ll 1$, the second term $\frac{r}{r+1} \cdot \mu_Y^{\frac{1}{r}}(y_i)$ in Eq. (5) can be negligible. Then, the rectangle has an approximate effective area $A_Y^a(y_i) = \mu_Y(y_i) \cdot s_i^a$, where s_i^a is an approximate effective width whose value is equal to the original width s_i. In this case, the approximate crisp value y_c^a is reduced as

$$
\begin{aligned}
y_c^a &= \frac{\sum_{i=1}^n A_Y^a(y_i) \cdot y_i}{\sum_{i=1}^n A_Y^a(y_i)} \\
&= \frac{\sum_{i=1}^n \mu_Y(y_i) \cdot s_i \cdot y_i}{\sum_{i=1}^n \mu_Y(y_i) \cdot s_i}.
\end{aligned}
$$ (9)

As can be seen in the simulation studies, the control performance of using y_c^a is not much degraded from that of using y_c^o. Further, this reduction is valuable when the proposed defuzzifier is implemented in hardware because the multiplication hardware can be saved greatly.

4. Determination of Defuzzification Weight Constants ω

The defuzzification weight constants ω are hard to determine by solving the linear algebraic equations since the distribution function of weight constants is not known. This paper proposes to use a genetic algorithm in order to determine the defuzzification weight constants ω. Genetic algorithms (GAs) [7] are iterative adaptive search and optimization procedures based on the natural selection and biological genetics.

GA works as follows. Let P be a population of N chromosomes. Let $P(0)$ and $P(t)$ be the randomly generated initial population and the population at time t, respectively. A new population $P(t + 1)$ is created by applying a set of genetic operations (reproduction, crossover, and/or mutation) over $P(t)$. Each chromosome in $P(t + 1)$ is reproduced in proportion to its fitness value at time t. Crossover recombines two chromosomes by cutting them at a random position and exchanging genetic materials in such a way that some of the first chromosome go to the second one and vice versa. Mutation changes some of the randomly selected gene values. The overall effect of GA is that a new population moves toward better solutions with higher values of fitness function. The genetic algorithm used in this paper is dissimilar to the one of [7] in that the genes of the chromosome are represented by the real numbers [9]. A detailed explanation of the genetic algorithm used here is given as follow.

(1) **Chromosome Representation** When the output variable consists of N fuzzy terms and the defuzzification weight constants are $\omega_i (i = 1, 2, \dots N)$, the chromosome is represented by a string of N consecutive real numbers.

(2) Initial Population Generation The defuzzification weight constants ω in the chromosome are obtained by randomly generating N real numbers within the interval of $[0.1]$. Initial population is obtained by repeating the above procedure M times.

(3) Fitness Function Assume that a typical example set $E = \{e^1, e^2, \ldots, e^T\}$. An example $e^t (t = 1, 2, \ldots, T)$ consists of an order set of m input values and n output values. When two input variables are x and ϕ and one output variable is θ, the training example e^t becomes $(e^t_x, e^t_\phi; e^t_\theta)$. All training examples E are applied to the fuzzy logic controller whose defuzzifier has the defuzzification weight constants ω corresponding to a specific chromosome. For each training example e^t, an error term ϵ^t can be calculated as $\epsilon^t = \sum_{i=1}^{S} (y^t(i) - y^t_c(i))^2$, where S, $y^t(i)$ and $y^t_c(i)$ are the number of steps to the goal, the true output value at the ith step of the tth example known from the training example set, and the crisp value at the ith step of the tth example obtained from Eq. (8), respectively. A total error E from all training examples is a sum of individual error from each training example as $E = [\sum_{t=1}^{T} (\epsilon^t)^2]^{\frac{1}{2}}$. Then, the fitness function F of each chromosome is defined as $F = \frac{1}{1 + k \cdot E}$, where k is an appropriate scaling constant.

(4) Genetic Operations A detailed explanation about the genetic operation used in determining the defuzzification weight constants ω is given as follows.

a) Reproduction A mixture of various selection methods is used to reproduce a new population. Firstly, the best chromosome based on the elitist selection is reproduced in the next generation. Secondly, the k-tournament method [8] is chosen with some modifications in order to reproduce some new chromosomes in the population. The method chooses a chromosome having the best fitness value among the k chromosomes randomly selected from the upper class of fitness values. The chromosomes P_1 and P_2 obtained by repeating the above procedure twice create a new chromosome C after applying the crossover and/or mutation operations explained later. New chromosome C will replace a certain chromosome P' having the worst fitness value among the k chromosomes randomly selected from the lower class of fitness values. The second reproduction procedure is repeated as many times as $pselect \times |P|$, where $|P|$ is the population size and $pselect$ is the percentage of chromosomes replaced. Thirdly, the reminiscent portion of the population set is filled by generating new chromosomes randomly similarly with the procedure of the initial population generation. The third reproduction is considered for avoiding the premature convergence. Fig. 4 shows the reproduction scheme used in this work.

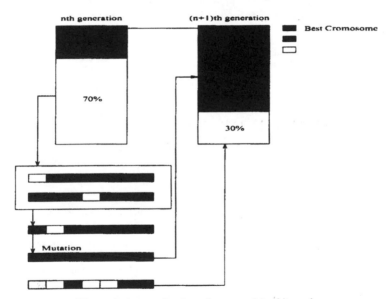

Figure 4. A reproduction scheme used in this work.

b) Crossover The crossover operation creates new defuzzification weight constants $\omega^c_i (i = 1, 2, \ldots, |N|)$ that are obtained by an average of two weight constants ω^1_i and ω^2_i in the chromosomes P_1 and P_2 as

$$\omega^c_i = \frac{\omega^1_i + \omega^2_i}{2}. \tag{10}$$

c) **Mutation** The mutation operation creates new defuzzification weight constants ω_i^m as

$$\omega_i^m = min(0, 1 - \omega_i).$$

(11)

The above equation implies that ω_i^m is smaller than ω_i by the amount of $|\omega_i - \frac{1}{2}|$ when ω_i is larger than $\frac{1}{2}$, vice versa. Min operator ensures that ω_i^m does not become negative.

5. Simulation Results and Discussion

The proposed various COG defuzzifiers are applied to the truck backer-upper control problem in order to compare the control performance with the conventional COG defuzzifier in terms of average tracing distance. The goal of the truck backer-upper control problem is to back a truck to a loading dock as quickly and precisely as possible. This control problem is a typical non-linear control problem that can not be solved by conventional control techniques.

Fig. 5 shows a model truck and a loading dock used in the truck backer-upper control problem. The position of the truck is precisely determined by (x, y, ϕ) where ϕ is an angle between the truck's onward direction and the x axis, and the tracking control of the truck is done by the θ where θ is an angle between the truck's onward direction and the axis of wheel. The approximate control dynamics of the truck backer-upper control problem is given as (See [10] for details.)

$$
\begin{aligned}
x(t+1) &= x(t) + cos[\phi(t) + \theta(t)] + sin[\theta(t)]sin[\phi(t)] \\
y(t+1) &= y(t) + sin[\phi(t) + \theta(t)] - sin[\theta(t)]cos[\phi(t)] \\
\phi(t+1) &= \phi(t) - sin^{-1}[2sin(\frac{\theta(t)}{b})],
\end{aligned}
$$

(12)

where b is the length of truck and $b = 4$ is taken in this work. If the distance between the truck and the loading dock is sufficiently great, it has only to back the truck straightforwardly once the truck comes close to near $x = 10$ and $\phi = 90°$. Thus, one variable y can be excluded from the fuzzy input variables (x, y, ϕ) for simplicity. So, the design problem of fuzzy controller for the truck backer-upper control problem is reconfigured as to back the truck at a certain position (x_0, ϕ_0) in the interval of $\{0 \leq x \leq 20, -90° \leq \phi \leq 270°\}$ to the loading dock at $x = 10, \phi = 90°$ as fast and precisely as possible.

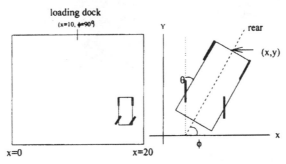

Figure 5. A model truck and a loading dock used in the truck backer-upper control problem.

Fig. 6 shows the fuzzy terms and the corresponding membership functions of the input and output variables when the shape constant n is 1. The input variable x, ϕ and the output variable θ consists of 5, 7, 7 partitions, respectively. Fig. 7 shows the fuzzy rule base used in this work. These two data come from the Wang and Mendel's paper [10].

Three kinds of different COG defuzzifiers are considered how much they improve the control performance compared with the conventional COG defuzzifier using Eq. (2). They are using the crisp value defined as an effective crisp value (Eq. (6)), an optimal crisp value (Eq. (8)), and an approximate crisp value (Eq. (9)), respectively. The control performance is evaluated by the average tracing distance of the model car starting from the 1,000 randomly selected initial values (x_i, ϕ_i). In the case of using an optimal crisp value, the defuzzification weight constants ω are obtained from the genetic algorithm by the 236 training example set that consist of the smooth trajectories from 14 initial positions to the loading dock (See from Table 1 to Table 14 in [10] for details.). The execution parameters of the genetic algorithm used for determining the defuzzification weight constants ω are given as the population size $M = 200$, the gene size $|\omega|$ = 7, the tournament size $k = 3$, $pselect = 0.7$, the crossover probability $P_c = 1.0$, the mutation probability $P_m = 0.001$, the number of generations $N = 2,000$. Fig. 8 shows the evolution curve of the fitness value generated from performing

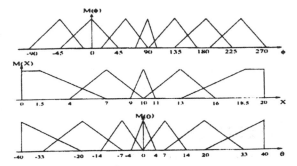

Figure 6. Membership functions used for truck backer-upper control.

		S2	S1	CE	B1	B2
	S3	S2	S3			
	S2	S2	S3	S3	S3	
	S1	B1	S1	S2	S3	S2
φ	CE	B2	B2	CE	S2	S2
	B1	B2	B3	B2	B1	S1
	B2		B3	B3	B3	B2
	B3				B3	B2

Figure 7. Fuzzy rule base used for truck backer-upper control.

the genetic algorithm. Values of the vertical axis represent the average of the fitness values among the chromosomes. It can be seen that the average fitness value is increased slowly as the number of generations is increased.

Table 1 shows the optimal defuzzification weight constants ω determined from the genetic algorithm. It can be seen that the global defuzzification weight constant w_{global} that is defined as the average of all weight constants becomes greater as the shape constant n is increased. This result seems to reflect the fact that the effective areas of the membership functions are decreased as the shape constant n becomes greater and the increase of the defuzzification weight constants ω are acting to compensate the reduction of the effective areas.

TABLE I
THE DEFUZZIFICATION WEIGHT CONSTANTS ω OBTAINED FROM THE GENETIC ALGORITHM.

Weights	w_1	w_2	w_3	w_4	w_5	w_6	w_7	w_{global}
r=0.5	0.705	0.652	0.618	0.618	0.633	0.615	0.554	0.628
r=1.0	0.732	0.693	0.652	0.652	0.669	0.651	0.605	0.605
r=2.0	0.761	0.712	0.682	0.676	0.679	0.697	0.631	0.691

Fig. 8 shows two trajectories of the model car starting from two different random starting points $\{(19.1, 0.0, 206°)$ and $(2.3, 0.0, -34.6°)\}$ when four different COG defuzzifiers are considered. It can be seen that the tracing distance is shorter in the order of the optimal COG(18 and 19 steps), the effective COG(19 and 20 steps), the approximate COG(21 and 21 steps), and the conventional COG(26 and 26 steps). Here, the first value in the parenthesis represents the tracing distance from the random starting point $\{(19.1, 0.0, 206°)$. In order to verify the generalization property of the proposed defuzzification methods, four different COG defuzzifiers were tested over the 1.000 randomly selected starting points.

Fig. 9 shows the average tracing distances of four different defuzzifiers(the shape constant n=2.0) obtained from the 1,000 starting points. It can bee seen that the average tracing distance is shorter in the order of the optimal COG(20.57 steps), the effective COG(20.69 steps), the approximate COG(21.41 steps), and the conventional COG(25.48 steps). This implies that the control performance is increased by 23.8 % on the average when the optimal COG defuzzifier and the conventional COG defuzzifier are compared.

Table 2 compares the average tracing distances among four different defuzzifiers when different shape constants n are considered. It can be seen that the control performance is better when the shape constant n is larger. This comes

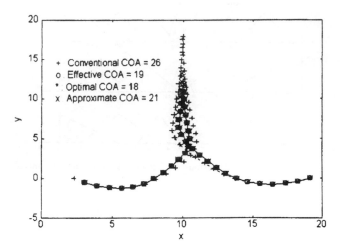

Figure 8. Trajectories of a model car from two random starting points $(19.1, 0.0, 206°)$ and $(2.3, 0.0, -34.6°)$.

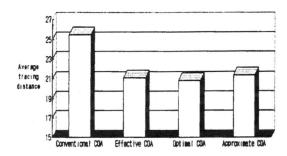

Figure 9. Average tracing distances of four different COG defuzzifiers.

from the fact when the effect of optimal defuzzification weight constants on the control performance is serious when the effective area is smaller (i.e., the shape constant n is larger.) and the overlapped area between two adjacent membership functions is diminished as the shape constant n is larger. There is no big difference of the control performance between the optimal COG defuzzifier and the effective COG defuzzifier on the average. But, the chance of non-convergent initial points in the effective COG defuzzifier is slightly greater than that of the optimal COG defuzzifier. In addition, the approximate COG defuzzifier shows an inconsiderable amount of degradation when compared with the optimal COG defuzzifier, but it only use the original widths of membership functions when the crisp value is calculated. This simplicity is very desirable for implementing the proposed defuzzifier in hardware.

6. Conclusion

This paper proposed new COG defuzzification methods that are intended to be applied for an accurate fuzzy logic controller. The conventional COG method considered the singleton's support values when a crisp value was calculated. So, the conventional method produced the wrong result that it calculated the same crisp value when the singleton's values were same but the widths of membership functions were different than each other. To correct this misuse, the concept of the effective area was introduced in calculating the crisp value. The effective area was equivalent to the area of the clipped membership value multiplied by the effective width. Since the effective width was known to be problem-

TABLE II

COMPARISON OF AVERAGE TRACING DISTANCES AMONG FOUR DIFFERENT COG DEFUZZIFIERS.

Shape Constant	Conventional COG	Effective COG	Optimal COG	Approximate COG
r=0.5	25.48 steps	21.11 steps	20.54 steps	21.41 steps
r=1.0	25.48 steps	20.89 steps	20.07 steps	21.41 steps
r=2.0	25.48 steps	20.69 steps	19.85 steps	21.41 steps

specific and dependent on shapes of the used membership function, the optimal defuzzification weight constants ω were determined by the genetic algorithm using a set of training examples.

The most important results of this work were (i) the control performance could be enhanced more than 20% by considering the optimal widths of membership functions when a crisp value is calculated, (ii) the optimal defuzzification weight constants ω could be determined by using the genetic algorithm easily, and (iii) the approximate COG defuzzification method that simply considered the widths of the membership functions resulted in a slightly degraded control performance than the optimal COG defuzzification method. However, the proposed optimal COG defuzzification method required the additional multipliers for calculating the effective area. To eliminate the use of multipliers, a new architecture of the COG defuzzifier using the stochastic logic circuits is developing now.

Appendix

Derivation of Eq.(4) :

From the Fig. 3. the area $A_Y(y_i)$ under the clipped membership function is the sum of the rectangular area(A_r) and the integrated area (A_i). And the parabolic equation $\mu_Y(y_i)$ can be represented by $\mu_Y(y_i) = m(y_i - \frac{s_i}{2})^n$ where the constant m is determined by the fact $\mu_Y(0) = 1$, i.e., $m(0 - \frac{s_i}{2})^r = 1$. Therefore, $m = \frac{1}{(-\frac{s_i}{2})^r}$ and $\mu_Y(y_i) = \frac{1}{(-\frac{s_i}{2})^r}(y_i - \frac{s_i}{2})^r = (1 - \frac{2y_i}{s_i})^r$. Also, an intersection point $P(y_i, \mu_i)$ can be determined by the fact that $\mu_i = \mu_Y(y_i)$, i.e., $(1 - \frac{2y_i}{s_i})^r = \mu_i$. Therefore, $y_i = \frac{s_i}{2} \cdot (1 - \mu_Y(y_i)^{\frac{1}{r}})$.

The rectangular area A_r becomes $A_r = 2 \cdot \mu_i \cdot y_i = 2 \cdot \mu_Y(y_i) \cdot y_i = s_i \cdot (1 - \mu_Y(y_i)^{\frac{1}{r}}) \cdot \mu_Y(y_i)$. The integrated area A_i becomes $A_i = 2 \cdot \int_{y_i}^{\frac{s_i}{2}} (1 - \frac{2y}{s_i})^r dy = 2[-\frac{s_i}{2} \cdot \frac{1}{r+1} \cdot (1 - \frac{2y}{s_i})^{r+1}]_{y_i}^{\frac{s_i}{2}} = \frac{s_i}{r+1}\mu_Y^{\frac{r+1}{r}}(y_i)$. Therefore, the total area $A_Y(y_i) = A_r + A_i = s_i \cdot (1 - \mu_Y^{\frac{1}{r}}(y_i)) \cdot \mu_Y(y_i) + \frac{s_i}{r+1} \cdot \mu_Y^{\frac{r+1}{r}}(y_i) = s_i \cdot \mu_Y(y_i) - s_i \cdot \frac{r}{r+1} \cdot \mu_Y^{\frac{r+1}{r}}(y_i) = s_i \cdot \mu_Y(y_i) \cdot (1 - \frac{r}{r+1} \cdot \mu_Y^{\frac{1}{r}}(y_i))$.

Furthermore, when $r > 1$, the same equation is derived through the similar one as the above procedure. This proves the derivation of Eq.(4).

REFERENCES

[1] Yager, R. R., and Filev D. P., 1993, SLIDE: A simple Adaptive Defuzzification Method, *IEEE Transactions on Fuzzy Systems*, 1, 69-78.
[2] Bastian. A.., 1995, Handling the nonlinearity of a fuzzy logic controller at the transition between rules, *Fuzzy Sets and Systems*, 71, 369-387.
[3] Ruiz. A., Gutiérrez. J., and Fernández, J., 1995, A Fuzzy Controller with an Optimized Defuzzification Algorithm, *IEEE Micro*, 15, 1-10.
[4] Lee, C. C., 1990, Fuzzy Logic in Control Systems: Fuzzy Logic Controller, *IEEE Transactions on System, Man, and Cybernetics*, 20, 404-435.
[5] Runkler, T. A., and Glesner, M., 1993, A set of Axioms for Defuzzification Strategies Towards a Theory of Rational Defuzzification Operators, *Second IEEE International Conference on Fuzzy Systems*, 2, 1161-1166.
[6] Miki, T., Matsumoto, H., Ohto, K., and Yamakawa, T., 1993, Silicon Implementation for a Novel High Speed Inference Engine: Mega-FLIPS Analog Fuzzy Processor, *Journal of Intelligent and Fuzzy Systems*, 1, 27-42.
[7] Goldberg, D. E., 1989, *Genetic Algorithms in Search, Optimization and Machine Learning*, Addison-Wesley Press.
[8] Goldberg. D. E., and Deb, K., 1991, A Comparative Analysis of Selection Schemes Used in Genetic Algorithms, Ed. Gregory J. E Rawlins, *Fundamental of Genetic Algorithms*, Morgan Kaufmann Publisher, 69-93.
[9] Davis, L., 1991, Hybridization and Numerical Representation, *Handbook of Genetic Algorithms*, Van Nostrand Reinhold, 61-71.
[10] Wang, L. X. and Mendel, J. M., *Generating Fuzzy Rules from Numerical Data, with Applications*, USC-SIPI Report 169, University of Southern California, USA.

Automatic Structuring of Unknown Dynamic Systems

Heikki Hyötyniemi

Helsinki University of Technology, Control Engineering Laboratory
Otakaari 5 A, FIN-02150 Espoo, Finland
Email: `heikki.hyotyniemi@hut.fi`

Keywords: dynamic systems, modeling, soft sensors, sparse coding

Abstract

In this paper, the novel algorithm called GGHA is applied to complex process structuring. The GGHA algorithm extracts linearly additive features from the data. It turns out that in the case of a dynamic system, these features are pulse responses. As a non-recurrent structure, the resulting feature model is of finite impulse response (FIR) type, whereas in a recurrent configuration, infinite impulse response (IIR) models can also be obtained. It is also shown how this kind of process models can be used for associative prediction tasks, so that different kinds of *soft sensors* can be realized. What is more, there exist concrete measures for the reliability of the obtained estimates.

1. Introduction

In today's industrial plants, the instrumentation makes it possible to collect more and more process measurement data. The data collecting devices deliver vast amounts of more or less relevant information about the process operation. There is a rapidly growing need of tools for mastering this complexity.

System identification usually means *parameter estimation*, that is, the structure of the process is assumed to be given beforehand, and the causal dependencies of the signals are assumed to be known. In an unstructured environment, however, the main problem is *structure identification*. There are not many tools for aiding in this task of finding structure from data—in the structuring phase, some 'intelligence' or 'understanding' is needed.

In [5], it is assumed that complex phenomena can be modeled by linearly additive *features*. In what follows, assume that the observed input vectors f can be expressed as weighted sums of N distinct features θ_i, where $1 \leq i \leq N$:

$$f = \sum_{i=1}^{N} \phi_i(f) \cdot \theta_i. \tag{1}$$

This can be presented conveniently in a matrix form

$$f = \theta \phi(f). \tag{2}$$

The features are the common characteristics of the input samples—these properties should be visible in the data correlation matrix, and, more specifically, in the eigenvectors of that matrix. Whereas the Principal Component Analysis (PCA) only extracts one set of eigenvectors of the data correlation matrix as a whole, the novel *GGHA algorithm* assumes that *various data distributions coexist*, extracting separate feature vector sequences when it is needed. This simple idea is applicable in very different cases (see [6] and [7]), and this paper shows that also in process modeling this approach is fruitful.

2. 'Feature basis' extraction

The GGHA algorithm is a generalization of the GHA algorithm that extracts the principal components of the input data matrix (for example, see [3]). The algorithm is based on self-organising maps of the Kohonen type [8]. There are i *nodes*, $1 \leq i \leq N$, with *prototype vectors* that are now denoted as $\hat{\theta}_i$. The new algorithm differs from the standard algorithm because the $n \times 1$ input vector f is iteratively modified and used as input many times, the number of iterations being $1 \leq m \leq N$. Additionally, the prototype vectors are normalised.

In the adaptation process, the prototype vectors become the feature vectors. The iterative process of adapting the self-organising feature network consists of the following steps:

1. Take the next input vector sample f.

2. Select the prototype vector with index c having the best (positive or negative) correlation with the (weighted) vector f:

$$c = \arg\max_{1 \leq i \leq N} \left\{ |\hat{\phi}_i| \right\} = \arg\max_{1 \leq i \leq N} \left\{ |\hat{\theta}_i^T W_{in} f| \right\}. \tag{3}$$

3. For each of the nodes i, apply the Kohonen type learning algorithm using the vector $\hat{\phi}_c \cdot f$ as input:

$$\hat{\theta}_i \leftarrow \hat{\theta}_i + \gamma \cdot h(i, c) \cdot \left(\hat{\phi}_c \cdot f - \hat{\theta}_i \right), \tag{4}$$

where the parameter $h(i, c)$ defines the *neighborhood relation* between the nodes i and c, and γ defines the adaptation rate as presented in [8]. More sophisticated algorithms for self-organization are derived in [4].

4. Normalise the feature estimate vectors:

$$\hat{\theta}_i \leftarrow \hat{\theta}_i / \sqrt{\hat{\theta}_i^T \hat{\theta}_i}. \tag{5}$$

5. Eliminate the contribution of the feature estimate number c by setting

$$f \leftarrow f - \hat{\phi}_c \cdot \hat{\theta}_c. \tag{6}$$

6. If the iteration limit m has not yet been reached, go back to Step 2, otherwise, go back to Step 1.

The parameter m reveals how many additive features are assumed to coexist in the input patterns. Normally there holds $1 < m < N$, and GGHA becomes a kind of a *sparse coding technique* (compare to [1]). The role of the (diagonal) weighting matrix W_{in} is to make different kinds of 'domain oriented' models possible: that means, if the weigth corresponding to some element of f is large, the algorithm will emphasise that variable, trying to explain it in terms of the other variables (see later).

3. Application example

In this example, five signals x_1 to x_5 of a process are measured (see Fig. 1). No information about the actual structure of the process producing the signals is assumed to be known (the simulation model is shown in Fig. 2). The 30×1 input vector f is constructed of the six latest samples of each of the five signals, that means, at time instant k the input vector structure is

$$f(k) = \left(\begin{array}{c} x(k) \\ \hline \vdots \\ \hline x(k-d) \end{array} \right) \quad \text{where} \quad x(\kappa) = \left(\begin{array}{c} x_1(\kappa) \\ \vdots \\ x_n(\kappa) \end{array} \right). \tag{7}$$

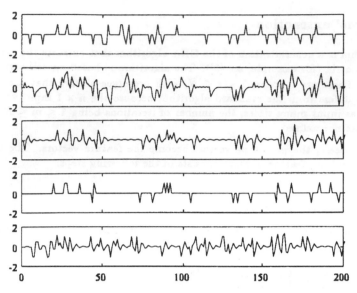

Figure 1. Sample sequences of the five process measurements

The symbols n and d denote the number of measured signals and the maximum delay, respectively (in this case, $n = 5$ and $d = 5$). The weighting matrix W_{in} is now identity matrix, that is, all measurements are equally emphasized. The self-organizing map is a two-dimensional grid of $4 \times 4 = 16$ nodes, and $m = 3$. The neighborhood function is Gaussian, the neighborhood radius changes gradually from 1.0 to 0.1.

The extracted 16 features are shown in Figs. 3, 4, 5, and 6. The vectors $\hat{\theta}_i$ are visualized by interpreting the 30 vector elements as successive samples of the five signals, all signals being plotted in the same frame. It turns out that the process features are now *pulse responses*, so that the process dynamics is presented as a set of Finite Impulse Response (FIR) models.

The meaning of the expression (1) becomes now clear: the model says that the measured signal combinations can be explained using scaled sums of pulse responses. Of course, the additivity of signals, or the linearity assumption is not valid in many practical applications, but this assumption offers a good basis for theoretical analyses. What is more, the nonlinearities can often be expressed as combinations of linearized models to sufficient degree.

The features can be interpreted in a natural way. For example, take the leftmost feature in Fig. 5, and denote it as ξ. Looking at the feature vector standing for ξ, it turns out that $\hat{\phi}_\xi$ correlates much with signal x_5 samples—in concrete terms, the connections between $\hat{\phi}_\xi(k)$ and $x_5(\kappa)$ for previous time instants κ can approximately be expressed as

$$\hat{\phi}_\xi(k) = 0.49 \cdot x_5(k-1) - 0.66 \cdot x_5(k-3) + 0.55 \cdot x_5(k-5). \tag{8}$$

This can most compactly be written using the *unit delay operator* q^{-1} (for example, see [12]) as

$$\hat{\phi}_\xi(k) = \left(0.49q^{-1} - 0.66q^{-3} + 0.55q^{-5}\right) x_5(k). \tag{9}$$

Using the theory of *balanced realizations*, this fifth order model can be reduced (see [2]). It turns out that if three of the five states are dropped, the resulting model approximation is

$$\hat{\phi}_\xi(k) \approx \frac{0.67q^{-1}}{1 + 0.65q^{-2}} \cdot x_5(k). \tag{10}$$

Comparing this model to the underlying process structure in Fig. 2, one can see that the dynamic nature of the autoregressive signal has been captured rather well (see the transfer function 4 in the figure). This

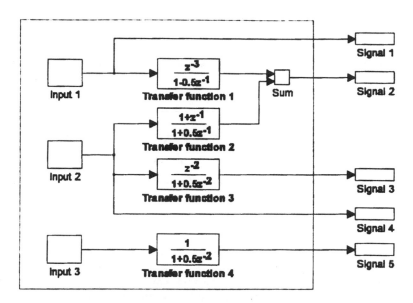

Figure 2. The actual 'black box' process producing the data. The inputs 1, 2, and 3 consist of unit pulses (positive or negative), pulse probabilities being about 35%

means that the behavior of the feature $\hat{\phi}_\xi(k)$ follows the behavior of the signal $x_5(k)$. Because of the dynamic nature of (10), the newest measurements of $x_5(k)$ are not relied on exclusively. That is why, the model can be used for filtering the signal if the measurement is noisy, or it can be used for prediction if the newest value of the signal is missing altogether.

It must be noted, however, that the above approximation is rather crude because of the truncated FIR model. This can also be seen when one calculates the *Hankel singular values* for the FIR model (9)— these values are 1.30, 0.97, 0.51, 0.31, and 0.25. Because the numeric magnitudes do not differ very much, reducing the model results in rather high approximation errors.

The algorithm mainly produced pulse responses in the above example, that means, the model become *input oriented,* where the effect of the input signals is clearly visible. However, in some applications, one would perhaps prefer an *output oriented* model, that would explain the (assumed) system outputs as functions of earlier inputs. As mentioned in the previous section, the structure of the feature model can be affected by weighting the signals appropriately — in this case, the output oriented model would be reached by specially emphasising the output signals in the weighting matrix W_{in}.

4. Using the network model

The converged network can be used as a model for the observations. To utilize this, the $n \times 1$ input pattern vectors f are presented as

$$f = \hat{\theta}\hat{\phi}(f) + e(f), \tag{11}$$

where $\hat{\theta}$ is an $n \times N$ matrix containing the feature estimates, $\hat{\phi}(f)$ is an $N \times 1$ vector containing the *loadings* or the contribution of the different features, and $e(f)$ stands for the *reconstruction error*. Now, minimizing the weighted sum of squared errors,

$$e^T(f)W_{out}e(f) = (f - \hat{\theta}\hat{\phi}(f))^T W_{out}(f - \hat{\theta}\hat{\phi}(f)), \tag{12}$$

gives the optimal loadings (assuming that all features are mutually additive)[1]

---

[1] The *Gauss-Markov estimate* [11] has the same form (13), even if the starting point is different: if W_{out}^{-1} is the covariance matrix of e, the formula gives the unbiased estimate for ϕ

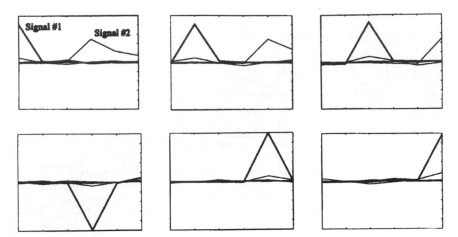

Figure 3. Features standing for correlations between signals 1 and 2, plotted with bold and normal style, respectively (only six samples of each signal is available, but the samples have been connected in the above frames). It is not a surprise that the best way to model dependencies between signals is through pulse responses! The algorithm does not 'know' that many of the input samples are just delayed copies of each other, and pulse responses are thus created separately for each time point. Note that the pulses and the responses may be inverted (see the fourth frame)—because of the assumed linearity of the processes, the signals can be arbitrarily scaled

$$\hat{\phi}^{opt}(f) = (\hat{\theta}^T W_{out}\hat{\theta})^{-1}\hat{\theta}^T W_{out} \cdot f. \tag{13}$$

For an *orthonormal* basis, so that $\hat{\theta}^T\hat{\theta} = I$, this reduces to a very simple form $\hat{\phi}(f) = \hat{\theta}^T f$, if also $W_{out} = I$. Even if the input signals in the experiment were sequences of distinct pulses (to make the features better distinguishable in the adaptation phase), the above formula can be used for feature extraction also if the excitation is continuous—the optimal $\hat{\phi}^{opt}$ values do not constitute a sparse code.

The role of the $n \cdot (d+1) \times n \cdot (d+1)$ positive definite weighting matrix W_{out}, normally of the form (see later)

$$W_{out} = \begin{pmatrix} W_0 & & \\ & \ddots & \\ & & W_d \end{pmatrix}, \tag{14}$$

is to emphasise the measurements appropriately. Setting some of the diagonal elements to zero, that measurement is not used for pattern matching—first calculating the loadings as in (13) and then constructing

$$\hat{f} = \hat{\theta}\hat{\phi}^{opt}(f), \tag{15}$$

gives an *associative prediction* of the unknown inputs using the FIR model library. For example, prediction of the new signal values when all previous signal values are available, can be accomplished by selecting

$$W_0 = \begin{pmatrix} 0 & & \\ & \ddots & \\ & & 0 \end{pmatrix} \quad \text{and} \quad W_i = \begin{pmatrix} 1 & & \\ & \ddots & \\ & & 1 \end{pmatrix} \tag{16}$$

for all other time indices $1 \le i \le d$ in (14). This way it is possible to construct different kinds of 'soft sensors' for the system. Figure 7 represents an experiment, where all the elements in W_{out} corresponding to the signal 2 are set to zero, so that

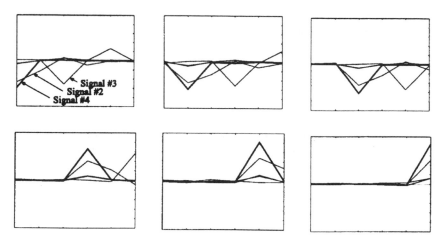

Figure 4. Features standing for correlations between signals 4, 2, and 3 (see the discussion in Fig. 3). Note that signals are scaled because of the feature vectors are always assumed to have unit length

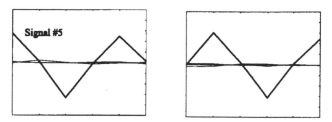

Figure 5. Features standing for signal 5. Note that there are too few feature vectors available for exactly modeling all of the independent signals. The algorithm has solved this problem by allocating *only two* features for the isolated signal 5. This signal oscillates with period of four time steps. Even if the dynamics is damped, the signal can best be approximated as a superposition of two sinusoidals with 90 degree phase difference

$$
W_i = \begin{pmatrix} 1 & & & & \\ & 0 & & & \\ & & 1 & & \\ & & & 1 & \\ & & & & 1 \end{pmatrix} \tag{17}
$$

for all $0 \leq i \leq d$. This means that the whole sequence of signal 2 values are unknown and should be reconstructed using only measurements of the other signals. In the figure, the reconstruction result is plotted against the real signal. The results look good.

Of course, the reconstruction of the signals cannot always be carried out. What is one of the best things about the proposed approach, is that there is a measure for the 'reconstructability' directly available: the condition number of $\hat{\theta}^T W_{out} \hat{\theta}$ tells us how reliable the results obtained by (13) are. For example, when reconstructing the signal 2 (see Fig. 7), the condition number is 3.1, while when reconstructing the independent signal 5 using only measurements of the signals 1, 2, 3, and 4, the condition number exceeds 20000!

5. Extensions of the model

One could wonder whether the modeling power of the data structure could be enhanced. One possibility is to manipulate the output of the network further—the vectors $\hat{\phi}^{opt}(k)$ can be used as input to yet another layer of the network. Even if the network structure is linear, so that two matrix operations can, in principle, be combined, the delayed signals enhance the expressive power of the multi-layer structure.

416

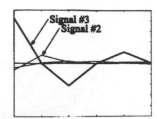 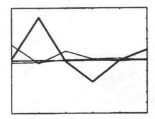

Figure 6. The last two features trying to 'explain the unexplained'. When the signal 4 passes the observation horizon of six time points, the signal 3 that is a delayed function of this input signal, still is visible in the observations. That is why, the 'tails' of the responses become modeled as autonomous autoregressive processes

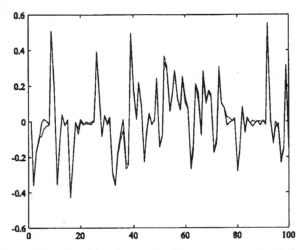

Figure 7. Reconstruction of the signal 2 when only the measurements of the signals 1, 3, 4, and 5 are available (the real, unmeasured signal sequence plotted with dotted line, and the reconstructed estimate plotted with solid line). It is interesting to note that the reconstruction errors are relatively large when the signal value is small—this is due to the fact that the pulse response models of the dynamics are truncated, so that the 'tails' of the responses are neglected. Having longer sequences of measurement data (and more nodes for features) would fix this problem

Qualitative enhancements can be obtained, if a *recurrent* network structure is implemented (Fig.8). Defining the augmented input vector

$$F(k) = \left(\frac{f(k)}{\dot{\Phi}(k-1)} \right) \quad \text{where} \quad \hat{\Phi}(k-1) = \left(\frac{\hat{\phi}^{opt}(F(k-1))}{\vdots}{\hat{\phi}^{opt}(F(k-d-1))} \right), \tag{18}$$

makes it possible to capture the auto-regressive characteristics of Infinite Impulse Response (IIR) models directly. However, the recurrence in the network structure often introduces stability problems, and to avoid this, some modifications are now needed. It turns out that it is wise to emphasise only the direct signal measurements, so that the estimated network output is

$$\hat{\phi}^{opt}(F(k)) = \left(\hat{\Theta}^T W_f \hat{\Theta} \right)^{-1} \hat{\Theta}^T W_f \cdot F(k), \tag{19}$$

where the $(n+N) \cdot (d+1) \times (n+N) \cdot (d+1)$ -dimensional weighting matrix W_f is now constructed as

$$W_f = \left(\begin{array}{c|c} I_{(d+1)n \times (d+1)n} & \\ \hline & 0_{(d+1)N \times (d+1)N} \end{array} \right). \tag{20}$$

The adaptation algorithm, of course, is modified so that the variables f and $\hat{\theta}$ are substituted for F and $\hat{\Theta}$, respectively. Additionally, Step 2 (selection of the 'winner') in the algorithm also needs to be calculated so that the weighting is taken into account:

$$c = \arg\max_{1 \le i \le N} \left\{ |\hat{\phi}_i| \right\} = \arg\max_{1 \le i \le N} \left\{ |\hat{\Theta}_i^T W_f F| \right\}. \tag{21}$$

The adaptation of this more complex network structure was carried out by continuing from the feature estimates that were received in the previous non-recurrent phase after augmenting the vectors appropriately. Figure 9 shows one view into the convergence results: the most significant connections between the three features that are shown in the upper part of Fig.3. It seems that the longer time a pulse has traversed, the less one counts on the direct measurements, and previous analysis results become more and more emphasised when judging whether a pulse is present or not.

In Figure 10, another case is presented, corresponding to Fig. 5. Denoting the features as shown in the figure, one can see that there holds

$$\begin{aligned} A_\xi(q^{-1})\hat{\phi}_\xi(k) &= B_\xi(q^{-1})\hat{\phi}_\zeta(k) + C_\xi(q^{-1})x_5(k) \\ A_\zeta(q^{-1})\hat{\phi}_\zeta(k) &= B_\zeta(q^{-1})\hat{\phi}_\xi(k) + C_\zeta(q^{-1})x_5(k), \end{aligned} \tag{22}$$

where the polynomials for feature ξ, for example, can be approximately written as

$$\begin{cases} A_\xi(q^{-1}) &= 1 + 0.32q^{-2} - 0.18q^{-4} + 0.08q^{-6} \\ B_\xi(q^{-1}) &= 0.37q^{-1} + 0.20q^{-3} + 0.17q^{-5} \\ C_\xi(q^{-1}) &= 0.38q^{-1} - 0.50q^{-3} + 0.42q^{-5}. \end{cases} \tag{23}$$

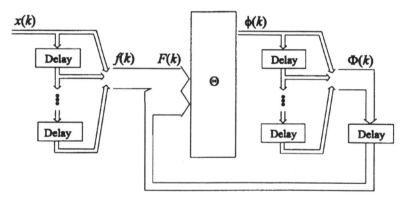

Figure 8. A recurrent network with virtually *infinite* number of layers

Eliminating $\hat{\phi}_\zeta(k)$ from (22) results in

$$\hat{\phi}_\xi(k) = \frac{B_\xi(q^{-1})C_\zeta(q^{-1}) + A_\zeta(q^{-1})C_\xi(q^{-1})}{A_\xi(q^{-1})A_\zeta(q^{-1}) - B_\xi(q^{-1})B_\zeta(q^{-1})} \cdot x_5(k). \tag{24}$$

This model can again be reduced. In principle, this model should be more accurate for presenting the dynamics of the signal $x_5(k)$ than the simpler expression (9) is. However, the truth is that the two available features carry too little information to reach good accuracy, however these features are utilised.

6. Discussion

Correlations can be assumed to be the basic building blocks of intelligent behavior—correlations between input data are features, and correlations between features are 'concepts', as presented in [5]. This experiment shows that when simple mathematical operations accumulate, the results can be rather conceptual:

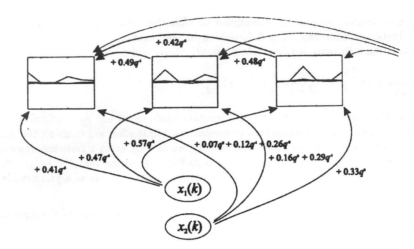

Figure 9. Some of the correlations between the features and measured signals

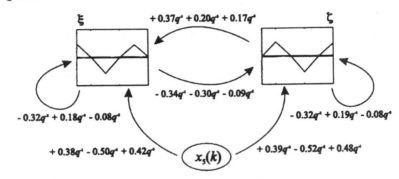

Figure 10. The correlations between the features that model the autoregressive signal $x_5(k)$

impulse responses, or *residence time distributions,* are still today one of the most widely used approaches to conceptualising the dynamics of complex industrial processes.

In [5], only the case of static, time-independent input data was discussed, and the question of *causal* relationships was not elaborated on. In dynamic processes, causality, or dependencies in time are fundamental—modeling correlations between signals in different time points has now been made possible simply by including previous signal samples together with the current ones in the input vector.

The role of the self-organising map as a basis of the GGHA algorithm is to assure that all prototype vectors get involved and utilised. The features are often rather far from each other (the 'pruning' of the input vector resembles the Gram–Schmidt orthogonalisation procedure), so that no actual 'map' of adjacent subspaces is created (as compared with the Adaptive Subspace SOM (ASSOM) method [9]). Because the goal of the adaptation is not to create an adjacency map, the adaptation process is rather fast also in large nets. However, because of the competitive learning, the final features are *not* strictly orthogonal, so that $\hat{\theta}_i^T \hat{\theta}_j$ is not necessarily zero for $i \neq j$. It can be argued that good features, like the pulse responses above, are not usually orthogonal.

Modeling of linear systems seems to be a very good application area of feature extraction methods. Usually, in other applications of GGHA, as in the examples presented in [6] and [7], all of the features cannot be combined freely. Now, on the other hand, the features or pulse responses are genuinely additive, and the whole feature map can be utilised simultaneously for presenting the signals. In this case the features span linear subspaces, whereas in other applications the features usually stand for distinct clusters. Distinct clusters with separate feature sequences can emerge also in dynamic processes, if there are some structural changes (for example, some valve is opened, or a pump is started), or if the operating point is changed.

As a final remark, it can be noted that there exist no explicit methods for finding the structure for data, and exhaustive search for the correct model structure generally cannot be avoided. The standard SISO (Single Input, Single Output) identification techniques do not necessarily work even if the process inputs and outputs were known. When various independent signals contribute, the SISO methods may give biased parameter estimates (for example, see [10]).

Acknowledgement

This work was financed by the Academy of Finland, and this support is gratefully acknowledged.

References

[1] Földiák, P., 1990: Forming sparse representations by local anti-Hebbian learning. *Biological Cybernetics*, **64**, No 2, 165–170.

[2] Glover, K., 1984: All optimal Hankel-norm approximations of linear multivariable systems and their L^∞-error bounds. *International Journal of Control*, **39**, No. 6, 1115–1193.

[3] Haykin, S., 1994: *Neural Networks. A Comprehensive Foundation*. Macmillan College Publishing, New York.

[4] Hyötyniemi, H., 1994: *Self-Organizing Artificial Neural Networks in Dynamic Systems Modeling and Control*. PhD Thesis, Helsinki University of Technology, Control Engineering Laboratory, Report 97.

[5] Hyötyniemi, H., 1995: Correlations—Building Blocks of Intelligence? In *Älyn ulottuvuudet ja oppihistoria (History and dimensions of intelligence)*, Finnish Artificial Intelligence Society, 199–226.

[6] Hyötyniemi, H., 1996: Constructing Non-Orthogonal Feature bases. *Proceedings of the International Conference on Neural Networks (ICNN'96)*, June 3–6, 1759–1764. Washington DC.

[7] Hyötyniemi, H., 1996: Text Document Classification with Self-Organising Maps. *Proceedings of STeP'96 (Finnish Artificial Intelligence Conference)*, Vaasa, Finland, August 20–23, 64–72. *Also available at* http://www.hut.fi/~hhyotyni/HH3/HH3.html

[8] Kohonen, T., 1984: *Self-Organization and Associative Memory*. Springer-Verlag, Berlin.

[9] Kohonen, T., 1995: *Self-Organizing Maps*. Springer-Verlag, Berlin.

[10] Ljung, L. and Söderström, T., 1983: *Theory and Practice of Recursive Identification*. MIT Press, Cambridge, Massachusetts.

[11] Luenberger, D.G., 1969: *Optimization by Vector Space Methods*. John Wiley & Sons, New York.

[12] Ogata, K., 1987: *Discrete-Time Control Systems*. Prentice-Hall International, Englewood Cliffs, New Jersey.

Takagi-Sugeno Fuzzy Control of Batch Polymerization Reactors

J. Abonyi[1], L. Nagy[2], F. Szeifert[3]

*Department of Chemical Engineering Cybernetics, University of Veszprém,
Veszprém, POB. 158, H-8201 Hungary
Email: [1]abonyij@kib.vein.hu, [2] nagyl@kib.vein.hu, [3] szeifert@kib.vein.hu*

Keywords: Fuzzy Control, PID Control, Batch Control

Abstract

The control problem focuses on the temperature control of a batch polymerization reactor. Achieving good temperature control is difficult because physical and chemical properties of the contents, such as mass, heat capacity, reaction rate and heat transfer coefficient vary from run to run and within a run.

Standard PID controllers are unable to perform satisfactory way over the entire range of the operation required, but the proposed Takagi-Sugeno fuzzy logic controllers (FLC) are shown to be capable of providing good overall system performance.

1. Introduction

It is a well-known fact that batch processes are gaining wider ground in chemical industries. The course is motivated first of all by the extraordinary flexibility of batch processes which allows quick adaptation to market demand and is made possible by rapid development of process control. Compared to continuos processes the control of processes is more difficult (complex sequential control task, operating conditions change with time, etc.).

The typical unit of batch systems in polymer, pharmaceutical and fine chemical industries is an autoclave with heating-cooling system, which uses directly cold and hot utility fluids of given temperatures available in the plant.

However, the many complications, nonlinearities and constrains make the control problem challenging. A further difficulty arises from the fact that vinyl polymerization is an auto-catalytic reaction. The problem becomes difficult to formulate succinctly and does not lend itself to mathematical formalism. Data and equations are provided for process hardware, reaction kinetics, heat transfer and reaction viscosity[1].

Several researches have addressed control problems in reactor systems of this type. Juba and Hamer [2], and Berber [3] provide an excellent overview on the challenges in batch reactor control and suggest several control strategies.

Davidson [4] proposes a control algorithm which combines knowledge of the process with the logic used by the skilled operator and compares the result with standard PI controllers.

Fuzzy control, a non mathematical control algorithm based on the set of decision rules has found applications in some industrial areas. Along with the practical successes observed with this control strategy in industry, there has also been some academic interest in fuzzy logic control. As a result, limitations of classical fuzzy control is extended, and a model based approach is suggested [5].

Examples for applications of fuzzy control to chemical reactors can be found in the literature [6].

In the following we shall use the concept of quasilinear fuzzy model [7] that based on Takagi-Sugeno-Kang reasoning method [8] to obtain a PI-like expression of the FLC to control of batch polymerization reactor.

2. The Control Problem

2.1. Process Description

A stirred tank reactor is used to produce expanded polystyrene in batch processes. Most processes consist of the following steps. An initial charge of prepolymer, surfactants, initiators, and monomer mixture is added to the reactor. After this initial stage the reaction mixture is heated to the temperature of the polymerization. The third stage called the impregnation stage, in which the blowing agent, commonly n-pentane is loaded to the reactor. Depending on the impregnation temperature and pressure minimum time must be allowed for diffusion of the blowing agent to reach the core of the beads and for complete monomer exhaustion [9]. The last step is the cooling down of the reaction mixture to the temperature required for further processing.

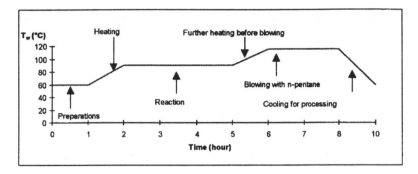

Figure 1. The prescribed temperature profile

2.2. Control Configuration

The scheme of the reactor system is summarized in Figure 2. The reactor temperature is maintained in desired value by adjusting the temperature for water recirculating thought the reactor jacket.

When the controller output is between 0 and 50%, the controller is in cooling mode. In cooling mode, the controller output is used to control the Sv valve.

For controller outputs between 50 and 100% the controller is in heating mode. The controller is used to throttle a control valve referred to as the steam valve (Sg). While the steam valve is open, medium pressure steam is injected directly into the recirculating water steam, thereby adding heat.

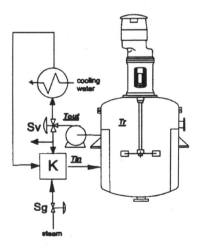

Figure 2. Scheme of the reactor system.

3. Structure of Fuzzy Controllers

3.1. Takagi and Sugeno's Fuzzy Model

In this paper we deal with Takagi and Sugeno's fuzzy model [10]. This fuzzy model can be formulated as the following form:

$$L^i: \quad IF \ x(1) \ is \ A_1^i \ and \ ... \ and \ x(n) \ is \ A_n^i \ THEN \quad y^i = a_0^i + a_1^i \cdot x(1) + ... + a_n^i \cdot x(n) \qquad (3.1.1)$$

where $L^i (i = 1, 2, ... l)$ denotes the i-th implication, l is the number of fuzzy implications, y^i is the output from the i-th implication, $a_p^i (p = 0, 1, ..., n)$ is consequent parameters, $x(1), ..., x(n)$ are the input variables, and A_p^i are fuzzy sets whose membership functions denoted by the same symbols as the fuzzy values. Given an input $\left(x(1), ..., x(n)\right)$, the final output of the fuzzy model is inferred by taking the weighted average of the y^i's:

$$y = \frac{\sum_{i=1}^{l} w^i y^i}{\sum_{i=1}^{l} w^i} \qquad (3.1.2)$$

where $w^i > 0$, and y^i is calculated for the input by consequent èquation of the i-th implications, and the weight w^i implies the overall truth value of premise of the i-th implication for input calculated as

$$w^i = \prod_{p=1}^{n} A_p^i \left(x(p)\right) \qquad (3.1.3)$$

3.2. PI -like Takagi-Sugeno Fuzzy Logic Controller

Several examples of realization of PID controllers by fuzzy control methods can be found in the literature [7,11,12,13,14].

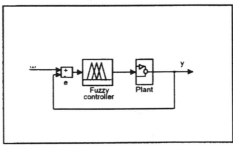

Figure 3. Scheme of the control loop (realized in MATLAB/Simulink)

The output of the conventional analog PI controller in the frequency s-domain is given by [15]

$$u_{PI}(s) = \left(K_p^c + \frac{K_I^c}{s}\right) e(s) \qquad (3.2.1)$$

where K_p^c and K_I^c are the proportional and integral gains, respectively, and $e(s)$ is the tracking error signal.

This equations can be transformed into the discrete form by applying the bilinear transformation $s = (2 / T)(z - 1) / (z + 1)$, where T > 0, is the sampling period, which results in the following form [16]:

$$u_{PI}(z) = \left(K_p^c - \frac{K_I^c T}{2} + \frac{K_I^c T}{1 - z^{-1}} \right) e(z) \qquad (3.2.2)$$

Letting:

$$K_p = K_p^c - \frac{K_I^c T}{2}; \qquad K_I = K_I^c T \qquad (3.2.3)$$

and taking the inverse z-transform, we have:

$$u_{PI}(kT) - u_{PI}(kT - T) = K_p \left[e(kT) - e(kT - T) \right] + K_i e(kT) \qquad (3.2.4)$$

From equation (3.2.4) can be obtained:

$$\Delta u_{PI}(kT) = K_p \Delta e(kT) + K_I e(kT) \qquad (3.2.5)$$

where

$$\Delta u_{PI}(kT) = u_{PI}(kT) - u_{PI}(kT - T) ,$$

$$\Delta e(kT) = e(kT) - e(kT - T) . \qquad (3.2.6)$$

From equation (3.1.1) and (3.2.5) the set of rules can be expressed with the following form:

$$L^i: \quad IF \ e(kT) \ is \ A_1^i \ and \ \Delta e(kT) \ is \ A_2^i \ and \ x(1) \ is \ B_1^i \ and \ ...$$
$$and \ x(n) \ is \ B_n^i \ THEN \quad \Delta u(kT)^i = a_1^i e(kT) + a_2^i \Delta e(kT) \qquad (3.2.7)$$

where $\Delta u(kT)^i$ denotes the output from the i-th implication, $a_p^i (p = 1, \ 2)$ are consequent parameters of the sub PI controller, $x(1), \ ...., \ x(n)$ are the auxiliary input variables (not necessary) , and A_p^i and B_p^i are fuzzy sets on $e(kT)$ and $\Delta e(kT)$ and on auxiliary input variables, whose membership functions denoted by the same symbols are the fuzzy values.

The final output of the fuzzy controller is inferred by taking the weighted average of the u^i 's:

$$u(kT) = u(kT - T) + \frac{\sum_{i=1}^{l} w^i \Delta u(kT)^i}{\sum_{i=1}^{l} w^i} \qquad (3.2.8)$$

where $w^i > 0$, and $u(kT)^i$ is calculated for the input by consequent equation of the i-th implications, and the weight w^i implies the overall truth value of premise of the i-th implication for input calculated as

$$w^i = A_1^i (e(kT)) \times A_2^i (\Delta e(kT)) \times \prod_{p=1}^{n} B_p^i (x(p)) \qquad (3.2.9)$$

3.3. Takagi-Sugeno Fuzzy supervised PI-control

In this section we describe the fuzzy supervisory PID control, where a fuzzy system is used to supervise a conventional PID controller. Several similar examples of fuzzy supervisory PID control can be found in literature [11].

424

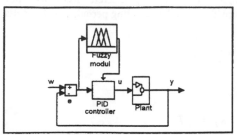

Figure 4. Scheme of the control loop

Our control algorithm is based on one of the early experimental evaluations of feedback controllers that was reported by Marroquin and Luyben [16] who developed a feedback PID controller.

Their approach to the nonlinear case was such that, the gain of the controllers K_p^c was increased depending on the error according to the equation bellow

$$K_p^c(kT) = K_{p0}^c \left(1 + b \cdot |e(kT)|\right) \tag{3.3.1}$$

where b is an adjustable parameter and K_{p0}^c is the base controller gain, when the error was zero (i.e. gain of the linear controller). Jutan and Uppal applied this algorithm in their study on simple feedback-feedforward control of a batch reactor [17].

From eq. (3.1.1) and (3.3.1) the rules of the supervising fuzzy system can be expressed with the following form:

$$L^i: \quad IF \ e(kT) \ is \ A_1^i \ and \ x(1) \ is \ B_1^i \ and \ ... $$
$$and \ x(n) \ is \ B_n^i \ THEN \quad K_p^{c\,i}(kT) = K_{p0}^{c\,i} + K_{p0}^{c\,i} \cdot b^i \cdot |e(kT)| \tag{3.3.2}$$

In place of (3.3.2) we used:

$$L^i: \quad IF \ e(kT) \ is \ A_1^i \ and \ x(1) \ is \ B_1^i \ and \ ... \ and \ x(n) \ is \ B_n^i \ THEN \quad K_p^{c\,i}(kT) = a_1^i + a_2^i \cdot |e(kT)| \tag{3.3.3}$$

where $K_p^{c\,i}$ denotes the output from the i-th implication, $a_p^i (p=1,2)$ are consequent parameters, $x(1),....x(n)$ are the auxiliary input variables (no necessary) , and A_1^i and B_p^i are fuzzy sets on $e(kT)$ and on auxiliary input variables, whose membership functions denoted by the same symbols are the fuzzy values.

The final output of the fuzzy system is inferred by taking the weighted average of the $K_p^{c\,i}$'s:

$$K_p^c(kT) = \frac{\sum_{i=1}^{l} w^i K_p^{c\,i}}{\sum_{i=1}^{l} w^i} \tag{3.3.4}$$

where $w^i > 0$, and $K_p^{c\,i}$ is calculated for the input by consequent equation of the i-th implications, and the weight w^i implies the overall truth value of premise of the i-th implication for input calculated as

$$w^i = A_1^i\big(e(kT)\big) \times \prod_{p=1}^{n} B_p^i\big(x(p)\big) \tag{3.3.5}$$

3.4. Modified PI - like Takagi-Sugeno Fuzzy Logic Controller

The role of the fuzzy supervising system can be expressed with the help of a PI-like Takagi-Sugeno fuzzy controller as well.

For simplicity from in the PI controller incremental form (3.4.1)

$$\frac{du}{dt} = K_p^c \left(\frac{de}{dt} + K_I^c \cdot e \right) \tag{3.4.1}$$

with discretization we obtain the following form:

$$\Delta u_{PI}(kT) = K_p \Delta e(kT) + K_I \cdot e(kT) \tag{3.4.2}$$

where $\quad K_p = K_p^c \qquad K_I = K_p^c \cdot K_I^c T$. $\tag{3.4.3}$

From (3.3.1) and (3.4.2) and (3.4.3) the discrete PI controller with varying gain can be expressed by

$$\Delta u_k = K_{p0}^c \cdot (1 + b \cdot |e(kT)|) \cdot \Delta e_k + K_{p0}^c \cdot (1 + b \cdot |e(kT)|) \cdot K_I^c \cdot T \cdot e(kT) . \tag{3.4.4}$$

The structure of the Takagi-Sugeno fuzzy controller is identical with the one described in (3.2.7) - (3.2.9)

$$L^i: \quad IF \ e(kT) \ is \ A_1^i \ and \ \Delta e(kT) \ is \ A_2^i \ and \ x(1) \ is \ B_1^i \ and \ ... \tag{3.4.5}$$
$$and \ x(n) \ is \ B_n^i \ THEN \quad \Delta u(kT)^i = a_1^i e(kT) + a_2^i \Delta e(kT)$$

but, $\quad a_1^i = K_{p0}^{c~i} (1 + b^i \cdot |e(kT)|) \cdot K_I^{c^i} T; \qquad a_2^i = K_{p0}^{c~i} (1 + b^i \cdot |e(kT)|) \cdot \tag{3.4.6}$

When we would like to imitate the controller described in section 3.3 and we do not use any auxiliary variables, the rules of the fuzzy controller become

$$L^i: \quad IF \ e(kT) \ is \ A_1^i \ THEN \quad \Delta u(kT)^i = a_1^i e(kT) + a_2^i \Delta e(kT) . \tag{3.4.7}$$

We designed the membership functions as illustrated in Figure 5.

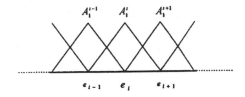

Figure 5. Membership functions of A_1^i

Obviously, by definition of the membership function, $A_1^i(e(kT)) + A_1^{i+1}(e(kT)) = 1$ when $e(kT) \in [e_i, e_{i+1}]$. Therefore, the output of the fuzzy controller can be simplified as

$$\Delta u = \sum_{i=1}^{l} \left(A_1^i(e(kT)) \right) \cdot \left(a_1^i \cdot e(kT) + a_2^i \Delta e(kT) \right) =$$
$$A_1^i(e(kT)) \left(a_1^i \cdot e(kT) + a_2^i \Delta e(kT) \right) + A_1^{i+1}(e(kT)) \left(a_1^{i+1} \cdot e(kT) + a_2^{i+1} \Delta e(kT) \right) = \tag{3.4.8}$$
$$\left(\frac{e_{i+1} - e(kT)}{e_{i+1} - e_i} \right) \cdot \left(a_1^i \cdot e(kT) + a_2^i \Delta e(kT) \right) + \left(\frac{e(kT) - e_i}{e_{i+1} - e_i} \right) \cdot \left(a_1^{i+1} \cdot e(kT) + a_2^{i+1} \Delta e(kT) \right)$$

By substituting (3.4.6) into (3.4.8) we get the output of the controller:

$$\Delta u = \left(\frac{e_{i+1} - e(kT)}{e_{i+1} - e_i} \right) \cdot \left(K_{p0}^{c\,i} (1 + b^j \cdot |e(kT)|) \cdot K_I^{c\,i} \cdot T \cdot e(kT) + K_{p0}^{c\,i} (1 + b^i \cdot |e(kT)|) \cdot \Delta e(kT) \right) +$$
$$\left(\frac{e(kT) - e_i}{e_{i+1} - e_i} \right) \cdot \left(K_{p0}^{c\,i+1} (1 + b^{i+1} \cdot |e(kT)|) \cdot K_I^{c\,i+1} \cdot T \cdot e(kT) + K_{p0}^{c\,i+1} (1 + b^{i+1} \cdot |e(kT)|) \cdot \Delta e(kT) \right) \qquad (3.4.9)$$

4. Simulation Results

4.1. Developing the Control Algorithms

In order to compare the control algorithms developed the following performance index was defined:

$$Q = \sum_{k=1}^{N} e(kT)^2 + \lambda \sum_{k=1}^{N} \Delta u(kT)^2 \qquad (4.1.1)$$

Where: $e(kT) = w(kT) - y(kT)$ the plan output error in the k.-th discrete time step,

$\Delta u(kT) = u(kT) - u(kT - T)$ the change of the control signal in the i.-th time step,

$N = \dfrac{t_{max}}{T_0}$ the maximal number of time steps, where T_0 is the sampling time,

λ weighting factor to maintain the dynamics of control signal (0.2).

The simulator was built using MATLAB and C based on a priori mathematical models according to [9,18,19,20].

All parameters of all controller types were determined by optimization with Sequential Quadratic Programming method with MATLAB *Optimization Toolbox's* constr function [22].

4.1. Conventional Solutions

Among the conventional solutions the PI controller has been examined first. Its performance index (Q) was 1.883. The performance index of optimal gain varying PI controller (3.3.1) was 1.633.

4.2. Applying PI -like Takagi-Sugeno Fuzzy Logic Controller

Among several possibilities we used the PI -type Takagi-Sugeno controller described in the 3.2 section without auxiliary variables in the following form:

$$L^{i,j}: \qquad IF \ e(kT) \ is \ A_1^i \ and \ \Delta e(kT) \ is \ A_2^j \ THEN \qquad \Delta u(kT)^{i,j} = a_1^i e(kT) + a_2^j \Delta e(kT) \qquad (4.2.1)$$

In the sequel $i = 1,2,...l$ and $j = 1,2,...l$ denotes the i-th and j-th membership functions on $e(kT)$ and $\Delta e(kT)$. We designed the membership functions as illustrated in Figure 5. and Figure 6.

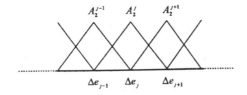

Figure 6. Membership functions of A_2^j

Therefore the output of the fuzzy controller from (3.2.9) can be expressed as

$$\Delta u(kT) = \frac{\sum_{i=1}^{l}\sum_{j=1}^{l}\left(A_1^i\left(e(kT)\right)\cdot A_2^j\left(\Delta e(kT)\right)\right)\cdot\left(a_1^i\cdot e(kT)+a_2^j\Delta e(kT)\right)}{\sum_{i=1}^{l}\sum_{j=1}^{l}\left(A_1^i\left(e(kT)\right)\cdot A_2^j\left(\Delta e(kT)\right)\right)} \tag{4.2.2}$$

Obviously, by definition of the membership function

$$\sum_{i=1}^{l}\sum_{j=1}^{l}\left(A_1^i\left(e(kT)\right)\cdot A_2^j\left(\Delta e(kT)\right)\right)=1 \tag{4.2.3}$$

when

$$e(kT)\in\left[e_i,e_{i+1}\right]\text{ and }\Delta e(kT)\in\left[\Delta e_j,\Delta e_{j+1}\right]. \tag{4.2.4}$$

So the controller output:

$$\Delta u = \left(\frac{e_{i+1}-e(kT)}{e_{i+1}-e_i}\right)\left(\frac{\Delta e_{j+1}-\Delta e(kT)}{\Delta e_{j+1}-\Delta e_j}\right)\cdot\left(a_1^i\cdot e(kT)+a_2^j\Delta e(kT)\right)+\left(\frac{e(kT)-e_i}{e_{i+1}-e_i}\right)\left(\frac{\Delta e_{j+1}-\Delta e(kT)}{\Delta e_{j+1}-\Delta e_j}\right)\cdot\left(a_1^{i+1}\cdot e(kT)+a_2^j\Delta e(kT)\right)+$$

$$\left(\frac{e_{i+1}-e(kT)}{e_{i+1}-e_i}\right)\left(\frac{\Delta e(kT)-\Delta e_j}{\Delta e_{j+1}-\Delta e_j}\right)\cdot\left(a_1^i\cdot e(kT)+a_2^{j+1}\Delta e(kT)\right)+\left(\frac{e(kT)-e_i}{e_{i+1}-e_i}\right)\left(\frac{\Delta e(kT)-\Delta e_j}{\Delta e_{j+1}-\Delta e_j}\right)\left(a_1^{i+1}\cdot e(kT)+a_2^{j+1}\Delta e(kT)\right) \tag{4.2.5}$$

In our simulation l was set to 3, and after the determination of the parameters of the three triangular membership function on $e(kT)$ and $\Delta e(kT)$ the 18 parameters of the fuzzy rules (nine a_1^i and nine a_2^j) were determined by optimization. So the performance index becomes 1.579.

4.3. Applying Takagi-Sugeno Fuzzy supervised PI-controller

The structure of the controller was described in 3.4 section, where l was set to 3.
After the determination of the parameters of the three triangular membership function on $e(kT)$, the 9 parameters of the fuzzy rules (three K_p^{ci} and three K_I^{ci} and three b^i) were set by optimization.

The performance of this algorithm with optimal parameters is shown by Figure 7., and the performance index (Q) was 1.509.

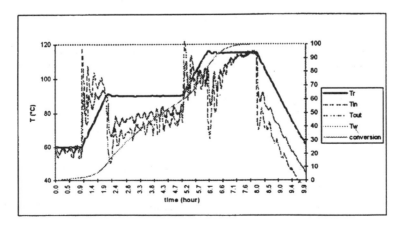

Figure 7. Performance of the optimal Fuzzy supervised PI -controller

428

5. Conclusion

In this study the batch polymerization reactors temperature control by Takagi-Sugeno fuzzy controller through the Expanded Polystyrene (EPS) process was investigated.

- In order to solve the problem the simulation program of the technology in MATLAB Simulink based on rigorous chemical engineering model was developed.
- The parameters of the optimal PI- and the gain varying PI- controllers according to a given cost function were determined .
- We established an optimal PI-type Takagi-Sugeno fuzzy controller,
- and the realisation (mimic) of the adaptive PI controller with the help of ANFIS [21] algorithm,
- as well the realisation and improvement by fuzzyfication of the adaptation algorithm.

As the figure bellow demonstrating the result and the derivation of algorithms show that we achieved 5-20% improvement of the performance index characterising the operation of the controller by fuzzy realization of the controllers.

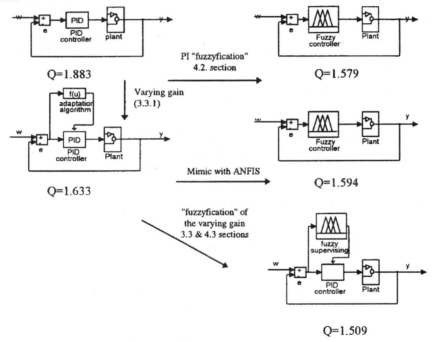

Figure 8. Scheme of the control algorithms

According to our experiences the conventional control solutions (PID algorithms, adaptation) can be realized by choosing the proper fuzzy structure (non equivalent connection) and can be improved by applying Takagi-Sugeno fuzzy models.

This type of applications can be a proper tool where the already existing linear control algorithm is to be replaced by a better, more effective nonlinear control algorithm.

References

[1] Chylla, R.W., Haase, R.W., 1993, Temperature Control of Semibatch Polymerisation Reactors, *Computers Chem. Eng*, **17**, 257-264

[2] Juba, M.R., Hamer, J.W., 1986, Progress and challenges in batch process control, *Proc. of CPC III,CHACE* Amsterdam, Netherlands.

[3] Berber, R.,1996, Control of Batch Reactors: Review, *Trans IChemE,* **74 Part A,** 3-20

[4] Davdison, R. S., 1989, Using Process Information to Control a Multipurpose Batch Chemical Reactor, *Computers chem. Engng,* **13,** 83-85

[5] Engell, S.,Hacketnhaler, T., 1995, Fuzzy control - An alternative to model based control ? *Methods on Model Based Process Control,* Kluwer Academic Publishers, Dordecht.

[6] King, P. J., Mamdani, E. H., 1977, The application of fuzzy control system to industrial processes, *Automatica,.* **13,** 235-242

[7] Filev, D. P., Yager, R.R., 1994, On the analysis of fuzzy logic controllers, *Fuzzy sets and System,* **68,** 39-66

[8] Sugeno, M., 1995, An Introductory Survey of Fuzzy Control, *Information Sciences,* **36,** 59-83

[9] Villalobos, M.A., Hamielec, A.E., Wood, P.E., 1993, Bulk and Suspension Polymerization in the Presence of n-pentane, *Journal of Applied Polymer Science,* **50,** 327-343

[10] Tanaka, K., Sugeno, M, 1992, Stability analysis and design of fuzzy control systems, *Fuzzy Sets and Systems,.* **42,** 135-156

[11] Driankov, D., Hellendom, H., 1993, *An Introduction to Fuzzy Control,* Springer-Verlag, London.

[12] Qiao W.Z, Mizumoto, M, 1996, PID type fuzzy controller and parameters adaptive method, *Fuzzy Sets and System,* **78,** 23-35

[13] Mizumoto, M, 1995, Realization of PID controls by fuzzy control methods, *Fuzzy Sets and Systems,* **70,** 171-182

[14] Mirsi, D, Malki, H.A., Chen, G., 1996, Design and analysis of a fuzzy proportional-integral-derivative controller, *Fuzzy Sets and Systems,* **79,** 297-314

[15] Äström, K., Wittenmark, B., 1990, *Computer Controlled Systems: Theory and Design,* Prentice-Hall International, Inc., London.

[16] Marroquin, G, Luyben, W.L, 1972, Experimental evaluation of nonlinear cascede controllers for batch reactors, *Ind. Eng. Chem. Fundam,* **11,** 552-556

[17] Jutan, A, Uppal, A.L, 1984, Combined feedforward-feedback servo control scheme tor an exothermic batch reactor, *Chem. Proc. Res. Dev.,* **23,** 597-602

[18] Cawthon, G.D., Knaebel, K.S., 1989, Optimalization of Semibatch Polymerisation Reactors. *Computers chem. Engng,* **13,** 63-72

[19] Nemeth, S. , Thyrion, F. C., 1985, Study of the Runaway Characteristics of Suspension Polymerisation of Styrene. *Chem. Eng. Technol,.* **18,** 315-323

[20] Choi, K.Y., Lei G.D., 1987, Modeling of Free-Radical Polymerisation of Styrene by Bifunctional Iniciators. *AIChE Journal.* **33,** 2067-2076

[21a] Gulley, N., Jang, J. -S. R., 1995, *Fuzzy Logic Toolbox User's Guide,* The Math Works Inc. Massachusetts.

[21b] Jang, J.-S.R., 1993, ANFIS: Adaptive-Network-based Fuzzy Inference Systems. *IEEE Transactions on Systems, Man and Cybernetics.* 23(3), 665-685

[22] Grace, A., 1992, *Optimization Toolbox User's Guide,* The Math Works Inc. Massachusetts.

Genetic Algorithms and Evolution Strategies Applied in Identification and Control: Case Study

L. S. Coelho[1], A. A. R. Coelho[2]

Federal University of Santa Catarina - Department of Automation and Systems
C.P. 476 - 88040.900 - Florianópolis - SC - BRAZIL
E-mail: [1] *lscoelho@lcmi.ufsc.br;* [2] *aarc@lcmi.ufsc.br*

Keywords: Genetic algorithms, Evolution strategies, Process identification, Process control, Practical application.

Abstract

In this paper the computational intelligence paradigm called evolutionary computation in process identification and control is utilized. The following methodologies are addressed: i) genetic algorithms, ii) hybrid algorithms composed by genetic algorithms with simulated annealing, and iii) evolution strategies. Experiments in identification were conducted in mono-tank level and temperature processes. Experimental control tests were evaluated in a non-linear level process, composed of coupled twin-tanks, which was submitted to reference change and load disturbance. In this control implementation, the evolutionary computation are utilized for tuning of the design parameters of a monovariable PID controller.

1. Introduction

Some non-conventional approaches in control, such as auto-tuning, predictive, and intelligent control have been proposed in literature [1], [2], [3]. This paper addresses the description, design and implementation of the computational intelligence methodologies in process control, with the purpose to deal with the restrictions and requirements in control system design. The computational intelligence (CI) paradigm has increased the robustness, flexibility and capacity of the algorithms in industrial process identification and control by incorporating notions form of the computer science, operational research and conventional control theory. The several efforts and recent developments by CI researches in intelligent control have presented good results.

The field of the intelligent control has embraced encourages by private and governmental investiments in the area, the field of intelligent control has come to embrace techniques, such as fuzzy logic, artificial neural networks, fuzzy neural networks, evolutionary computation, simulated annealing, pattern recognition, expert systems, intelligent hybrid systems, and chaos theory [1], [2]. In this paper, one of the CI paradigms called evolutionary computation (EC) is utilized. EC paradigm finds its main application area in optimization, design, scheduling and routing problems, being particularly significant in aerospace and automotive industry, telecommunications and chemical engineering [3], [4]. Several papers have provided overviews and surveys of EC and it is possible to find the features that are particularly appropriate for control engineering applications in academia and industry [3], [4], [5], [6], [7], [8].

The purpose of this paper is to utilize the EC in process identification and control. The following methodologies are addressed: i) genetic algorithms, ii) hybrid algorithms composed by genetic algorithms with simulated annealing, and iii) evolution strategies [7]. The paper is organized as follows. In section 2, the evolutionary computation paradigms called genetic algorithms and evolution strategies are described. In section 3, the description of mono-tank level and temperature processes, and the experimental results in identification these processes are shown and analysed. The description of a non-linear level process, composed of coupled twin-tanks and simulation results are treated in section 4, and the conclusions are presented in section 5.

2. Evolutionary Computation

The EC constitutes a goal option to be used instead of conventional methodologies in search and optimization. EC imitates a rudimentary and simplified model of nature as an adaptive procedure of search and optimization, which enables computational implementations. The EC suggests a mechanism in which a population P(t) of individuals (solutions) intends to improve, in average, the general performance with respect to a given problem.

The EC includes a crescent number of the methodologies, from which the more importance are: genetic algorithms, evolution strategies, evolutionary programming, genetic programming, and classifier systems [9], [10]. In Figure 1 a basic pseudocode of the EC methodologies is shown.

```
generation ← 0
initialization (P(t))
evaluate of the population fitness (P(t))
while the stop criterion of P(t) is not obtained
{        generation ← generation + 1
         P(t) = selection (P(t-1))
         crossover (P(t))
         mutation (P(t))
         evaluate of the population fitness (P(t))  }
```

Figure 1. Basic pseudocode of the EC methodologies

2.1. Genetic Algorithms

The genetic algorithms (GA's) are global optimization techniques based on genetics and natural selection mechanisms and are characterized by being robust and powerful in complex and irregular search space. The aptitude of the individual to the environment is obtained according to the evaluation of an objective function which utilizes random selection, crossover and chromosomes mutation. A GA is based on following steps:

i) initialize a population of chromosomes (solutions);
ii) generate the phenotype and evaluate the fitness for each population chromosome;
iii) select the chromosomes according to the selection strategy;
iv) realize the crossover and mutation operations;
v) originate a new population;
vi) repeat the steps (ii) up to (v) so that the stop condition is satisfied.

In this paper the floating point representation of the chromosomes is used because it is more adequate for the treatment of restrictions problems and it presents more precision and efficiency in computing time when compared with the binary representation. The evolutionary operators utilized in this work are non-uniform mutation and arithmetic crossover proposed by Michalewicz [11].

2.2. Hybrid GA's with Simulated Annealing

The simulated annealing algorithm (SA) is a variation of hill-climbing algorithms where the objective is cost function minimization (energy level) and based on the physical process in which a substance is taken to a higher level of energy and subsequently it is gradually cooled down until it reaches a lower state of energy. The GA's are robust in applications where the global searches are adequated for treatment of the problem in question. Meanwhile, for local searches generally do not present an adequate behavior. GA's can become more efficient if a fine adjust for local searches through the SA algorithm is down. The SA procedure is applied to GA (GASA) in every generation after the crossover and mutation operations through a little random perturbation (low temperature factor and a fast annealing factor) on the best individuals in the population [12], [13].

2.3. Evolution Strategies

The evolution strategies (ES's) were developed for solving optimization technical problems in engineering and, nowadays, constitute relevant computational algorithms in parameter optimization problems. The ES's are techniques which emphasize the behavior connection between the populations generated, more than that of genetic association. The first ES to be developed was the ES-(1+1), at the Technical University of Berlin, by I. Rechenberg and H.-P. Schwefel [14]. The ES's with multimembers are divided into ES-$(\mu+\lambda)$ which suggest that μ fathers produce λ offsprings, after the μ fathers and the λ offsprings compete for survival, on the contrary of the ES-(μ,λ) which suggest λ offsprings compete for survival and the μ fathers are all replaced in every generation [15]. In this work the ES-$(\mu+\lambda)$ and ES-(μ,λ) are addressed. The two employed variants of these ES's utilize the mutation operator with: i) individual step size control, which enable the treatment of the problems inadequately scheduled with relation to a bigger sensibility of the inherent parameters of mutation

432

operator, and ii) correlated variations and parameter adaptation scheme, where the algorithm is complex, robust, and appropriate for effective of the correlated mutations in individual phenotype level [16], [17].

The ES's consist of techniques with similar features of the GA's. However, it adopts biological characteristics of the poligeny and pleiotropy. The differences between ES's and GA's is that the ES's uses the mutation operator as main operator, it allows self-adaptation characteristics through standard deviation and covariances, and it employs deterministic (extintive) selection. In the usual implementation of the GA's, the main element is the crossover operator, which does not present mechanisms for auto-adaptation and has probabilistic (preservative) selection [7], [10], [15]. In Table 1 the EC paradigms utilized in process identification and control are shown.

Table 1. EC paradigms utilized in process identification and control

no.	techniques (population)	d	selection	crossover	mutation
1	GA (30)	-	roulette wheel	[2]$p_r = 0.8$	[1]$p_m = 0.1$
2	GASA (30)	-	roulette wheel	[2]$p_r = 0.8$	[1]$p_m = 0.1$
3	GASA (30)	-	breeder	[2]$p_r = 0.8$	[1]$p_m = 0.1$
4	ES-(1+1)	1	-	-	yes
5	ES-(5+25)	1	-	-	yes
6	ES-(5+25)	1	-	[3]	yes
7	ES-(1,29)[4]	1	-	-	correlated
8	ES-(1+29)[5]	1	-	-	correlated
conventions					
[1]	non-uniform mutation with b=2				
[2]	arithmetic crossover				
[3]	discrete crossover (x) and intermediary arithmetic (σ)				
[4]	individual step size control				
[5]	correlated variations and parameters adaptation scheme				

3. Experiments in Process Identification

Theoretical and practical aspects for system identification have received great attention in the last decades and the literature presents a large number of conventional and advanced techniques for industrial process modeling. In this paper the estimation algorithm based on EC is evaluated in two processes: mono-tank level and temperature [17]. The parametric mathematical model utilized in the parameter estimation has the following expression:

$$A(z^{-1})y(t) = B(z^{-1})u(t-d) + v(t) \tag{1}$$

where

$$A(z^{-1}) = 1 + a_1 z^{-1} + a_2 z^{-2} + \ldots + a_{na} z^{-na} \tag{2}$$

$$B(z^{-1}) = b_1 z^{-1} + b_2 z^{-2} + b_3 z^{-3} + \ldots + b_{nb} z^{-nb} \tag{3}$$

and y(t), u(t), d and v(t) are output, input, time delay and noise, respectively. The noise is a normal random number sequence with zero mean and unitary variance. The objective of EC methodologies is to achieve the parameters of the $A(z^{-1})$ and $B(z^{-1})$ polynomials. The validation of the estimated discrete models is based on the comparative study between real response of the process and identified response. The estimated parameters of the models are adequate if the error criterions applied in the estimation provide minimum values for each experimentation. The error criterions considered are:

i) Integral Square-Error (ISE)

$$ISE = \sum_{k=1}^{N} [y(k) - \hat{y}(k)]^2 \tag{4}$$

and, ii) Relative Prediction-Error Criterion (RPE)

$$RPE = \frac{\sum_{k=1}^{N} [y(k) - \hat{y}(k)]^2}{\sum_{k=1}^{N} [y(k)]^2} \tag{5}$$

where y(k) is the measured output, ŷ(k) is the estimated output and N is the number of samples acquired in each experiment (stored in data files). The evaluation function or fitness F, dopted for process identification and control is defined as

$$F(ISE) = \frac{\alpha}{1 + ISE} \qquad (6)$$

where α is the factor scaled and it is adjusted in $\alpha = 10^6$. Two control systems are utilized to evaluate the EC algorithms were designed in the Department of Automation and Systems (UFSC/DAS/LCP). The mono-tank level and temperature processes are present in many industrial plants. Further information of the processes are available in Coelho [18] and http://lcp.lcmi.ufsc.br.

3.1. Mono-Tank Level Process

The schematic diagram of the level process is shown in Figure 2. The experimental prototype of the level system consists of a 29cmx12cmx28cm rectangular glass tank and with a 10 liters capacity. At the bottom of the process there is a reservoir capable of holding approximatelly 10 liters of the fluid. The water is pumped into the tank from the reservoir by a small DC electric pump. The desired level is measured by a potentiometer/float attached to the top of the tank. A change in the resistance of the level sensor is converted to an analog voltage by a Wheatstone bridge with zero adjustment. An operational amplifier, implemented in the gain configuration (span adjust), is utilized to calibrate the level from 0 to +5 volts (level range from 5 cm to 18 cm). The level plant is submitted to an input and, through a data aquisition board connected to the computer, 1500 samples of the process signals are collected. The sampling period is one second.

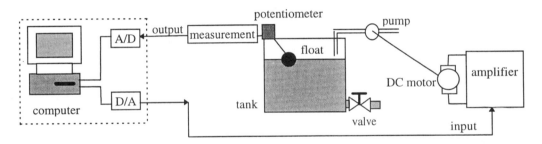

Figure 2. Mono-Tank Level Process

3.2. Temperature Process

The experimental diagram for the temperature process is shown in Figure 3. The temperature prototype, in laboratorial scale, consists of a rectangular metalic tank with dimension 15cm x 13cm x 23cm and capacity of 4.5 liters. There are also the following components: a temperature sensor (PT-100) connected to a conditioning circuit (to provide scaling and amplification of the sensor signal) to operate in the range from 0 to +5 volts (to measure temperatures from 25°C to 90°C); and an industrial electric resistance of 750 watts. The plant is submitted to an input and 300 samples of the process are collected. The sampling time is 5 seconds.

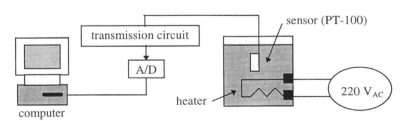

Figure 3. Temperature Process.

3.3. Experimental Results

The EC methodologies applied in process identification have the configuration presented in Figure 4.

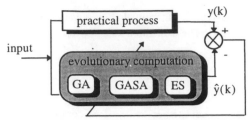

Figure 4. Configuration in process identification through EC methodologies

Table 2 presents EC paradigms, refering to the first and second order modeling (linearization) for the mono-tank level process. Figure 5 shows the evolution of ISE in 10 experiments (average) for mono-tank level process identification.

Table 2. Mono-tank level process identification by EC methodologies

no.	na	nb	a_1	a_2	b_1	b_2	ISE	RPE
1	1	1	-0.9958	-	0.0039	-	15.2700	5.83×10^{-4}
2	1	2	-0.9959	-	0.0882	-0.0843	14.2665	5.45×10^{-4}
3	1	2	-0.9957	-	-0.3115	0.3155	14.3508	5.48×10^{-4}
4	2	2	-0.5019	-0.4919	-0.2013	0.2071	16.0045	6.11×10^{-4}
5	*1*	*2*	*-0.9958*	*-*	*0.8305*	*-0.8266*	*14.1431*	*5.40×10^{-4}*
6	1	2	-0.9958	-	-0.1732	0.1771	14.1528	5.41×10^{-4}
7	1	2	-0.9958	-	0.5878	-0.5839	14.3694	5.49×10^{-4}
8	1	2	-0.9958	-	0.5878	-0.5839	14.3694	5.49×10^{-4}

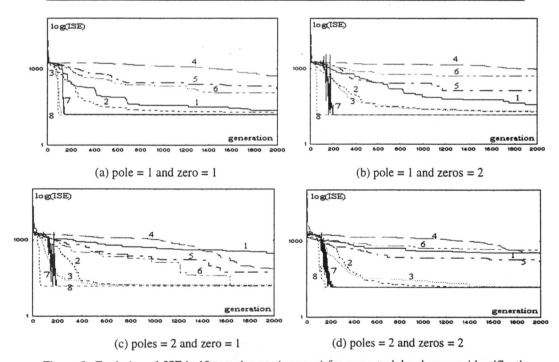

(a) pole = 1 and zero = 1

(b) pole = 1 and zeros = 2

(c) poles = 2 and zero = 1

(d) poles = 2 and zeros = 2

Figure 5. Evolution of ISE in 10 experiments (average) for mono-tank level process identification

The analysis of Table 2 suggests that best result was acquired with ES-(5+25) algorithm in mono-tank level process identification for one pole (a_1=-0.9958) and two zeros (b_1=0.8305, b_2=-0.8266). The measured and estimated responses of mono-tank level process in open-loop by ES-(5+25) is shown in Figure 6.

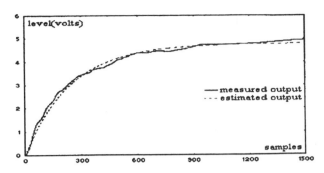

Figure 6. Measured and estimated responses of mono-tank level process in open-loop by ES-(5+25)

Table 3 presents EC paradigms, refering to the first and second order modeling (linearization) for the temperature process with and without time delay estimation.

Table 3. Temperature process identification by EC methodologies

(a) without time delay estimation

no.	na	nb	a_1	a_2	b_1	b_2	ISE	RPE
1	1	1	-0.9957	-	0.0053	-	9.2633	1.84×10^{-3}
2	1	1	-0.9959	-	0.0052	-	9.2374	1.83×10^{-3}
3	1	1	-0.9958	-	0.0052	-	9.0960	1.80×10^{-3}
4	1	1	-0.9959	-	0.0052	-	9.4590	1.88×10^{-3}
5	1	1	-0.9958	-	0.0052	-	9.0960	1.80×10^{-3}
6	1	1	-0.9958	-	0.0052	-	9.0960	1.80×10^{-3}
7	1	1	-0.9958	-	0.0052	-	9.0960	1.80×10^{-3}
8	1	1	-0.9958	-	0.0052	-	9.0960	1.80×10^{-3}

(b) with time delay estimation

no.	na	nb	d	a_1	a_2	b_1	b_2	ISE	RPE
1	2	2	9	-0.1732	-0.8230	-1.861	1.8701	114.202	2.27×10^{-2}
2	1	1	3	-0.9949	-	0.0059	-	3.9704	7.88×10^{-4}
3	1	1	7	-0.9952	-	0.0058	-	3.9533	7.84×10^{-4}
4	1	1	2	-0.9956	-	0.0052	-	9.0182	1.79×10^{-3}
5	1	2	2	-0.9959	-	0.1623	-0.1599	10.4556	2.07×10^{-3}
6	1	1	3	-0.9950	-	0.1501	-0.1456	12.0432	2.39×10^{-3}
7	1	1	2	-0.9950	-	0.0059	-	3.9561	7.85×10^{-4}
8	1	1	2	-0.9950	-	0.0058	-	3.9200	7.78×10^{-4}

The analysis of Table 3 suggests that best result was acquired with ES-(1+29) algorithm with correlated variations and parameters adaptation scheme in mono-tank level process identification for one pole ($a_1 = -0.9950$) and one zero ($b_1 = 0.0058$). The measured and estimated responses of temperature process in open-loop by ES-(1+29) is shown in Figure 7.

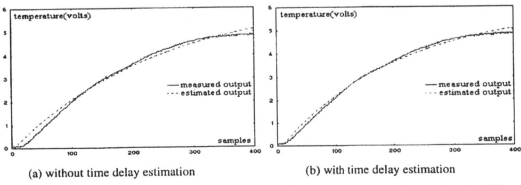

(a) without time delay estimation　　　　　(b) with time delay estimation

Figure 7. Measured and estimated responses of temperature process in open-loop by EC

436

4. Experiments in Process Control

In this section the description and the main features of the twin-tanks level process utilized and the tuning evaluation of intelligent PID controller through EC (PID-EC) are treated. The system of the non-linear level regulation is composed by two coupled tanks and it is illustrated in Figure 8 (proposed by Li & Ng [19]). The equations which characterize the open-loop dynamic are

$$A_1\dot{h}_1 = u - a_1c_1\sqrt{2g(h_1 - h_2)} \tag{7}$$

$$A_2\dot{h}_2 = a_1c_1\sqrt{2g(h_1 - h_2)} - a_2c_2\sqrt{2g(h_2 - h_0)} \tag{8}$$

$$y = h_2(t - \tau) \tag{9}$$

where the tanks area are $A_1 = A_2 = 97$ cm^2; the apertures area are $a_1 = 0.396$ cm^2 and $a_2 = 0.395$ cm^2; the constant discharge are $c_1 = 0.53$ and $c_2 = 0.63$; the apertures height is $h_0 = 3$ cm, and the gravity acceleration is $g = 981$ cm/s^2. The time delay included in the system is $\tau = 6$ units of time [13], [19].

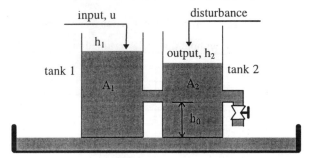

Figure 8. Level regulation system composed by twin coupled tanks

The experimental tests are conducted to analyse the stability of the techniques in parameters tuning of PID control [20]. The process is submitted to reference change in the controller tuning phase, and it is subject to load disturbance in the test phase of the robustness of the control algorithms. The cost function J(u), for minimization, is given by equation (10) with w=0.1 adopted, where

$$J(u) = \sum_{k=1}^{N} k(e(k)^2 + w\Delta u(k)^2) \tag{10}$$

The experiments realized in the non-linear system of level regulation aim to maintain the liquid level of the tank 2 during 600 iterations, in each one of the three references proposed of 9 cm, 12 cm and 5 cm, respectively, with a minimum overshoot and zero steady-state error. In respct of the controller design with intelligent configuration it must be mentioned that the controllers are configured and optimized only for reference change not for load disturbance. The liquid input control signal of the tank 1 is restricted to a range of [0;33.3 cm^3/s] and the output of the plant is calculated by the Runge-Kutta of the 4^a order method [7], [19]. In Table 1 the EC algorithms utilized in tuning of conventional and intelligent PID control are shown. Figure 9 presents the PID-EC control configuration.

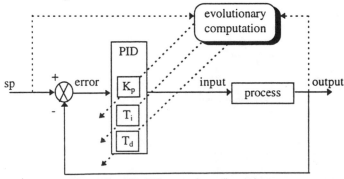

Figure 9. PID-EC control configuration.

Load disturbance about 25% of the maximum control (8.325 cm³/s) are applied in tank 2, at time instant 300 to 600, 900 to 1200, and 1500 to 1800, after the controller design (tuning phase), for analysis of the robustness, sensibility, and actuation of PID-EC controller for situation not foreseen in the design of the controller. The best result obtained by PID-EC, analysing GA, GASA and ES paradigms, is given by ES-(1+29) with correlated variations and parameters adaptation scheme. The obtained tuning parameters are K_p=16.68, T_i=0.32, T_d=0.02, and J(u)=127104 with 20 iterations (generations) and are presented in Table 4.

Table 4. Best result obtained for the PID-EC with ES-(1+29)

ES-(1+29) with correlated variations and parameters adaptation scheme					
ts_1	150	ts_2	309	ts_3	152
tp_1	162	tp_2	323	tp_3	164
o_1	0.52	o_2	0 12	o_3	117
Δu_1	114.4	Δu_2	4.17	Δu_3	109.4
conventions					
ts_k	rise time				
tp_k	peak time				
$o_k\%$	percentage of overshoot				
Δu_k	control signal change				
k = (1,2,3)	k=1 (0 to 600 samples); k=2 (600 to 1200 samples); k=3 (1200 to 1800 samples)				

Figure 10 presents the closed-loop response of PID-EC controller more adequate in the end of the tuning phase (ES-(1+29)), obtained subjected to reference change and load disturbance (test phase).

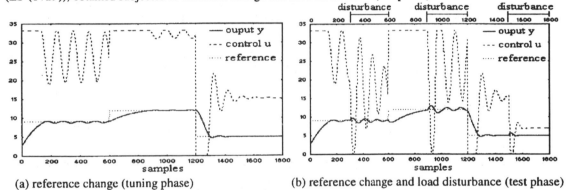

(a) reference change (tuning phase)　　　　　(b) reference change and load disturbance (test phase)

Figure 10. Output and control for reference change via PID-EC.

5. Conclusions

In this paper the computational intelligence paradigm called evolutionary computation in process identification and control is utilized. The following methodologies are addressed: i) genetic algorithms, ii) hybrid algorithms composed by genetic algorithms with simulated annealing, and iii) evolution strategies. Experiments in identification were conducted in mono-tank level and temperature processes. The experimental results showed that the EC techniques was appropriate in process identification. Experimental control tests were conducted in a non-linear level process, composed of coupled twin-tanks, which was submitted to reference change and load disturbance. The simulation results showed that the tuning of the PID controller via GA, GASA, and ES provided an appropriate performance for the controlled process, being an alternative to the conventional tuning techniques for the controllers. The drawbacks of the EC is that each algorithm generation, all population of the solutions (controller parameters) need to be evaluated for fitness function calculation. The ES methodology with correlated mutations was the one which presented the best performance in convergency speed, due to its self-adaptation features and second level learning in process identification and control. However, this technique presents computational complexity similar to the other evolutionary methods, notwithstanding with greatest algorithmic complexity and improved mathematical base [10], [15], [16], [17].

The obtained results open perspectives for studies that incompass the analysis of ES with SA hybrid, evolutionary self-adaptation mechanisms of the control parameters in GA, the treatment of the applications in linearization, identification and tuning of the controllers so that they can be applied in monovariable and multivariable processes. However it is required the analysis of stability, convergence and robustness of these design techniques, which should be further investigated in future researches. Other possibilities are techniques combining EC with paradigms of the computational intelligence, such as fuzzy logic, artificial neural networks, pattern recognition and expert systems [1], [2], [5], [6], [11]. In spite of the high computational complexity, the IC algorithms behaved adequately to aid in the configuration in process identification, and design and tuning in process control. So, they offer perspectives of combining IC, self-tuning and adaptive approaches for development of an efficient hybrid system for application in complex industrial processes.

References

[1] Levine, W. S. (ed.), 1996, *The Control Handbook*. IEEE Press and CRC Press, Inc.

[2] Katayama, R., Kajitani, Y., Kuwata, K., and Nishida, Y., 1993, Developing Tools and Methods for Applications for Incorporating Neuro, Fuzzy and Chaos Technology, *Computer Ind. Engng.*, 24(4), pp.579-592.

[3] RayChaudhuri, T., Hamey, L. G. C., and Bell, R. D., 1996, From Conventional Control to Autonomous Intelligent Methods, *IEEE Control Systems Magazine*, vol.16, no.5, pp. 78-84.

[4] EvoNews, 1996, *Newsletter of the Evonet Network of Excellence in Evolutionary Computation*, European Commision's ESPRIT IV Programme, Issue 2.

[5] Chipperfiled, A., and Fleming, P., 1996, Evolutionary Algorithms for Control Engineering, *13th World Congress of IFAC*, San Francisco, USA, pp. 181-186.

[6] Alander, J. T., 1995, *An Indexed Bibliography of Genetic Algorithms in Control*, Report 94-1, Univ. of Vaasa, Dep. of Information Technology and Production Economics, Finland.

[7] Coelho, L. S., 1997, *Computational Intelligence Methodologies in Process Identification and Control: Fuzzy, Evolutionary, and Neural Approaches*, Msc Thesis in Computer Science, Federal University of Santa Catarina, Florianópolis, SC, Brazil (in Portuguese).

[8] Varsek, A., Urbanic, T., and Filipic, B., 1993, Genetic Algorithms in Controller Design and Tuning, *IEEE Trans. Syst. Man and Cybernetics*, vol. 23, no. 5, pp. 1330-1339.

[9] Goldberg, D. E., 1989, *Genetic Algorithms in Search, Optimization, and Machine Learning*. Addison-Wesley Publishing Company.

[10] Soucek, B., and The IRIS Group, 1992, *Dynamic, Genetic, and Chaotic Programming: the Sixth Generation*. John Wiley & Sons, Inc.

[11] Michalewicz, Z., 1992, *Genetic Algorithms + Data Structures = Evolution. Programs*, Springer-Verlag.

[12] Ghoshray, S., Yen, K. K., and Andrian, J., 1995, Modified Genetic Algorithms by Efficient Unification with Simulated Annealing, *Int. Conf. on Artificial Neural Networks and Genetic Algorithms*, Alés, France, pp. 487-490.

[13] Li, Y., Tan, K. C., and Ng, K. C., 1995, *Performance Based Linear Control System Design by Genetic Evolution with Simulated Annealing*, CSC 95017, Centre for Systems and Control, Univ. of Glasgow, U.K.

[14] Rechenberg, I., 1973, *Evolutionsstrategie Optimierung technischer Systeme nach Prinzipien der biologischen Evolution*, Fromman-Holzboog, Stuttgart.

[15] Bäck, T., and Schwefel, H. -P., 1993, An Overview of Evolutionary Algorithms for Parameter Optimization, *Evolutionary Computation*, 1(1), pp. 1-23.

[16] Ostermeier, A., Gawelczyk, A., and Hansen, N., 1994, A Derandomized Approach to Self-Adaptation of Evolution Strategies, *Evolutionary Computation*, 2(4), pp. 369-380.

[17] Rudolph, G., 1992, On Correlated Mutations in Evolution Strategies, in R. Männer and B. Manderick (eds.), *Proc. Parallel Problem Solving from Nature 2*, Amsterdam: North-Holland, pp. 105-114.

[18] Coelho, A. A. R., 1995, Laboratory Experiments for Education in Process Control, *Workshop on Control Education and Technology Transfer Issues*, Curitiba, PR, Brazil, pp. 133-138.

[19] Li, Y., and Ng, N. G., 1995, *Uniform Approach to Model-Based Fuzzy Control System Design and Structural Optimization*, CSC 95007, Centre for Systems and Control, Univ. of Glasgow, U.K.

[20] Åström, K. J., and Hägglund, T., 1996, PID Control, in Levine, W. S. [1], pp. 198-209.

Part 9: Summary of Discussions

Summary of Discussions

The papers presented in the eight parts of this book comprise of the papers presented in the 2nd On-line World Conference on Soft Computing in Engineering Design and Manufacturing (WSC2). This part summarises the discussions about the papers during the conference. The summary is divided into eight sections according to the eight sessions in the conference, and most of the sections have been prepared by the session facilitators. The name of the person who has prepared the summary is given at the beginning of each section.

1. Evolutionary Computing

Prepared by: Ivanoe De Falco, Italy

This session concerned Evolutionary Computing. It consisted of six papers, plus adequate links to fifteen more papers which were actually contained in other sessions, yet had some interest for this session too. As facilitator of this session I have to say that discussion was really interesting in quality and noticeable in quantity: a very high number of questions and answers were posted by email. This, on the one hand, showed the success of the WSC2 concept, and on the other hand suggested that the papers in the session were of high technical relevance.

The paper, "Fitness Causes Bloat" by W. B. Langdon and R. Poli, raised lot of interest during the discussion. This paper also obtained the Second Prize in the Best Paper Award. Among the questions, a very interesting one was raised by Bentley (University College of London, UK) who wondered whether there is a feature of natural evolution that introduces an indirect parsimony pressure to slow or stop the growth of genetic material. Further, he asked the authors how much the degree of bloat varies in living creatures, and why. Langdon's answer was 'yes', there is some incentive in nature to keep the genetic material as short as possible (for example, bacteria which reproduce rapidly have shorter chromosomes, and viruses even less). Moreover, he pointed out that objective function in natural evolution is far from static, and that since each DNA base requires energy there is some incentive to keep the genetic material as short as possible. Gordon Robinson (University of Southampton, UK) reported that almost 100% of the DNA of bacteria and viruses is expressed, whereas in eukaryotes the majority of DNA is not, and there are huge differences in genome length which are not reflected in the apparent complexity of the organism. He guessed that this could indicate the presence of strong evolutionary pressure on genome length. He also wondered to what extent the differences in the selection pressures on genome size are due to differences in rates of reproduction between prokaryotes and eukaryotes. His opinion was that most evolutionary algorithms are more like in the eukaryotic case in that the cost in CPU time and memory of maintaining the additional genetic material will usually be small relative to the cost of evaluating the objective function. He finally raised the question of whether to encourage or suppress the bloat. As a response, the authors reported the argument of accumulation of neutral changes to justify that some form of redundancy is in practice necessary to evolution. However, they also added, in simple forms such redundancy appears to make subsequent evolution more difficult. Then they introduced the question of whether multi-cellular organisms' genomes are still increasing in size, or if they have reached some form of equilibrium. Bentley again asked whether entities that must constantly keep changing and updating their DNA over few generations require shorter DNA. He argued that if there is so little bloat in viruses perhaps there is a real advantage in minimising bloat. GAs would probably benefit from this idea. De Falco (IRSIP-CNR, Italy) asked whether there is some relation between bloat and complexity defined according to Kolmogorov. Langdon's answer was affirmative, and he added that bloat highlights a difficulty, since their work involves the generation of huge programs with the same fitness and identical behaviour. De Falco again suggested the idea of giving different fitness values to programs showing different lengths, even if they do the same things; this is an idea coming out from Kolmogorov Complexity; he provided the authors with a web address where a paper by himself and his colleagues about Genetic Programming, Kolmogorov Complexity and Program Fitness depending on Length can be found. Mario Koeppen (Fraunhofer IPK, Germany) wondered whether the authors' hypothesis, that the number of representations with equal fitness increases with the length, can be estimated, and what would happen if a No-operation node was allowed in the search. The authors mentioned that, statistically, exact methods are impractical. A biased estimate could be obtained by randomly generating programs from ones for which one already knows the fitness, and looking at the distribution of lengths and fitness. This is biased since it depends on the starting point chosen. Another feasible approach is to use a Genetic Programming population, which introduces multiple biases. At this point Koeppen suggested that an additional reason for bloat might be the maximum number of incoming connections to a node. He has asked whether there is a difference in bloat, if one uses only node functions with at most three inputs, rather than four. He said that in later stages of the run the probability of getting nodes with higher inputs increases, hence the trees are getting a more complex structure. Even though this is not bloat, it could, in his opinion, enhance the effects of bloating. The authors' reply has been that they do not remember any explicit study between function set arity and bloat, though they agree on the comment. Finally, James Foster (University of Idaho, USA) affirmed that there is no correlation between complexity of organism and size of genome, and that there is no correlation between complexity of organisms and the amount of expressed genetic material (he has reported that this is called in biology the C value paradox). He also claimed for parsimony pressure in some organisms like Prokaryotes. For instance, viruses copy their genetic material quickly, and so they have no introns, further they show some form of data compression.

The paper "An Overview of Evolutionary Computing for Multimodal Function Optimisation" (by R. Roy and I. C. Parmee) provoked an interesting question from Bill Langdon, who noticed that the technique proposed by authors, i.e. ARTS, tends to favour crossover between different niches. Since it has been supposed that children generated by recombination between very different parents are likely to have a poor fitness ("lethals"), Langdon asked whether this has been found true by the authors in their experiments. Their answer was that this takes place in the early generations, where some very weak individuals appear, but as the run progresses, such individuals reduce more and more.

As regards the paper "Evolution of Cellular-automaton Based Associative Memories" (by M. Chady and R. Poli), De Falco remarked that the problem cited therein, i.e. the pattern association, can be solved also by Hopfield Networks, so he was curious as to whether the results obtained by authors with Cellular Automata are competitive with those obtained by ANNs. The answer was that, as regards memory capacity, CA performs better than Hopfield networks on the problem, the difference being that Hopfield tested his network with completely random patterns, while they have used a fixed set of training patterns. As regards the search time, ANNs are faster than CA which needs to run a lengthy GA to find the required weights. They are not faster in retrieving the stored patterns, since they are recurrent, so they need some time to relax into some stable state after an input pattern has been presented. One more question by De Falco was about the termination condition for the algorithms: why stop the search as soon as, after generation 50, in one generation both the best and the average fitness values do not improve? Could not this simply mean that a local optimum has been found? The answer has been that this comes from preliminary experiments, which showed that population tends to converge long before generation 50 and further search rarely produced significant improvements.

Also the paper "Simple implementation of Genetic Programming by Column Tables" (by V. Kvasnicka and Jiri Posphical) received several questions: De Falco asked about application of the proposed method to real problems, and about results compared to classical GP methods. The authors said they had used the method on the real problem of finding correlation of data describing chemical compounds against their anti AIDS activity. The results have been not better than those found by multilinear (and partly polinomial) regression. As regards convergence times, the authors do not expect the column tables to converge to a better solution faster than classical GP for simple model tasks, though they expect this to take place for complicated functions. R. Poli (University of Birmingham, UK) pointed out that the basic idea is not new, since he has been working on something similar during the last two years, designing the Parallel Distributed GP, which can evolve graph-like programs. He provided the authors with web reference to the paper. The authors, after looking at it, replied that they have been stimulated, just like Poli, by feed-forward neural networks. However, the two approaches are not really the same, since Poli's is aimed on hardware, it is not really general, since edges can exist only between adjacent layers. They claimed column tables are useful for coding acyclic oriented graphs.

Bapi's paper "Triune brain inspired unifying view of Intelligent Computation" provoked one question by De Falco about whether his quite formal description of the model could somehow lead to the practical definition of a system suitable for a real industrial problem. Bapi pointed out that the main aim of the problem-solving process is not only to find a solution, but to make the solution process explicit, in the sense that the logic behind solution steps is transparent and that the solution process can be repeated with ease for a similar problem. As a practical example, Bapi proposed the pole-balancing control problem. He reported that some methods, like linear controllers, work well for small pole angles and low speed of falling, but, when the linear controller fails, the system needs to switch to a heuristic controller, like GAs, Fuzzy Methods, or others. What is still missing is the integration step among them. He added that the soft-computing methods do not exploit the knowledge of linear range and of some already known things. He claimed there should be a way of incorporating prior-knowledge into the methods, rather than letting it discover again "from scratch" by the tool. He gave out the web address of a paper about a discussion on how this can be done in a neuro-resistive grid.

As a conclusion, it seems to me that the particular structure of the WSC2 Conference allowed a wide and deep discussion among researchers from many different places in the world, giving origin to a fruitful exchange of ideas and experiences, in a quantity probably higher than allowed by a "classical" Conference, and with the advantages of time and money savings.

2. Neural Networks

Prepared by: Stephen Potter, UK

Kovacs, T., XCS Classifier System Reliably Evolves Accurate, Complete and Minimal Representations for Boolean Functions. NetQ (netq@spo.rowland.org) noted that the Subset-Extraction method for obtaining the optimal population, the classifiers of a candidate [O] are tested to see if they form a complete non-overlapping map of the input/action space, and wondered whether there is not a tractability problem with that method - the subset-extraction method itself is plausibly tractable, but within it, this completeness test would seem not to be tractable (i.e., how do the authors test for completeness in a way that does not require testing [O] against every possible input?) In reply Tim Kovacs (T.Kovacs@cs.bham.ac.uk) said that the test for completeness of [O] is integrated with the rest of the subset extraction algorithm in the following manner: when an element of [P] is added to the candidate [O], its condition is converted into line segments in the input/action space of the problem. This is done by converting those input strings that the classifier's condition would match into ranges of base 10 numbers. (For example, 00000# matches both 000000 and 000001, which are binary for 0 and 1 respectively. 0 and 1 are adjacent integers and so are considered a single line segment and written: [0-1]. Similarly, 0000#0 converts to two integers, and are represented by two points [0-0] and [2-2], while ###### converts to [0-63].) Each time an element of [P] is added to [O], its line segment(s) is added to a list of line segments for that action (separate lists for each action do not have to be maintained, since two classifiers cannot overlap if they specify different actions). If any two line segments in a list overlap, e.g. as [12-20] and [0-16] do, then the population is overlapping. If any two line segments are adjacent, they are combined into a single segment, e.g. [0-1] and [2-3] combine into [0-4]. If a point is reached at which there is a single line segment in each list of line segments, then a complete non-overlapping candidate [O] is known to exist. So it is not necessary to test [O] against each possible input. Chris Gathercole (chrisg@harlequin.co.uk) complemented the author on his approach before remarking that XCS doesn't allow for chained classifiers where the action component of one classifier can match the condition component of other classifiers, etc., before eventually producing an action to be passed out to the environment. Would the XCS idea work with chained reasoning? How would the credit assignment work? Tim Kovacs concurred that in some other classifier systems, classifiers can post their actions to an internal message list, and then other classifiers can respond to these messages if their conditions match them. However, XCS has no internal message list and thus this kind of rule chaining cannot occur. This makes XCS reactive - its response is determined only by the current input to the system. He went on to say that Wilson suggested adding temporary memory to his earlier ZCS system by using a register of bits. In this scheme, classifier actions have two parts, one of which specifies the action for the system to take, while the other optionally sets bits in the register. Conditions are also composed of two parts, one of which must match the contents of the register while the other must match environmental input in order to fire. Cliff and Ross actually implemented this system for ZCS, and the same approach should work for XCS. Credit assignment would remain the same under this scheme, i.e. accurate classifiers are rewarded while inaccurate ones are not. The register-setting behaviour of a classifier would not affect its fitness calculation (other than indirectly) - fitness is based on the accuracy of the condition/output-action pair. Finally, he remarked that it should be possible to generalise this dual-action idea so that classifiers can post messages to an internal message list (as well as specifying an action for the system to take).

Tsui K. and Plumbley, M., have designed Neural Networks using a Genetic Rule-based System. S. E. Potter (enssep@bath.ac.uk) asked the authors where do the network configuration rules come from and how have they decided that a particular condition should be met with a particular response. In addition, he wondered whether the authors thought it possible to extend their method to include rules for adding units to the network, and

444

indeed, adding layers. K. C. Tsui (tsuikc@info.bt.co.uk) replied that the classifier rule used consists of a number of conditions and an action. The training of the MaCS involves presentation of input-output patterns that the neural network will see. The response to restructure the neural network is then decided. The system is assisted by the hill-climbing search built into the MaCS system. In response to the final point, the author answered that as connection matrix is used to represent the neural network, adding neurons or layers of neurons are indeed difficult. Inclusion of these new features would affect the decoding scheme of the classifier action, which is undesirable. The experience the authors have with the two-spiral problem is that MaCS is able to find a network smaller than that it had been given at the beginning of the search. A possible direction is to evolve grammar rules (such as those used by Kitano and Gruau) that indirectly manipulate the network structure. The authors have had some success along this line, but the work is ongoing.

Blumenstein, M. and Verma, B., presented a Neural Network for Real-World Postal Address Recognition. S. E. Potter (enssep@bath.ac.uk) asked what size bit matrix the authors had used for representing the character inputs to the system and how that size was determined. Michael Blumenstein (M.Blumenstein@eas.gu.edu.au) replied that the character matrix size varied from one experiment to another and that it was definitely important to have a size which allowed for good resolution and kept important detail, while not being too large to slow the training of the ANN. Matrix sizes of 15x23 and 19x24 for the hand-written characters were used, and for the printed characters a matrix of 13x12 was used. S. E. Potter remarked that the results show that the RBF networks performed relatively poorly in comparison with the back-propagation networks, and asked the authors if they knew of any reasons for this to be so. He went on to note that such a network, when embedded within a knowledge-based system, may provide a powerful recognition tool. Michael Blumenstein concurred that the RBF network's results were indeed unsatisfactory, but noted that they improved as the number of hidden units was increased. Time restrictions had prevented the authors from conducting further experiments with the RBF network, but in general the back-propagation algorithm allowed the ANN to generalise far better with a smaller number of hidden units, whilst the RBF's linear training method simply did not offer a good recognition rate with a low number of hidden units (possibly attributable to the difficult handwriting used for training and testing). In response to the final remark, he agreed that for problems such as zipcode recognition, an integrated approach is indeed a very viable option and that for further research a database or lexicon of zipcodes and also street names, etc. will be used.

Zhang, B. T. and Joung, J. G., presented how to Evolve Neural trees for Heart Rate Prediction. S. E. Potter (enssep@bath.ac.uk) asked the authors how they had arrived at the fitness measure given in section 3 of the paper, and why it is necessary to include an explicit penalty for complex networks, as the decrease in prediction performance as complexity increases would seem to supply an evolutionary pressure toward more parsimonious networks. Byoung-Tak Zhang (btzhang@comp.snu.ac.kr) responded that the penalty term in the fitness function was indeed used to evolve parsimonious networks, based on the theory that a parsimonious neural network tends to have better generalisation performance than a complex network, all other factors being equal. S. E. Potter then asked whether interconnections between neural trees, other than at the output-input interface, were permitted, and asked the authors to describe in greater detail the steps involved in training the network weights between generations. Byoung-Tak Zhang replied that although there are no explicit connections between the individual neural trees, they can share sub-trees as building blocks which are stored in a common library (more detail on this matter may be found in the paper by Zhang, Ohm, and Muehlenbein to appear in Evolutionary Computation, Vol. 5, No. 3, 1997). In reply to the second question, the author responded as follows. The newly generated networks undergo a hill-climbing search to train the weights. Each hill-climbing step involves generating a new weight vector, W', by weight mutation of the current weight vector, W. Then W' replaces W, but only if W' is fitter (having less error on the training examples) than W. The hill-climbing steps are repeated until the step count reaches a limit or a fixed number of consecutive steps generate no improvement.

3. Fuzzy Logic

Prepared by: Wallace E. Kelly, III, USA

The following are edited excerpts of the discussion which took place in the Technical Session on Fuzzy Logic. Because the entire discussion could not be included here, it is the intent of the session facilitator to highlight the themes of the discussion without removing too much of the context in which these comments were made.

Industrial Application of Chaos Engineering
by Tadashi Iokibe, Meidensha Corporation, Japan

Kelly: In your comparison with ARX model for forecasting water usage, you say, "when the non-linear short-term prediction method ... is applied, absolutely no data other than one line of observed time series data is needed." Were the seven "neighboring data vectors" taken from recently recorded observed time series data?

Iokibe: No, the recently recorded observed time series data is one of the components of Z(T). Z(T) is the data vector lately reconstructed in a multidimensional state space using Takens' embedding theorem. Neighbouring data vectors X(i) are selected by the Euclidean distance from Z(T).

Kelly: Does the database from which you pull the neighbouring data vectors grow as the system is observed?

Iokibe: Yes, the database is updated automatically when the new time series datum is observed.

Kelly: How does the local fuzzy reconstruction method deal with the situation in which the proximity vectors, X(i), are all concentrated on one side of the state space surrounding the data vector, Z(T)? Without doing my own simulations, it would seem that the prediction Z(T+s) would be biased towards that direction in state space. Would you expect an ARX, or similar linear model, to outperform the local fuzzy reconstructor, in this case?

Iokibe: The relation between Z(T) and X(i) are projected on the two dimensional space which are the components of the multidimensional state space. The membership functions are defined in the two dimensional space. The universe of discourse of the membership function is defined as the double scale of the length | X(i) - Z(T) | which is projected on the two dimensional space. So even if X(i) are concentrated one side, you can make a fuzzy inference. I have already examined with same time series, and the correlation coefficient as a prediction performance of the local fuzzy reconstruction method and ARIMA model with Kalman filter are 0.986, 0.974 respectively. When the non-linearity of the time series is increased or the prediction step is larger, the prediction performance by the local fuzzy reconstruction method will be higher than that by other linear methods. The reason is that the local reconstruction method is a non-linear method.

A Fuzzy Control Course on Internet
Jan Jantzen and Mariagrazia Dotoli
Technical University of Denmark, Lyngby, Denmark and DEE-Politecnico di Bari, Bari, Italy

Kelly: You mention that the textbook material was specifically formatted to fit the course. Do you think such a course could be taught using a "standard" classroom textbook?

Jantzen: Yes, I think standard textbooks can be used in this sort of course. You would have to build the lessons around the textbook, and it would probably be best if the order of lessons follow the organisation of the textbook. You could use the Internet lessons as a supplement to a traditional course with lectures and slides; for example you could have a 20 minute introduction in a lecture theatre, and then take the students to the terminals afterwards to do an Internet lesson. In the fuzzy control course, however, the textbook is just a photocopied, soft cover book that I have written myself. Therefore, I have full control over it. I can add material and change the organisation over time, instead of depending on a publisher. For example, it may be necessary in the future to split the course into an introductory course and an advanced course; then the book has to be changed accordingly.

An Iterative Two-phase Approach to Fuzzy System Design
F. Abbattista, G. Castellano, and A. M. Fanelli
Universita degli Studi di Bari, Bari, Italy

Kelly identified a mistake in the paper which was then corrected by the authors, and a revised paper was submitted.

FUZOS - Fuzzy Operating System Support for Information Technology
A. B. Patki, G. V. Raghunathan, and Azar Khurshid
Department of Electronics, Government of India, India

Kelly: You interpret the AND operator as the pessimistic operator and are implementing it with min. Technically, I agree with you that AND should be the intersection of two sets, and be implemented with a T-norm. However, I suspect that given this command:
c:>dir large files and average files

..most users would expect to see BOTH the large files and the average files. Of course, you and me both understand the set interpretation of AND, but has there been any discussion about how most users might interpret this command?

Khurshid: It must be quite clear that the fuzzy AND and fuzzy OR connectives are not similar to normal Boolean connectives. As explained in the paper, it is difficult to implement the meanings of fuzzy AND and OR for every kind of user by simple S and T Norms. In the corporate decision making environment, AND may not be interpreted as independent satisfaction of all conditions, but a fuzzy interpretation of them, such that the results may only partially satisfy the conditions. The inclusion of these types of queries involving 'and/or' from same domain for the FUZOS has not been felt as a useful feature by some users (with IT management background) who participated in performance testing (section 5). However, the same have been included for completeness as a OS level support for tool, utilities, and application building. For further clarifications about operator behaviour in information handling, the reader is referred to reference number 8 in the paper.

Ho: I found that your title Fuzzy OS is a big exaggeration of a system which can be best described as a Fuzzy Command Line Interface. I cannot see how it can give any OS support in terms of process scheduling, memory management or file system management. The paper starts out with a very good point about the deficiency of computer systems. We have an explosion of information and we do not know how to manage it efficiently. Finding relevant information on a computer system is not easy and finding the RIGHT information is even harder. However, your approach (from your examples of FUZOS, "dir large and average files") does not seem to solve the problem. Corporate computing is more interested in asking the computer to generate the latest report on the corporate assets or profits. They are not interested in large or average files. "Given a set of sales figures, can the computer generate a summary of the profit and loss or the buying tendency of the consumers?" is probably more relevant to the corporate world. In order to solve the problem, I am not sure your approach on the file level is sufficient or relevant. I think content or knowledge representation of data is needed in order to provide some sort of support to the type of query that corporate world is interested in. In my opinion, the search and indexing engines that are available on the World Wide Web are better suited to locate information for us based on a set of keywords. Perhaps applying your fuzzy interface technology to those search engines may result in a better IT system.

Khurshid: The title of the paper is "FUZOS -- Fuzzy Operating System Support for Information Technology" and not "Fuzzy OS". The reader should not get an impression that either the title or the contents of the paper discuss the Fuzzy Operating System, covering all its facets, in its entirety. A full length discussion on the FUZOS is beyond the scope of any single paper. References [3,4] report some of the aspects of relevance. The authors are aware of the special requirements of corporate computing and envisage the need for Workstation and OS level system software support from soft computing for IT managers. FUZOS is an effort towards system software with emphasis on Corporate Computing. Application products including associated utilities running on "Workstations for IT Managers" is a separate issue. Dr. Ho states that content or knowledge representation of data will be useful for Corporate Computing. Similarly, he mentions that there exists a scope for improving the existing search and indexing engines on WWW. These issues typically fall under the category of Application Software Modules for Corporate Computing running on Workstations for IT Managers. The authors strongly feel that FUZOS research background could be very well utilised for developing such applications. The transition from consumer product market applications of fuzzy systems to the information technology sector expects certain pre-requisites in the form of availability of fuzzy processing hardware and associated software support. With the availability of new chip-sets designed to carry out fuzzy logic computations, entirely new applications will emerge that were hitherto considered not practical. Internet explosion and the associated security and integrity requirements for intranets of enterprises are leading to more openings for the soft computing applications. However, the major limitation in achieving effective and sustained solution is the lack of availability of Operating System level support. The crucial issue is not whether FUZOS is comprised of 'smart commands' or has characteristics of 'command line interface', but the opening of a new avenue towards realising the full commercial potential of fuzzy technology in the Information Technology area by incorporating the support at the operating system level.

Hypertrapezoidal Fuzzy Membership Functions for Decision Aiding
Wallace E. Kelly, III and John H. Painter
Texas A&M University, College Station, TX, USA

While no discussion regarding this paper took place during the WSC2 conference, the authors welcome comments and questions via wkelly@tamu.edu and painter@tamu.edu.

4. Genetic Algorithms

Prepared by: Kang, Myung-Ju, Korea

This session covered genetic algorithms. A genetic algorithm is simply a series of steps for solving a problem and a problem-solving method that uses genetics as its model of problem solving. In recent years, the field of genetic algorithms has grown and developed in all areasincluding economy, industry and communication. In this session, 6 papers on genetic algorithms were presented. The main points of discussions raised by each paper are as follows.

1. A Genetic Method for Evolutionary Agents in a Competitive Environment.

This paper described an n-BDD to investigate co-evolutional environment. The main points of discussion were what kinds of problems are better suited for evolution using a n-BDD than a binary bit-string, and that a n-BDD would have a bias towards the earlier bits in the input string.

2. Empirically-Derived Population Size and Mutation Rate Guidelines for a Genetic Algorithm with Uniform Crossover

This paper described the uniform crossover method and an empirically-derived populations size. This was the most discussed paper in this session, and also won the best presentation award. The main point of discussion for this paper is the guidelines for population sizes, mutation rates and the string length.

3. Analysis of various evolutionary algorithms and the classical dumped least squares in the Optimisation of the doublet

This paper described the optimisation of the doublet with various algorithms including dumped least squares, adaptive steady state genetic algorithm without duplicates, (1+1)evolutionary strategy, (μ, λ) evolutionary strategy with or without recombination.

4. Genetic Programming with One-Point Crossover

This paper described one-point crossover where the same crossover point is selected in both parent programs.

5. Evolutionary Tabu Search for Geometric Primitive Extraction

This paper was an invited paper and described evolutionary tabu search (ETS) that genetic algorithm and tabu search algorithm were combined. The main point of discussion for this paper was the comparison with other search algorithms and whether the feature of GA was used.

6. Parallel Genetic Algorithms in the Optimization of Composite Structures

This paper described PGA in the optimisation of composite structures. Especially, examples of the successful use of PGA to design composite structures, such as, energy-absorbing laminated beams, airfoils with tailored bending-twisting coupling, and flywheel structures, described.

This session covered genetic algorithms including a genetic method for evolutionary agents, empirically-derived population size and mutation rate guidelines, evolutionary tabu search, and so on. In this session, 6 papers were presented and all authors replied all their questions eagerly. I think that these papers will help the researchers to enhance their studies.

5. Decision Support, Constraints and Optimisation

Prepared by: Rajkumar Roy, UK

This session included seven papers, and was one of the most discussed sessions in the conference. Discussions about individual papers are summarised below:

I. De Falco, A. Della Cioppa, A. Iazzetta and E. Tarantino: Mijn Mutation Operator for Aerofoil Design Optimisation

Vladislav B. Valkovsky opened the discussion by asking for the authors' view on 'prediction of sets of probable solutions by analysis of input or intermediate data and use of the information to guide optimisation'. Ivan de

Falco presented his view with caution. He recognises the need for using hybrid systems to solve specific industrial optimisation problems faster. He reminded the community by saying 'the real challenge for us in the field is, in my opinion, to find such new models that are able to i) explain evolution and ii) solve real problems, rather than find hybrids or smart ideas to solve specific problems'. Stephen Rudlof observed that when Mijn is applied to the left most bit, it behaves the same way as Bit Flip Mutation. On the issue of whether 'the left most variables in the string might have a higher probability of undergoing macromutation', Ivan de Falco reported on additional experiments that are not included in the paper. Ivan reported that the additional experiments show the maximum number of bits affected by Mijn-EA at a time is linearly proportional to the logarithm of the number of fitness evaluations.

Marian Mach: Design Problems with Soft Linear Constraints

In reply to a question by Vladislav Valkovsky on whether prediction about probable solutions should be used to guide an optimisation, Marian Mach described it similar to incorporating domain knowledge in a search. He argues that the use of domain knowledge can transform a weak algorithm into a strong one. In this way the algorithm can be made very efficient for a limited range of problems.

P.J. Bentley, J.P. Wakefield: Finding Acceptable Solutions in the Pareto-Optimal Range using Multiobjective Genetic Algorithms

Barry O'Sullivan raised an issue about handling constrained search spaces. He pointed out that due to the presence of constraints between variables, there can be a considerable degree of epistasis. Peter Bentley mentioned the difficulty with 'hard' constraints, where certain combinations of parameter values are 'forbidden'. But he mentioned about 'soft' constraints which can be implemented using a penalty on the fitness value. William Crossley raised an issue about whether the objective is to obtain 'a set of Pareto-Optimal (P-O) solutions' or a single 'acceptable' P-O solution. Peter Bentley replied by saying 'The problem with generating a range of P-O solutions is that not all P-O solutions are particularly good or acceptable solutions'. If you are not aware of the kind of distribution of P-O solutions that your method generates, you may end up with a GA that consistently evolves a range of P-O, but still very poor, solutions'. The authors used a concept of 'importance' to guide the GA, and make it favour those solutions which satisfy the objectives with increased relative importance in the problem. Thus by using distinct user-definable biases (SWGR) the authors have obtained a single 'acceptable' P-O solution.

P. J. Baron, R. B. Fisher, F. Mill, A. Sherlock, A. L. Tuson: A Voxel-based Representation for Evolutionary Shape Optimisation: A case-study of a problem-centred approach to operator design

Peter Bentley expressed his doubts about the 'scalability' of the low-level enumeration of the design space using Voxel-based representation. He mentioned that 'the number of parameters needed for even simple approximations of designs would actively inhibit evolution by the GA to good solutions'. And of course this representation also suffers from 'holes' and other such problems'. Andrew Tuson, one of the authors, recognises the problem but believes by improving the GA they can handle such cases. He proposes to use domain knowledge to improve the search capability, but not sure how to implement it. He mentioned that parametric design is probably better suited for complex problems. However, for a smaller problem, a less restricted representation may be preferred as it could look in the space of 'novel' solutions.

MJ Kang and CG Han: A comparison of crossover for rural postman problem with time windows

In reply to a question from Vladislav Valkovsky, Myung-Ju Kang reported that they are also studying Simulated Annealing and Neural Networks for their problem. They are going to compare the results with that from GA.

R. K. Pant & C.M.Kalker-kalkman: On Generating Optimum Configurations of Commuter Aircraft using Stochastic Optimisation Methods

C. M. Kalker-Kalkman mentioned that when she incorporates few good feasible solutions in the initial population of her GA, it converges faster. The discussion was taken further by William A. Crossley when he raised the issue of how to select the value of cooling rate in Simulated Annealing. R. K. Pant replied by saying 'the appropriate value has to be determined by carrying out a few trial runs'. He mentioned that they reject only very poor solutions in a GA population while use a penalty function for rest of the poor solutions. The authors used a variable rate of mutation, and stressed the usefulness of their interactive interface. The interface allows them to investigate the effect of the mutation rates on the achieved solutions easily. R. K. Pant also mentioned that '10% mutation rate is not being recommended as general practice'.

6. Engineering Design

Prepared by: John Greene, South Africa

The discussion involved three papers - Generic Evolutionary Design by Bentley and Wakefield, Evolving Digital Logic Circuits on Xilinx 6000 Family FPGAs by Fogarty, Millar and Thomson and Hard vs Soft Computing issues in Configuration Design by Potter, Chawdhry and Cutter. The first and the third papers raised questions focusing on the broader issues of aims and methodology while the second elicited interest in a number of matters of detail.

Dr Roy commented on the great flexibility of the generic evolutionary system proposed, but questioned whether a system of such great generality was relevant and usable in real life. The authors replied that their aim had been precisely to explore the feasibility of a generic system which could be used for a wide range of design tasks with minimum set-up time. They pointed out the modular nature of the system and the way that it could easily be specialised for various design tasks by the selection of a suitable subset of modules. They claimed that their work had demonstrated that a single system was in principle capable of evolving many different types of designs. They conceded that the route to practical use might well be further modularisation, with the development of industry-specific 'libraries' of evaluation modules operating around a common software core. In response to a question about the elimination of infeasible crossover-generated solutions, the authors stated that there was no separate 'culling' operation - the system was purely 'fitness' driven, fine-tuning solutions until they became acceptable. They also pointed out that the evolutionary process was driven by function alone, not form. In view of the fact that trial solutions had a finite lifetime and good solutions could be lost, it was asked whether a separate record was kept of good solutions. The authors replied that the steady-state GA at the heart of the system had the advantage of preserving all good solutions. However, with 'noisy' fitness functions, a 'lucky' individual could dominate and perhaps become immortal, thereby corrupting the evolutionary process. For this reason a lifespan limiter was set which drastically reduced fitness after a pre-defined number of generations. To ensure that this affected only the 'lucky' trial solutions the number of generations was set in excess of the 6 or 7 generations which was the typical lifespan of a 'normal' solution.

Poli and Greene expressed interest in the second paper but requested clarification on a number of points of detail. Poli, for example, asked how syntactical correctness was preserved, as well as the avoidance of non-local connections as required by the Xilinx 60000. In response, the authors expanded on the details of their implementation, pointing out, for example, that a constraint on local connectivity was implemented in the form of a 'levels back' parameter (which also had the effect of ensuring that all trial solutions were valid combination circuits by preventing illegitimate feedback connections). Solutions generated by simple crossover would always be legitimate. The authors also stated in reply to questions that the grid-size had been determined by trial and error, starting with one which could accommodate a conventional solution and gradually reducing it to improve the efficiency. They said that thus far the behaviour of the Xilinx chip had been simulated but the planned the use of in-circuit evolution, and were hoping for a significant speedup. In response to queries about the GA used and computational effort involved, they stated that (after unsuccessfully trying much larger sizes) they were using a population of 100, running the GA for thousands of generations. Many of the runs produced 100% functionally-correct solutions. The evolution of a 1-bit adder required 2000 generations (about 20 minutes on a Sun Workstation), a 2-bit adder 10 000 generations and a 3-bit adder 90 000 generations. A 2-bit multiplier required 60 000 generations. A 3-bit multiplier and 4-bit adder had also been successfully evolved. The GA uses tournament solution with elitism, with an additional probability-of-accepting-the-winner parameter (0.7) There is 100% crossover and 1-5% of the genes are mutated each generation.

On the third paper, Greene requested clarification regarding the "function recognition network", and questioned whether a radical functional decomposition at the level of 'one component one function' was always possible (mad whether it led to the best designs). The authors replied that basic functionality was provided by the template mechanism, the templates spanning the domain of virtually all hydraulic systems. The function recognition network performs a fine-tuning by subset selection of domain-dependent abstractions. Dynamic aspects were handled by the use of a time-delay neural network (TDNN) approach which essentially converted a dynamic description into a static one. It remained an open question as to whether this was the best approach to take. Regarding the functional decomposition issue the authors stated that they did consider groups of components where such groups jointly achieve a basic function. However, they conceded, the independent choices of component groups could serve to mask effects (benign or otherwise) which could result from interaction, and that this too was an ongoing research issue.

7. Scheduling, Manufacturing and Robotics

Prepared by: Masahiro Tanaka, Japan

This section consisted of 6 papers, where 4 papers out of 6 were based on the genetic algorithm. The application area spread in the diversity of fields in scheduling, manufacturing and robotics. This fact probably made the authors difficult to discuss others' papers. Only one question was raised from other persons but the facilitator. The followings are the papers and the summary of the discussions (if any). I have added some personal view to some papers.

B. Visweswaran and D.K. Anvekar: A Genetic Algorithm Based Hybrid Channel Allocation Scheme

This paper developed an application of GA to the hybrid channel allocation scheme in mobile cellular communication systems. This kind of application seems to be timely, but the example was not necessarily suitable; they used GA for the problem with only 32 alternatives in the solution space. In this case, an exhaustive search is surely superior to the GA application. As the author replied, the GA based HCA will converge faster than a random or exhaustive search as the size of the sample space increases because this is the common feature of the GA application. However, I think that it must be further validated experimentally whether it conveniently works for the problems when the sample size is large.

I.P. Norenkov and E.D. Goodman: Solving Scheduling Problems via Evolutionary Methods for Rule Sequence Optimization

Paper by Norenkov and Goodman used GA to solve the scheduling problem for the flow shop scheduling problem. In most of GA applications in scheduling, the chromosome expresses the phenotypic representations directly. But it is clear that various heuristics supplementary algorithm needs to be invoked to satisfy the constraints, otherwise it is very hard for the schedule to be feasible. In this paper, the i-th gene does not represent the number or the server but expresses the number of the heuristic to be used. This kind of approach will widen the GA application fields.

L. Hazra, H. Kato and T. Kuroda: Estimation of Geometrical Parameters of Drill Point by Combining Genetic Algorithm and Gradient Method

The paper by Hazra et al. deals with the measurement problem by GA. It is very interesting to try to estimate the unmeasurable values by using GA, where the values of the geometrical parameters were to be identified. First GA was used and after the convergence the steepest gradient method was used. The GA was said to have been the floating point GA, where the real value was used as the chromosome. Two questions were raised. One question was whether the floating point GA efficiently worked or not. If not, the result is almost the same as the one only using the gradient algorithm. Another question was whether the greedy algorithm with local search at every evaluation of GA would work the same or not. Although there was no reply from the authors, the greedy algorithm will generally give better result. The problem is probably the computation time for this.

M. Tanaka, M. Yamada and O. Matsuo: Application-Based Time Tabling by Genetic Algorithm

Tanaka et al. developed a new concept for the timetabling problems using GA. In the traditional timetabling problem, the interest was mainly devoted to the constraint satisfaction as well as to optimise the efficiency of the table, where the jobs to be assigned to the table have no will. But they often have the desire; for example, in the school timetable, the lectures to be assigned are delivered by the tutors who get tired by consecutive lectures and sometimes want to be allocated only in the morning. Considering such features, they used application slips on which the applicants write their preferable time slots for each lecture. The chromosome of GA expresses the order of the application slips to be used for the assignment. Each application slips are used so that the higher preferences are tried to be allocated. In this sense, this paper is similar to the one by Norenkov and Goodman. These papers show how important it is to use GA indirectly to the scheduling with heavy constraints that needs the heuristics.

F. Michaud: Adaptability by Behaviour Selection and Observation for Mobile Robots

The paper by Michaud considered that adaptability of mobile robots by behaviour selection and observation. To make the robots more adaptable to real-world environments, it is good to dynamically select behaviours that control the actions of the system. It also demonstrated the actual interesting experimental result, but unfortunately there was no discussion for this paper during the conference.

D. T. Pham and A. B. Chan: A Novel Self-Organising Neural Network for Control Chart Pattern Recognition

The paper by Pham and Chan developed a new neural network that is used for the pattern recognition of the control chart. Control chart is a time series data to know whether a process is functioning correctly or not. The neural network they developed has stability-elasticity dilemma. This means the network is a growing feature in nature. The fundamental idea for this neural network employs that of "Adaptive Resonance Theory (ART)" by Carpenter and Grossberg, although the learning method is different from this.

8. Dynamic Systems, Identification and Control

Prepared by: Andreas Nürnberger, Germany

D. Kim and I.-H. Cho: An Optimal COG Defuzzification Method for a Fuzzy Logic Controller

In this paper, D. Kim and I.-H. Cho presented an approach to optimise the defuzzification method of a fuzzy logic controller using a genetic algorithm. The discussion focused on problems occurring by the use of 'weighted rules' and on further optimising approaches for the presented methods.

The authors proposed the use of an approximate COG (Centre Of Gravity) defuzzifier as the defuzzification method in a fuzzy logic controller. This COG defuzzifier is defined in equation (9). This equation corresponds to defuzzification methods of optimising approaches using 'weighted rules' (for example those used in the ARIC-model [1, 5]). Many researchers suggested this kind of defuzzification equation for the more accurate computation of the crisp value, in spite of problems which may occur if the derived rulebase has to be interpreted [5]. (Different weights can be assigned to the same fuzzy set as the consequence of different rules. This will yield different meanings (linguistic terms).). It should be emphasised that the 'weighted rules' approach as presented in this paper is motivated by the mathematical exact computation of the area defined by the clipped membership function in order to eliminate the misuse of the conventionally simplified COG defuzzifiers [3][1]. This mathematical exact COG defuzzifier is defined in equation (8). The primary goal of the authors' research is to implement the fuzzy logic controller in hardware. Because additional multipliers are required, it is difficult to implement equation (8) directly in hardware. Therefore, the authors used the approximated form defined by equation (9). Further publications present some improvements which resolve the problem caused by equation (8) with stochastic computing methods [2, 3]. Since the defuzzifier defined in equation (9) is just the same as in defuzzification methods used in optimising approaches which use 'weighted rules', the presented algorithm can be directly compared to them. In table 2 the authors presented the results of control performance using equation (8) and equation (9) (last two columns). The results illustrate that the exact COG defuzzifier is hardly better than the approximate COG defuzzifier. These results support the authors' choice of equation (9), because it significantly reduces hardware complexity at a small loss of control performance.

Another approach to optimise the COG defuzzification is to use a genetic algorithm for the Optimization of the width s_i in equation (9). The authors have developed an algorithm for the Optimization of the width, which extends the results presented in this paper. They added another parameter (shape constant r) to their defuzzification equation and derived a learning algorithm that is based on back-propagation. Furthermore, they constructed a method based on a genetic algorithm to find the optimal COG's parameters (centre y_i, width s_i, and shape constant r). Finally, they suggested a new co-adaptation method that combines learning and evolution to reduce the computation time for the calculation of the optimal values. According to the authors, simulation results have shown that the co-adaptation scheme is a very promising technique to solve any kind of Optimization problems. These results will be presented in future publications. Another approach that uses the co-adaptation scheme in codebook design is presented in [4].

J. Abonyi, L. Nagy and F. Szeifert: Takagi-Sugeno Fuzzy Control of Batch Polymerization Process

In this paper, J. Abonyi, L. Nagy and F. Szeifert show that conventional control solutions for batch Polymerization reactors can be improved by applying Tagaki-Sugeno fuzzy models [6]. Comments concerned

---

[1] The cited publications [3, 4] are also available via anonymous ftp from the ftp-server vlsi.donga.ac.kr. (Files ifsa.ps.gz [3] and ieeetnn.ps.gz [4] in directory /pub)

the presented simulation results in section 4. The main bias that the simulation results all have in common is that the input intervals for the tracking error signal e(kt) (and (e(kt)) are partitioned by three triangular membership functions. Another possibility to decrease the performance index is to refine the partitioning. The authors argued that the measure of change will be negligible and that the Optimization and simulation time will increase strongly. For instance, when increasing the number of membership functions from 3 to 5, the number of parameters to be determined increases to 30 compared to 18 in the original example. Therefore, the authors suggest choosing only a coarse partitioning to reduce execution time.

References

[1] Berenji, H.R., 1992, A Reinforcement Learning-Based Architecture for Fuzzy Logic Control, *Int. Journal of Approximate Reasoning*, **6**, 267-292.

[2] Kim, Dajin and Cho, In-Hyun, 1997, A Dividerless COG Defuzzifier with an Efficient Searching of Moment Equilibrium Point, *Proceedings of the seventh International Fuzzy System Association World Congress (IFSA '97), Prague, Academia, Prague*, 3, pp. 428-433.

[3] Kim, Dajin and Cho, In-Hyun, 1997, An Accurate and Cost-Effective COG Defuzzifier Without the Multiplier and the Divider, *Fuzzy Sets and Systems*, Elsevier Science B.V., Amsterdam (to appear)

[4] Kim, Dajin and Ahn, Sunha, 1997, Co-adaption of Self-Organizing Maps by Evolution and Learning for An Optimal VQ Codebook, *IEEE Transactions on Neural Networks* (submitted)

[5] Nauck, Detlef; Klawonn, Frank; Kruse, Rudolf, 1997, *Foundations of Neuro-Fuzzy Systems*, John Wiley & Sons, Inc., New York, Chichester, et.al. (to appear)

[6] Tanaka, K., and Sugeno, M., 1992, Stability Analysis and Design of Fuzzy Control Systems, *Fuzzy Sets and Systems*, Elsevier Science B.V., Amsterdam, **42**, 135-156.

Keyword Index

List of the Reviewers

Prof. Erik Goodman, Michigan State University, USA
Prof. Peter Sackett, Cranfield University, UK
Prof. Terry Fogarty, Napier University, UK
Prof. M. A. Faruqi, Indian Institute of Technology, Kharagpur, INDIA
Dr. Kay Hameyer, Katholieke Universiteit Leuven, BELGIUM
Dr. I. S. Fan, Cranfield University, UK
Prof. V. A. Zarubin, SSAU, RUSSIA
Dr. Ram Sriram, NIST, USA
Dr. Riccardo Poli, The University of Birmingham, UK
Mr. G. V. Raghunathan, DOE, INDIA
Dr. Michael Lee, BISC, University of California, USA
Mr. Hossein Soltan, Cranfield University, UK
Prof. Okyay Kaynak, Bogazici University, TURKEY
Dr. Raju Bapi, University of Plymouth, UK
Ir. C. M. Kalker-Kalkman, University of Technology, Delft, THE NETHERLANDS
Prof. John Gero, University of Sydney, AUSTRALIA
Dr. Peter Cowley, Rolls Royce plc., UK
Dr. Frank Hoffmann, University of Kiel, GERMANY
Dr. Tadashi Iokibe, Meidensha Corporation, JAPAN
Mr. Steve G. Thornton, British Steel plc., UK
Dr. Ivanoe De Falco, IRSIP-CNR, ITALY
Prof. SongDe MA, Institute of Automation, CHINA
Prof. Sin-Horng Chen, National Chiao-Tung Univ., TAIWAN
Prof. Asim Roy, Arizona State University, USA
Dr. Rajkumar Roy , Cranfield University, UK
Prof. Takeshi Furuhashi, Nagoya University, JAPAN
Dr. Pravir Chawdhry, Univ. of Bath, UK
Prof. R. K. Pant, Indian Institute of Technology, Bombay, INDIA
Dr. George Bilchev, Advanced Research & Tecnologies, BT Laboratories, UK

P.K. Chawdhry, R. Roy and R.K. Pant (Eds)

Soft Computing in Engineering Design and Manufacturing

Springer

P.K. Chawdhry, PhD
School of Mechanical Engineering, University of Bath, Bath BA2 7AY, UK

R. Roy, PhD
The CIM Institute, Cranfield University, Cranfield, Bedford MK43 0AL, UK

R.K. Pant, PhD
Department of Aerospace Engineering, Indian Institute of Technology,
Powai, Mumbai 400 076, India

ISBN 3-540-76214-0 Springer-Verlag Berlin Heidelberg New York

British Library Cataloguing in Publication Data
Soft computing in engineering design and manufacturing
   1.Engineering design - Congresses   2.Engineering - Data processing - Congresses   3.Artificial intelligence -
   Congresses
   I.Chawdhry, P.   II.Roy, R.   III.Pant, R.
   620'.0042028563
   ISBN 3540762140

Library of Congress Cataloging-in-Publication Data
Soft computing in engineering design and manufacturing / P. Chawdhry, R. Roy and R. Pant, eds.
     p.   cm.
   Includes bibliographical references.
   ISBN 3-540-76214-0   (alk. paper)
   1. Soft computing.   2. Computer-aided engineering.   I. Chawdhry, P. (Pravir), 1958-
II. Roy, R. (Rajkumar), 1966-   III. Pant, R. (Raj), 1962-
QA76.9.S63S63   1998
670'.285'63- -dc21   97-31959

Typesetting: Camera ready by editors
Printed and bound by Professional Book Supplies Ltd., Abingdon, Oxfordshire
69/3830-543210 Printed on acid-free paper